Operator Theory: Advances and Applications
Volume 222

Founded in 1979 by Israel Gohberg

Harry Dym
Mauricio C. de Oliveira
Mihai Putinar
Editors

Mathematical Methods in Systems, Optimization, and Control

Festschrift in Honor of J. William Helton

 Birkhäuser

Editors
Harry Dym
Department of Mathematics
Weizmann Institute of Science
Rehovot, Israel

Mauricio C. de Oliveira
Department of Mechanical and
  Aerospace Engineering
University of California San Diego
La Jolla, CA
USA

Mihai Putinar
Department of Mathematics
University of California
Santa Barbara, CA
USA

ISBN 978-3-0348-0410-3   ISBN 978-3-0348-0411-0 (eBook)
DOI 10.1007/978-3-0348-0411-0
Springer Basel Heidelberg New York Dordrecht London

Library of Congress Control Number: 2012943938

Printed on acid-free paper

Springer Basel AG is part of Springer Science+Business Media (www.birkhauser-science.com)

Dedicated to Bill Helton

on the occasion of his 65th birthday.
Photo by Matthew James

Preface

This volume is dedicated to John W. Helton, also known as Bill, on the occasion of his 65th birthday. It includes, biographical information, personal notes and articles by many of Bill's friends and collaborators.

The first part of the volume begins with a list of Bill's publications and a number of short notes by close friends. This is followed by a transcript of two public addresses delivered at the conference dinner during a three day workshop held at UCSD from October 2nd to the 4th, 2010, to celebrate Bill's birthday. The first is a survey of Bill's contributions to mathematics and engineering, by Miroslav Krstic. The second is an essay on Bill's early work, by Jim Agler.

The editors and most of the authors of this volume have had the privilege to know and work closely with Bill. His unbounded curiosity, original ideas, and generosity have marked many of us. Entire chapters of modern operator theory and control theory of linear systems of differential equations have been shaped by his thoughts. The nineteen expository articles in the second part of this volume illustrate his remarkable impact. Subjects include interpolation, Szegö limit theorems, Nehari problems, trace formulas, systems and control theory, convexity, matrix completion problems, linear matrix inequalities and much more.

Bill's mathematical talent was discovered and directed by the famous mathematical pedagogue Robert L. Moore at UT Austin. From those early days, Bill preserved the experimental attitude and a Socratic approach to research, as displayed throughout his career. This might help explain his success at talking to and doing mathematics with engineers, physicists and all humans with an inclination to listen or respond to his endless questions. For Bill, every mathematical problem is interesting, while solving it in a continuous dialogue is the most natural approach – the numerous experts in mathematical education (in such an expansive state nowadays) have much to learn from Bill. Another explanation might be on his unreserved kindness and his friendly manner. Under the guise of an absent-minded gentleman, Bill hides a lucid, focused search for the mathematical truth. His openness to dialogue and willingness to share ideas is an inspiration to all of us.

December 2011

Maurício de Oliveira
Harry Dym
Mihai Putinar

Acknowledgement

The editors would like to thank the many friends and colleagues that contributed to and helped to bring this volume to life. We would like to thank specially RUTH WILLIAMS, for her help in mining information about Bill and about his friends while keeping this project a secret for more than a year, and MIROSLAV KRSTIC and JIM AGLER, for allowing us to reproduce their public addresses in this volume. We thank MIROSLAV KRSTIC, ROBERT BITMEAD, WILLIAM McENEANEY, and MATTHEW JAMES for organizing the workshop in honor of Bill at UCSD, where the idea for this book was born. Special thanks go to MARINA ROBENKO, for taking excellent care of Bill and his friends during the workshop.

Contents

x　　　　　　　　　　　　　　　Contents

Part I

Biographical and Personal Notes

Operator Theory:
Advances and Applications, Vol. 222, 3–12
© 2012 Springer Basel

Helton's Bibliography

[1] J. Agler, J.W. Helton, S. McCullough, and L. Rodman. Positive semidefinite matrices with a given sparsity pattern. In *Proceedings of the Victoria Conference on Combinatorial Matrix Analysis (Victoria, BC, 1987)*, volume 107, pages 101–149, 1988.

[2] J. Agler, J.W. Helton, and M. Stankus. Classification of hereditary matrices. *Linear Algebra Appl.*, 274:125–160, 1998.

[3] F.N. Bailey, J.W. Helton, and O. Merino. Alternative approaches in frequency domain design of single loop feedback systems with plant uncertainty. *Internat. J. Robust Nonlinear Control*, 7(3):265–277, 1997.

[4] F.N. Bailey, J.W. Helton, and O. Merino. Performance functions and sublevel sets for frequency domain design of single loop feedback control systems with plant uncertainty. *Internat. J. Robust Nonlinear Control*, 7(3):227–263, 1997.

[5] J.A. Ball, V. Bolotnikov, J.W. Helton, L. Rodman, and I.M. Spitkovsky, editors. *Topics in operator theory. Volume 1. Operators, matrices and analytic functions*, volume 202 of *Operator Theory: Advances and Applications*, Basel, 2010. Birkhäuser Verlag. A tribute to Israel Gohberg on the occasion of his 80th birthday.

[6] J.A. Ball, V. Bolotnikov, J.W. Helton, L. Rodman, and I.M. Spitkovsky, editors. *Topics in operator theory. Volume 2. Systems and mathematical physics*, volume 203 of *Operator Theory: Advances and Applications*, Basel, 2010. Birkhäuser Verlag. A tribute to Israel Gohberg on the occasion of his 80th birthday.

[7] J.A. Ball, V. Bolotnikov, J.W. Helton, L. Rodman, and I.M. Spitkovsky. The XIXth International Workshop on Operator Theory and its Applications. II. In *Topics in operator theory. Volume 2. Systems and mathematical physics*, volume 203 of *Oper. Theory Adv. Appl.*, pages vii–ix. Birkhäuser Verlag, Basel, 2010.

[8] J.A. Ball, Y. Eidelman, J.W. Helton, V. Olshevsky, and J. Rovnyak, editors. *Recent advances in matrix and operator theory*, volume 179 of *Operator Theory: Advances and Applications*, Basel, 2008. Birkhäuser Verlag.

[9] J.A. Ball, C. Foias, J.W. Helton, and A. Tannenbaum. Nonlinear interpolation theory in H^∞. In *Modelling, robustness and sensitivity reduction in control systems (Groningen, 1986)*, volume 34 of *NATO Adv. Sci. Inst. Ser. F Comput. Systems Sci.*, pages 31–46. Springer, Berlin, 1987.

[10] J.A. Ball, C. Foias, J.W. Helton, and A. Tannenbaum. On a local nonlinear commutant lifting theorem. *Indiana Univ. Math. J.*, 36(3):693–709, 1987.

[11] J.A. Ball, C. Foias, J.W. Helton, and A. Tannenbaum. A Poincaré-Dulac approach to a nonlinear Beurling-Lax-Halmos theorem. *J. Math. Anal. Appl.*, 139(2):496–514, 1989.

[12] J.A. Ball and J.W. Helton. Nonnormal dilations, disconjugacy and constrained spectral factorization. *Integral Equations Operator Theory*, 3(2):216–309, 1980.

[13] J.A. Ball and J.W. Helton. Subinvariants for analytic mappings on matrix balls. *Analysis*, 1(3):217–226, 1981.

[14] J.A. Ball and J.W. Helton. Factorization results related to shifts in an indefinite metric. *Integral Equations Operator Theory*, 5(5):632–658, 1982.

[15] J.A. Ball and J.W. Helton. Lie groups over the field of rational functions, signed spectral factorization, signed interpolation, and amplifier design. *J. Operator Theory*, 8(1):19–64, 1982.

[16] J.A. Ball and J.W. Helton. A Beurling-Lax theorem for the Lie group U(m, n) which contains most classical interpolation theory. *J. Operator Theory*, 9(1):107–142, 1983.

[17] J.A. Ball and J.W. Helton. Beurling-Lax representations using classical Lie groups with many applications. II. GL(n, **C**) and Wiener-Hopf factorization. *Integral Equations Operator Theory*, 7(3):291–309, 1984.

[18] J.A. Ball and J.W. Helton. Linear fractional parameterizations of matrix function spaces and a new proof of the Youla-Jabr-Bongiorno parameterization for stabilizing compensators. In *Mathematical theory of networks and systems (Beer Sheva, 1983)*, volume 58 of *Lecture Notes in Control and Inform. Sci.*, pages 16–23. Springer, London, 1984.

[19] J.A. Ball and J.W. Helton. Beurling-Lax representations using classical Lie groups with many applications. III. Groups preserving two bilinear forms. *Amer. J. Math.*, 108(1):95–174 (1986), 1986.

[20] J.A. Ball and J.W. Helton. Beurling-Lax representations using classical Lie groups with many applications. IV. GL(n, **R**), U*($2n$), SL(n, **C**), and a solvable group. *J. Funct. Anal.*, 69(2):178–206, 1986.

[21] J.A. Ball and J.W. Helton. Interpolation problems of Pick-Nevanlinna and Loewner types for meromorphic matrix functions: parametrization of the set of all solutions. *Integral Equations Operator Theory*, 9(2):155–203, 1986.

[22] J.A. Ball and J.W. Helton. Shift invariant manifolds and nonlinear analytic function theory. *Integral Equations Operator Theory*, 11(5):615–725, 1988.

[23] J.A. Ball and J.W. Helton. Shift invariant subspaces, passivity, reproducing kernels and H^∞-optimization. In *Contributions to operator theory and its applications (Mesa, AZ, 1987)*, volume 35 of *Oper. Theory Adv. Appl.*, pages 265–310. Birkhäuser, Basel, 1988.

[24] J.A. Ball and J.W. Helton. Factorization and general properties of nonlinear Toeplitz operators. In *The Gohberg anniversary collection, Vol. II (Calgary, AB, 1988)*, volume 41 of *Oper. Theory Adv. Appl.*, pages 25–41. Birkhäuser, Basel, 1989.

[25] J.A. Ball and J.W. Helton. H^∞ control for nonlinear plants: connections with differential games. In *Proceedings of the 28th IEEE Conference on Decision and Control, Vol. 1-3 (Tampa, FL, 1989)*, pages 956–962, New York, 1989. IEEE.

[26] J.A. Ball and J.W. Helton. Interconnection of nonlinear causal systems. *IEEE Trans. Automat. Control*, 34(11):1132–1140, 1989.

[27] J.A. Ball and J.W. Helton. Nonlinear H^∞ control theory: a literature survey. In *Robust control of linear systems and nonlinear control (Amsterdam, 1989)*, volume 4 of *Progr. Systems Control Theory*, pages 1–12. Birkhäuser Boston, Boston, MA, 1990.

[28] J.A. Ball and J.W. Helton. Inner-outer factorization of nonlinear operators. *J. Funct. Anal.*, 104(2):363–413, 1992.

[29] J.A. Ball and J.W. Helton. Nonlinear H^∞ control theory for stable plants. *Math. Control Signals Systems*, 5(3):233–261, 1992.

[30] J.A. Ball and J.W. Helton. Viscosity solutions of Hamilton-Jacobi equations arising in nonlinear H_∞-control. *J. Math. Systems Estim. Control*, 6(1):22 pp. (electronic), 1996.

[31] J.A. Ball, J.W. Helton, M. Klaus, and L. Rodman, editors. *Current trends in operator theory and its applications*, volume 149 of *Operator Theory: Advances and Applications*, Basel, 2004. Birkhäuser Verlag.

[32] J.A. Ball, J.W. Helton, M. Klaus, and L. Rodman. IWOTA 2002 and recent achievements and new directions in operator theory and applications. In *Current trends in operator theory and its applications*, volume 149 of *Oper. Theory Adv. Appl.*, pages ix–xxii. Birkhäuser, Basel, 2004.

[33] J.A. Ball, J.W. Helton, and C.H. Sung. Nonlinear solutions of Nevanlinna-Pick interpolation problems. *Michigan Math. J.*, 34(3):375–389, 1987.

[34] J.A. Ball, J.W. Helton, and M. Verma. A *J*-inner-outer factorization principle for the H^∞ control problem. In *Recent advances in mathematical theory of systems, control, networks and signal processing, I (Kobe, 1991)*, pages 31–36. Mita, Tokyo, 1992.

[35] J.A. Ball, J.W. Helton, and M.L. Walker. H^∞ control for nonlinear systems with output feedback. *IEEE Trans. Automat. Control*, 38(4):546–559, 1993.

[36] E. Basor and J.W. Helton. A new proof of the Szegő limit theorem and new results for Toeplitz operators with discontinuous symbol. *J. Operator Theory*, 3(1):23–39, 1980.

[37] A. Ben-Artzi, J.W. Helton, and H.J. Woerdeman. Non-minimal factorization of nonlinear systems: the observable case. *Internat. J. Control*, 63(6):1069–1104, 1996.

[38] E.A. Bender, W.J. Helton, and L.B. Richmond. Asymptotics of permutations with nearly periodic patterns of rises and falls. *Electron. J. Combin.*, 10:Research Paper 40, 27 pp. (electronic), 2003.

[39] C.I. Byrnes and J.W. Helton. Cascade equivalence of linear systems. *Internat. J. Control*, 44(6):1507–1521, 1986.

[40] J.F. Camino, J.W. Helton, and R.E. Skelton. Solving matrix inequalities whose unknowns are matrices. *SIAM J. Optim.*, 17(1):1–36 (electronic), 2006.

[41] J.F. Camino, J.W. Helton, R.E. Skelton, and J. Ye. Matrix inequalities: a symbolic procedure to determine convexity automatically. *Integral Equations Operator Theory*, 46(4):399–454, 2003.

[42] G. Craciun, J.W. Helton, and R.J. Williams. Homotopy methods for counting reaction network equilibria. *Math. Biosci.*, 216(2):140–149, 2008.

[43] M.C. de Oliveira and J.W. Helton. Computer algebra tailored to matrix inequalities in control. *Internat. J. Control*, 79(11):1382–1400, 2006.

[44] M.C. de Oliveira, J.W. Helton, S.A. McCullough, and M. Putinar. Engineering systems and free semi-algebraic geometry. In *Emerging applications of algebraic geometry*, volume 149 of *IMA Vol. Math. Appl.*, pages 17–61. Springer, New York, 2009.

[45] R.G. Douglas and J.W. Helton. Inner dilations of analytic matrix functions and Darlington synthesis. *Acta Sci. Math. (Szeged)*, 34:61–67, 1973.

[46] H. Dym, J.M. Greene, J.W. Helton, and S.A. McCullough. Classification of all noncommutative polynomials whose Hessian has negative signature one and a noncommutative second fundamental form. *J. Anal. Math.*, 108:19–59, 2009.

[47] H. Dym and J.W. Helton. The matrix multidisk problem. *Integral Equations Operator Theory*, 46(3):285–339, 2003.

[48] H. Dym, J.W. Helton, and S. McCullough. The Hessian of a noncommutative polynomial has numerous negative eigenvalues. *J. Anal. Math.*, 102:29–76, 2007.

[49] H. Dym, J.W. Helton, and O. Merino. Multidisk problems in H^∞ optimization: a method for analysing numerical algorithms. *Indiana Univ. Math. J.*, 51(5):1111–1159, 2002.

[50] H. Dym, W. Helton, and S. McCullough. Irreducible noncommutative defining polynomials for convex sets have degree four or less. *Indiana Univ. Math. J.*, 56(3):1189–1231, 2007.

[51] J.W. Evans and J.W. Helton. Applications of Pick-Nevanlinna interpolation theory to retention-solubility studies in the lungs. *Math. Biosci.*, 63(2):215–240, 1983.

[52] B.A. Francis, J.W. Helton, and G. Zames. $\mathcal{H}^\infty$-optimal feedback controllers for linear multivariable systems. *IEEE Trans. Automat. Control*, 29(10):888–900, 1984.

[53] B.A. Francis, J.W. Helton, and G. Zames. H^∞-optimal feedback controllers for linear multivariable systems. In *Mathematical theory of networks and systems (Beer Sheva, 1983)*, volume 58 of *Lecture Notes in Control and Inform. Sci.*, pages 347–362. Springer, London, 1984.

[54] I. Gohberg, J.W. Helton, and L. Rodman, editors. *Contributions to operator theory and its applications*, volume 35 of *Operator Theory: Advances and Applications*, Basel, 1988. Birkhäuser Verlag.

[55] F.A. Grünbaum, J.W. Helton, and P. Khargonekar, editors. *Signal processing. Part II*, volume 23 of *The IMA Volumes in Mathematics and its Applications*. Springer-Verlag, New York, 1990. Control theory and applications.

[56] D.M. Hay, J.W. Helton, A. Lim, and S. McCullough. Non-commutative partial matrix convexity. *Indiana Univ. Math. J.*, 57(6):2815–2842, 2008.

[57] F.J. Helton and J.W. Helton. Scattering theory and non linear systems. In *Category theory applied to computation and control (Proc. First Internat. Sympos., San Francisco, Calif., 1974)*, pages 164–169. Lecture Notes in Comput. Sci., Vol. 25. Springer, Berlin, 1975.

[58] H. Helton. Operators with a representation as multiplication by $\times$ on a Sobolev space. In *Hilbert space operators and operator algebras (Proc. Internat. Conf., Tihany, 1970)*, pages 279–287. (loose errata) Colloq. Math. Soc. János Bolyai, 5. North-Holland, Amsterdam, 1972.

[59] J.W. Helton. Unitary operators on a space with an indefinite inner product. *J. Functional Analysis*, 6:412–440, 1970.

[60] J.W. Helton. Jordan operators in infinite dimensions and Sturm Liouville conjugate point theory. *Bull. Amer. Math. Soc.*, 78:57–61, 1971.

[61] J.W. Helton. Operators unitary in an indefinite metric and linear fractional transformations. *Acta Sci. Math. (Szeged)*, 32:261–266, 1971.

[62] J.W. Helton. Infinite dimensional Jordan operators and Sturm-Liouville conjugate point theory. *Trans. Amer. Math. Soc.*, 170:305–331, 1972.

[63] J.W. Helton. Passive network realization using abstract operator theory. *IEEE Trans. Circuit Theory*, CT-19:518–520, 1972.

[64] J.W. Helton. The characteristic functions of operator theory and electrical network realization. *Indiana Univ. Math. J.*, 22:403–414, 1972/73.

[65] J.W. Helton. Continuity of linear fractional transformations on an operator algebra. *Proc. Amer. Math. Soc.*, 40:217–218, 1973.

[66] J.W. Helton. Discrete time systems, operator models, and scattering theory. *J. Functional Analysis*, 16:15–38, 1974.

[67] J.W. Helton. Book Review: Indefinite inner product spaces. *Bull. Amer. Math. Soc.*, 81(6):1028–1030, 1975.

[68] J.W. Helton. A spectral factorization approach to the distributed stable regular problem; the algebraic Riccati equation. *SIAM J. Control Optimization*, 14(4):639–661, 1976.

[69] J.W. Helton. Systems with infinite-dimensional state space: the Hilbert space approach. *Proc. IEEE*, 64(1):145–160, 1976. Recent trends in system theory.

[70] J.W. Helton. An operator algebra approach to partial differential equations; propagation of singularities and spectral theory. *Indiana Univ. Math. J.*, 26(6):997–1018, 1977.

[71] J.W. Helton. Orbit structure of the Möbius transformation semigroup acting on H^∞ (broadband matching). In *Topics in functional analysis (essays dedicated to M.G. Kreĭn on the occasion of his 70th birthday)*, volume 3 of *Adv. in Math. Suppl. Stud.*, pages 129–157. Academic Press, New York, 1978.

[72] J.W. Helton. The distance of a function to H^∞ in the Poincaré metric; electrical power transfer. *J. Funct. Anal.*, 38(2):273–314, 1980.

[73] J.W. Helton. Broadbanding: gain equalization directly from data. *IEEE Trans. Circuits and Systems*, 28(12):1125–1137, 1981.

[74] J.W. Helton. Book Review: Minimal factorization of matrix and operator functions. *Bull. Amer. Math. Soc. (N.S.)*, 6(2):235–238, 1982.

[75] J.W. Helton. A conference report. *Integral Equations Operator Theory*, 5(1):605–607, 1982.

[76] J.W. Helton. Non-Euclidean functional analysis and electronics. *Bull. Amer. Math. Soc. (N.S.)*, 7(1):1–64, 1982.

[77] J.W. Helton. Optimization, engineering, and a more general Corona theorem. In *Operators and function theory (Lancaster, 1984)*, volume 153 of *NATO Adv. Sci. Inst. Ser. C Math. Phys. Sci.*, pages 39–40. Reidel, Dordrecht, 1985.

[78] J.W. Helton. Worst case analysis in the frequency domain: the H^∞ approach to control. *IEEE Trans. Automat. Control*, 30(12):1154–1170, 1985.

[79] J.W. Helton. Optimization over spaces of analytic functions and the corona problem. *J. Operator Theory*, 15(2):359–375, 1986.

[80] J.W. Helton. Book review. *Integral Equations Operator Theory*, 10(5):751–753, 1987. Operator theory: Advances and Applications Operator theory and systems: edited by I. Gohberg and M.A. Kaasshoek, Birkhäuser Verlag, Basel-Boston-Stuttgart 1986. OT 19.

[81] J.W. Helton. A numerical method for computing the structured singular value. *Systems Control Lett.*, 10(1):21–26, 1988.

[82] J.W. Helton. Factorization of nonlinear systems. In *The Gohberg anniversary collection, Vol. II (Calgary, AB, 1988)*, volume 41 of *Oper. Theory Adv. Appl.*, pages 311–328. Birkhäuser, Basel, 1989.

[83] J.W. Helton. Beyond commutant lifting. In *Operator theory: operator algebras and applications, Part 1 (Durham, NH, 1988)*, volume 51 of *Proc. Sympos. Pure Math.*, pages 219–224. Amer. Math. Soc., Providence, RI, 1990.

[84] J.W. Helton. Optimal frequency domain design vs. an area of several complex variables. In *Robust control of linear systems and nonlinear control (Amsterdam, 1989)*, volume 4 of *Progr. Systems Control Theory*, pages 33–59. Birkhäuser Boston, Boston, MA, 1990.

[85] J.W. Helton. Two topics in systems engineering: frequency domain design and nonlinear systems. In *H_∞-control theory (Como, 1990)*, volume 1496 of *Lecture Notes in Math.*, pages 106–140. Springer, Berlin, 1991.

[86] J.W. Helton. Some adaptive control problems which convert to a "classical" problem in several complex variables. *IEEE Trans. Automat. Control*, 46(12):2038–2043, 2001.

[87] J.W. Helton. "Positive" noncommutative polynomials are sums of squares. *Ann. of Math. (2)*, 156(2):675–694, 2002.

[88] J.W. Helton. Manipulating matrix inequalities automatically. In *Mathematical systems theory in biology, communications, computation, and finance (Notre Dame, IN, 2002)*, volume 134 of *IMA Vol. Math. Appl.*, pages 237–256. Springer, New York, 2003.

[89] J.W. Helton and J.A. Ball. The cascade decompositions of a given system vs. the linear fractional decompositions of its transfer function. *Integral Equations Operator Theory*, 5(3):341–385, 1982.

[90] J.W. Helton, J.A. Ball, C.R. Johnson, and J.N. Palmer. *Operator theory, analytic functions, matrices, and electrical engineering*, volume 68 of *CBMS Regional Conference Series in Mathematics*. Published for the Conference Board of the Mathematical Sciences, Washington, DC, 1987.

[91] J.W. Helton and R.E. Howe. Integral operators: commutators, traces, index and homology. In *Proceedings of a Conference Operator Theory (Dalhousie Univ., Halifax, N.S., 1973)*, pages 141–209. Lecture Notes in Math., Vol. 345, Berlin, 1973. Springer.

[92] J.W. Helton and R.E. Howe. Traces of commutators of integral operators. *Acta Math.*, 135(3-4):271–305, 1975.

[93] J.W. Helton and R.E. Howe. A bang-bang theorem for optimization over spaces of analytic functions. *J. Approx. Theory*, 47(2):101–121, 1986.

[94] J.W. Helton and M.R. James. On the stability of the information state system. *Systems Control Lett.*, 29(2):61–72, 1996.

[95] J.W. Helton and M.R. James. *Extending H^∞ control to nonlinear systems*, volume 1 of *Advances in Design and Control*. Society for Industrial and Applied Mathematics (SIAM), Philadelphia, PA, 1999. Control of nonlinear systems to achieve performance objectives.

[96] J.W. Helton, M.R. James, and W.M. McEneaney. Reduced-complexity nonlinear H^∞ control of discrete-time systems. *IEEE Trans. Automat. Control*, 50(11):1808–1811, 2005.

[97] J.W. Helton, V. Katsnelson, and I. Klep. Sign patterns for chemical reaction networks. *J. Math. Chem.*, 47(1):403–429, 2010.

[98] J.W. Helton, I. Klep, and R. Gomez. Determinant expansions of signed matrices and of certain Jacobians. *SIAM J. Matrix Anal. Appl.*, 31(2):732–754, 2009.

[99] J.W. Helton, I. Klep, and S. McCullough. Analytic mappings between noncommutative pencil balls. *J. Math. Anal. Appl.*, 376(2):407–428, 2011.

[100] J.W. Helton, I. Klep, and S. McCullough. Proper analytic free maps. *J. Funct. Anal.*, 260(5):1476–1490, 2011.

[101] J.W. Helton, I. Klep, S. McCullough, and N. Slinglend. Noncommutative ball maps. *J. Funct. Anal.*, 257(1):47–87, 2009.

[102] J.W. Helton, F.D. Kronewitter, W.M. McEneaney, and M. Stankus. Singularly perturbed control systems using non-commutative computer algebra. *Internat. J. Robust Nonlinear Control*, 10(11-12):983–1003, 2000. George Zames commemorative issue.

[103] J.W. Helton, D. Lam, and H.J. Woerdeman. Sparsity patterns with high rank extremal positive semidefinite matrices. *SIAM J. Matrix Anal. Appl.*, 15(1):299–312, 1994.

[104] J.W. Helton, J.B. Lasserre, and M. Putinar. Measures with zeros in the inverse of their moment matrix. *Ann. Probab.*, 36(4):1453–1471, 2008.

[105] J.W. Helton and D.E. Marshall. Frequency domain design and analytic selections. *Indiana Univ. Math. J.*, 39(1):157–184, 1990.

[106] J.W. Helton, D.P. McAllaster, and J.A. Hernandez. Non-commutative harmonic and subharmonic polynomials. *Integral Equations Operator Theory*, 61(1):77–102, 2008.

[107] J.W. Helton and S. McCullough. Convex noncommutative polynomials have degree two or less. *SIAM J. Matrix Anal. Appl.*, 25(4):1124–1139 (electronic), 2004.

[108] J.W. Helton, S. McCullough, and M. Putinar. Matrix representations for positive noncommutative polynomials. *Positivity*, 10(1):145–163, 2006.

[109] J.W. Helton, S. McCullough, and M. Putinar. Strong majorization in a free *-algebra. *Math. Z.*, 255(3):579–596, 2007.

[110] J.W. Helton, S. McCullough, M. Putinar, and V. Vinnikov. Convex matrix inequalities versus linear matrix inequalities. *IEEE Trans. Automat. Control*, 54(5):952–964, 2009.

[111] J.W. Helton and S.A. McCullough. A Positivstellensatz for non-commutative poly-nomials. *Trans. Amer. Math. Soc.*, 356(9):3721–3737 (electronic), 2004.

[112] J.W. Helton, S.A. McCullough, and M. Putinar. A non-commutative Positivstel-lensatz on isometries. *J. Reine Angew. Math.*, 568:71–80, 2004.

[113] J.W. Helton, S.A. McCullough, and M. Putinar. Non-negative hereditary polyno-mials in a free ∗-algebra. *Math. Z.*, 250(3):515–522, 2005.

[114] J.W. Helton, S.A. McCullough, and V. Vinnikov. Noncommutative convexity arises from linear matrix inequalities. *J. Funct. Anal.*, 240(1):105–191, 2006.

[115] J.W. Helton and O. Merino. Optimal analytic disks. In *Several complex variables and complex geometry, Part 2 (Santa Cruz, CA, 1989)*, volume 52 of *Proc. Sympos. Pure Math.*, pages 251–262. Amer. Math. Soc., Providence, RI, 1991.

[116] J.W. Helton and O. Merino. Conditions for optimality over H^∞. *SIAM J. Control Optim.*, 31(6):1379–1415, 1993.

[117] J.W. Helton and O. Merino. Novel approach to accelerating Newton's method for sup-norm optimization arising in H^∞-control. *J. Optim. Theory Appl.*, 78(3):553–578, 1993.

[118] J.W. Helton and O. Merino. A fibered polynomial hull without an analytic selection. *Michigan Math. J.*, 41(2):285–287, 1994.

[119] J.W. Helton and O. Merino. *Classical control using H^∞ methods.* Society for In-dustrial and Applied Mathematics (SIAM), Philadelphia, PA, 1998. Theory, opti-mization, and design.

[120] J.W. Helton and O. Merino. *Classical control using H^∞ methods.* Society for Indus-trial and Applied Mathematics (SIAM), Philadelphia, PA, 1998. An introduction to design.

[121] J.W. Helton, O. Merino, and T.E. Walker. Algorithms for optimizing over analytic functions. *Indiana Univ. Math. J.*, 42(3):839–874, 1993.

[122] J.W. Helton, O. Merino, and T.E. Walker. Optimization over analytic functions whose Fourier coefficients are constrained. *Integral Equations Operator Theory*, 22(4):420–439, 1995.

[123] J.W. Helton, O. Merino, and T.E. Walker. H^∞ optimization with plant uncertainty and semidefinite programming. *Internat. J. Robust Nonlinear Control*, 8(9):763–802, 1998.

[124] J.W. Helton and J. Nie. Sufficient and necessary conditions for semidefinite repre-sentability of convex hulls and sets. *SIAM J. Optim.*, 20(2):759–791, 2009.

[125] J.W. Helton and J. Nie. Semidefinite representation of convex sets. *Math. Program.*, 122(1, Ser. A):21–64, 2010.

[126] J.W. Helton, S. Pierce, and L. Rodman. The ranks of extremal positive semidefinite matrices with given sparsity pattern. *SIAM J. Matrix Anal. Appl.*, 10(3):407–423, 1989.

[127] J.W. Helton and M. Putinar. Positive polynomials in scalar and matrix variables, the spectral theorem, and optimization. In *Operator theory, structured matrices, and dilations*, volume 7 of *Theta Ser. Adv. Math.*, pages 229–306. Theta, Bucharest, 2007.

[128] J.W. Helton and J.V. Ralston. The first variation of the scattering matrix. *J. Differential Equations*, 21(2):378–394, 1976.

[129] J.W. Helton, R. Rashidi Far, and R. Speicher. Operator-valued semicircular elements: solving a quadratic matrix equation with positivity constraints. *Int. Math. Res. Not. IMRN*, (22):Art. ID rnm086, 15, 2007.

[130] J.W. Helton and L. Rodman. Signature preserving linear maps of Hermitian matrices. *Linear and Multilinear Algebra*, 17(1):29–37, 1985.

[131] J.W. Helton and L. Rodman. Vandermonde and resultant matrices: an abstract approach. *Math. Systems Theory*, 20(2-3):169–192, 1987.

[132] J.W. Helton and L. Rodman. Correction: "Vandermonde and resultant matrices: an abstract approach". *Math. Systems Theory*, 21(1):61, 1988.

[133] J.W. Helton and L.A. Sakhnovich. Extremal problems of interpolation theory. *Rocky Mountain J. Math.*, 35(3):819–841, 2005.

[134] J.W. Helton and D.F. Schwartz. Polynomially contractible sets. *Complex Variables Theory Appl.*, 10(4):333–348, 1988.

[135] J.W. Helton, D.F. Schwartz, and S.E. Warschawski. Local optima in H^∞ produce a constant objective function. *Complex Variables Theory Appl.*, 8(1-2):65–81, 1987.

[136] J.W. Helton and A. Sideris. Frequency response algorithms for H_∞ optimization with time domain constraints. *IEEE Trans. Automat. Control*, 34(4):427–434, 1989.

[137] J.W. Helton, P.G. Spain, and N.J. Young. Tracking poles and representing Hankel operators directly from data. *Numer. Math.*, 58(6):641–660, 1991.

[138] J.W. Helton and M. Stankus. Computer assistance for "discovering" formulas in system engineering and operator theory. *J. Funct. Anal.*, 161(2):289–363, 1999.

[139] J.W. Helton, M. Stankus, and J.J. Wavrik. Computer simplification of formulas in linear systems theory. *IEEE Trans. Automat. Control*, 43(3):302–314, 1998.

[140] J.W. Helton and M. Tabor. On classical and quantal Kolmogorov entropies. *J. Phys. A*, 18(14):2743–2749, 1985.

[141] J.W. Helton and M. Tabor. On the classical support of quantum mechanical wavefunctions. *Phys. D*, 14(3):409–415, 1985.

[142] J.W. Helton and V. Vinnikov. Linear matrix inequality representation of sets. *Comm. Pure Appl. Math.*, 60(5):654–674, 2007.

[143] J.W. Helton and A.E. Vityaev. Analytic functions optimizing competing constraints. *SIAM J. Math. Anal.*, 28(3):749–767, 1997.

[144] J.W. Helton, M.L. Walker, and W. Zhan. H^∞ control of systems with compensators which read the command signal. *Internat. J. Control*, 64(1):9–27, 1996.

[145] J.W. Helton and J.J. Wavrik. Rules for computer simplification of the formulas in operator model theory and linear systems. In *Nonselfadjoint operators and related topics (Beer Sheva, 1992)*, volume 73 of *Oper. Theory Adv. Appl.*, pages 325–354. Birkhäuser, Basel, 1994.

[146] J.W. Helton and F. Weening. Some systems theorems arising from the Bieberbach conjecture. *Internat. J. Robust Nonlinear Control*, 6(1):65–82, 1996.

[147] J.W. Helton and M.A. Whittlesey. Global uniqueness tests and performance bounds for H^∞ optima. *SIAM J. Control Optim.*, 42(1):363–380 (electronic), 2003.

[148] J.W. Helton and H.J. Woerdeman. Symmetric Hankel operators: minimal norm extensions and eigenstructures. *Linear Algebra Appl.*, 185:1–19, 1993.

[149] J.W. Helton and N.J. Young. Approximation of Hankel operators: truncation error in an H^∞ design method. In *Signal Processing, Part II*, volume 23 of *IMA Vol. Math. Appl.*, pages 115–137. Springer, New York, 1990.

[150] J.W. Helton and A.H. Zemanian. The cascade loading of passive Hilbert ports. *SIAM J. Appl. Math.*, 23:292–306, 1972.

[151] J.W. Helton and W. Zhan. An inequality governing nonlinear H_∞ control. *Systems Control Lett.*, 22(3):157–165, 1994.

[152] J.W. Helton and W. Zhan. Piecewise Riccati equations and the bounded real lemma. *Internat. J. Robust Nonlinear Control*, 7(8):741–757, 1997.

[153] J.W. Helton, Jr. *Invariant subspaces of certain commuting families of operators on linear spaces with an indefinite inner product.* ProQuest LLC, Ann Arbor, MI, 1968. Thesis (Ph.D.) – Stanford University.

[154] W. Helton, J. Rosenthal, and X. Wang. Matrix extensions and eigenvalue completions, the generic case. *Trans. Amer. Math. Soc.*, 349(8):3401–3408, 1997.

[155] E.A. Jonckheere and J.W. Helton. Power spectrum reduction by optimal Hankel norm approximation of the phase of the outer spectral factor. *IEEE Trans. Automat. Control*, 30(12):1192–1201, 1985.

[156] M.L. Walker and J.W. Helton. High frequency inverse scattering and the Luneberg-Kline asymptotic expansion. *IEEE Trans. Antennas and Propagation*, 40(4):450–453, 1992.

[157] S. Yuliar, M.R. James, and J.W. Helton. Dissipative control systems synthesis with full state feedback. *Math. Control Signals Systems*, 11(4):335–356, 1998.

Operator Theory:
Advances and Applications, Vol. 222, 13–15
© 2012 Springer Basel

Personal Notes

Just as the control systems field owes a debt of gratitude to Bill Helton for introducing it to techniques that have enabled the development of H_∞ control theory and several other sub-disciplines of control and optimization, the UCSD control community is tremendously indebted to Bill for being a beacon of excellence in control theory and mathematics at UC San Diego and whose presence has facilitated the formation of the new control engineering program at UCSD in the late 1990s. Thank you Bill for always being a thoughtful, constructive, and unselfish colleague and friend.

MIROSLAV KRSTIC, Department of Mechanical and Aerospace Engineering
University of California San Diego, La Jolla, CA 92093-0411, USA
e-mail: krstic@ucsd.edu

* * *

Bill and I have been good friends for more than 45 years. We have shared many experiences, some funny, some serious. In the summer of 1968 when I decided that part of my PhD thesis was hopelessly wrong, I went to Bill to sort things out. He put his feet up, stared at the pages for a while, and explained that everything would be fine if I added a few lines.

You may have noticed that Bill likes to film things. Over the years he has found ever smaller, more easily concealed cameras to do this. He has probably filmed you. I would like to report one of his early cinematic efforts.

For a time when we were graduate students five of us rented a small green stucco house in Menlo Park. Our landlord, Gracie – we called it Gracie Mansion – conveniently lived a good distance away near Santa Cruz. She left the maintenance of Gracie Mansion to us... with predicable results. Bill decided to make a film of our efforts to clean the place up. The resulting 5 minute silent, "Cleaning the House," built to a fine dramatic climax as one of the house cleaners flushed himself down the drain. If it had had wider distribution, it might now be a surrealist classic. I fear that it has been lost, and write this as proof of its (one time) existence.

JAMES RALSTON, Department of Mathematics
University of California, Los Angeles, CA 90095-1555, USA
e-mail: ralston@math.ucla.edu

I first met Bill in the winter of 1971, when I was on leave from Berkeley and spending time on the east coast. We met during a visit of a few days I made to Stony Brook and we must have discussed mathematics, but I don't remember any of that. What I do remember is that we spent some time batting a ball around a handball court. (Bill was much better at it than I was.)

In the fall of 1971 I was the referee for the A.M.S. Transactions of one of Bill's early papers, "Infinite-dimensional Jordan operators and Sturm-Liouville conjugate point theory." I was very impressed by the paper and strongly recommended it for publication. Bill came to know I was the referee because of what happened next.

In the spring of 1972 Bill was pessimistic about his prospects for tenure at Stony Brook, due to the situation there at the time. He decided he had best get on the job market, and he had the thought that the (anonymous) referee of his paper might serve as one of his references. He contacted the Transactions editor, who contacted me, and I agreed. I thus played a bit part in Stony Brook's loss and UCSD's gain.

In spring 1972 Bill's migration toward systems theory was well underway. In a letter to me dated April 17, 1973, he first explained the murky tenure picture at Stony Brook, and went on to write: "I've spent the year learning engineering systems theory which at some levels is almost straight operator theory. Some of the best functional analysis (Krein, Livsic) has come from engineering institutes and I'm beginning to see why. Such collaboration does not exist in this country and I would very much like to go to a place with strong engineers where an engineering and operator theory group might be formed." As the saying goes, the rest is history.

Despite Bill's best efforts, I never became strongly immersed in the systems theory viewpoint – my purist instincts kept leading me elsewhere. Our different perspectives notwithstanding, I learned a great deal of mathematics from Bill over the years, both through our frequent interactions at conferences and through studying some of his papers.

No operator theorist can help but be strongly influenced by Bill, if not directly, then by the way his vision and his leadership have transformed our subject.

DONALD SARASON, Department of Mathematics
University of California, Berkeley, CA 94720-3840, USA
e-mail: sarason@math.berkeley.edu

* * *

I came to UCSD in November, 1996 with the assignment to build a systems and control group at UCSD. I was impressed with the courage of UCSD to fund the whole group at once, to try to create an immediate impact, rather than the traditional university approach of adding a junior faculty every few years. I expected this to be a difficult challenge, and a very lonely one, building the program from scratch. Then like the Lone Ranger, help arrived from across the campus. Bill Helton attended endless seminars and dinners and BBQs and convinced candidates that there was a control-friendly math department on campus. Quality people came, and I am very pleased with the recruiting results. But Bill Helton contributed just as much toward the development of a world-class systems and control group at UCSD.

I express my sincere appreciation to Bill for his generous contributions of time and wisdom and friendship during this build-up phase. I have been told by the administration that the creation of the systems and control program was a significant milestone in the growth of the still very young UCSD. Bill Helton was partially responsible for this event. I appreciate Ruth as well, for the many BBQs, the patience to tolerate Bill's abundant engineering activities, and the parking lot for sabbatical cars. While Bill's math skills are impecible, he has a rare human quality that dwarfs all other attributes. He is a kind and humble person who lets his work do his talking. He presents to the world a strength of character as strong as his strength of skill. He is one of my favorite people on this planet. I summarize with a Limmerick.

> I once knew a mathematician
> Whose skills would scare a magician
> But the engineers he helped
> Are greatly in his debt
> For the friendship he added in addition.

ROBERT SKELTON, Department of Mechanical and Aerospace Engineering
University of California San Diego, La Jolla, CA 92093-0411, USA
e-mail: bobskelton@ucsd.edu

Operator Theory:
Advances and Applications, Vol. 222, 17–21
© 2012 Springer Basel

After Dinner Biographical Notes

*As presented by Miroslav Krstic at the gala
dinner during UCSD Bill's Festricht*

Basic Career Facts About Bill Helton

Bill Helton was born in November 1944 in Jacksonville, Texas. His father was an oral surgeon initially in the US Army and later in private practice. His mother was a drama major at Northwestern University and worked in her husband's oral surgery practice. Bill finished high school at Alamo Heights, TX, at the age of 16. He received his undergraduate degree from the University of Texas at Austin at the age of 19 and completed his PhD at Stanford in 1968 at the age of 23. After receiving his PhD, Bill was on the faculty at SUNY Stony Brook until 1973. He moved to UCSD in 1974. He has held the title of Distinguished Professor of Mathematics since the mid-1990s. Throughout his career Bill has made a remarkable number of breakthroughs which led to new fields. I will review them briefly.

H_∞ Engineering

What dominated control engineering until the late 1970s was control based on mean square specifications, namely, H_2 control. Bill Helton became a household name in control engineering by introducing tools that led to the development of powerful methods dealing with worst case frequency-domain specifications, namely, H_∞ control. Mathematically, this amounts to approximation in the supremum norm with functions analytic in the disk or right half-plane.

Originally the system techniques which now dominate H_∞ engineering were developed to solve the classical broadband impedance matching problem of electric circuit theory. Bill's breakthrough was in solving the MIMO problem in these three papers [H78a], [H80], [H82], with the results first appearing in engineering conferences from 1976 to 1978:

[H78a] J.W. Helton: "Orbit structure of the Möbius transformation semigroup action on H_∞ (broadband matching)," *Adv. in Math. Suppl. Stud.*, 3 Academic Press, New York (1978), 129–197.

[H80]   J.W. Helton: "The distance from a function to H_∞ in the Poincaré metric; electrical power transfer," *J. Functional Analysis*, 38 (1980), No. 2, 273–314.

[H82] J.W. Helton: "Non-Euclidean functional analysis and electronics," *Bull. AMS*, 7 (1982), No.1, 1–64.

Bill solved the impedance matching problem by converting it to problems solved using the Adamjan-Arov-Krein Theorem (AAK), the Commutant Lifting Theory of Foias-Nagy-Sarason, and Nehari Theory. A genuine scientific revolution followed, largely through Bill's interactions with George Zames, who propelled H_∞ into the area of control, and with Patrick Dewilde and Tom Kailath, who introduced these techniques to signal processing.

In addition to providing impetus to others, Bill and collaborators, most notably Joe Ball, worked out substantial amounts of the early input-output theory of H_∞ control. This was published mostly in mathematics articles, however, the control paper by Francis-Helton-Zames [FrHZ84] won an outstanding paper award from the IEEE Control Systems Society.

[FrHZ84] B.A. Francis, J.W. Helton and G. Zames: "H_∞ optimal feedback controllers for linear multivariable systems," *IEEE Trans. Auto. Control*, AC-29 (1984), No. 10, 888–990.

Linear Matrix Inequalities vs Convexity

Some of Bill's more recent work resolved open questions in convex optimization, more specifically semidefinite programming through LMIs. A first set of contributions revolves around the question of what classes of convex sets can be represented as LMIs?

1) With Victor Vinnikov, Bill developed a test which a convex set defined by polynomial inequalities must satisfy in order to have an LMI representation, and showed that not all such sets can pass the test. The test is also a sufficient condition in two dimensions, and helped P.A. Parrilo, A.S. Lewis and M.V. Ramana settle the 1958 Lax conjecture;

2) With Jiawang Nie, Bill then went to show that nonempty compact convex sets defined by polynomial inequalities satisfying a technical regularity condition can be lifted into an LMI, that is represented as an LMI with extra variables.

A second area of Bill's effort is in convexity and positivity of noncommutative functions.

1) Bill and Scott McCullough showed that every convex noncommutative polynomial has degree two or less;

2) Bill, Scott McCullough and Victor Vinnikov showed that every convex noncommutative rational function has an LMI representation, with further refinements with Harry Dym and Scott McCullough;

3) Bill, Scott McCullough and Mihai Putinar provided a noncommutative form of Positivstellensatz by extending the result that every positive noncommutative polynomial is a sum-of-squares.

Other Areas that Bill Opened

Engineering vs Operator Model Theory vs Lax-Phillips Scattering

Bill's papers [H74], [H76b] were devoted to showing that the core of operator model theory, Lax-Phillips Scattering Theory, and linear engineering systems theory are essentially equivalent.

[H74]   Helton, Discrete time systems, operator models and scattering theory, *J. Functional Analysis*, (1974), 15–38.

[H76b] Helton: "Systems with infinite-dimensional state space; the Hilbert space approach," *Proc. IEEE* (1976), 145–160.

Independently Paul Fuhrmann and Patrick Dewilde observed something similar. The conference MTNS was originally set up to explore such interplays between operator theory and system theory.

In related early work, one of Bill's papers with Ron Douglas [DH73a] provided a result which, when specialized to rational functions, says that any energy dissipating circuit can be realized using a lossless circuit attached to a unit resistor.

[DH73a] R.G. Douglas and J.W. Helton: "The precise theoretical limits of causal Darlington synthesis," *Trans. IEEE Circuit Theory*, CT 20 No. 3 (1973).

Noncommutative Geometry

Bill's 1976 paper with Roger Howe [HH76] revealed a structure in an area which became one of the two cornerstones of non-commutative geometry.

[HH76] J.W. Helton and R. Howe, Traces of commutators of integral operators, *Acta Math.*, (1976), 272–305.

This paper influenced Alain Connes in his invention of noncommutative differential geometry, which was later used in string theory.

By the 1980s noncommutative geometry had become one of the main branches of operator theory.

Spectral Representations of Operators

In this paper from the beginning of the 1970s [H72], aged barely over 25, Bill described which operators are equivalent to Multiplication on a Sobolev Space.

[H72]   J.W. Helton: "Infinite-dimensional Jordan operators and Sturm-Liouville conjugate point theory," *Trans. Amer. Math. Soc.*, 170 (1972), 305–331.

He then wrote down a lift of these to simpler operators. The surprise here was that the lifting theory for a class of seemingly abstract operators defined algebraically turns out to generalize classical ODE theorems. This was the forerunner of the hereditary operator theory.

Frequency Domain Optimization

By the mid-1980s, Bill had moved on to a worst case frequency domain theory for more general problems, including those in several complex variables.

1. Bill and Roger Howe [HHo86] showed that when worst case performance of a system is optimized the resulting performance does not depend on frequency.
2. Bill and Don Marshall [HMar90] showed that for one complex variable (SISO systems) the optimal solution is unique.

[HHo86]   J.W. Helton and R. Howe: "A bang-bang theorem for optimization over spaces of analytic functions," *J. Approx. Theory*, 47 (1986), No. 2, 101–121.

HMar90 11 J.W. Helton and D. Marshall: "Frequency domain design and analytic selections," *Indiana Univ. Math. Journal*, 39 (Spring 1990), No. 1, 157–184.

In addition, Bill with Orlando Merino provided 2nd-order convergent numerical algorithms and diagnostics for use with practically any other algorithm.

Extending H_∞ Control to Nonlinear Systems

In the early 1980's Bill began wondering if H_∞ theory would extend in some form to nonlinear systems and with Joe Ball, Ciprian Foias, and Allen Tannenbaum [BFHTa87] laid out the basic problem and gave suitable first solutions.

[BFHTa87] J.A. Ball, C. Foias, J.W. Helton, and A. Tannenbaum. Nonlinear interpolation theory in H_∞. In *Modelling, robustness and sensitivity reduction in control systems* (Groningen, 1986), volume 34 of *NATO Adv. Sci. Inst. Ser. F Comput. Systems Sci.*, 31–46. Springer, Berlin, 1987.

Then he and Joe Ball provided state space results for stable systems. A couple of years later came the breakthrough of Doyle-Glover-Kargonekaar-Francis for linear systems which identified H_∞ control with differential games. This signaled the direction that should be pursued in the nonlinear case and many people made significant contributions, including Tamer Basar, Arjan van der Schaft, Alessandro Astolfi, Alberto Isidori and Matthew James. Bill and Matthew James provided results on asymptotics of the state estimation equations.

Noncommutative Computer Algebra

In the early 1990s Bill produced a package, NCAlgebra which is the main tool in Mathematica for doing noncommutative computer algebra. It is currently maintained and expanded by Bill, Mauricio de Oliveira and Mark Stankus. It is aimed at doing computations with matrices and operators without writing them in terms of all of their entries.

Bill has also done computational work on noncommutative computer algebra algorithms for systems (with Mauricio de Oliveira) and for solving inequalities in matrix unknowns (with Juan Camino and Bob Skelton).

Interdisciplinary Conferences

The Mathematical Theory of Networks and Systems (MTNS) is one of the main conferences today in the mathematics of systems. Bill was one of the founders and remains to this day one of the driving forces on the Steering Committee.

A related ongoing conference is the International Workshop on Operator Theory and Applications, IWOTA. The first IWOTA was organized in 1981 by Bill in the week before MTNS, with the idea of giving pure mathematicians a place where they could speak and then encouraging them to learn engineering by going to the MTNS. This association with the MTNS has worked continuously for nearly 30 years, with the main strategic decisions being made by Israel Gohberg (President) and the Vice Presidents, Bill Helton and Rien Kaashoek.

Operator Theory:
Advances and Applications, Vol. 222, 23–28

Bill's Early Work

*As presented by Jim Agler at the gala
dinner during UCSD Bill's Festricht*

Hi everybody. My name is Jim Agler. I have known Bill for 28 years now, and, well frankly, can think of no one that has influenced my mathematical perspectives more than Bill. Bill also was enormously helpful when I was still a young mathematician in helping me to establish my career. For those gifts Bill, I am very grateful.

Now those of you who know me, know well my capacity for flakiness, and also my tendency for reclusiveness. When I heard via the grapevine that Bill was going to have a birthday bash, my first thought was:

"Oh my God, this could involve my actually having to do something."

What I needed was plausible deniability. Now, I've never been an easy person to contact, not doing the Email, and after a bit of reflection, I conceived what I thought would be an iron-clad plan to be able to deny I ever knew there was a birthday party for Bill. I would not answer the phone nor would I talk to any of my collaborators (all of whom of course know Bill) until after the birthday bash was over.

So secure in the knowledge that I had escaped having to go to Bill's birthday bash, I was sitting in my back yard a couple of weeks ago, studying Bill's latest beautiful paper analyzing the feasibility of eliminating commutative algebra from the high school math curriculum, when suddenly, a pigeon landed on my bistro table and started drinking my coffee. It had a note attached to its leg:

"Jim, Bill has a birthday party... can you say a few words about the early work at the banquet? Ruth."

Damn... I should have known my plan was not going to work on Ruth... resigning myself to my fate, I composed the return message:

"Ruth, of course I would be willing to talk about my early work at Bill's banquet... would an hour be enough material or do you want more? Jim"

I attached the message to the pigeon, who by that time had finished the rest of my coffee, and sent it packing. I then went inside to get more coffee, returned to my bistro table, and recommenced my study of Bill's latest paper. Apparently, if you lay off all of the 7th grade algebra teachers in California, there are just enough funds freed up to preserve the off-scale pay structures for all the math departments in the UC System.

An hour later the pigeon returned, I hastily ripped a few pages from Bill's paper and stuffed them into the top of my coffee cup and began to read a new message.

"Jim, You Worthless Drongo. Sometimes I wonder whether you could find a grand piano in a one roomed house. I don't want you to talk about YOUR early work... I want you to talk about BILL's early work. Can you prove to me that you are not as useless as an ashtray on a motorbike and get on that now. Ruth... P.S. 10 minutes should be plenty of time. "

Mike Crandall

Well I hadn't wanted to mention to Ruth that I hadn't a clue what Bill's early work had been, so, I thought "I know, I'll ask Mike. Surely, a guys thesis advisor is the one to ask about early work". I got him on the phone and after a few pleasantries I ask him "Mike, what can you tell me about Bill's early work?" After a very long pause, he replies:

"Hmm... lets see now... Bill's work... uhh... well actually I can't recall any off hand... now I do have a vivid memory of that brilliant paper that you wrote, Jim, what was it called now... oh yea..."On the operators of the form multiplication by x on Sobolev space". Our analysis group at Stanford used to hang around late into the night reading and rereading that paper. We desperately wanted to hire you but, oddly, no matter how hard we tried were never able to get in touch with you. Anyway, if you find out anything about how Bill's career panned out, let me know, he was a nice guy."

Roger Howe

I next resolved to ask Roger. I remembered that as a graduate student at Indiana University, some of the other students and me had formed a seminar and studied a beautiful paper in #345 that Roger had written with Bill, *Integral operators: commutators, traces, index, and homology, cohomology, K, KK, and KKK theory, Hilbert modules, Cantor sets, the rings of Saturn, Tokamaks, and why BDF is kids stuff*. Even absorbing the title of that paper raised our IQs 10 points, and after a month of hard work, even though we had gotten only a third of the way through the Introduction, we all had clearly become much smarter than our professors. Surely, as an early collaborator of Bill's, Roger could give me some good pointers on the details of Bill's early work. I got Roger on the phone and after a few pleasantries, I asked him

"Roger, what can you tell me about Bill's early work?"

"Absolutely first rate Jim. As I am sure you know, they haven't always had things like the Email. But you see, back then Jim, they didn't even have things like the AMS TeX, everything had to be typed up by hand on primitive machines called typewriters. Bill used to love spending hours and hours trying different spacings and fonts, correcting typos, and fussing over the subtleties of the notations we were using. Unbelievably, he could differentiate between 18 brands of white-out by smell alone. Bill's work on our paper *Integral operators* was so outstanding that I think if you have a look at the old #345 you will agree with me that it is by far the most beautiful paper in the volume. Of his early period, I can't really tell you much more. You might ask Ron tho, I have a vague memory that Bill might have helped Larry, Ron and Peter write up BDF, he was much sought after by everyone as a collaborator."

Ron Douglas

Now, having asked Ron questions before, I realized that this was going to be like trying to take a drink from a fire hose. I had no choice tho, since I knew Ruth was not going to be satisfied with what I had learned so far about Bill's early work. Steeling myself, I gave him a call and when Bunny picked up, I could hear somebody shouting in the background and what sounded very much like things being deliberately smashed. "Jim, thank God you called. I am really worried about Ron. He has been rampaging through the house for the last hour, ranting about all kinds of horrible things. Something about how BDF was supposed to have really been DBF, and how his career has been ruined by a typographical error. Plus, he's been saying the most awful things about Bill Helton. What exactly is BDF anyway... its not some kind of kinky sex is it?" Just then, over the telephone, I heard what sounded like a bomb going off in the distance, followed by a loud crash and then what sounded like dishes being thrown against a wall. I hurriedly, reassured Bunny as best I could and then hung up.

Joe

Well, I hadn't wanted to bother Joe, he's always so busy. And I knew that after our conversation I was going to have to have a stiff one to clear my brain. But if you need to know something about operator theory, Joe is the one to ask. I get him on the phone and ask:

"Joe, can you give me some insights into Bill's early work, I mean from before you started working with him on those m-symmetric operators and indefinite metrics?"... An hour later, Joe was still deeply immersed in explaining how, based on his analysis of over a thousand references in the literature, he had deduced that John Doyle had actually stolen the idea of μ-synthesis from Thomas Edison. Interrupting him as gently as I could at this point, I asked, " Joe, but what about Bill's early work?"

Joe answered, "Jim, sorry, I gotta go give a lecture now in the EE department on what Euclid had to say on the robust stabilization of systems with fuzzy feedback. Why don't you look in Bill's Lecture Notes from the 85 CBMS" and then abruptly hung up.

Thank God for Joe. I should have thought of Bill's CBMS lecture notes right away. Surely, I would find practically all of the early work referenced there. Over the years, on many occasions, maybe at Christmas, or when Bill would get a big promotion, or when Ruth would force him to come to my birthday parties, he would wrap up a copy of those CBMS notes and present it to me as a gift.

Backyard again

I gathered a copy of the notes from my office, got my coffee, went out to the backyard, and had just settled down at my bistro table to have a look-see, when a pigeon flew in. I hastily balanced the lecture notes on the top of my coffee cup and then detached a message from the pigeon's leg.

"Jim, How is it going? You aren't going to let me down are you?, Ruth."

Well, while I was reading this message, the pigeon, not being able to get to my coffee had become enraged and knocked over my coffee cup spilling coffee all over Bill's Lecture Notes. It had then gone into a frenzy, ripping and shredding the pages with its beak and claws, all the while constantly pooping all over everything. But this was not the worst of it... when I started to salvage at least the cover page of the notes, I noticed that the coffee had dissolved the "J. William" and in its place I could clearly make out a "Robert E."... the upper case H had turned into a lower case k and much to my amazement, in the space before the H was a clearly visible smudge of white-out. Rubbing it, I discovered a capital S. To my horror, I realized that Bill was not actually the author of these lecture notes. Rather, the author was some guy I had never heard of, a Robert E. Skelton. I had my graduate student do the Google to him and discovered that he was an engineer who had had a promising career until all the references to his early work had mysteriously disappeared in a short amount of time. Fortunately, his career has gotten back on track recently due to his receiving a great deal of positive publicity in the national press due to his being the victim of a number of high profile lawsuits from the toy companies that make Legos, Tinkertoys, Lincoln Logs and Erector Sets. Apparently, he wrote an influential article arguing that the current building systems in place were having a harmful effect on our children and advocating that all children's construction sets should have a minimum content of 90% string-like materials by weight.

In any case, I quickly realized that my plan to base my analysis of Bill's early work on these notes was seriously flawed.

I sent the pigeon back to Ruth with the following message:

"Ruth, I am so sorry, I haven't been able to find anything out about Bill's early work. I have let you down. Jim cry/cry"

As you can imagine, it was with considerable trepidation that I awaited the pigeon's return. While waiting, I had the presence of mind to go inside and make several cups of coffee which I then laid out on my Bistro table. An hour later, the pigeon returned. As before it had a message attached to its left leg but now there was a cd dangling from its right leg. The message read:

"Jim, I swear you have roos loose in your top paddock. Attached is a cd with everything you will need to complete the task I assigned to you. Ruth"

A label on the cd read "insert into the cd drive of your computer." Upon insertion the disc took control of my computer and installed something called "Outlook" which I learned is Microsoft's way of letting you do the Email. It had already sent a message to a Miss MacGillicutty, apparently Bill's 7th grade algebra teacher. An hour later, I received my first piece of the Email. It read:

> Dear Professor Agler,
>
> It was very exciting for me to receive an email from such a famous and important mathematician. My students and I are closely following (as best we can) your current work on Carathéodory-Julia theory on bidiscs and are pulling for you to crack the Loewner problem in n-variables. Just yesterday, one of my students turned in an extra credit report, a biography of your life. I found it a fascinating read. Keep up the good work... you are an awesome role model for my 12 year olds.
>
> You ask after one of my former students, Bill Helton, what his work was like in my class. Even after 50 years I can remember that he displayed absolutely no aptitude for algebra. Through some combination of slowness, laziness, and willfulness he was unable to master even the simplest rules of algebra. For example, even after a year of drill he had not learned that the simplification of x plus y squared is x squared plus 2xy plus y squared. He would always tediously multiply everything out and then leave the xy and yx terms ungrouped. By the end of the year there were so many holes in his algebra background that I resolved to hold him back. But at the last moment, I felt sorry for him and let him go on to 8th grade math. He's not causing you any trouble is he? If so just let me know. As I feel sort of responsible, I would be willing to read him the riot act.
>
> Sincerely, Miss McGillicutty

So in one deft stroke, Ruth had not only gotten me up and running with the Email, but had also helped me to ferret out some key information about Bill's early work. I plan to continue my research into Bill's early career with an eye to eventually starting Volume 1 of what will be a 3 volume biography of his career. If you have any knowledge of his early work please e-mail me.

Thank You & Happy Birthday Bill

Conclusion: So what can we say we have learned from Bill's Early Work. Clearly, Bill is a great and famous mathematician... that is a given. Now normally, the early work of a great and famous mathematician is widely remembered and widely regarded to have been seminal... that is how you know the mathematician is great and is deservedly famous. In Bill's case, since Bill is a great and famous mathematician, yet, the early work seems to have been largely forgotten, one can only conclude that whatever it was it must have been truly great indeed.

Operator Theory:
Advances and Applications, Vol. 222, 29–30
© 2012 Springer Basel

List of Contributors

Agler, Jim; Department of Mathematics University of California San Diego La
Jolla, CA 92093-0112, USA, e-mail: `jagler@ucsd.edu`

Ball, Joseph A.; Department of Mathematics, Virginia Tech, Blacksburg,
VA 24061-0123, USA, e-mail: `joball@math.vt.edu`

Basor, Estelle; American Institute of Mathematics, Palo Alto, CA 94306, USA,
e-mail: ebasor@aimath.org

Dewilde, Patrick; TU München, Institute for Advanced Study,
e-mail: p.dewilde@me.com

Douglas, Ronald G.; Department of Mathematics, Texas A&M University, College
Station, TX, USA, e-mail: `rdouglas@math.tamu.edu`

Dym, Harry; Department of Mathematics, The Weizmann Institute of Science,
Rehovot 76100, Israel, e-mail: `harry.dym@weizmann.ac.il`

Eschmeier, Jörg; Fachrichtung Mathematik, Universität des Saarlandes,
Saarbrücken, Germany, e-mail: `eschmei@math.uni-sb.de`

Fang, Quanlei; Department of Mathematics & Computer Science, CUNY-BCC,
Bronx, NY 10453, USA, e-mail: `fangquanlei@gmail.com`

Frazho, A.E.; Department of Aeronautics and Astronautics, Purdue University,
West Lafayette, IN 47907, USA, e-mail: `frazho@ecn.purdue.edu`

Fuhrmann, P.A.; Department of Mathematics, Ben-Gurion University of the
Negev, Beer Sheva, Israel, e-mail: `fuhrmannbgu@gmail.com`,

Georgiou, Tryphon T.; Department of Electrical and Computer Engineering,
University of Minnesota, Minneapolis, MN 55455, USA,
e-mail: `tryphon@umn.edu`

Harrison, Martin; Department of Mathematics, University of California Santa
Barbara, Santa Barbara, CA 93106, USA, e-mail: `martin@math.ucsb.edu`

Helmke, U.; Universität Würzburg, Institut für Mathematik, Würzburg, Germany,
e-mail: `helmke@mathematik.uni-wuerzburg.de`

Helton, J. William; Department of Mathematics, University of California San
Diego, USA, e-mail: `helton@math.ucsd.edu`

Howe, Roger; Yale University, Mathematics Dept., PO Box 208283, New Haven, CT 06520-8283, USA, e-mail: howe@math.yale.edu

James, M.R.; ARC Centre for Quantum Computation and Communication Technology, Research School of Engineering, Australian National University, Canberra, ACT 0200, Australia, e-mail: Matthew.James@anu.edu.au

Kaashoek, M.A.; Department of Mathematics, VU University Amsterdam, De Boelelaan 1081a, NL-1081 HV Amsterdam, The Netherlands, e-mail: m.a.kaashoek@vu.nl

Klep, Igor; Department of Mathematics, The University of Auckland, New Zealand, e-mail: igor.klep@auckland.ac.nz

McCullough, Scott: Department of Mathematics, University of Florida, Gainesville, FL 32611-8105, USA, e-mail: sam@math.ufl.edu

Nie, Jiawang; Department of Mathematics, University of California, 9500 Gilman Drive, La Jolla, CA 92093, USA, e-mail: njw@math.ucsd.edu

Plaumann, Daniel; Fachbereich Mathematik und Statistik, Universität Konstanz, D-78457 Konstanz, Germany, e-mail: Daniel.Plaumann@uni-konstanz.de

Rodman, L.; Department of Mathematics, College of William and Mary, Williamsburg, VA, USA, e-mail: lxrodm@math.wm.edu

Sturmfels, Bernd; Dept. of Mathematics, University of California, Berkeley, CA 94720, USA, e-mail: bernd@math.berkeley.edu

Takyarm Mir Shahrouz; Department of Electrical and Computer Engineering, University of Minnesota, Minneapolis, MN 55455, USA, e-mail: shahrouz@umn.edu

Tannenbaum, Allen; Departments of Electrical & Computer and Biomedical Engineering, Boston University, Boston, MA, USA, e-mail: tannenba@bu.edu

Tannenbaum, Emmanuel; Department of Chemistry, Ben-Gurion University of the Negev, Israel, e-mail: emmanuel.tannenbaum@gmail.com

ter Horst, S.; Department of Mathematics, Utrecht University, P.O. Box 80010, NL-3508 TA Utrecht, The Netherlands, e-mail: s.ter.horst@uu.nl

Vinnikov, Victor; Department of Mathematics, Ben-Gurion University of the Negev, Beer-Sheva, Israel, 84105, e-mail: vinnikov@math.bgu.ac.il

Vinzant, Cynthia; Department of Mathematics, University of Michigan, Ann Arbor, MI 48109, USA, e-mail: vinzant@umich.edu

Woerdeman, H.J.; Department of Mathematics, Drexel University, Philadelphia, PA, USA, e-mail: hugo@math.drexel.edu

Young, N.J.; Department of Pure Mathematics, Leeds University, Leeds LS2 9JT, England *and* School of Mathematics and Statistics, Newcastle University, Newcastle upon Tyne NE1 7RU, England, e-mail: N.J.Young@leeds.ac.uk

Part II

Technical Papers

Operator Theory:
Advances and Applications, Vol. 222, 33–41
© 2012 Springer Basel

The Carathéodory-Julia Theorem and the Network Realization Formula in Two Variables

Jim Agler

Abstract. Harry Dym and Donald Sarason have given independent treatments of the classical Carathéodory-Julia Theorem on the disc using operator-theoretic methods. Here, I describe some joint work with John McCarthy and Nicholas Young in which similar operator-theoretic methods are used to generalize the classical theorem to the bidisc.

Mathematics Subject Classification. 32A40, 32A70, 47B32, 47N70, 93B15.

Keywords. Network realization formula, Carathéodory–Julia theorem, bidisc, models.

1. Introduction

In the early 70's Bill wrote a seminal paper [8] pointing out that the network realization formula from electrical engineering and the characteristic function of the Sz. Nagy-Foias model theory for contractions on Hilbert space were intimately related. This paper, along with many other novel perspectives of Bill, influenced a generation of young operator theorists, including myself, to take a life long interest in the mathematics of engineering. I learned the network realization formula for the first time in a seminar Bill gave on H^∞-control theory in 1985, shortly after I received my appointment at UCSD. I have many fond memories of those exciting times, and, over the years, the realization formula has played a recurring role in my research.

In this note I would like to describe some recent joint work with John McCarthy and Nicholas Young [3] in which the network realization formula leads the way to the discovery of some new results on the Carathéodory-Julia Theorem in two variables.

2. One variable realizations and models

Let $\mathbb{C}$ denote the complex numbers, $\mathbb{D} = \{z \in \mathbb{C} \mid |z| < 1\}$, and $\mathbb{T} = \{z \in \mathbb{C} \mid |z| = 1\}$. We let $\mathcal{S}$ denote the Schur class, i.e., the set of functions ϕ that are defined and analytic on $\mathbb{D}$ that satisfy the inequality $|\phi(\lambda)| \leq 1$ for all $\lambda \in \mathbb{D}$.

Definition 1. Let ϕ be a function on $\mathbb{D}$. We say a 4-tuple (A, B, C, D) is a realization for ϕ if there exists a Hilbert space, $\mathcal{M}$, such that the 2×2 block matrix,

$$V = \begin{bmatrix} A & B \\ C & D \end{bmatrix},$$

is a well-defined isometry acting on $\mathcal{M} \oplus \mathbb{C}$, and

$$\phi(\lambda) = D + C\lambda(1 - A\lambda)^{-1}B \tag{1}$$

for all $\lambda \in \mathbb{D}$.

As a consequence of the following theorem, realizations, as defined in the above definition, can be a powerful tool for penetrating the function theoretic properties of functions in the Schur class.

Theorem 2. *Let ϕ be function defined on $\mathbb{D}$. $\phi \in \mathcal{S}$ if and only if ϕ has a realization.*

In order to use realizations to study functions, it turns out that it is efficacious to strip out some of the "noise" in the realization. This is done by constructing what is called a model.

Definition 3. Let ϕ be a function on $\mathbb{D}$. By a model for ϕ is meant an ordered pair, $(\mathcal{M}, u)$, where $\mathcal{M}$ is a Hilbert space, $u : \mathbb{D} \to \mathcal{M}$ and

$$1 - \overline{\phi(\mu)}\phi(\lambda) = (1 - \overline{\mu}\lambda)\langle u_\lambda, u_\mu \rangle \tag{2}$$

for all $\lambda, \mu \in \mathbb{D}$.

Armed with this notion of a model, it is possible to transform Theorem 2 into the following result.

Theorem 4. *Let ϕ be function defined on $\mathbb{D}$. $\phi \in \mathcal{S}$ if and only if ϕ has a model.*

Proof. The theorem will follow from Theorem 2 if we can show that ϕ has a realization if and only if ϕ has a model. First suppose that ϕ has a realization as described in Definition 1. Set $u_\lambda = (1 - A\lambda)^{-1}B(1)$ so that

$$A(\lambda u_\lambda) + B(1) = u_\lambda. \tag{3}$$

Also, note that (1) implies that

$$C(\lambda u_\lambda) + D(1) = \phi(\lambda). \tag{4}$$

Combining (3) and (4) yields that

$$V \begin{pmatrix} \lambda u_\lambda \\ 1 \end{pmatrix} = \begin{pmatrix} u_\lambda \\ \phi(\lambda) \end{pmatrix}. \tag{5}$$

Hence, as V is an isometry,

$$\left\langle \begin{pmatrix} u_\lambda \\ \phi(\lambda) \end{pmatrix}, \begin{pmatrix} u_\mu \\ \phi(\mu) \end{pmatrix} \right\rangle$$

$$= \left\langle V \begin{pmatrix} \lambda u_\lambda \\ 1 \end{pmatrix}, V \begin{pmatrix} \mu u_\mu \\ 1 \end{pmatrix} \right\rangle$$

$$= \left\langle \begin{pmatrix} \lambda u_\lambda \\ 1 \end{pmatrix}, \begin{pmatrix} \mu u_\mu \\ 1 \end{pmatrix} \right\rangle,$$

which unravels to

$$\langle u_\lambda, u_\mu \rangle + \phi(\lambda)\overline{\phi(\mu)} = \langle \lambda u_\lambda, \mu u_\mu \rangle + 1. \tag{6}$$

As (6) implies (2), it follows that ϕ has a model.

To prove the converse which shall employ a "lurking isometry" argument. Assume that ϕ has a model as in Definition 3. (2) implies that (6) holds, which in turn implies that

$$\left\langle \begin{pmatrix} u_\lambda \\ \phi(\lambda) \end{pmatrix}, \begin{pmatrix} u_\mu \\ \phi(\mu) \end{pmatrix} \right\rangle = \left\langle \begin{pmatrix} \lambda u_\lambda \\ 1 \end{pmatrix}, \begin{pmatrix} \mu u_\mu \\ 1 \end{pmatrix} \right\rangle \tag{7}$$

for all $\lambda, \mu \in \mathbb{D}$. (7) suggests that we attempt to define an isometric linear operator $V : \mathcal{M} \oplus \mathbb{C} \to \mathcal{M} \oplus \mathbb{C}$ with the property that formula (5) holds for all $\lambda \in \mathbb{D}$. This construction will always succeed by simply extending (5) by linearity provided one is willing to enlarge $\mathcal{M}$ sufficiently. Once V is so defined, one represents V as a 2×2 block matrix,

$$V = \begin{bmatrix} A & B \\ C & D \end{bmatrix},$$

and obtains the equations (3) and (4) from (5). Solving (3) for u_λ and then substituting into (4) yields that (1) holds. This shows that ϕ has a realization.   □

3. The Carathéodory-Julia Theory in One Variable

Let ϕ be a Schur function of norm one. The Carathéodory-Julia Theory is actually a number of related results that describe the geometric behavior of ϕ when viewed as a mapping from $\mathbb{D}$ into $\mathbb{D}$ near a point τ on the boundary of $\mathbb{D}$ where ϕ attains its norm. We apologize in advance to any experts in function theory for the somewhat "kinky" description of Carathéodory's and Julia's results that we give here. We are setting up so that the step to two variables will be easily digestible.

One way to understand the results, is to see that they correspond to saying that an a priori very weak regularity condition on ϕ at τ is actually equivalent to a very strong regularity condition at τ.

For $\phi \in \mathcal{S}$, we define J_ϕ, *the Julia quotient of* ϕ, to be the function defined on $\mathbb{D}$ by the formula,

$$J_\phi(\lambda) = \frac{1 - |\phi(\lambda)|}{1 - |\lambda|}. \tag{8}$$

Before continuing, we remark that the quotient originally introduced by Julia in [9] was defined for Pick functions in the upper half-plane. When transformed to the Schur class it takes the form, $J_\phi^{\text{True}}(\lambda) = \frac{1-|\phi(\lambda)|^2}{1-|\lambda|^2}$. However, fortunately, for $\phi \in \mathcal{S}$, $J_\phi(\lambda)$ and $J_\phi^{\text{True}}(\lambda)$ bound each other on $\mathbb{D}$. While J_ϕ^{True} is the usual quotient studied by function theorists, generalizing it to several variables requires a lot of annoying subtleties involving invariant metrics.

Now consider the following increasingly stringent conditions on the behavior of ϕ at a point $\tau \in \mathbb{T}$.

I $\liminf\limits_{\substack{\lambda\to\tau \\ \lambda\in\mathbb{D}}} J_\phi(\lambda) < \infty$.

II $J_\phi(\lambda)$ is non-tangentially bounded at τ.

III ϕ is directionally differentiable at τ and $|\phi(\tau)|= 1$.

IV ϕ is non-tangentially differentiable at τ and $|\phi(\tau)|= 1$.

V ϕ is non-tangentially continuously differentiable at τ and $|\phi(\tau)|= 1$.

Condition I above is self explanatory. Condition III asserts that there exists a complex number with modulus one, denoted $\phi(\tau)$, such that whenever $\delta \in \mathbb{C}$ and $\tau + t\delta \in \mathbb{D}$ for sufficiently small positive t,

$$\lim_{t\to 0+} \frac{\phi(\tau + t\delta) - \phi(\tau)}{t} \quad \text{exists.} \tag{9}$$

The other conditions use the geometric notion of "non-tangential" approach to a boundary point. For $S \subseteq \mathbb{D}$ and $\tau \in \mathbb{T}$ we say that S *approaches τ non-tangentially*, $S \xrightarrow{\text{nt}} \tau$, if $\tau \in S^-$ and there exists a constant c such that

$$|\lambda - \tau| \leq c(1-|\lambda|) \tag{10}$$

for all $\lambda \in S$. We then say that a property holds "non-tangentially" at τ if it holds on every set that approaches τ non-tangentially. For example, condition II above is simply saying that ϕ is bounded on every set in $\mathbb{D}$ that approaches τ non-tangentially. Condition IV says that there exist complex numbers denoted $\phi(\tau)$ and $\phi'(\tau)$ such that

$$\lim_{\substack{\lambda\to\tau \\ \lambda\in S}} \frac{\phi(\lambda) - \phi(\tau) - \phi'(\tau)(\lambda - \tau)}{|\lambda - \tau|} = 0 \tag{11}$$

whenever $S \subseteq \mathbb{D}$ and $S \xrightarrow{\text{nt}} \tau$. Finally, Condition V is saying that in addition,

$$\lim_{\substack{\lambda\to\tau \\ \lambda\in S}} \phi'(\lambda) = \phi'(\tau), \tag{12}$$

whenever $S \subseteq \mathbb{D}$ and $S \xrightarrow{\text{nt}} \tau$.

Theorem 5. *Let $\tau \in \mathbb{T}$ and let $\phi \in \mathcal{S}$. Then conditions I–V are equivalent.*

Carathéodory essentially gave the first proof of Theorem 5 in 1929 using geometric methods [6]. Since then, treatments of various aspects of the theorem from virtually every possible point of view have been developed. Of particular interest to us here, are the Hilbert space proofs that I implies IV that were discovered independently by Harry Dym [7] and Donald Sarason [12]. Both approaches involved developing structures based on the use of de Branges spaces that have great power in the study of Schur functions generally.

Fortunately, if one is only interested in the Carathéodory-Julia theory, then these discoveries of Harry and Don, when stripped of the technicalities that are necessitated through the use of de Branges spaces, can be stated in a very simple way by using the notion of a model as defined in Definition 3. The following two propositions both give very precise relationships between the local function-theoretic behavior of ϕ at the point τ and the Hilbert space behavior of u near τ.

Proposition 6. *Let $\phi \in \mathcal{S}$ and let $(\mathcal{M}, u)$ be a model for ϕ. If $\tau \in \mathbb{T}$, then $J_\phi(\lambda)$ is non-tangentially bounded at τ if and only if u_λ is non-tangentially bounded at τ.*

Proposition 7. *Let $\phi \in \mathcal{S}$ and let $(\mathcal{M}, u)$ be a model for ϕ. If $\tau \in \mathbb{T}$, then ϕ is non-tangentially differentiable at τ if and only if u_λ is non-tangentially continuous at τ, i.e., there exists a vector $x \in \mathcal{M}$ such that*

$$\lim_{\substack{\lambda \to \tau \\ \lambda \in S}} u_\lambda = x$$

whenever $S \subseteq \mathbb{D}$ and $S \overset{\mathrm{nt}}{\to} \tau$.

Once Propositions 6 and 7 are obtained, it is apparent that the equivalence of II–IV is equivalent to the following fact.

Fact 8. *Let $\phi \in \mathcal{S}$, let $(\mathcal{M}, u)$ be a model for ϕ, and let $\tau \in \mathbb{T}$. If u_λ is non-tangentially bounded at τ, then u_λ is non-tangentially continuous at τ.*

Proofs of the propositions and the fact all can be worked out using elementary, follow your nose, Hilbert space arguments. This gives that II–IV are equivalent. The full loop of equivalences I–V is then completed by proving that I implies II (this uses (2) and the Cauchy-Schwarz Inequality), proving that IV implies V (this uses (1) and a resolvent estimate on $(1 - A\lambda)^{-1}$), and finally, closing the loop with the triviality that V implies I.

4. Two variable realizations and models

How does it work in two variables? Let $\mathbb{C}^2 = \mathbb{C} \times \mathbb{C}$, $\mathbb{D}^2 = \mathbb{D} \times \mathbb{D}$, and $\mathbb{T}^2 = \mathbb{T} \times \mathbb{T}$. We let $\mathcal{S}_2$ denote the Schur class of the bidisc, i.e., the set of functions ϕ that are defined and analytic on $\mathbb{D}^2$ that satisfy the inequality $|\phi(\lambda)| \leq 1$ for all $\lambda \in \mathbb{D}^2$. Just as in one variable it is possible to realize Schur functions.

Definition 9. Let ϕ be a function on $\mathbb{D}^2$. We say a 4-tuple (A, B, C, D) is a realization for ϕ if there exists a decomposed Hilbert space, $\mathcal{M} = \mathcal{M}_1 \oplus \mathcal{M}_2$, such that the 2×2 block matrix,

$$V = \begin{bmatrix} A & B \\ C & D \end{bmatrix},$$

is a well-defined isometry acting on $\mathcal{M} \oplus \mathbb{C}$, and

$$\phi(\lambda) = D + C\lambda(1 - A\lambda)^{-1}B \tag{13}$$

for all $\lambda \in \mathbb{D}^2$. Here, in formula (13), for $\lambda = (\lambda_1, \lambda_2) \in \mathbb{D}^2$, we view λ as an operator on $\mathcal{M}$ via the formula,

$$\lambda = \lambda_1 P_{\mathcal{M}_1} + \lambda_2 P_{\mathcal{M}_2}$$

where $P_{\mathcal{M}_1}$ and $P_{\mathcal{M}_2}$ denote the orthogonal projections of $\mathcal{M}$ onto $\mathcal{M}_1$ and $\mathcal{M}_2$.

As a consequence of the following theorem from [2], two variable realizations, as defined in the above definition, can be used to study the function theoretic properties of functions in the two variable Schur class. I think it would be fair to say that were it not for the strong influence of Bill, it never would have occurred to me to formulate such a theorem.

Theorem 10. *Let ϕ be function defined on $\mathbb{D}^2$. $\phi \in \mathcal{S}_2$ if and only if ϕ has a realization.*

Now, it turns out that realizations in two variables have a lot more "noise" than one variable realizations. Accordingly, the notion of a model has correspondingly more power than in one variable.

Definition 11. Let ϕ be a function on $\mathbb{D}^2$. By a model for ϕ is meant an ordered pair, $(\mathcal{M}, u)$, where $\mathcal{M} = \mathcal{M}_1 \oplus \mathcal{M}_2$ is a decomposed Hilbert space, $u : \mathbb{D}^2 \to \mathcal{M}$ and

$$1 - \overline{\phi(\mu)}\phi(\lambda) = \langle (1 - \mu^*\lambda)u_\lambda, u_\mu \rangle \tag{14}$$

for all $\lambda, \mu \in \mathbb{D}^2$. In (14), λ and μ are interpreted as operators on $\mathcal{M}$ as in Definition 9.

Once one has this notion of a model for elements of $\mathcal{S}_2$, mimicking the proof of Theorem 4 yields the following theorem.

Theorem 12. *Let ϕ be function defined on $\mathbb{D}^2$. $\phi \in \mathcal{S}_2$ if and only if ϕ has a model.*

5. The Carathéodory-Julia theorem in two variables

In [3], John McCarthy, Nicholas Young and myself used two variable models and realizations to probe the Carathéodory-Julia theory on the bidisc. Our first discovery was that a straightforward generalization of Proposition 6 obtains. To state this result we need to generalize both the Julia quotient and the notion of nontangential approach to the setting of the bidisc.

First notice that the expression, $1- |\lambda|$, which appears in the denominator on the right side of equation (8), can be identified geometrically with the distance from λ to the boundary of $\mathbb{D}$. Letting $\partial(\mathbb{D}^2)$ denote the boundary of $\mathbb{D}^2$ (i.e., the set $(\mathbb{D} \times \mathbb{T}) \cup (\mathbb{T} \times \mathbb{D}) \cup \mathbb{T}^2$), we thereby see that a natural way to interpret the expression in two variables is to consider the quantity $\mathrm{dist}(\lambda, \partial(\mathbb{D}^2))$. As

$$\mathrm{dist}(\lambda, \partial(\mathbb{D}^2)) = \min\{1- |\lambda_1|, 1- |\lambda_2|\},$$

we are led to defining the Julia quotient on the bidisc by the formula,

$$J_\phi(\lambda) = \frac{1- |\phi(\lambda)|}{\min\{1- |\lambda_1|, 1- |\lambda_2|\}}. \tag{15}$$

Now notice that the expression $1- |\lambda|$ also appears on the right-hand side of inequality (10). Accordingly, if $S \subseteq \mathbb{D}^2$ and $\tau \in \mathbb{T}^2$ we say that S *approaches* τ *non-tangentially*, $S \overset{\mathrm{nt}}{\to} \tau$, if $\tau \in S^-$ and there exists a constant c such that

$$|\lambda - \tau| \leq c(\min\{1- |\lambda_1|, 1- |\lambda_2|\}) \tag{16}$$

for all $\lambda \in S$. Following Section 3, we then say that a property of a Schur function on the bidisc holds non-tangentially at a point $\tau \in \mathbb{T}^2$ if the property holds for every set $S \subseteq \mathbb{D}^2$ such that S approaches τ non-tangentially. Finally, we can state our generalization of Proposition 6 to two variables.

Proposition 13. *Let $\phi \in \mathcal{S}_2$ and let $(\mathcal{M}, u)$ be a model for ϕ. If $\tau \in \mathbb{T}^2$, then $J_\phi(\lambda)$ is non-tangentially bounded at τ if and only if u_λ is non-tangentially bounded at τ.*

How does Proposition 7 generalize? Evidently, by analogy with (11) we want to say that ϕ *is non-tangentially differentiable at* τ if there exist a complex number denoted $\phi(\tau)$ and a vector in $\mathbb{C}^2$ denoted $\nabla\phi(\tau)$ such that

$$\lim_{\substack{\lambda \to \tau \\ \lambda \in S}} \frac{\phi(\lambda) - \phi(\tau) - \nabla\phi(\tau) \cdot (\lambda - \tau)}{|\lambda - \tau|} = 0 \tag{17}$$

whenever $S \subseteq \mathbb{D}^2$ and $S \overset{\mathrm{nt}}{\to} \tau$. The following analog of Proposition 7 then obtains.

Proposition 14. *Let $\phi \in \mathcal{S}_2$ and let $(\mathcal{M}, u)$ be a model for ϕ. If $\tau \in \mathbb{T}^2$, then ϕ is non-tangentially differentiable at τ if and only if u_λ is non-tangentially continuous at τ.*

We now come to an interesting discovery. It turns out that Fact 8 does not generalize to two variables. A counter-example is provided by the simple function

$$\phi(\lambda) = \frac{2\lambda_1\lambda_2 - \lambda_1 - \lambda_2}{2 - \lambda_1 - \lambda_2}. \tag{18}$$

As a consequence we are led to the discovery that there actually two types of "carapoints" in two variables, B-points and C-points.

Definition 15. Let $\phi \in \mathcal{S}_2$, $\tau \in \mathbb{T}^2$, and let $(\mathcal{M}, u)$ be a model for ϕ. We say τ is a carapoint for ϕ if the two variable analog of condition I from Section 3 obtains, i.e.,

$$\liminf_{\substack{\lambda \to \tau \\ \lambda \in \mathbb{D}^2}} J_\phi(\lambda) < \infty.$$

We say that τ is a B-point for ϕ if u_λ is non-tangentially bounded at τ and we say that τ is a C-point for ϕ if u_λ is non-tangentially continuous at τ.

We now are ready to generalize Theorem 5 to the bidisc. Note that we have given each of the regularity conditions I–V from Section 3 an unambiguous meaning in two variables.

Theorem 16. *Let $\tau \in \mathbb{T}^2$ and let $\phi \in \mathcal{S}_2$. Then conditions I, II, and III are equivalent and conditions IV and V are equivalent. Furthermore I–III hold if and only if τ is a B-point for ϕ and IV–V hold if and only if τ is a C-point for ϕ.*

6. Conclusion

Over the years, one of the recurrent and productive themes in Bill's approach toward mathematics has been the identification and exploitation of fruitful synergies between ideas in mathematics and engineering. In this note, I have described a concrete example of how, through an insight originally put forth by Bill, the network realization formula has pointed the way to the discovery and proof of new results in several complex variables. Many other applications of the realization formula to function theory have been discovered in the last 25 years. For example, it has played a role in the generalization of the classical Nevanlinna-Pick Theorem to several variables [1] and when Cayley transformed to the upper half-plane, in the generalization of Loewner's beautiful characterization of the matrix monotone functions [10] to severable variables [5]. Most recently in [4], a new proof with sharp bounds was obtained for the Oka Extension Theorem [11] using a version of the realization formula on polyhedrons in $\mathbb{C}^n$.

References

[1] J. Agler. Some interpolation theorems of Nevanlinna-Pick type. Preprint, 1988.

[2] J. Agler. On the representation of certain holomorphic functions defined on a polydisc. In *Operator Theory: Advances and Applications, Vol. 48*, pages 47–66. Birkhäuser, Basel, 1990.

[3] J. Agler, J.E. McCarthy, and N.J. Young. A Carathéodory theorem for the bidisk using Hilbert space methods. To appear.

[4] J. Agler, J.E. McCarthy, and N.J. Young. On the representation of holomorphic functions defined on polyhedra. To appear.

[5] J. Agler, J.E. McCarthy, and N.J. Young. Operator monotone functions and Loewner functions of several variables. To appear.

[6] C. Carathéodory. Über die Winkelderivierten von beschränkten analytischen Funktionen. *Sitzunber. Preuss. Akad. Wiss.*, pages 39–52, 1929.

[7] H. Dym. *J-Contractive Matrix Functions, Reproducing Kernel Hilbert Spaces and Interpolation*. Number 71 in CBMS Lecture Notes. American Mathematical Society, Providence, 1989.

[8] J.W. Helton. The characteristic functions of operator theory and electrical network realization. *Ind. Univ. Math. Jour.*, 22(5):403–414, 1972.

[9] G. Julia. Extension nouvelle d'un lemme de Schwarz. *Acta Math.*, 42:349–355, 1920.

[10] K. Löwner. Über monotone Matrixfunktionen. *Math. Z.*, 38:177–216, 1934.

[11] K. Oka. Domaines convexes par rapport aux fonctions rationelles. *J. Sci. Hiroshima Univ.*, 6:245–255, 1936.

[12] D. Sarason. *Sub-Hardy Hilbert spaces in the unit disk*. University of Arkansas Lecture Notes, Wiley, New York, 1994.

Jim Agler
Department of Mathematics
University of California San Diego
La Jolla, CA 92093-0112, USA
e-mail: jagler@ucsd.edu

Operator Theory:
Advances and Applications, Vol. 222, 43–71
© 2012 Springer Basel

Nevanlinna-Pick Interpolation via Graph Spaces and Kreĭn-space Geometry: A Survey

Joseph A. Ball and Quanlei Fang

Dedicated to Bill Helton, a dear friend and collaborator

Abstract. The Grassmannian/Kreĭn-space approach to interpolation theory introduced in the 1980s gives a Kreĭn-space geometry approach to arriving at the resolvent matrix which parametrizes the set of solutions to a Nevanlinna-Pick interpolation or Nehari-Takagi best-approximation problem. We review the basics of this approach and then discuss recent extensions to multivariable settings which were not anticipated in the 1980s.

Mathematics Subject Classification. 47A57.

Keywords. Kreĭn space, maximal negative and maximal positive subspaces, graph spaces, projective space, Beurling-Lax representations.

1. Introduction

We take this opportunity to update the Grassmannian approach to matrix- and operator-valued Nevanlinna-Pick interpolation theory introduced in [24]. It was a privilege for the first-named current author to be a participant with Bill Helton in the development of all these operator-theory ideas and their connections with Kreĭn-space projective geometry and engineering applications (in particular, circuit theory and control). Particularly memorable was the eureka moment when Bill observed that our J-Beurling-Lax representer was the same as the Adamjan-Arov-Kreĭn "resolvent matrix" Θ parameterizing all solutions of a Nehari-Takagi problem. This gave us an alternative way of constructing and understanding the origin of such resolvent matrices, and provided a converse direction for Bill's earlier results on orbits of matrix-function linear-fractional maps [50].

The present paper is organized as follows. Following this Introduction, in Section 2 we review the Grassmannian approach to the basic bitangential Sarason interpolation problem, including an indication of how the simplest bitangential

matrix Nevanlinna-Pick interpolation problem is included as a special case. We
also highlight along the way where some additional insight has been gained over
the years. In Section 3 we show how a reformulation of the problem as a bitangential operator-argument interpolation problem leads to a set of coordinates which
leads to state-space realization formulas for the Beurling-Lax representer, i.e., the
resolvent matrix providing the linear-fractional parametrization for solutions of the
interpolation problem. The rational case of this construction essentially appears in
the book [17] while the general operator-valued case is more recent (see [30]). The
final Section 4 surveys extensions of the Grassmannian method to more general
settings, with the main focus on the results from [45] where it is shown that the
Grassmannian approach applies to left-tangential operator-argument interpolation
problems for contractive multipliers on the Drury-Arveson space (commuting variables) and on the Fock space (noncommuting variables).

2. The Sarason bitangential interpolation problem via the Grassmannian approach

We formulate the *bitangential Sarason* (**BTS**) interpolation problem as follows.
Given an input Hilbert space $\mathcal{U}_I$ and an output Hilbert space $\mathcal{U}_O$, we let $H^\infty_{\mathcal{L}(\mathcal{U}_I,\mathcal{U}_O)}$
denote the space of bounded holomorphic functions on the unit disk $\mathbb{D}$ with values
in the space $\mathcal{L}(\mathcal{U}_I,\mathcal{U}_O)$ of bounded linear operators between $\mathcal{U}_I$ and $\mathcal{U}_O$. We let
$\mathcal{S}(\mathcal{U}_I,\mathcal{U}_O)$ denote the *Schur class* consisting of the elements of $H^\infty_{\mathcal{L}(\mathcal{U}_I,\mathcal{U}_O)}$ with
infinity norm over the unit disk at most 1. For a general coefficient Hilbert space
$\mathcal{U}$, an element B of $H^\infty_{\mathcal{L}(\mathcal{U})}$ is said to be *two-sided inner* if the nontangential strong-
limit boundary-values $B(\zeta)$ of B on the unit circle $\mathbb{T}$ are unitary operators on $\mathcal{U}$
for almost all $\zeta \in \mathbb{T}$. The *data set* for a bitangential Sarason interpolation problem
$\mathfrak{D}_{\mathrm{BTS}}$ consists of a triple (S_0, B_I, B_O) where S_0 is a function in $H^\infty_{\mathcal{L}(\mathcal{U}_I,\mathcal{U}_O)}$, and B_I
and B_O are two-sided inner functions with values in $\mathcal{L}(\mathcal{U}_I)$ and $\mathcal{L}(\mathcal{U}_O)$ respectively.
Then we formulate the bitangential Sarason interpolation problem as follows:

Problem BTS (Bitangential Sarason Interpolation Problem): Given a data set
$\mathfrak{D}_{\mathrm{BTS}} = (S_0, B_I, B_O)$ as above, find $S \in \mathcal{S}(\mathcal{U}_I,\mathcal{U}_O)$ so that the function $Q :=
B_O^{-1}(S - S_0)B_I^{-1}$ is in $H^\infty_{\mathcal{L}(\mathcal{U}_I,\mathcal{U}_O)}$.

By way of motivation, let us consider the special case where $\mathcal{U}_I = \mathbb{C}^{n_I}$ and
$\mathcal{U}_O = \mathbb{C}^{n_O}$ are finite dimensional and where for simplicity we assume that $\det B_I$
and $\det B_O$ are finite Blaschke products of respective degrees n_I and n_O. Let
us also assume that all zeros of $\det B_I$ and of $\det B_O$ are simple (but possibly
overlapping). Then it is not hard to see that the **BTS** interpolation problem is
equivalent to a *bitangential Nevanlinna-Pick* (**BTNP**) interpolation problem which
we now describe. We suppose that we are given nonzero row vectors $x_1, \ldots, x_{n_I}$
of size $1 \times n_O$, row vectors $y_1, \ldots, y_{n_O}$ of size $1 \times n_I$, distinct points $z_1, \ldots, z_{n_O}$ in
$\mathbb{D}$ (the zeros of $\det B_O$), together with nonzero column vectors $u_1, \ldots, u_{n_I}$ of size
$n_I \times 1$, column vectors $v_1, \ldots, v_{n_I}$ of size $n_O \times 1$, and distinct points $w_1, \ldots, w_{n_I}$ in

$\mathbb{D}$ (the zeros of $\det B_I$, possibly overlapping with the z_i's), together with complex numbers ρ_{ij} for any pair of indices i, j such that $z_i = w_j =: \xi_{ij}$. The bitangential Nevanlinna-Pick problem then is:

Problem BTNP (Bitangential Nevanlinna-Pick interpolation problem): *Given a data set* $\mathfrak{D} = \mathfrak{D}_{\mathrm{BTNP}}$ *given by*

$$\mathfrak{D} = \{(x_i, y_i, z_i) \text{ for } i = 1, \dots, n_O, (u_j, v_j, w_j) \text{ for } j = 1, \dots, n_I, \xi_{ij} \text{ for } z_i = w_j\}$$

as described above, find a matrix Schur-class function $S \in \mathcal{S}(\mathbb{C}^{n_I}, \mathbb{C}^{n_O})$ *so that* S *satisfies the collection of interpolation conditions:*

$$x_i S(z_i) = y_i \text{ for } i = 1, \dots, n_O,$$
$$S(w_j) u_j = v_j \text{ for } j = 1, \dots, n_I, \text{ and}$$
$$x_i S'(\xi_{ij}) u_j = \rho_{ij} \text{ for } i, j \text{ such that } z_i = w_j =: \xi_{ij}. \tag{1}$$

We remark that it is Donald Sarason [72] who first made this connection between the operator-theoretic **BTS** interpolation problem and the classical point-by-point **BTNP** interpolation problem for the scalar case.

We now present the solution of the **BTS** problem as originally presented in [24, 26]. In addition to the function spaces $H^\infty_{\mathcal{L}(\mathcal{U}_I, \mathcal{U}_O)}$ already introduced above, let us now introduce the spaces of vector-valued functions $L^2_{\mathcal{U}}$ (measurable $\mathcal{U}$-valued functions on $\mathbb{T}$ which are norm-square integrable) and its subspace $H^2_{\mathcal{U}}$ consisting of those $L^2_{\mathcal{U}}$-functions with vanishing Fourier coefficients of negative index; as is standard, we can equivalently view $H^2_{\mathcal{U}}$ as holomorphic $\mathcal{U}$-valued functions f on the unit disk $\mathbb{D}$ for which the 2-norm over circles of radius r centered at the origin are uniformly bounded as r increases to 1. The space $L^2_{\mathcal{U}}$ comes equipped with the bilateral shift operator M_z of multiplication by the coordinate functions z (on the unit circle):

$$M_z \colon f(z) \mapsto z f(z).$$

When restricted to $H^2_{\mathcal{U}}$, we get the unilateral shift (of multiplicity equal to $\dim \mathcal{U}$: information not encoded in the notation M_z). For F a function in $H^\infty_{\mathcal{L}(\mathcal{U}_I, \mathcal{U}_O)}$, there is an associated multiplication operator

$$M_F \colon f(z) \mapsto F(z) f(z)$$

mapping $H^2_{\mathcal{U}_I}$ into $H^2_{\mathcal{U}_O}$ and intertwining the respective shift operators: $M_F M_z = M_z M_F$. More generally, we may consider M_F as an operator from $L^2_{\mathcal{U}_I}$ into $L^2_{\mathcal{U}_O}$ which intertwines the respective bilateral shift operators; in this setting we need not restrict F to $H^\infty_{\mathcal{L}(\mathcal{U}_I, \mathcal{U}_O)}$ but may allow $F \in L^\infty_{\mathcal{L}(\mathcal{U}_I, \mathcal{U}_O)}$. A key feature of this correspondence between functions and operators is the correspondence of norms: given $F \in H^\infty_{\mathcal{L}(\mathcal{U}_I, \mathcal{U}_O)}$, the operator norm of M_F is the same as the supremum norm (over the unit disk or over the unit circle) of the function F:

$$\|M_F\|_{\mathrm{op}} = \|F\|_\infty := \sup\{\|F(z)\| \colon z \in \mathbb{D}\} = \text{ess-sup}\{\|F(\zeta)\| \colon \zeta \in \mathbb{T}\}.$$

Let us suppose that we are given a data set $\mathfrak{D}_{\mathrm{BTS}} = (S_0, B_I, B_O)$ for a BTS problem as above. We introduce the space $\mathcal{K} = L^2_{\mathcal{U}_O} \oplus B_I^{-1} H^2_{\mathcal{U}_I}$ (elements of which will be written as column vectors $\begin{bmatrix} f \\ g \end{bmatrix}$ with $f \in L^2_{\mathcal{U}_O}$ and $g \in B_I^{-1} H^2_{\mathcal{U}_I}$). We use the signature matrix $J_{\mathcal{K}} := \begin{bmatrix} I_{\mathcal{U}_O} & 0 \\ 0 & -I_{\mathcal{U}_I} \end{bmatrix}$ to define a Kreĭn-space inner product on $\mathcal{K}$:

$$\left[\!\!\left[\begin{bmatrix} f \\ B_I^{-1}g \end{bmatrix}, \begin{bmatrix} f \\ B_I^{-1}g \end{bmatrix}\right]\!\!\right]_{\mathcal{K}} := \left\langle J_{\mathcal{K}} \begin{bmatrix} f \\ B_I^{-1}g \end{bmatrix}, \begin{bmatrix} f \\ B_I^{-1}g \end{bmatrix}\right\rangle_{L^2}$$

$$= \|f\|^2_{L^2} - \|g\|^2_{H^2} \text{ for } \begin{bmatrix} f \\ B_I^{-1}g \end{bmatrix} \in \mathcal{K}.$$

We note that a *Kreĭn space* is simply a linear space $\mathcal{K}$ equipped with an indefinite inner product $[\cdot, \cdot]$ with respect to which $\mathcal{K}$ has an orthogonal decomposition $\mathcal{K} = \mathcal{K}_+ \oplus \mathcal{K}_-$ with $\mathcal{K}_+$ a Hilbert space in the $[\cdot, \cdot]$-inner product and $\mathcal{K}_-$ a Hilbert space in the $-[\cdot, \cdot]$-inner product; good references for more complete information are the books [8, 34]. We then consider the subspace $\mathcal{M}$ of $\mathcal{K}$ completely determined by the data set $\mathfrak{D}_{\text{BTS}} = (S_0, B_I, B_O)$:

$$\mathcal{M} := \mathcal{M}_{S_0, B_I, B_O} = \begin{bmatrix} B_O & S_0 B_I^{-1} \\ 0 & B_I^{-1} \end{bmatrix} \begin{bmatrix} H^2_{\mathcal{U}_O} \\ H^2_{\mathcal{U}_I} \end{bmatrix} \tag{2}$$

Then one checks that the function $S \in L^\infty_{\mathcal{L}(\mathcal{U}_I, \mathcal{U}_O)}$ is a solution of the **BTS** problem if and only if its graph $\mathcal{G} := \begin{bmatrix} M_S \\ I \end{bmatrix} B_I^{-1} H^2_{\mathcal{U}_I}$ satisfies:

1. $\mathcal{G}$ *is a subspace of* $\mathcal{M}_{S_0, B_I, B_O}$ (and hence also is a subspace of $\mathcal{K}$),
2. $\mathcal{G}$ *is a negative subspace of* $\mathcal{K}$, i.e., $[g, g]_{\mathcal{K}} \leq 0$ for all $g \in \mathcal{G}$, and, moreover, $\mathcal{G}$ *is maximal* with respect to this property: if $\mathcal{N}$ is a negative subspace of $\mathcal{K}$ with $\mathcal{G} \subset \mathcal{N}$, then $\mathcal{G} = \mathcal{N}$, and
3. $\mathcal{G}$ *is shift-invariant*, i.e., whenever $g \in \mathcal{G}$ then the vector function $\tilde{g}$ given by $\tilde{g}(z) = zg(z)$ is also in $\mathcal{G}$.

Let us verify each of these conditions in turn:

(1): If $S = S_0 + B_O Q B_I$ where $Q \in H^\infty_{\mathcal{L}(\mathcal{U}_I, \mathcal{U}_O)}$, then

$$\begin{bmatrix} M_S \\ I \end{bmatrix} B_I^{-1} H^2_{\mathcal{U}_I} = \begin{bmatrix} M_{S_0} + M_{B_O} M_Q M_{B_I} \\ I \end{bmatrix} B_I^{-1} H^2_{\mathcal{U}_I}$$

$$\subset \begin{bmatrix} B_O \\ 0 \end{bmatrix} H^2_{\mathcal{U}_I} + \begin{bmatrix} S_0 B_I^{-1} \\ B_I^{-1} \end{bmatrix} H^2_{\mathcal{U}_I} \text{ (since } M_Q : H^2_{\mathcal{U}_I} \to H^2_{\mathcal{U}_O})$$

$$= \mathcal{M}_{S_0, B_I, B_O}.$$

(2): If $S \in \mathcal{S}(\mathcal{U}_I, \mathcal{U}_O)$, then by the remarks above it follows that $\|M_S\|_{\text{op}} \leq 1$. This is enough to imply that $\mathcal{G}$ is $\mathcal{K}$-maximal negative.

(3): Due to the intertwining properties of M_z mentioned above, we have

$$M_z \begin{bmatrix} M_S \\ I \end{bmatrix} M_{B_I}^{-1} H^2_{\mathcal{U}_I} = \begin{bmatrix} M_S \\ I \end{bmatrix} M_{B_I}^{-1} M_z H^2_{\mathcal{U}_I} \subset \begin{bmatrix} M_S \\ I \end{bmatrix} B_I^{-1} H^2_{\mathcal{U}_I}$$

from which we see that $\mathcal{G}$ is invariant under M_z.

Conversely, one can show that if $\mathcal{G}$ is *any* subspace of $\mathcal{K}$ satisfying conditions (1), (2), (3) above, then $\mathcal{G}$ has the form $\mathcal{G} = \begin{bmatrix} M_S \\ I \end{bmatrix} B_I^{-1} H^2_{\mathcal{U}_I}$ with S a solution of the **BTS** problem. Indeed, condition (2) forces $\mathcal{G}$ to be the graph space $\mathcal{G} =$

$[^X_I] B_I^{-1} H^2_{\mathcal{U}_I}$ for a contraction operator $X \colon B_I^{-1} H^2_{\mathcal{U}_I} \to L^2_{\mathcal{U}_O}$. Condition (3) then forces X to be a multiplier $X = M_S$ for some $S \in L^\infty_{\mathcal{U}_I, \mathcal{U}_O}$ and $\|X\| \leq 1$ implies that $\|S\|_\infty \leq 1$. Finally, condition (1) then forces S to be of the form $S = S_0 + B_O K B_I$ with $K \in H^\infty_{\mathcal{L}(\mathcal{U}_I, \mathcal{U}_O)}$ from which we see that S is solution of the BTS problem.

Elementary Kreĭn-space geometry implies that, if there exists a $\mathcal{G}$ satisfying conditions (1) and (2), then necessarily the orthogonal complement of $\mathcal{M}$ inside $\mathcal{K}$ with respect to the indefinite Kreĭn-space inner product must be a positive subspace:

$$\mathcal{P} := \mathcal{P}_{S, B_I, B_O} = \mathcal{K} \ominus_J \mathcal{M}_{S, B_I, B_O} \text{ is a positive subspace,} \qquad (3)$$

i.e., $[p, p]_\mathcal{K} \geq 0$ for all $p \in \mathcal{P}$.

We conclude that *the subspace $\mathcal{P} := \mathcal{P}_{S_0, B_I, B_O}$ being a positive subspace is a necessary condition for the existence of solutions to the* BTS *Problem.* More explicitly one can work out that positivity of $\mathcal{P}$ in (3) is equivalent to contractivity of the Sarason model operator:

$$\|T_{S_0, B_I, B_O}\| \leq 1 \text{ where } T_{S_0, B_I, B_O} = P_{L^2_{\mathcal{U}_O} \ominus B_O H^2_{\mathcal{U}_O}} M_{S_0}|_{B_I^{-1} H^2_{\mathcal{U}_I}}. \qquad (4)$$

In terms of the **BTNP** formulation, condition (3) translates to positive semidefiniteness of the associated Pick matrix $\Lambda_{\mathfrak{D}_{\mathrm{BTNP}}}$:

$$\Lambda_{\mathfrak{D}_{\mathrm{BTNP}}} := \begin{bmatrix} \Lambda_I & (\Lambda_{OI})^* \\ \Lambda_{OI} & \Lambda_O \end{bmatrix} \geq 0 \qquad (5)$$

where

$$\Lambda_I = \begin{bmatrix} \dfrac{u_i^* u_j - v_i^* v_j}{1 - \overline{w}_i w_j} \end{bmatrix}, \quad [\Lambda_{OI}]_{ij} = \begin{cases} \dfrac{x_i v_j - y_i u_j}{w_j - z_i} & \text{for } z_i \neq w_j, \\ \rho_{ij} & \text{for } z_i = w_j \end{cases}, \quad \Lambda_O = \begin{bmatrix} \dfrac{x_i x_j^* - y_i y_j^*}{1 - z_i \overline{z}_j} \end{bmatrix}.$$

To prove sufficiency of any of the three equivalent conditions (3), (4), (5), we must be able to show that solutions of the **BTS** problem exist when $\mathcal{P}$ is a positive subspace. Let us therefore suppose that the subspace $\mathcal{P} := \mathcal{P}_{S, B_I, B_O}$ is a positive subspace of $\mathcal{K}$. Then *any* subspace $\mathcal{G}$ contained in $\mathcal{M}_{S_0, B_I, B_O}$ which is maximal as a negative subspace of $\mathcal{M}_{S_0, B_I, B_O}$ is also maximal as a negative subspace of $\mathcal{K}$ (i.e., $\mathcal{M}_{S_0, B_I, B_O}$-*maximal negative implies $\mathcal{K}$-maximal negative*) and hence $\mathcal{G}$ satisfies conditions (1) and (2). The rub is to find such a $\mathcal{G}$ which also satisfies the shift-invariance condition (3).

It is at this point that we make a leap of faith and assume what is called in [9] the *Beurling-Lax Axiom: there exists a (bounded) J-unitary function $\Theta(z)$ so that*

$$\mathcal{M}_{S_0, B_I, B_O} = \Theta \cdot H^2_{\mathcal{U}} \qquad (6)$$

for some appropriate Kreĭn space $\mathcal{U}$. Thus we assume that $\mathcal{U}$ has a Kreĭn-space inner product induced by a fundamental decomposition $\mathcal{U} = \mathcal{U}_+ \oplus \mathcal{U}_-$ with $\mathcal{U}_+$ a Hilbert space and $\mathcal{U}_-$ an anti-Hilbert space. More concretely, we simply take $\mathcal{U}_+$ and $\mathcal{U}_-$ to be Hilbert spaces and the Kreĭn-space inner product on $\mathcal{U} = \mathcal{U}_+ \oplus \mathcal{U}_-$ is given by

$$[[\begin{smallmatrix} u_+ \\ u_- \end{smallmatrix}], [\begin{smallmatrix} u_+ \\ u_- \end{smallmatrix}]]_{\mathcal{U}} = \|u_+\|^2_{\mathcal{U}_+} - \|u_-\|^2_{\mathcal{U}_-}.$$

The J-unitary property of Θ means that the values $\Theta(\zeta)$ of Θ are J-unitary for almost all ζ in the unit circle $\mathbb{T}$ (as a map between Kreĭn coefficient spaces $\mathcal{U}$ and $\mathcal{U}_O \oplus \mathcal{U}_I$ with the inner product induced by $J_K = \begin{bmatrix} I_{\mathcal{U}_O} & 0 \\ 0 & -I_{\mathcal{U}_I} \end{bmatrix}$). It then follows that without loss of generality we may take $\mathcal{U}_+ = \mathcal{U}_O$ and $\mathcal{U}_- = \mathcal{U}_I$. The crucial point is that then the operator M_Θ of multiplication by Θ *is a Kreĭn-space isomorphism* between $H^2_{\mathcal{U}}$ ($\mathcal{U} = \mathcal{U}_O \oplus \mathcal{U}_I$) and $\mathcal{M}_{S_0,B_I,B_O}$, i.e., M_Θ maps $H^2_{\mathcal{U}}$ one-to-one and onto $\mathcal{M}_{S_0,B_I,B_O}$, preserves the respective Kreĭn-space inner products:

$$[\Theta u, \Theta u]_{\mathcal{K}} = [u, u]_{\mathcal{U}},$$

and simultaneously *intertwines the respective shift operators*:

$$M_\Theta M_z = M_z M_\Theta.$$

It turns out that if condition (3) holds, then any such J-unitary representer Θ for $\mathcal{M}$ is actually J-*inner*, i.e., Θ has meromorphic pseudocontinuation to the unit disk $\mathbb{D}$ such that the values $\Theta(z)$ are J-contractive at all points of analyticity z inside the unit disk:

$$J - \Theta(z)^* J \Theta(z) \geq 0 \text{ for } z \in \mathbb{D}, \quad \Theta \text{ analytic at } z.$$

Under the assumption that we have such a representation (6) for $\mathcal{M}_{S_0,B_I,B_O}$, we can complete the solution of the **BTS** problem (under the assumption that the subspace $\mathcal{P}_{S_0,B_I,B_O}$ is a positive subspace) as follows. Since $M_\Theta \colon H^2_{\mathcal{U}} \to \mathcal{M}_{S_0,B_I,B_O}$ is a Kreĭn-space isomorphism, all the Kreĭn-space geometry is preserved. Thus a subspace $\mathcal{N}$ of $H^2_{\mathcal{U}}$ is maximal negative as a subspace of $H^2_{\mathcal{U}}$ if and only if its image $M_\Theta \mathcal{N} = \Theta \cdot \mathcal{N}$ is maximal negative as a subspace of $\mathcal{M}_{S_0,B_I,B_O}$. Moreover, since $M_z M_\Theta = M_\Theta M_z$, we see that $\mathcal{N}$ is shift-invariant in $H^2_{\mathcal{U}}$ if and only if its image $\Theta \cdot \mathcal{N}$ is a shift-invariant subspace of $\mathcal{M}_{S_0,B_I,B_O}$. From the observations made above, under the assumption that the subspace $\mathcal{P}_{S_0,B_I,B_O}$ is positive, getting a subspace $\mathcal{G}$ to satisfy conditions (1) and (2) in the Grassmannian reduction of the **BTS** problem is the same as getting $\mathcal{G} \subset \mathcal{M}_{S_0,B_I,B_O}$ to be maximal negative as a subspace of $\mathcal{M}_{S_0,B_I,B_O}$. We conclude that $\mathcal{G}$ meets all three conditions (1), (2), (3) in the Grassmannian reduction of the **BTS** problem if and only if $\mathcal{G} = \Theta \cdot \mathcal{N}$ where $\mathcal{N}$ is maximal negative as a subspace of $H^2_{\mathcal{U}}$ and is shift invariant. But these subspaces are easy: they are just subspaces of the form $\mathcal{N} = \begin{bmatrix} M_G \\ I \end{bmatrix} H^2_{\mathcal{U}_I}$ where G is in the Schur class $\mathcal{S}(\mathcal{U}_I, \mathcal{U}_O)$. We conclude that S solves the **BTS** problem if and only the graph $\mathcal{G}_S = \begin{bmatrix} M_S \\ I \end{bmatrix} B_I^{-1} H^2_{\mathcal{U}_I}$ satisfies

$$\begin{bmatrix} M_S \\ I \end{bmatrix} B_I^{-1} \cdot H^2_{\mathcal{U}_I} = \Theta \cdot \begin{bmatrix} M_G \\ I \end{bmatrix} H^2_{\mathcal{U}_I}$$

$$= \begin{bmatrix} \Theta_{11} G + \Theta_{12} \\ \Theta_{21} G + \Theta_{22} \end{bmatrix} \cdot H^2_{\mathcal{U}_I}. \tag{7}$$

Next note that the operator $M_\Theta \begin{bmatrix} M_G \\ I \end{bmatrix}$, as the composition of injective maps, is injective as an operator acting on $H^2_{\mathcal{U}_O}$. We claim that the bottom component $M_{\Theta_{21} G + \Theta_{22}}$ is already injective. Indeed, if $(\Theta_{21} G + \Theta_{22}) h = 0$ for some nonzero

$h \in H^2_{\mathcal{U}_I}$, then $\left[\begin{smallmatrix} (\Theta_{11}G + \Theta_{12})h \\ 0 \end{smallmatrix} \right]$ would be a strictly positive element of the negative subspace $\Theta \left[\begin{smallmatrix} G \\ I \end{smallmatrix} \right] \cdot H^2_{\mathcal{U}_I}$, a contradiction. Thus $M_{\Theta_{21}G + \Theta_{22}}$ must be injective as claimed. From the identity of bottom components in (7), we see that multiplication by $\Theta_{21}G + \Theta_{22}$ maps $H^2_{\mathcal{U}_I}$ onto $B_I^{-1}H^2_{\mathcal{U}_I}$. We conclude that the function $K := B_I(\Theta_{21}G + \Theta_{22})$ and its inverse are in $H^\infty_{\mathcal{L}(\mathcal{U}_I)}$. Then we may rewrite (7) as

$$
\begin{aligned}
\begin{bmatrix} S \\ I \end{bmatrix} B_I^{-1} \cdot H^2_{\mathcal{U}_I} &= \begin{bmatrix} (\Theta_{11}G + \Theta_{12})(\Theta_{21}G + \Theta_{22})^{-1} \\ I \end{bmatrix} B_I^{-1}K \cdot H^2_{\mathcal{U}_I} \\
&= \begin{bmatrix} (\Theta_{11}G + \Theta_{12})(\Theta_{21}G + \Theta_{22})^{-1} \\ I \end{bmatrix} B_I^{-1} \cdot H^2_{\mathcal{U}_I}.
\end{aligned}
$$

Thus for each $h \in B_I^{-1}H^2_{\mathcal{U}_I}$ there is an element h' of $B_I^{-1}H^2_{\mathcal{U}_I}$ such that

$$
\begin{bmatrix} S \\ I \end{bmatrix} h = \begin{bmatrix} (\Theta_{11}G + \Theta_{12})(\Theta_{21}G + \Theta_{22})^{-1} \\ I \end{bmatrix} h'.
$$

Equality of the bottom components forces $h = h'$ and then equality of the top components for all h leads to the linear-fractional parametrization for the set of solutions of the **BTS** problem: S *solves the* **BTS** *Problem if and only if S has the form*

$$
S = (\Theta_{11}G + \Theta_{12})(\Theta_{21}G + \Theta_{22})^{-1} \tag{8}
$$

for a uniquely determined $G \in \mathcal{S}(\mathcal{U}_I, \mathcal{U}_O)$. In this way we arrive at the linear-fractional parametrization of the set of all solutions appearing in the work of Nevanlinna [68] for the classical Nevanlinna-Pick interpolation problem and in the work of Adamjan-Arov-Kreĭn [1] in the context of the Nehari-Takagi problem.

Remark 1. We note that the derivation of the linear-fractional parametrization (8) used essentially only coordinate-free Kreĭn-space geometry. It is also possible to arrive at this parametrization without any appeal to Kreĭn-space geometry via working directly with properties of J-inner functions: see, e.g., [17] where a winding number argument plays a key role, and [39] for an alternative reproducing-kernel method.

All the success of the preceding paragraphs is predicated on the validity of the so-called Beurling-Lax Axiom (6). Validity of the Beurling-Lax Axiom requires at a minimum that the subspace $\mathcal{M}_{S_0, B_I, B_O}$ be a Kreĭn space in the indefinite inner product inherited from $\mathcal{K}$. Unlike the Hilbert space case, this is not automatic (see, e.g., [8, Section 1.7]). We say that the subspace $\mathcal{M}$ of the Kreĭn space $\mathcal{K}$ is *regular* if it is the case that $\mathcal{M}$ is itself a Kreĭn space with inner product inherited from $\mathcal{K}$; an equivalent condition is that $\mathcal{K}$ decomposes as an orthogonal (in the Kreĭn-space inner product) direct sum $\mathcal{K} = \mathcal{M} \oplus \mathcal{M}^{[\perp]}$ (where $\mathcal{M}^{[\perp]}$ is the orthogonal complement of $\mathcal{M}$ inside $\mathcal{K}$ with respect to the Kreĭn-space inner product). For the Nevanlinna-Pick problem involving only finitely many interpolation conditions, regularity of $\mathcal{M}$ is automatic under the condition that the solution of the interpolation problem is not unique (completely indeterminate in the language of some authors). Nevertheless, even when the subspace $\mathcal{M} = \mathcal{M}_{S_0, B_I, B_O}$ is regular in $\mathcal{K} = \begin{bmatrix} L^2_{\mathcal{U}_O} \\ B_I^{-1} H^2_{\mathcal{U}_I} \end{bmatrix}$, it can happen that only a weakened version of the Beurling-Lax Axiom holds. The following is one of the main results from [24] (see [58] for extensions to shift-invariant subspaces contractively included in $H^2_{\mathcal{U}}$).

Theorem 2. *Suppose that $\mathcal{M}$ is a regular subspace of $L^2_{\mathcal{U}}$ (where $L^2_{\mathcal{U}}$ is considered to be a Kreĭn space in the indefinite inner product induced by the Kreĭn-space inner product on the space of constants $\mathcal{U} = \mathcal{U}_O \oplus \mathcal{U}_I$). Then there exists a multiplier Θ with values in $\mathcal{L}(\mathcal{U})$ such that*

1. *$M_{\Theta^{\pm 1}} : \mathcal{U} \to L^2_{\mathcal{U}}$,*
2. *$\Theta(\zeta)^* J \Theta(\zeta) = J$ for almost all $\zeta \in \mathbb{T}$ (where $J = \begin{bmatrix} I_{\mathcal{U}_O} & 0 \\ 0 & -I_{\mathcal{U}_I} \end{bmatrix}$),*
3. *the densely defined operator $M_\Theta P_{H^2_{\mathcal{U}}} M_{\Theta^{-1}} = M_\Theta P_{H^2_{\mathcal{U}}} J M_{\Theta^*} J$ extends to define a bounded J-orthogonal projection operator on $L^2_{\mathcal{U}}$, and*
4. *the space $\mathcal{M}$ is equal to the closure of $\Theta \cdot \mathcal{P}^+_{\mathcal{U}}$, where $\mathcal{P}^+_{\mathcal{U}}$ is the space of analytic trigonometric polynomials $p(\zeta) = \sum_{k=0}^n u_k \zeta^k$ with coefficients u_k in $\mathcal{U}$ ($n = 0, 1, 2, \dots$).*

Conversely, whenever Θ is a multiplier satisfying conditions (1), (2), and (3) and the subspace $\mathcal{M}$ is defined via (4), then $\mathcal{M}$ is a regular subspace of $L^2_{\mathcal{U}}$ (with J-orthogonal projection onto $\mathcal{M}$ along $\mathcal{M}^{[\perp]}$ given by the bounded extension of $M_{\Theta^{-1}} P_{H^2_{\mathcal{U}}} M_\Theta$ onto all of $L^2_{\mathcal{U}}$).

This illustrates a general phenomenon in the Kreĭn-space setting in contrast with the Hilbert-space setting: there is no reason why unitary operators need be bounded. The moral of the story is: the Beurling-Lax Axiom does hold in case $\mathcal{M}_{S_0, B_I, B_O}$ is a regular subspace of $\mathcal{K}$, but only with in general densely defined and unbounded Beurling-Lax representer Θ. This technical detail in turn complicates the Kreĭn-space geometry argument given above leading to the existence and parametrization of the set of all solutions of the **BTS** problem under the necessary condition (3) that the subspace $\mathcal{P}_{S_0, B_I, B_O}$ be a positive subspace. This point was handled in [24] (and revisited in [29]) via an approximation argument using the fact that bounded functions are dense in any shift-invariant subspace of H^2.

Here we use an idea from [45] based on an ingredient from the approach of Dym [39] to obtain a smoother derivation of the linear-fractional parametrization even for the case where the Beurling-Lax representer may be unbounded. The following lemma proves to be helpful.

Lemma 3 (see [45, Lemma 2.3.1]). *Let $\mathcal{K}$ be a Kreĭn space and let $\mathcal{M}$ be a regular subspace of $\mathcal{K}$ such that $\mathcal{M}^{[\perp]}$ is a positive subspace. If $\mathcal{G}$ is a maximal negative subspace of $\mathcal{K}$, then the following are equivalent:*

1. *$\mathcal{G} \subset \mathcal{M}$.*
2. *$P_{\mathcal{M}} \mathcal{G}^{[\perp]}$ is a positive subspace, where $P_{\mathcal{M}}$ is the J-orthogonal projection of $\mathcal{K}$ onto $\mathcal{M}$.*

Now we suppose that $\mathcal{P}_{S_0, B_I, B_O}$ is a positive subspace (as is necessary for solutions to the **BTS** problem to exist) and that $S \in \mathcal{S}(\mathcal{U}_I, \mathcal{U}_O)$ is a solution. Thus

$\mathcal{G} = \begin{bmatrix} S \\ I \end{bmatrix} B_I^{-1} \cdot H_{\mathcal{U}_I}^2$ is maximal negative and contained in $\mathcal{M}_{S_0,B_I,B_O}$. According to the lemma, this means that $P_{\mathcal{M}}\mathcal{G}^{[\perp]}$ is a positive subspace. By the result of Theorem 2 we know that $P_{\mathcal{M}} = M_\Theta J P_{H_{\mathcal{U}}^2} M_{\Theta^*} J$ (formally unbounded but having bounded extension to the whole space). Also, an elementary computation gives

$$\mathcal{G}^{[\perp]} = \begin{bmatrix} I \\ P_{B_I^{-1}H_{\mathcal{U}_I}^2} M_{S^*} \end{bmatrix} L_{\mathcal{U}_O}^2.$$

Thus the condition (2) in Lemma 3 becomes

$$\left\langle J M_\Theta J P_{H^2} M_{\Theta^*} J \begin{bmatrix} I \\ P_{B_I^{-1}H_{\mathcal{U}_I}^2} M_{S^*} \end{bmatrix} f, \begin{bmatrix} I \\ P_{B_I^{-1}H_{\mathcal{U}_I}^2} M_{S^*} \end{bmatrix} f \right\rangle_{L^2 \oplus B_I^{-1}H^2} \geq 0 \qquad (9)$$

for all $f \in L_{\mathcal{U}_O}^2$. Since the range of M_Θ is contained in $\mathcal{M}$ which in turn is contained in $\begin{bmatrix} L_{\mathcal{U}_O}^2 \\ B_I^{-1}H_{\mathcal{U}_I}^2 \end{bmatrix}$, we see that the projection $P_{B_I^{-1}H_{\mathcal{U}_I}^2}$ in (9) is removable. We can then rewrite (9) as

$$\left\langle \begin{bmatrix} I & -M_S \end{bmatrix} M_\Theta J P_{H^2} M_{\Theta^*} \begin{bmatrix} I \\ -M_{S^*} \end{bmatrix} f, f \right\rangle \geq 0.$$

Restricting to an appropriate dense domain and writing F in place of M_F for multiplication operators for simplicity, we arrive at the operator inequality

$$0 \leq \begin{bmatrix} \Theta_{11} - S\Theta_{21} & \Theta_{12} - S\Theta_{22} \end{bmatrix} J P_{H^2} \begin{bmatrix} \Theta_{11}^* - \Theta_{21}^* S^* \\ \Theta_{12}^* - \Theta_{22}^* S^* \end{bmatrix}$$
$$= (\Theta_{11} - S\Theta_{21}) P_{H^2} (\Theta_{11} - S\Theta_{21})^* - (\Theta_{12} - S\Theta_{22}) P_{H^2} (\Theta_{12} - S\Theta_{22})^*. \qquad (10)$$

It is a consequence of the Commutant Lifting Theorem (in this form actually a version of the Leech Theorem – see [70]) that (10) implies that there is a Schur-class function written as $-G \in \mathcal{S}(\mathcal{U}_I, \mathcal{U}_O)$ so that

$$\Theta_{12} - S\Theta_{22} = (\Theta_{11} - S\Theta_{21})(-G).$$

It is now a straightforward matter to solve for S in terms of G to arrive at

$$S = (\Theta_{11}G + \Theta_{12})(\Theta_{21}G + \Theta_{22})^{-1}. \qquad (11)$$

Conversely the steps are reversible: for any Schur-class function $G \in \mathcal{S}(\mathcal{U}_I, \mathcal{U}_O)$, the formula (11) leads to a solution S of the **BTS** problem. In this way we arrive at the linear-fractional parametrization (8) for the set of all solutions of the **BTS** problem even in the case where $\mathcal{M}$ is regular but its Beurling-Lax representer Θ is not bounded.

Remark 4. The case where $\mathcal{M}$ is regular is only a particular instance of the so-called "completely indeterminate case" where solutions of the **BTS** problem exist having norm strictly less than 1. In this case there is still a linear-fractional parametrization of the set of all solutions of the form (8) even though the associated interpolation subspace $\mathcal{M}_{S_0,B_I,B_O}$ is not a regular subspace of $\mathcal{K}$; see [5].

3. State-space realization of the J-Beurling-Lax representer

Various authors ([40, 17]), perhaps beginning with Nudelman [69], have noticed that the detailed interpolation conditions (1) can be written more compactly in aggregate form as

$$\frac{1}{2\pi i} \int_{\mathbb{T}} (zI - Z)^{-1} X S(z) \, dz = Y,$$

$$\frac{1}{2\pi i} \int_{\mathbb{T}} S(z) U (zI - A)^{-1} \, dz = V, \tag{12}$$

$$\frac{1}{2\pi i} \int_{\mathbb{T}} (zI - Z)^{-1} X S(z) U (zI - A)^{-1} \, dz = \Gamma, \tag{13}$$

where the collection of seven matrices $\mathfrak{D}_{\text{BTOA}} = (U, V, A, Z, X, Y, \Gamma)$ (the label **BTOA** refers to the *bitangential operator-argument* interpolation problem which is described below) is given by

$$X = \begin{bmatrix} x_1 \\ \vdots \\ x_{n_O} \end{bmatrix}, \quad Y = \begin{bmatrix} y_1 \\ \vdots \\ y_{n_O} \end{bmatrix}, \quad Z = \begin{bmatrix} z_1 & & \\ & \ddots & \\ & & z_{n_O} \end{bmatrix},$$

$$U = \begin{bmatrix} u_1 & \cdots & u_{n_I} \end{bmatrix}, \quad V = \begin{bmatrix} v_1 & \cdots & v_{n_I} \end{bmatrix}, \quad A = \begin{bmatrix} w_1 & & \\ & \ddots & \\ & & w_{n_O} \end{bmatrix},$$

$$[\Gamma]_{ij} = \begin{cases} \frac{x_i v_j - y_i u_j}{w_j - z_i} & \text{if } w_j \neq z_i, \\ \rho_{ij} & \text{if } w_j = z_i \end{cases} \quad \text{for } 1 \leq i \leq n_O, \ 1 \leq j \leq n_I. \tag{14}$$

The interpolation conditions expressed in this form (13) make sense even if the matrices A and Z, while maintaining spectrum inside the unit disk, have more general Jordan canonical forms (i.e., are not diagonalizable); in this way we get a compact way of expressing higher-order bitangential interpolation conditions. By expanding the resolvent operators inside the contour integrals in Laurent series, it is not hard to see that we can rewrite the interpolation/moment conditions in (13) in the form

$$P_{H^{2\perp}_{\mathcal{U}_O}} M_S \widehat{\mathcal{O}}^b_{U,A} = \widehat{\mathcal{O}}^b_{V,A}, \quad \widehat{\mathcal{C}}^b_{Z,X} M_S|_{H^2_{\mathcal{U}_I}} = \widehat{\mathcal{C}}^b_{Z,Y}, \quad \widehat{\mathcal{C}}^b_{Z,X} P_{H^2_{\mathcal{U}_O}} M_S \widehat{\mathcal{O}}^b_{U,A} = \Gamma \tag{15}$$

where $\widehat{\mathcal{O}}^b_{U,A} \colon \mathbb{C}^{n_I} \to H^{2\perp}_{\mathcal{U}_I}$ and $\widehat{\mathcal{O}}^b_{V,A} \colon \mathbb{C}^{n_I} \to H^{2\perp}_{\mathcal{U}_O}$ are the *backward-time observation operators* given by

$$\widehat{\mathcal{O}}^b_{U,A} \colon x \mapsto U(zI - A)^{-1} x = \sum_{n=1}^{\infty} (U A^{n-1} x) z^{-n},$$

$$\widehat{\mathcal{O}}^b_{V,A} \colon x \mapsto V(zI - A)^{-1} x = \sum_{n=1}^{\infty} (V A^{n-1} x) z^{-n},$$

and where $\widehat{\mathcal{C}}^b_{Z,X} \colon H^2_{\mathcal{U}_I} \to \mathbb{C}^{no}$, $\widehat{\mathcal{C}}^b_{Z,Y} \colon H^2_{\mathcal{U}_I} \to \mathbb{C}^{no}$ are the *backward-time control operators* given by

$$\widehat{\mathcal{C}}^b_{Z,X} \colon f(z) = \sum_{n=0}^{\infty} f_n z^n \mapsto \sum_{n=0}^{\infty} Z^n X f_n, \quad \widehat{\mathcal{C}}^b_{Z,Y} \colon g(z) = \sum_{n=0}^{\infty} g_n z^n \mapsto \sum_{n=0}^{\infty} Z^n Y g_n.$$

The terminology is suggested from the following connections with linear systems. Given a discrete-time state-output linear system running in backwards time with specified initial condition at time $n = 0$

$$\begin{aligned} x(n) &= Ax(n+1) \\ y(n) &= Cx(n+1) \end{aligned}, \quad x(0) = x_0, \tag{16}$$

the resulting output string $\{y(n)\}_{n=-1,-2,\dots}$ is given by

$$y(-n) = CA^{n-1}x_0 \text{ for n=1,2, \dots .}$$

It is natural to let $\mathcal{O}^b_{C,A}$ denote the *time-domain backward-time observation operator* given by

$$\mathcal{O}^b_{C,A} \colon x \mapsto \{y(n)\}_{n=-1,-2,\dots} = \{CA^{-n-1}x\}_{n=-1,-2,\dots}.$$

Upon taking Z-transform $\{y(n)\} \mapsto \widehat{y}(z) = \sum_{n\in\mathbb{Z}} y(n)z^n$, we arrive at the *frequency-domain backward-time observation operator* $\widehat{\mathcal{O}}^b_{C,A}$ given by

$$\widehat{\mathcal{O}}^b_{C,A} \colon x \mapsto \widehat{y}(z) = \sum_{n=1}^{\infty}(CA^{n-1}x)z^{-n} = C(zI - A)^{-1}x.$$

In these computations we assumed that the matrix A has spectrum inside the disk; we conclude that $C(zI - A)^{-1}x \in H^{2\perp}_{\mathcal{U}_O}$ when viewed as a function on the circle; note that $C(zI - A)^{-1}x$ is rational with all poles inside the disk and vanishes at infinity.

Similarly, given a discrete-time input-state linear system running in backwards time

$$x(n) = Zx(n+1) + Xu(n+1) \tag{17}$$

where we assume that $x(n) = 0$ for $n \geq N$ and $u(n) = 0$ for all $n > N$ for some large N, solving the recursion successively for $x(N-1)$, $x(N-2)$, $\dots$, $x(0)$ leads to the formula

$$x(0) = \sum_{k=0}^{\infty} Z^k X u(k) = \begin{bmatrix} X & ZX & Z^2X & \cdots \end{bmatrix} \begin{bmatrix} u(0) \\ u(1) \\ u(2) \\ \vdots \end{bmatrix}.$$

As Z by assumption has spectrum inside the unit disk, the matrix

$$\begin{bmatrix} X & ZX & Z^2X & \cdots \end{bmatrix},$$

initially defined only on input strings having finite support, extends to the space of all $\mathcal{U}_I$-valued ℓ^2 input-strings $\ell^2_{\mathcal{U}_I}$. It is natural to define the *frequency-domain backward-time control operator* $\mathcal{C}^b_{Z,X}$ by

$$\mathcal{C}^b_{Z,X}: \{u(n)\}_{n\geq 0} \mapsto \begin{bmatrix} X & ZX & Z^2X & \cdots \end{bmatrix} \begin{bmatrix} u(0) \\ u(1) \\ u(2) \\ \vdots \end{bmatrix}.$$

Application of the inverse Z-transform to $\{u(n)\}_{n=0,1,2,\ldots}$ then leads us to the *frequency-domain backward-time control operator* $\widehat{\mathcal{C}}^b_{Z,X}: H^2_{\mathcal{U}_I} \to \mathbb{C}^{n_O}$ given by

$$\widehat{\mathcal{C}}^b_{Z,X}: u(z) = \sum_{n=0}^{\infty} u(n)z^n \mapsto \sum_{n=0}^{\infty} Z^n X u(n).$$

The next step is to observe that conditions (15) make sense even if the data set $\mathfrak{D}_{\text{BTOA}}$ does not consist of matrices. Instead, we now view X,Y,Z,U,V,A,Γ as operators

$$X: \mathcal{U}_O \to \mathcal{X}_L, \quad Y: \mathcal{U}_I \to \mathcal{X}_L, \quad Z: \mathcal{X}_L \to \mathcal{X}_L,$$
$$U: \mathcal{X}_R \to \mathcal{U}_I, \quad V: \mathcal{X}_R \to \mathcal{U}_O, \quad A: \mathcal{X}_R \to \mathcal{X}_R, \quad \Gamma: \mathcal{X}_R \to \mathcal{X}_L. \quad (18)$$

Note that when the septet (X,Y,Z,U,V,A,Γ) is of the form as in (14), then the Sylvester equation

$$\Gamma A - Z\Gamma = XV - YU. \quad (19)$$

is satisfied. To avoid degeneracies, it is natural to impose some additional controllability and observability assumptions. The full set of admissibility requirements is as follows.

Definition 5. Given a septet of operators $\mathfrak{D}_{\text{BTOA}} := (X,Y,Z,U,V,A,\Gamma)$ as in (18), we say that $\mathfrak{D}_{\text{BTOA}}$ is *admissible* if the following conditions are satisfied:

1. (X,Z) is a *stable exactly controllable input pair*, i.e., $\widehat{\mathcal{C}}^b_{Z,X}$ defines a bounded operator from $H^2_{\mathcal{U}_I}$ into $\mathcal{X}_L$ with range equal to the whole space $\mathcal{X}_L$.

2. (U,A) is a *stable exactly observable output pair*, i.e., $\widehat{\mathcal{O}}^b_{U,A}$ maps the state space $\mathcal{X}_R$ into $H^{2\perp}_{\mathcal{U}_I}$ and is bounded below:

$$\|\widehat{\mathcal{O}}^b_{U,A}x\|^2_{H^{2\perp}_{\mathcal{U}_I}} \geq \delta\|x\|^2_{\mathcal{X}_R} \text{ for some } \delta > 0.$$

3. The operator Γ is a solution of the Sylvester equation (19).

We can now formulate the promised *bitangential operator-argument interpolation problem*.

Problem BTOA (Bitangential Operator Argument Interpolation Problem): *Given an admissible operator-argument interpolation data set $\mathfrak{D}_{\text{BTOA}}$ as described in Definition 5, find a function S in the Schur class $\mathcal{S}(\mathcal{U}_I,\mathcal{U}_O)$ which satisfies the interpolation conditions (15).*

It can be shown that there is a bijection between **BTS** data sets $\mathfrak{D}_{\mathrm{BTS}} = \{S_0, B_I, B_O\}$ and admissible **BTOA** data sets $\mathfrak{D}_{\mathrm{BTOA}}$ (18) so that the corresponding interpolation problems **BTS** and **BTOA** have exactly the same set of solutions. For the rational matrix-valued case, details can be found in [17] (see Theorem 16.9.3 there); the result for the general case can be worked out using these ideas and the results from [30].

Let us now suppose that $\mathfrak{D}_{\mathrm{BTS}} = \{S_0, B_I, B_O\}$ and $\mathfrak{D}_{\mathrm{BTOA}}$ (18) are equivalent in this sense. Then the subspace $\mathcal{M}_{S_0, B_I, B_O}$ (2) is the subspace of $L^2_{\mathcal{U}_O} \oplus B_I^{-1}H^2_{\mathcal{U}_I}$ spanned by the graph spaces $\left[\begin{smallmatrix}S\\I\end{smallmatrix}\right] B_I^{-1} \cdot H^2_{\mathcal{U}_I}$ of solutions S of the **BTS** interpolation problem. Hence this same subspace can be expressed as the span $\mathcal{M}_{\mathrm{BTOA}}$ of the graph spaces of all solutions S of the **BTOA** interpolation problem. One can work out that $\mathcal{M}_{\mathrm{BTOA}}$ can be expressed directly in terms of the data set $\mathfrak{D}_{\mathrm{BTOA}}$ as:

$$\mathcal{M}_{\mathrm{BTOA}} = \left\{ \widehat{\mathcal{O}}^b_{[\begin{smallmatrix}V\\U\end{smallmatrix}],A} x + \begin{bmatrix} f_+ \\ f_- \end{bmatrix} : x \in \mathcal{X}_R, \begin{bmatrix} f_+ \\ f_- \end{bmatrix} \in H^2_{\mathcal{U}_O \oplus \mathcal{U}_I} \right.$$
$$\left. \text{such that } \widehat{\mathcal{C}}^b_{Z,[X\ -Y]} \begin{bmatrix} f_+ \\ f_- \end{bmatrix} = \Gamma x \right\}. \tag{20}$$

Remark 6. For the representation of general shift-invariant subspaces in terms of null-pole data developed in [30], the coupling operator Γ in general is only a closed (possibly unbounded) operator with dense domain in $\mathcal{X}_R$. In the context of the **BTOA** interpolation problem as we have here, from the last of the interpolation conditions (15) we see that Γ is bounded whenever the **BTOA** interpolation problem has solutions. Therefore for the discussion here we may avoid the complications of unbounded Γ and always assume that Γ is bounded.

By the analysis of the previous section, we see that parametrization of the set of all solutions of the **BTOA** interpolation problem follows from a J-Beurling-Lax representation for the subspace $\mathcal{M}_{\mathrm{BTOA}}$ (20) as in Theorem 2. Toward this end we have the following result; we do not go into details here but the main ingredients can be found [30] (see Corollary 6.4, Theorem 6.5 and Theorem 7.1 there).

Theorem 7. *Let $\mathfrak{D}_{\mathrm{BTOA}}$ be an admissible bitangential operator-argument interpolation data set as in Definition 5 and let $\mathcal{M}_{\mathrm{BTOA}}$ be the associated shift-invariant subspace as in* (20). *Then:*

1. *$\mathcal{M}_{\mathrm{BTOA}}$ is regular as a subspace of the Kreĭn space $L^2_{\mathcal{U}_O} \oplus B_I^{-1} \cdot H^2_{\mathcal{U}_I}$, or equivalently, as a subspace of the Kreĭn space $L^2_{\mathcal{U}_O} \oplus L^2_{\mathcal{U}_I}$ (both with the indefinite inner product induced by $J = \left[\begin{smallmatrix} I_{\mathcal{U}_O} & 0 \\ 0 & -I_{\mathcal{U}_I} \end{smallmatrix}\right]$) if and only if the operator*

$$\Lambda_{\mathrm{BTOA}} := \begin{bmatrix} -(\widehat{\mathcal{O}}^b_{[\begin{smallmatrix}V\\U\end{smallmatrix}],A})^* J \widehat{\mathcal{O}}^b_{[\begin{smallmatrix}V\\U\end{smallmatrix}],A} & \Gamma^* \\ \Gamma & \widehat{\mathcal{C}}^b_{Z,[X\ -Y]} J (\widehat{\mathcal{C}}^b_{Z,[X\ -Y]})^* \end{bmatrix} : \begin{bmatrix} \mathcal{X}_R \\ \mathcal{X}_L \end{bmatrix} \to \begin{bmatrix} \mathcal{X}_R \\ \mathcal{X}_L \end{bmatrix}$$
$$\tag{21}$$

   is invertible.

2. *The subspace*

$$\mathcal{P}_{\mathrm{BTOA}} := \mathcal{K} \ominus_J \mathcal{M}_{\mathrm{BTOA}} \ where \ \mathcal{K} = \begin{bmatrix} \widehat{\mathcal{O}}^b_{V,A} \\ \widehat{\mathcal{O}}^b_{U,A} \end{bmatrix} \mathcal{X}_R \oplus \begin{bmatrix} H^2_{\mathcal{U}_O} \\ H^2_{\mathcal{U}_I} \end{bmatrix}$$

  is a positive subspace if and only the **BTOA** *Pick matrix* Λ_{BTOA} *as in* (21) *is positive semidefinite.*

3. *Assume that* Λ_{BTOA} *is invertible. Then a Beurling-Lax representer* Θ *for* $\mathcal{M}_{\mathrm{BTOA}}$ *has bidichotomous realization*

$$\Theta(z) = \begin{bmatrix} V \\ U \end{bmatrix} (zI - A)^{-1} \mathbf{B}_- + \mathbf{D} + z \begin{bmatrix} X^* \\ Y^* \end{bmatrix} (I - zZ^*)^{-1} \mathbf{B}_+. \qquad (22)$$

  where the operators appearing in (22) *not specified in the data set* $\mathfrak{D}_{\mathrm{BTOA}}$, *namely* $\mathbf{B}_-$, $\mathbf{B}_+$, *and* $\mathbf{D}$, *are constructed so that the operator*

$$\begin{bmatrix} \mathbf{B}_- \\ \mathbf{B}_+ \\ \mathbf{D} \end{bmatrix} : \begin{bmatrix} \mathcal{U}_O \\ \mathcal{U}_I \end{bmatrix} \to \begin{bmatrix} \mathcal{X}_R \\ \mathcal{X}_L \\ \begin{bmatrix} \mathcal{U}_O \\ \mathcal{U}_I \end{bmatrix} \end{bmatrix}$$

  is a J-unitary isomorphism from $\begin{bmatrix} \mathcal{U}_O \\ \mathcal{U}_I \end{bmatrix}$ *onto* $Ker \ \Psi \subset \mathcal{X}_R \oplus \mathcal{X}_L \oplus \begin{bmatrix} \mathcal{U}_O \\ \mathcal{U}_I \end{bmatrix}$, *where*

$$\Psi = \begin{bmatrix} \Gamma & -Z\widehat{\mathcal{C}}^b_{Z,[X \ -Y]} J (\widehat{\mathcal{C}}^b_{Z,[X \ -Y]})^* & [-X \ Y] \\ -A^* (\widehat{\mathcal{O}}^b_{[\begin{smallmatrix} V \\ U \end{smallmatrix}],A})^* J \widehat{\mathcal{O}}^b_{[\begin{smallmatrix} V \\ U \end{smallmatrix}]} & -\Gamma^* & [-V^* \ U^*] \end{bmatrix},$$

  and where $\mathcal{X}_R \oplus \mathcal{X}_L \oplus \begin{bmatrix} \mathcal{U}_O \\ \mathcal{U}_I \end{bmatrix}$ *carries the indefinite inner product induced by the selfadjoint operator*

$$\mathcal{J} := \begin{bmatrix} (\mathcal{O}^b_{[\begin{smallmatrix} V \\ U \end{smallmatrix}],A})^* J \widehat{\mathcal{O}}^b_{[\begin{smallmatrix} V \\ U \end{smallmatrix}],A} & 0 & 0 \\ 0 & \widehat{\mathcal{C}}^b_{Z,[X \ -Y]} J (\widehat{\mathcal{C}}^b_{Z,[X \ -Y]})^* & 0 \\ 0 & 0 & J \end{bmatrix}.$$

  In case Λ_{BTOA} *in* (15) *is also positive definite, then* Θ *parametrizes all solutions of the* **BTOA** *interpolation problem via the formula* (8) *with free parameter* $G \in \mathcal{S}(\mathcal{U}_I, \mathcal{U}_O)$.

Remark 8. The idea for the derivation of the formula (22) for the Beurling-Lax representer Θ for the subspace $\mathcal{M}_{\mathrm{BTOA}}$ in Theorem 7 goes back to [24]: Θ, when viewed as an operator from the Kreĭn space of constant functions $\mathcal{U}_O \oplus \mathcal{U}_I$ into the Kreĭn space of functions $L^2_{\mathcal{U}_O} \oplus L^2_{\mathcal{U}_I}$ is a Kreĭn-space isomorphism from $\mathcal{U}_O \oplus \mathcal{U}_I$ to the wandering subspace $\mathcal{L} := M_z(\mathcal{M}_{\mathrm{BTOA}})^{[\perp]} \cap \mathcal{M}_{\mathrm{BTOA}}$. Similar state-space realizations hold for affine Beurling-Lax representations (or the *Beurling-Lax Theorem for the Lie group* $GL(n, \mathbb{C})$ in the terminology of [25]). Here one is given a pair of subspaces $(\mathcal{M}, \mathcal{M}^\times)$ such that $\mathcal{M}$ is forward shift invariant, $\mathcal{M}^\times$ is backward shift invariant, $\mathcal{M}$ and $\mathcal{M}^\times$ form a direct-sum decomposition for $\mathcal{L}^2_{\mathcal{U}}$, and one seeks an invertible operator function Θ on the circle $\mathbb{T}$ so that roughly [1] $\mathcal{M} = \Theta \cdot H^2_{\mathcal{U}}$ and $\mathcal{M}^\times = \Theta \cdot H^{2\perp}_{\mathcal{U}}$. State-space implementations for the Beurling-Lax representer Θ where $\mathcal{M}$ and $\mathcal{M}^\times$ are assumed to have representations of the form (20) are worked out in [30] (see also [17, Theorem 5.5.2] and [16] for the rational matrix-valued case).

Remark 9. When we consider the result of Theorem 7 for the case of matricial data, arguably the solution is not as explicit as one would like; one must find a J-orthonormal basis for a certain finite-dimensional regular subspace of $L^2_{\mathcal{U}_O \oplus \mathcal{U}_I}$. One can explain this as follows. In this general setting, no assumptions are made on the locations of the poles (i.e., the spectrum of A inside the unit disk and the reflection of the spectrum of Z to outside the disk) and zeros (i.e., the spectrum of Z and the reflection of the spectrum of A to outside the disk) in the extended complex plane; hence there is no global chart with respect to which one can set up coordinates. This issue can be resolved in several ways. For example, one could specify a point ζ_0 on the unit circle at which no interpolation conditions are specified, and demand that $\Theta(\zeta_0)$ be some given J-unitary matrix (e.g., $I_{\mathcal{U}_O \oplus \mathcal{U}_I}$: see Theorem 7.5.2 in [17]); however in the case of infinite-dimensional data it is possible for Z and A to have spectrum including the whole unit circle thereby making this approach infeasible. Alternatively, one might assume that both A and Z are invertible (no interpolation conditions at the point 0) in which case Theorem 7.1.7 in [17] gives a more explicit formula for Θ. A difficulty for numerical implementation of the formulas is the challenge of inverting the Pick matrix in these formulas; this difficulty was later addressed by adapting the use of fast recursive algorithms for the inversion of structured matrices by Olshevsky and his collaborators (see, e.g., [61, 62]).

3.1. A special case: left tangential operator-argument interpolation

We now discuss the special case of Theorem 7 where there are only left tangential interpolation conditions present (so the right-side state space $\mathcal{X}_R = \{0\}$ is trivial). In this case the bitangential operator-argument interpolation data set $\mathfrak{D}_{\mathrm{BTOA}}$ consisting of seven operators collapses to a left tangential operator-argument data set $\mathfrak{D}_{\mathrm{LTOA}}$ consisting of only three operators

$$\mathfrak{D}_{\mathrm{LTOA}} = \{X, Y, Z\} \text{ where } X \colon \mathcal{U}_O \to \mathcal{X}_L, \quad Y \colon \mathcal{U}_I \to \mathcal{X}_L, \quad Z \colon \mathcal{X}_L \to \mathcal{X}_L,$$

the interpolation problem collapses to just the second of the conditions (15) which can be written also in more succinct left-tangential operator-argument form

$$(\widehat{XS})^{\wedge L}(Z) := \sum_{n=0}^{\infty} Z^n X S_n = Y \tag{23}$$

---

[1]In general the same technicality occurs here as occurs for the case where $\mathcal{M}^\times = \mathcal{M}^{[\perp]}$ as described in Theorem 2 above: one can only require that $\Theta^{\pm 1}\mathcal{U} \subset L^2_{\mathcal{U}}$, $\mathcal{M}$ is equal to the closure of $\Theta \cdot \mathcal{P}^+_{\mathcal{U}}$ and that $\mathcal{M}^\times$ is the closure of $\Theta \cdot z^{-1}\overline{\mathcal{P}^+_{\mathcal{U}}}$, where $\overline{\mathcal{P}^+_{\mathcal{U}}}$ are the complex conjugates of $\mathcal{U}$-valued analytic trigonometric polynomials, i.e., the antianalytic trigonometric polynomials.

(here S_n ($n = 0, 1, 2, \ldots$) are the Taylor coefficients of S: $S(z) = \sum_{n=0}^{\infty} S_n z^n$ for $z \in \mathbb{D}$). The shift-invariant subspace $\mathcal{M}_{\mathrm{BTOA}}$ collapses to the left-tangential version

$$
\mathcal{M}_{\mathrm{LTOA}} = \left\{ \begin{bmatrix} f_+ \\ f_- \end{bmatrix} : \widehat{\mathcal{C}}^b_{Z, [X \ -Y]} \begin{bmatrix} f_+ \\ f_- \end{bmatrix} = 0 \right\}
$$
$$
= \mathrm{Ker}\, \widehat{\mathcal{C}}^b_{Z, [X \ -Y]} \subset H^2_{\mathcal{U}_O \oplus \mathcal{U}_I}, \tag{24}
$$

while the solution criterion $\Lambda_{\mathrm{BTOA}} \geq 0$ collapses to

$$
\Lambda_{\mathrm{LTOA}} := \widehat{\mathcal{C}}^b_{Z, [X \ -Y]} J (\widehat{\mathcal{C}}^b_{Z, [X \ -Y]})^* \geq 0.
$$

In the regular case (which we now assume), we have in addition that Λ_{LTOA} is invertible. It follows that $H^2_{\mathcal{U}_O \oplus \mathcal{U}_I} \ominus_J \mathcal{M}_{\mathrm{LTOA}}$ is given by

$$
H^2_{\mathcal{U}_O \oplus \mathcal{U}_I} \ominus_J \mathcal{M}_{\mathrm{LTOA}}
$$
$$
= \mathrm{Ran}\, J \left(\widehat{\mathcal{C}}^b_{Z, [X \ -Y]} \right)^* = \mathrm{Ran}\, \widehat{\mathcal{O}}^f_{\begin{bmatrix} X^* \\ Y^* \end{bmatrix}, Z^*} := \left\{ \begin{bmatrix} X^* \\ Y^* \end{bmatrix} (I - zZ^*)^{-1} x : x \in \mathcal{X}_L \right\}.
$$

To simplify the notation let us introduce the quantities

$$
C = \begin{bmatrix} X^* \\ Y^* \end{bmatrix}, \quad A = Z^* \tag{25}
$$

so that we may write $\widehat{\mathcal{O}}^f_{C,A}$ rather than the more cumbersome $\widehat{\mathcal{O}}^f_{\begin{bmatrix} X^* \\ Y^* \end{bmatrix}, Z^*}$ and write simply $\mathcal{M}$ for $\mathcal{M}_{\mathrm{LTOA}}$ and $\mathcal{M}^{[\perp]}$ for $H^2_{\mathcal{U}_O \oplus \mathcal{U}_I} \ominus_J \mathcal{M}_{\mathrm{LTOA}}$. Then the regularity of $\mathcal{M}$ and the positivity of Λ_{LTOA} can be expressed as

$$
\Lambda_{\mathrm{LTOA}} = (\widehat{\mathcal{O}}^f_{C,A})^* J \widehat{\mathcal{O}}^f_{C,A} > 0.
$$

If we impose the positive-definite inner product induced by Λ_{LTOA} on $\mathcal{X}_L$, then the map

$$
\iota : x \mapsto \mathcal{O}^f_{C,A} x \tag{26}
$$

is a Kreĭn-space isomorphism between $\mathcal{X}_L$ and $\mathcal{M}^{[\perp]}$. If we set

$$
K(z, w) = C(I - zA)^{-1} \Lambda^{-1} (I - \overline{w} A^*)^{-1} C^* \tag{27}
$$

(with $\Lambda = \Lambda_{\mathrm{LTOA}}$), then one can use the J-unitary property of the map ι (26) to compute, for $f(z) = (\widehat{\mathcal{O}}^f_{C,A} x)(z) = C(I - zA)^{-1} x$, $w \in \mathbb{D}$ and $u \in \mathcal{U}_O \oplus \mathcal{U}_I$,

$$
\langle Jf, K(\cdot, w) u \rangle_{H^2_{\mathcal{U}_O \oplus \mathcal{U}_I}} = \langle \Lambda x, \Lambda^{-1} (I - \overline{w} A^*)^{-1} C^* u \rangle_{\mathcal{X}_L}
$$
$$
= \langle C(I - wA)^{-1} x, u \rangle_{\mathcal{U}_O \oplus \mathcal{U}_I} = \langle f(w), u \rangle_{\mathcal{U}_O \oplus \mathcal{U}_I}
$$

from which we see that $K(z, w)$ is the *reproducing kernel* for the space $\mathcal{M}^{[\perp]}$. On the other hand, if we construct $\begin{bmatrix} \mathbf{B} \\ \mathbf{D} \end{bmatrix}$ so that

$$
\begin{bmatrix} A & \mathbf{B} \\ C & \mathbf{D} \end{bmatrix} \begin{bmatrix} \Lambda^{-1} & 0 \\ 0 & J \end{bmatrix} \begin{bmatrix} A^* & C^* \\ \mathbf{B}^* & \mathbf{D}^* \end{bmatrix} = \begin{bmatrix} \Lambda^{-1} & 0 \\ 0 & J \end{bmatrix}, \tag{28}
$$

and set

$$
\Theta(z) = \mathbf{D} + zC(I - zA)^{-1} \mathbf{B},
$$

then Θ is J-inner with associated kernel $K_\Theta(z, w)$ satisfying

$$K_\Theta(z, w) := \frac{J - \Theta(z)J\Theta(w)^*}{1 - z\overline{w}} = C(I - zA)^{-1}\Lambda^{-1}(I - \overline{w}A^*)^{-1}C^* = K(z, w)$$

where $K(z, w)$ is as in (27). From this it is possible to show that the closure of $\Theta \cdot (H_{\mathcal{U}}^2)_0$ is exactly $(\mathcal{M}^{[\perp]})^{[\perp]} = \mathcal{M}$, i.e., the J-Beurling-Lax representer for $\mathcal{M}$ can be constructed in this way. To make the construction of $[\begin{smallmatrix} \mathbf{B} \\ \mathbf{D} \end{smallmatrix}]$, note that solving (28) for $\mathbf{B}$ and $\mathbf{D}$ amounts to solving the J-Cholesky factorization problem

$$\begin{bmatrix} \mathbf{B} \\ \mathbf{D} \end{bmatrix} J \begin{bmatrix} \mathbf{B}^* & \mathbf{D}^* \end{bmatrix} = \begin{bmatrix} \Lambda^{-1} & 0 \\ 0 & J \end{bmatrix} - \begin{bmatrix} A \\ C \end{bmatrix} \Lambda^{-1} \begin{bmatrix} A^* & C^* \end{bmatrix}. \tag{29}$$

An amusing exercise is to check that this recipe is equivalent to that in Theorem 7 when specialized to the case where $\mathcal{X}_R = \{0\}$.

4. Extensions and generalizations of the Grassmannian method

The CBMS monograph [51] and the survey article [9] mention various adaptations of the Grassmannian method to other sorts of interpolation and extension problems. We note that the Treil-Volberg commutant lifting theorem [73] used the Grassmannian setup as described here but then used the Iokhvidov-Ky Fan fixed-point theorem rather than a Beurling-Lax representation theorem. We also mention the Grassmannian version of the abstract band method (including the Takagi version where one seeks a solution in a generalized Schur class (the kernel $K_S(z, w) = [I - S(z)S(w)^*]/(1 - z\overline{w})$ is required to have a most some number κ of negative squares rather than to be a positive kernel)) worked out in [55]. Also the Grassmannian approach certainly influenced the theory of time-varying interpolation developed in [18, 19, 20]; see also [41, 42] for an approach to time-varying interpolation using the isometric-completion and unitary-coupling method of Katsnelson-Kheifets-Yuditskii [56]. Moreover, one can argue that the Grassmannian approach to interpolation, in particular the point of view espoused in [27], foreshadowed the behavioral formulation and solution of the H^∞-control problem (see [60, 74]). Here we discuss some more recent extensions of the Grassmannian method to several variable contexts.

4.1. Interpolation problems for multipliers on the Drury-Arveson space

A multivariable generalization of the Szegő kernel much studied of late is the positive kernel

$$k_d(\boldsymbol{\lambda}, \boldsymbol{\zeta}) = \frac{1}{1 - \langle \boldsymbol{\lambda}, \boldsymbol{\zeta} \rangle}$$

on $\mathbb{B}^d \times \mathbb{B}^d$, where $\mathbb{B}^d = \{\boldsymbol{\lambda} = (\lambda_1, \dots, \lambda_d) \in \mathbb{C}^d : \langle \boldsymbol{\lambda}, \boldsymbol{\lambda} \rangle < 1\}$ is the unit ball of the d-dimensional Euclidean space $\mathbb{C}^d$ and $\langle \boldsymbol{\lambda}, \boldsymbol{\zeta} \rangle$ is the standard inner product in $\mathbb{C}^d$. The reproducing kernel Hilbert space $\mathcal{H}(k_d)$ associated with k_d is called the Drury-Arveson space (also denoted as H_d^2) and acts as a natural multivariable analogue of the Hardy space H^2 of the unit disk. The many references on this topic include [38, 6, 7, 2, 31, 44, 48, 59].

For $\mathcal{Y}$ an auxiliary Hilbert space, we consider the tensor product Hilbert space $\mathcal{H}_{\mathcal{Y}}(k_d) := \mathcal{H}(k_d) \otimes \mathcal{Y}$ whose elements can be viewed as $\mathcal{Y}$-valued functions in $\mathcal{H}(k_d)$. Then $\mathcal{H}_{\mathcal{Y}}(k_d)$ has the following characterization:

$$\mathcal{H}_{\mathcal{Y}}(k_d) = \left\{ f(\boldsymbol{\lambda}) = \sum_{\mathbf{n} \in \mathbb{Z}_+^d} f_{\mathbf{n}} \boldsymbol{\lambda}^{\mathbf{n}} : \|f\|^2 = \sum_{\mathbf{n} \in \mathbb{Z}_+^d} \frac{\mathbf{n}!}{|\mathbf{n}|!} \cdot \|f_{\mathbf{n}}\|_{\mathcal{Y}}^2 < \infty \right\}. \tag{30}$$

Here and in what follows, we use standard multivariable notations: for multi-integers $\mathbf{n} = (n_1, \ldots, n_d) \in \mathbb{Z}_+^d$ and points $\boldsymbol{\lambda} = (\lambda_1, \ldots, \lambda_d) \in \mathbb{C}^d$ we set

$$|\mathbf{n}| = n_1 + n_2 + \cdots + n_d, \qquad \mathbf{n}! = n_1! n_2! \ldots n_d!, \qquad \boldsymbol{\lambda}^{\mathbf{n}} = \lambda_1^{n_1} \lambda_2^{n_2} \ldots \lambda_d^{n_d}. \tag{31}$$

For coefficient Hilbert spaces $\mathcal{U}$ and $\mathcal{Y}$, the operator-valued Drury-Arveson Schur-multiplier class $\mathcal{S}_d(\mathcal{U}, \mathcal{Y})$ is defined to be the space of functions S holomorphic on the unit ball $\mathbb{B}^d$ with values in the space of operators $\mathcal{L}(\mathcal{U}, \mathcal{Y})$ such that the multiplication operator

$$M_S \colon f(\boldsymbol{\lambda}) \to S(\boldsymbol{\lambda}) \cdot f(\boldsymbol{\lambda})$$

maps $\mathcal{H}_{\mathcal{U}}(k_d)$ contractively into $\mathcal{H}_{\mathcal{Y}}(k_d)$, or equivalently, the associated multivariable de Branges-Rovnyak kernel

$$K_S(\boldsymbol{\lambda}, \boldsymbol{\zeta}) := \frac{I - S(\boldsymbol{\lambda}) S(\boldsymbol{\zeta})^*}{1 - \langle \boldsymbol{\lambda}, \boldsymbol{\zeta} \rangle} \tag{32}$$

should be a positive kernel.

Let $\mathbf{A} = (A_1, \ldots, A_d)$ be a commutative d-tuple of bounded, linear operators on the Hilbert space $\mathcal{X}$. If $C \in \mathcal{L}(\mathcal{X}, \mathcal{Y})$, then the pair $(C, \mathbf{A})$ is said to be *output-stable* if the associated observation operator

$$\widehat{\mathcal{O}}_{C,\mathbf{A}} \colon x \mapsto C(I - \lambda_1 A_1 - \cdots - \lambda_d A_d)^{-1} x$$

maps $\mathcal{X}$ into $\mathcal{H}_{\mathcal{Y}}(k_d)$, or equivalently (by the closed graph theorem), the observation operator is bounded. Just as in the single-variable case (see (16)), there is a system-theoretic interpretation for this operator, but now in the context of multidimensional systems (see [12] for details). We can then pose the Drury-Arveson space version of the left-tangential operator-argument interpolation (**LTOA**) problem formulated in Subsection 3.1 by replacing the unit disk $\mathbb{D}$ by the unit ball $\mathbb{B}^d$ and the Schur class $\mathcal{S}(\mathcal{U}, \mathcal{Y})$ by the Drury-Arveson Schur-multiplier class $\mathcal{S}_d(\mathcal{U}, \mathcal{Y})$.

Problem LTOA (Left Tangential Operator-Argument Interpolation Problem): *Let $\mathcal{U}_I, \mathcal{U}_O$ and $\mathcal{X}$ be Hilbert spaces and suppose that we are given the data set $(\mathbf{Z}, X, Y)$ with $\mathbf{Z} = (Z_1, \ldots, Z_d)$ with each $Z_i \in \mathcal{L}(\mathcal{X})$, $X \in \mathcal{L}(\mathcal{U}_O, \mathcal{X})$, $Y \in \mathcal{L}(\mathcal{U}_I, \mathcal{X})$ such that $(\mathbf{Z}, X)$ is an input stable pair, i.e., $(X^*, \mathbf{Z}^*)$ is an output stable pair. Find $S \in \mathcal{S}_d(\mathcal{U}_I, \mathcal{U}_O)$ such that*

$$\left(\widehat{\mathcal{O}}_{X^*, \mathbf{Z}^*} \right)^* M_S = \left(\widehat{\mathcal{O}}_{Y^*, \mathbf{Z}^*} \right)^*, \tag{33}$$

or equivalently,

$$(\widehat{XS})^{\wedge L}(\mathbf{Z}) = Y, \tag{34}$$

where the multivariable left tangential operator-argument point-evaluation is given by

$$(\widehat{XS})^{\wedge L}(\mathbf{Z}) = \sum_{n\in\mathbb{Z}_+^d} \mathbf{Z}^n X S_n.$$

Here $S(z) = \sum_{n\in\mathbb{Z}_+^d} S_n z^n$ is the multivariable Taylor series for S and we use the commutative multivariable notation

$$\mathbf{Z}^n = Z_1^{n_1} \cdots Z_d^{n_d} \text{ for } n = (n_1,\ldots,n_d) \in \mathbb{Z}_+^d.$$

We note that this and related interpolation problems were studied in [11] by using techniques from reproducing kernel Hilbert spaces, Schur-complements and isometric extensions borrowed from the work of [39, 57, 56]; here we show how the problem can be handled via the Grassmannian approach.

As a motivation for this formalism, we consider a simple example: take $\mathcal{U}_I =$

$$\mathcal{U}_O = \mathbb{C}, \; X = \begin{bmatrix} 1 \\ 1 \\ \vdots \\ 1 \end{bmatrix}, Y = \begin{bmatrix} w_1 \\ w_1 \\ \vdots \\ w_N \end{bmatrix}, \; \mathbf{Z} = (Z_1,\ldots,Z_d) \text{ with } Z_j = \begin{bmatrix} \lambda_j^{(1)} & & & \\ & \lambda_j^{(2)} & & \\ & & \ddots & \\ & & & \lambda_j^{(N)} \end{bmatrix},$$

where $j = 1,\ldots,d$ and where $\lambda^{(i)} = (\lambda_1^{(i)},\ldots,\lambda_d^{(i)}) \in \mathbb{B}^d$ for $i = 1,\ldots,N$. Then the **LTOA** problem collapses to Nevanlinna-Pick-type interpolation problem for Drury-Arveson space multipliers, as studied in [66, 4, 37, 31]: *for given points* $\lambda^{(1)},\ldots,\lambda^{(N)}$ *in the ball* $\mathbb{B}^d$ *and given complex numbers* $w_1,\ldots,w_N$, *find* $S \in \mathcal{S}_d$ *so that*

$$S(\lambda^{(i)}) = w_i \qquad \text{for} \quad i = 1,\cdots,N.$$

We transform the problem to projective coordinates (following the Grassmannian approach) as follows. We identify the Drury-Arveson-space multiplier $S \in \mathcal{S}_d$ with its graph to convert the nonhomogeneous interpolation conditions to homogeneous interpolation conditions for the associated subspaces (i.e., we projectivize the problem). Then one checks that the function $S \in \mathcal{S}_d$ is a solution of the **LTOA** problem if and only if its graph $\mathcal{G} := \begin{bmatrix} M_S \\ I \end{bmatrix} \mathcal{H}(k_d) \subset \begin{bmatrix} \mathcal{H}(k_d) \\ \mathcal{H}(k_d) \end{bmatrix}$ satisfies:

1. $\mathcal{G}$ *is a subspace of* $\mathcal{M} = \{f \in \begin{bmatrix} \mathcal{H}(k_d) \\ \mathcal{H}(k_d) \end{bmatrix} : [\,1 \; -w_i\,]\, f(\lambda^i) = 0 \quad for \quad i = 1,\ldots,N\}$

   (and hence also is a subspace of the Kreĭn space $\mathcal{K} = \begin{bmatrix} \mathcal{H}(k_d) \\ \mathcal{H}(k_d) \end{bmatrix}$ with $J = \begin{bmatrix} I_{\mathcal{H}(k_d)} & \\ & -I_{\mathcal{H}(k_d)} \end{bmatrix}$),

2. $\mathcal{G}$ *is maximal negative in* $\mathcal{K}$ *and*

3. $\mathcal{G}$ *is* M_{λ_k} *invariant for* $k = 1,\ldots,d$.

Conversely, if $\mathcal{G}$ as a subspace of $\begin{bmatrix} \mathcal{H}(k_d) \\ \mathcal{H}(k_d) \end{bmatrix}$ satisfies $(1),(2),(3)$, then $\mathcal{G}$ is in the form of $\begin{bmatrix} M_S \\ I \end{bmatrix} \mathcal{H}(k_d)$ for a solution S of the interpolation problem. Thus the **LTOA** interpolation problem translates to the problem of finding subspaces $\mathcal{G}$ of $\begin{bmatrix} \mathcal{H}(k_d) \\ \mathcal{H}(k_d) \end{bmatrix}$ which satisfy the conditions (1), (2), (3) above.

For the general **LTOA** problem, the analysis is similar. One can see that $S \in \mathcal{S}_d(\mathcal{U}_I, \mathcal{U}_O)$ solves the **LTOA** problem if and only if its graph $\mathcal{G} := [\begin{smallmatrix} S \\ I \end{smallmatrix}] \cdot \mathcal{H}_{\mathcal{U}_I}(k_d)$ satisfies the following conditions:

1. $\mathcal{G} \subset \mathcal{M}$ where

$$\mathcal{M} = \left\{ f \in \mathcal{H}_{\mathcal{U}_O \oplus \mathcal{U}_I}(k_d) : \left(\begin{bmatrix} X & -Y \end{bmatrix} f \right)^{\wedge L}(\mathbf{Z}) = \mathbf{0} \right\}, \tag{35}$$

where $\mathcal{H}_{\mathcal{U}_O \oplus \mathcal{U}_I}(k_d) = \begin{bmatrix} \mathcal{H}_{\mathcal{U}_O}(k_d) \\ \mathcal{H}_{\mathcal{U}_I}(k_d) \end{bmatrix}$.

2. $\mathcal{G}$ is J-maximal negative subspace of $\mathcal{H}_{\mathcal{U}_O \oplus \mathcal{U}_I}(k_d)$, where $J = I_{\mathcal{H}_{\mathcal{U}_O}(k_d)} \oplus -I_{\mathcal{H}_{\mathcal{U}_I}(k_d)}$.

3. $\mathcal{G}$ is invariant under $M_{\lambda_k}, k = 1, 2, \ldots, d$.

Just as in the single-variable case, we see that a necessary condition for solutions to exist is that the analogue of (3) holds:

$$\mathcal{P} := \mathcal{H}_{\mathcal{U}_O \oplus \mathcal{U}_I}(k_d) \ominus_J \mathcal{M} \text{ is a positive subspace of } \mathcal{H}_{\mathcal{U}_O \oplus \mathcal{U}_I}(k_d). \tag{36}$$

Given that (36) holds, we see that solutions $\mathcal{G}$ of (1), (2), (3) above amount to subspaces $\mathcal{G}$ of $\mathcal{M}$ which are maximal negative as subspaces of $\mathcal{M}$ ($\mathcal{M}$-maximal negative) and which are shift invariant. These in turn can be parametrized if $\mathcal{M}$ has a suitable J-Beurling-Lax representer. For the Hilbert space setting ($J = I$), there is a Beurling-Lax representation theorem (see [6, 59, 48, 12, 15]): *given a closed shift-invariant subspace $\mathcal{M}$ of $\mathcal{H}_{\mathcal{U}}(k_d)$, there is a suitable Hilbert space $\mathcal{U}'$ and a Schur-class multiplier $\mathcal{S}_d(\mathcal{U}', \mathcal{U})$ so that the orthogonal projection $P_{\mathcal{M}}$ of $\mathcal{H}_{\mathcal{U}}(k_d)$ onto $\mathcal{M}$ is given by $P_{\mathcal{M}} = M_\Theta (M_\Theta)^*$.* Unlike the single-variable case ($d = 1$), in general one cannot take M_Θ to be an isometry, but rather, M_Θ is only a *partial isometry*.

An analogous result holds in the J-setting as follows, as can be seen by following the construction sketched in Subsection 3.1 for the single-variable case. In general we say that an operator T between two Kreĭn spaces $\mathcal{K}'$ and $\mathcal{K}$ is a (possibly unbounded) *Kreĭn-space partial isometry* if $T^{[*]}T$ and $TT^{[*]}$ (where $T^{[*]}$ is the adjoint of T with respect to the Kreĭn-spaces indefinite inner products) are bounded J-self-adjoint projection operators on $\mathcal{K}'$ and $\mathcal{K}$ respectively.

Theorem 10 (see [45, Theorem 3.3.2]). *Suppose that $\mathcal{M}$ is a regular subspace of $\mathcal{H}_{\mathcal{U}_O \oplus \mathcal{U}_I}(k_d)$. Then there is a coefficient Kreĭn space $\mathcal{E}$ and a (possibly unbounded) Drury-Arveson multiplier Θ so that M_Θ is a (possibly unbounded) Kreĭn-space partial isometry with final projection operator (the bounded extension of $M_\Theta J_\mathcal{E} M_\Theta^* J$) equal to the J-orthogonal projection of $\mathcal{H}_{\mathcal{U}_O \oplus \mathcal{U}_I}(k_d)$ onto $\mathcal{M}$. In case condition (36) holds, then one can take $\mathcal{E}$ to have the form $(\mathcal{U}_{O,\text{aug}} \oplus \mathcal{U}_O) \oplus \mathcal{U}_I$ with $J_\mathcal{E} = I_{\mathcal{U}_{0,\text{aug}} \oplus \mathcal{U}_O} \oplus -I_{\mathcal{U}_I}$ for a suitable augmentation Hilbert space $\mathcal{U}_{0,\text{aug}}$.*

*If $\mathcal{M}$ comes from a **LTOA** interpolation problem as in (35), then condition (36) holds if and only if*

$$\Lambda := \left(\widehat{\mathcal{O}}_{\left[\begin{smallmatrix} X^* \\ Y^* \end{smallmatrix} \right], \mathbf{z}^*} \right)^* J \widehat{\mathcal{O}}_{\left[\begin{smallmatrix} X^* \\ Y^* \end{smallmatrix} \right], \mathbf{z}^*} \geq 0. \tag{37}$$

*Then $\mathcal{M}$ is regular if and only if Λ is strictly positive and then the set of all solutions S of the **LTOA** interpolation problem is given by formula (8) where now the free parameter G sweeps the Drury-Arveson Schur class $\mathcal{S}_d(\mathcal{U}_I, \mathcal{U}_{O,\mathrm{aug}} \oplus \mathcal{U}_O)$. Moreover, a realization formula for the representer Θ is given by*

$$\Theta(\boldsymbol{\lambda}) = \mathbf{D} + C(I - \lambda_1 A_1 - \cdots - \lambda_d A_d)^{-1}(\lambda_1 \mathbf{B}_1 + \cdots + \lambda_d \mathbf{B}_d)$$

where the nonbold components of the matrix

$$\mathbf{U} = \begin{bmatrix} A & \mathbf{B} \\ C & \mathbf{D} \end{bmatrix} = \begin{bmatrix} A_1 & \mathbf{B}_1 \\ \vdots & \vdots \\ A_d & \mathbf{B}_d \\ C & \mathbf{D} \end{bmatrix}$$

are given by

$$A = \begin{bmatrix} Z_1^* \\ \vdots \\ Z_d^* \end{bmatrix}, \quad C = \begin{bmatrix} X^* \\ Y^* \end{bmatrix}$$

while the bold components are given via solving the J-Cholesky factorization problem:

$$\begin{bmatrix} \mathbf{B} \\ \mathbf{D} \end{bmatrix} J \begin{bmatrix} \mathbf{B}^* & \mathbf{D}^* \end{bmatrix} = \begin{bmatrix} \oplus_{k=1}^d \Lambda^{-1} & 0 \\ 0 & J \end{bmatrix} - \begin{bmatrix} A \\ C \end{bmatrix} \Lambda^{-1} \begin{bmatrix} A^* & C^* \end{bmatrix}.$$

Remark 11. The major new feature in the multivariable setting compared to the single-variable case is that Θ is only a (possibly unbounded) partial J-isometry rather a J-unitary map. Nevertheless, there is still a correspondence (8) between maximal negative subspaces in the model (or parameter) Kreĭn space $(\mathcal{U}_{O,\mathrm{aug}} \oplus \mathcal{U}_O) \oplus \mathcal{U}_I$ and $\mathcal{M}$-maximal negative subspaces of $\mathcal{M} \subset H_{\mathcal{U}_O \oplus \mathcal{U}_I}(k_d)$, but with the price that the solution S no longer uniquely determines the associated free parameter G. Roughly, what makes this work is that the construction guarantees that $\mathrm{Ker}\, M_\Theta$ is necessarily a positive subspace of $\mathcal{H}_{(\mathcal{U}_{O,\mathrm{aug}} \oplus \mathcal{U}_O) \oplus \mathcal{U}_I}(k_d)$. Verification of this correspondence for the unbounded case can be done analogously to the single-variable case by use of the Drury-Arveson-space Leech theorem which in turn follows from the Commutant Lifting Theorem for the Drury-Arveson-spaces multipliers (see [64, 31, 36]).

4.2. Interpolation problems for multianalytic functions on the Fock space

Recently there has been much interest in noncommutative function theory and associated multivariable operator theory and multidimensional system theory, spurred on by a diverse collection of applications too numerous to mention in any depth here. Let us just point out that there are at least three points of view: (1) formal power series in freely noncommuting indeterminates [21, 65, 64, 32, 33, 13], (2) functions in d noncommuting operators acting on some fixed infinite-dimensional separable Hilbert space [22, 23, 10], and (3) functions of d $N \times N$-matrix arguments where the size $N = 1, 2, 3, \ldots$ is arbitrary [3, 54, 43, 52, 53].

We restrict our discussion here to the noncommutative version of the Drury-Arveson Schur class, elements of which first appeared in the work of Popescu [63] as the characteristic functions of row contractions. This Schur class consists of formal power series in a set of noncommuting indeterminates which define contractive multipliers between (unsymmetrized) vector-valued Fock spaces. To introduce this setting, let $\{1, \ldots, d\}$ be an alphabet consisting of d letters and let $\mathcal{F}_d$ be the associated free semigroup generated by the letters $1, \ldots, d$ consisting of all words γ of the form $\gamma = i_N \cdots i_1$, where each $i_k \in \{1, \ldots, d\}$ and where $N = 1, 2, \ldots$. For $\gamma = i_N \cdots i_1 \in \mathcal{F}_d$ we set $|\gamma| := N$ to be the *length* of the word γ. Multiplication of two words $\gamma = i_N \cdots i_1$ and $\gamma' = j_{N'} \cdots j_1$ is defined via concatenation:

$$\gamma\gamma' = i_N \cdots i_1 j_{N'} \cdots j_1.$$

The empty word $\emptyset$ is included in $\mathcal{F}_d$ and acts as the unit element for this multiplication; by definition $|\emptyset| = 0$. We let $z = (z_1, \ldots, z_d)$ be a d-tuple of freely noncommuting indeterminates with associated noncommutative formal monomials $z^\gamma = z_{i_N} \cdots z_{i_1}$ if $\gamma = i_N \cdots i_1 \in \mathcal{F}_d$.

For a Hilbert space $\mathcal{U}$, we define the associated *Fock space* $H^2_{\mathcal{U}}(\mathcal{F}_d)$ to consist of formal power series in the set of noncommutative indeterminates $z = (z_1, \ldots, z_d)$

$$\widehat{u}(z) = \sum_{\gamma \in \mathcal{F}_d} u(\gamma) z^\gamma$$

satisfying the square-summability condition on the coefficients:

$$\sum_{\gamma \in \mathcal{F}_d} \|u(\gamma)\|^2_{\mathcal{U}} < \infty.$$

Given two coefficient Hilbert spaces $\mathcal{U}$ and $\mathcal{Y}$, we define the noncommutative Schur class $\mathcal{S}_{nc,d}(\mathcal{U}, \mathcal{Y})$ to consist of formal power series with operator coefficients

$$S(z) = \sum_{\gamma \in \mathcal{F}_d} S_\gamma z^\gamma$$

such that the noncommutative multiplication operator

$$M_S \colon \widehat{u}(z) = \sum_{\gamma \in \mathcal{F}_d} u(\gamma) z^\gamma \mapsto S(z) \cdot \widehat{u}(z) := \sum_{\gamma \in \mathcal{F}_d} \left(\sum_{\alpha, \beta \in \mathcal{F}_d \colon \alpha\beta = \gamma} S_\alpha u(\beta) \right) z^\gamma$$

defines a contraction operator from $H^2_{\mathcal{U}}(\mathcal{F}_d)$ into $H^2_{\mathcal{Y}}(\mathcal{F}_d)$.

One can view elements S of the noncommutative Schur class $\mathcal{S}_{nc,d}(\mathcal{U}, \mathcal{Y})$ as defining functions of d noncommuting arguments and then set up noncommutative analogues of Nevanlinna-Pick interpolation problems as follows. Given a (not necessarily commutative) d-tuple of bounded operators $\mathbf{A} = (A_1, \ldots, A_d)$ on a

Hilbert space $\mathcal{X}$ together with an output operator $C\colon \mathcal{X} \to \mathcal{Y}$, let us say that the output-pair $(C, \mathbf{A})$ is *output stable* if the noncommutative observation operator

$$\widehat{\mathcal{O}}^{nc}_{C,\mathbf{A}} \colon x \mapsto C(I - z_1 A_1 - \cdots - z_d A_d)^{-1} x = \sum_{\gamma \in \mathcal{F}_d} (C\mathbf{A}^\gamma x) z^\gamma$$

maps $\mathcal{X}$ into the Fock space $H^2_{\mathcal{Y}}(\mathcal{F}_d)$; here we use the noncommutative multivariable notation:

$$\mathbf{A}^\gamma = A_{i_N} \cdots A_{i_1} \text{ if } \gamma = i_N \cdots i_1 \in \mathcal{F}_d \text{ with } A^\emptyset = I_{\mathcal{X}}.$$

If $(\mathbf{Z} = (Z_1, \ldots, Z_d), X)$ is a multivariable input-pair (so Z_j acts on a state space $\mathcal{X}$ and X is an input operator mapping an input space $\mathcal{U}_I$ into $\mathcal{X}$) such that the output-pair $(X^*, \mathbf{Z}^* = (Z_1^*, \ldots, Z_d^*))$ is output-stable, then $\widehat{\mathcal{O}}^{nc}_{X^*, \mathbf{Z}^*}$ maps $\mathcal{X}$ boundedly into $H^2_{\mathcal{U}_I}(\mathcal{F}_d)$ and hence its adjoint $\left(\widehat{\mathcal{O}}^{nc}_{X^*, \mathbf{Z}^*}\right)^*$ maps $H^2_{\mathcal{U}_I}(\mathcal{F}_d)$ boundedly into $\mathcal{X}$: in this case we say that the input pair $(\mathbf{Z}, X)$ is *input-stable*. We can use such operators to define interpolation conditions on a noncommutative Schur-class function.

Problem ncLTOA (noncommutative Left Tangential Operator Argument Interpolation Problem): Let $\mathcal{U}_I$, $\mathcal{U}_O$, $\mathcal{X}$ be Hilbert spaces. Suppose that we are given the data set $(\mathbf{Z}, X, Y)$ with $\mathbf{Z} = (Z_1, \ldots, Z_d)$ with each $Z_j \in \mathcal{L}(\mathcal{X})$, $X \in \mathcal{L}(\mathcal{U}_O, \mathcal{X})$, $Y \in \mathcal{L}(\mathcal{U}_I, \mathcal{X})$ such that $(\mathbf{Z}, X)$ is a stable input pair. Find $S \in \mathcal{S}_{nc,d}(\mathcal{U}_I, \mathcal{U}_O)$ such that

$$\left(\widehat{\mathcal{O}}^{nc}_{X^*, \mathbf{Z}^*}\right)^* M_S = \left(\widehat{\mathcal{O}}^{nc}_{Y^*, \mathbf{Z}^*}\right)^*, \tag{38}$$

or equivalently,

$$(\widehat{XS})^{\wedge L, nc}(\mathbf{Z}) = Y \tag{39}$$

where the noncommutative left tangential operator-argument point-evaluation is given by

$$(\widehat{XS})^{\wedge L, nc}(\mathbf{Z}) = \sum_{\gamma \in \mathcal{F}_d} \mathbf{Z}^{\gamma^\top} X S_\gamma \text{ if } S(z) = \sum_{\gamma \in \mathcal{F}_d} S_\gamma z^\gamma.$$

Here we use the notation $\gamma^\top$ for the *transpose* of the word γ: $\gamma^\top = i_1 \cdots i_N$ if $\gamma = i_N \cdots i_1$.

Problems of this sort have been studied in the literature, e.g., in [66, 35, 10]. The solution of the **ncLTOA** interpolation problem via the Grassmannian approach proceeds in a completely analogous fashion as in the commutative case. In this setting, the shift-invariant subspaces are subspaces of $H^2_{\mathcal{U}_O \oplus \mathcal{U}_I}(\mathcal{F}_d)$ which are invariant under the right creation operators

$$R_{z_k} \colon f(z) = \sum_{\gamma \in \mathcal{F}_d} f_\gamma z^\gamma \mapsto f(z) z_k = \sum_{\gamma \in \mathcal{F}_d} f_\gamma z^{\gamma \cdot k}$$

for $k = 1, \ldots, d$. We view $H^2_{\mathcal{U}_O \oplus \mathcal{U}_I}(\mathcal{F}_d)$ is a Kreĭn space in the indefinite inner product induced by $J = \begin{bmatrix} I_{\mathcal{U}_O} & 0 \\ 0 & -I_{\mathcal{U}_I} \end{bmatrix}$. Graph spaces $\mathcal{G} = \begin{bmatrix} M_S \\ I \end{bmatrix} H^2_{\mathcal{U}_I}(\mathcal{F}_d)$ of solutions

S of the **ncLTOA** interpolation problem are characterized by the condition: $\mathcal{G}$ *is an $H^2_{\mathcal{U}_O \oplus \mathcal{U}_I}(\mathcal{F}_d)$-maximal negative subspace of*

$$\mathcal{M} := \left\{ f \in H^2_{\mathcal{U}_O \oplus \mathcal{U}_I}(\mathcal{F}_d) \colon \left(\begin{bmatrix} X & -Y \end{bmatrix} f \right)^{\wedge L, \mathrm{nc}}(\mathbf{Z}) = 0 \right\} \tag{40}$$

which is also shift-invariant. The Pick matrix condition

$$H^2_{\mathcal{U}_O \oplus \mathcal{U}_I} \ominus_J \mathcal{M} \text{ is a positive subspace} \tag{41}$$

is necessary for solutions to exist; conversely, if (41) holds, then it suffices to look for any shift-invariant subspace $\mathcal{G}$ contained in $\mathcal{M}$ ($\mathcal{M}$ as in (40)) which is maximal negative as a subspace of $\mathcal{M}$. Such subspaces $\mathcal{G} = \begin{bmatrix} M_S \\ I \end{bmatrix} \cdot H^2_{\mathcal{U}_I}(\mathcal{F}_d)$ can be parametrized via the linear-fractional formula (8) (where now the free parameter G is in the noncommutative Schur class $\mathcal{S}_{\mathrm{nc},d}(\mathcal{U}_I, \mathcal{U}_{O,\mathrm{aug}} \oplus \mathcal{U}_O)$) if there is a suitable J-Beurling-Lax representation for $\mathcal{M}$. For the case $J = I$, such Beurling-Lax representations (with M_Θ isometric rather than merely partially isometric) have been known for some time (see [63, 67]); we note that the paper [13] derives the $J = I$ Beurling-Lax theorem for the Fock-space setting from the point of view which we have here, where the shift-invariant subspace $\mathcal{M}$ is presented as the kernel of an operator of the form $\left(\widehat{\mathcal{O}}^{\mathrm{nc}}_{C,A} \right)^*$. Adaptation of this construction to the J-case (with the complication that M_Θ, while J-isometric, may be unbounded) is carried out in [45]. The following theorem summarizes the results for solving the **ncLTOA** interpolation problem via the Grassmannian approach.

Theorem 12. *Suppose that $\mathcal{M}$ is a regular subspace of $H^2_{\mathcal{U}_O \oplus \mathcal{U}_I}(\mathcal{F}_d)$. Then there is a coefficient Kreĭn space $\mathcal{E}$ and a (possibly unbounded) noncommutative Schur-class multiplier S so that M_S is a (possibly unbounded) Kreĭn-space isometry with the bounded extension of $M_\Theta J_\mathcal{E} M_\Theta^* J$ equal to the (bounded) J-orthogonal projection of $H^2_{\mathcal{U}_O \oplus \mathcal{U}_I}(\mathcal{F}_d)$ onto $\mathcal{M}$. In case condition (41) holds, then one can take $\mathcal{E}$ to have the form $(\mathcal{U}_{O,\mathrm{aug}} \oplus \mathcal{U}_O) \oplus \mathcal{U}_I$ with $J_\mathcal{E} = I_{\mathcal{U}_{O,\mathrm{aug}} \oplus \mathcal{U}_O} \oplus -I_{\mathcal{U}_I}$.*

*If $\mathcal{M}$ comes from a **ncLTOA** interpolation problem as in (40), then condition (41) holds if and only if*

$$\Lambda := \left(\widehat{\mathcal{O}}^{\mathrm{nc}}_{\begin{bmatrix} X^* \\ Y^* \end{bmatrix}, \mathbf{z}^*} \right)^* J \widehat{\mathcal{O}}^{\mathrm{nc}}_{\begin{bmatrix} X^* \\ Y^* \end{bmatrix}, \mathbf{z}^*} \geq 0. \tag{42}$$

*Then $\mathcal{M}$ is regular if and only if Λ is strictly positive and then the set of all solutions S of the **ncLTOA** interpolation problem is given by formula (8) where now the free parameter G is in the noncommutative Schur class $\mathcal{S}_{\mathrm{nc},d}(\mathcal{U}_I, \mathcal{U}_{O,\mathrm{aug}} \oplus \mathcal{U}_O)$. Moreover, a realization formula for the representer Θ is given by*

$$\Theta(z) = \mathbf{D} + C(I - z_1 A_1 - \cdots - z_d A_d)^{-1}(z_1 \mathbf{B}_1 + \cdots + z_d \mathbf{B}_d)$$

where the associated colligation matrix

$$\mathbf{U} = \begin{bmatrix} A & \mathbf{B} \\ C & \mathbf{D} \end{bmatrix} = \begin{bmatrix} A_1 & \mathbf{B}_1 \\ \vdots & \vdots \\ A_d & \mathbf{B}_d \\ C & \mathbf{D} \end{bmatrix}$$

is constructed via the same recipe as given in Theorem 10, the one distinction now being that the d-tuple $\mathbf{Z} = (Z_1, \ldots, Z_d)$ *is no longer assumed to be commutative.*

We note that not all multivariable interpolation problems succumb to the Grassmannian/Beurling-Lax approach. Indeed, the lack of a Beurling theorem in the polydisk setting (see, e.g., [71]) is the tipoff to the more complicated structures that one can encounter. To get state-space formulas for solutions as we are getting here, one must work with the Schur-Agler class rather than the Schur class; moreover, without imposing additional apparently contrived moment conditions, it is often impossible to get a single linear-fractional formula which parametrizes the set of all solutions; for a recent survey we refer to [28].

References

[1] V.M. Adamjan, D.Z. Arov, and M.G. Kreĭn, *Infinite block Hankel matrices and their connection with interpolation problem*, Amer. Math. Soc. Transl. (2) **111**, 133–156 [Russian original: 1971].

[2] J. Agler and J.E. McCarthy, *Complete Nevanlinna-Pick kernels*, J. Funct. Anal., **175** (2000), 111–124.

[3] D. Alpay and D. Kaliuzhnyĭ-Verbovetzkiĭ, *Matrix-J-unitary non-commutative rational formal power series*, in: The State Space Method Generalizations and Applications (eds. D. Alpay and I. Gohberg), pp. 49–113, **OT 161** Birkhäuser Verlag, Basel, 2006.

[4] A. Arias and G. Popescu, *Noncommutative interpolation and Poisson transforms*, Israel J. Math. **115** (2000), 205–234.

[5] D.Z. Arov, *γ-generating matrices, J-inner matrix-functions and related extrapolation problems*, Teor. Funktsii Funktsional. Anal. i Prilozhen., I, **51** (1989), 61–67; II, **52** (1989), 103–109; III, **53** (1990), 57–64; English transl., J. Soviet Math. I, **52** (1990), 3487–3491; II, **52** (1990), 3421–3425; III, **58** (1992), 532, 537.

[6] W. Arveson, *Subalgebras of C*-algebras. III. Multivariable operator theory*, Acta Math. **181** (1998), no. 2, 159–228.

[7] W. Arveson, *The curvature invariant of a Hilbert module over* $\mathbb{C}[z_1, \ldots, z_d]$, J. Reine Angew. Math. **522** (2000), 173–236.

[8] T.Y. Azizov, I.S. Iokhidov, *Linear Operators in Spaces with an Indefinite Metric*, Wiley & sons Ltd. Chichester, 1989.

[9] J.A. Ball, *Nevanlinna-Pick interpolation: generalizations and applications*, in: Recent Results in Operator Theory Vol. I (eds. J.B. Conway and B.B. Morrel), Longman Scientific and Tech., Essex, 1988, pp. 51–94.

[10] J.A. Ball and V. Bolotnikov, *Interpolation in the noncommutative Schur-Agler class*, J. Operator Theory **58** (2007), no. 1, 83–126.

[11] J.A. Ball and V. Bolotnikov, *Interpolation problems for Schur multipliers on the Drury-Arveson space: from Nevanlinna-Pick to Abstract Interpolation Problem*, Integral Equations and Operator Theory **62** (2008), 301–349.

[12] J.A. Ball, V. Bolotnikov and Q. Fang, *Multivariable backward-shift invariant subspaces and observability operators*, Multidimens. Syst. Signal Process **18** (2007), 191–248.

[13] J.A. Ball, V. Bolotnikov and Q. Fang, *Schur-class multipliers on the Fock space: de Branges-Rovnyak reproducing kernel spaces and transfer-function realizations*, in: Operator Theory, Structured Matrices, and Dilations, Tiberiu Constantinescu Memorial Volume, Theta Foundation in Advance Mathematics, 2007, 85–114.

[14] J.A. Ball, V. Bolotnikov and Q. Fang, *Transfer-function realization for multipliers of the Arveson space*, J. Math. Anal. Appl. **333** (2007), no. 1, 68–92.

[15] J.A. Ball, V. Bolotnikov and Q. Fang, *Schur-class multipliers on the Arveson space: de Branges-Rovnyak reproducing kernel spaces and commutative transfer-function realizations*, J. Math. Anal. Appl. **341** (2008), no. 1, 519–539.

[16] J.A. Ball, N. Cohen and A.C.M. Ran, *Inverse spectral problems for regular improper rational matrix functions*, in: Topics in interpolation theory of rational matrix-valued functions (ed. I. Gohberg) **OT 33**, Birkhäuser Verlag, Basel, 1988, pp. 123–173.

[17] J.A. Ball, I. Gohberg and L. Rodman, *Interpolation of Rational Matrix Functions*, **OT 45** Birkhäuser Verlag, Basel-Boston, 1990.

[18] J.A. Ball, I. Gohberg and M.A. Kaashoek, *Nevanlinna-Pick interpolation for time-varying input-output maps: the discrete case*, in: Time-Variant Systems and Interpolation (ed. I. Gohberg), **OT 56** Birkhäuser Verlag, Basel, 1992, pp. 1–51.

[19] J.A. Ball, I. Gohberg and M.A. Kaashoek, *Bitangential interpolation for input-output operators of time-varying systems: the discrete time case*, in: New Aspects in Interpolation and Completion Theories (ed. I. Gohberg), **OT 64** Birkhäuser Verlag, Basel, 1993, pp. 33–72.

[20] J.A. Ball, I. Gohberg and M.A. Kaashoek, *Two-sided Nudelman interpolation for input-output operators of discrete time-varying systems*, Integral Equations and Operator Theory **21** (1995), 174–211.

[21] J.A. Ball, G. Groenewald and T. Malakorn, *Structured noncommutative multidimensional linear systems*, SIAM J. Control and Optimization **44** no. 4 (2005), 1474–1528.

[22] J.A. Ball, G. Groenewald and T. Malakorn, *Conservative structured noncommutative multidimensional linear systems*, in: The State Space Method Generalizations and Applications (eds. D. Alpay and I. Gohberg), pp. 179–223, **OT 161** Birkhäuser Verlag, Basel, 2006.

[23] J.A. Ball, G. Groenewald and T. Malakorn, *Bounded real lemma for structured noncommutative multidimensional linear systems and robust control*, Multidimensional Systems and Signal Processing **17** (2006), 119–150.

[24] J.A. Ball and J.W. Helton, *A Beurling-Lax theorem for the Lie group $U(m, n)$ which contains most classical interpolation*, J. Operator Theory **9** (1983), 107–142.

[25] J.A. Ball and J.W. Helton, *Beurling-Lax representations using classical Lie groups with many applications II: $GL(n, \mathbb{C})$ and Wiener-Hopf factorization*, Integral Equations and Operator Theory **7** (1984), 291–309.

[26] J.A. Ball and J.W. Helton, *Interpolation problems of Pick-Nevanlinna and Loewner types for meromorphic matrix functions: parametrization of the set of all solutions*, Integral Equations and Operator Theory **9** (1986), 155–203.

[27] J.A. Ball and J.W. Helton, *Shift invariant subspaces, passivity, reproducing kernels and H^∞-optimization*, in: Contributions to Operator Theory and its Applications (eds. J.W. Helton and L. Rodman), **OT 35** Birkhäuser Verlag, Basel, 1988, pp. 265–310.

[28] J.A. Ball and S. ter Horst, *Multivariable operator-valued Nevanlinna-Pick interpolation: a survey*, in: Operator Algebras, Operator Theory and Applications (eds. J.J. Grobler, L.E. Labuschagne, and M. Möller), pages 1–72, **OT 195** Birkhäuser, Basel-Berlin, 2009.

[29] J.A. Ball, K.M. Mikkola and A.J. Sasane, *State-space formulas for the Nehari-Takagi problem for nonexponentially stable infinite-dimensional systems*, SIAM J. Control and Optimization **44** (2005) no. 2, 531–563.

[30] J.A. Ball and M.W. Raney, *Discrete-time dichotomous well-posed linear systems and generalized Schur-Nevanlinna-Pick interpolation*, Complex Analysis and Operator Theory **1** (2007), 1–54.

[31] J.A. Ball, T.T. Trent and V. Vinnikov, *Interpolation and commutant lifting for multipliers on reproducing kernel Hilbert spaces*, in: Operator Theory and Analysis (eds. H. Bart, I. Gohberg and A.C.M. Ran), pp. 89–138, **OT 122**, Birkhäuser, Basel, 2001.

[32] J.A. Ball and V. Vinnikov, *Formal reproducing kernel Hilbert spaces: the commutative and noncommutative settings*, in: Reproducing Kernel Spaces and Applications (ed. D. Alpay), pp. 77–134, **OT 143** Birkhäuser Verlag, Basel, 2003.

[33] J.A. Ball and V. Vinnikov, *Lax-Phillips scattering and conservative linear systems: A Cuntz-algebra multidimensional setting*, Memoirs of the American Mathematical Society, Volume **178** Number 837, American Mathematical Society, Providence, 2005.

[34] J. Bognár, *Indefinite Inner Product Spaces*, Springer-Verlag, Berlin-New York, 1974.

[35] T. Constantinescu and J.L. Johnson, *A note on noncommutative interpolation*, Can. Math. Bull. **46** (2003) no. 1, 59–70.

[36] K.R. Davidson and T. Le, *Commutant lifting for commuting row contractions*. Bull. Lond. Math. Soc. **42** (2010) no. 3, 506–516.

[37] K.R. Davidson and D.R. Pitts, *Nevanlinna-Pick interpolation for non-commutative analytic Toeplitz algebras*, Integral Equations & Operator Theory **31** (1998) no. 2, 401–430.

[38] S.W. Drury, *A generalization of von Neumann's inequality to the complex ball*, Proc. Amer. Math. Soc. **68** (1978), 300–304.

[39] H. Dym, *J Contractive Matrix Functions, Reproducing Kernel Hilbert Spaces and Interpolation*, CBMS Regional Conference series **71**, American Mathematical Society, Providence, 1989.

[40] H. Dym, *Linear fractional transformations, Riccati equations and bitangential interpolation, revisited*, in: Reproducing Kernel Spaces and Applications (ed. D. Alpay), pp. 171–212, **OT 143** Birkhäuser Verlag, Basel, 2003.

[41] H. Dym and B. Freydin, *Bitangential interpolation for upper triangular operators*, in: Topics in Interpolation Theory (eds. H. Dym, B. Fritzsche, V. Katsnelson, and B. Kirstein), pp. 105–142, **OT 95**, Birkhäuser, Basel, 1997.

[42] H. Dym and B. Freydin, *Bitangential interpolation for triangular operators when the Pick operator is strictly positive*, in: Topics in Interpolation Theory (eds. H. Dym, B. Fritzsche, V. Katsnelson, and B. Kirstein), pp. 143–164, **OT 95**, Birkhäuser, Basel, 1997.

[43] H. Dym, J.W. Helton, and S. McCullough, *The Hessian of a noncommutative polynomial has numerous negative eigenvalues*, J. Anal. Math. **102** (2007), 29–76.

[44] J. Eschmeier and M. Putinar, *Spherical contractions and interpolation problems on the unit ball*, J. Reine Angew. Math. **542** (2002), 219–236.

[45] Q. Fang, *Multivariable Interpolation Problems*, PhD dissertation, Virginia Tech, 2008.

[46] C. Foias and A.E. Frazho, *The Commutant Lifting Approach to Interpolation Problems*, **OT 44** Birkhäuser Verlag, Basel-Boston, 1990.

[47] C. Foias, A.E. Frazho, I. Gohberg, and M.A. Kaashoek, *Metric Constrained Interpolation, Commutant Lifting and Systems*, **OT 100** Birkhäuser Verlag, Basel-Boston, 1998.

[48] D.C. Greene, S. Richter, and C. Sundberg, *The structure of inner multipliers on spaces with complete Nevanlinna-Pick kernels*, J. Funct. Anal. **194** (2002), 311–331.

[49] P.R. Halmos, *Shifts on Hilbert space*, J. für die Reine und Angewandte Math. **208** (1961), 102–112.

[50] J.W. Helton, *Orbit structure of the Möbius transformation semigroup acting on H^∞ (broadband matching)*, Topics in Functional Analysis (essays dedicated to M.G. Kreĭn on the occasion of his 70th birthday), pp. 129–157, Advances in Math. Suppl. Stud., **3**, Academic Press, New York-London, 1978.

[51] J.W. Helton, J.A. Ball, C.R. Johnson, and J.N. Palmer, *Operator Theory, Analytic Functions, Matrices, and Electrical Engineering*, CBMS Regional Conference Series **68**, American Mathematical Society, Providence, 1987.

[52] J.W. Helton, I. Klep, and S. McCullough, *Proper analytic free maps*, J. Funct. Anal. **260** (2011) no. 5, 1476–1490.

[53] J.W. Helton, I. Klep, and S. McCullough, *Analytic mappings between noncommutative pencil balls*, J. Math. Anal. Appl. **376** (2011) no. 2, 407–428.

[54] J.W. Helton, S.A. McCullough, and V. Vinnikov, *Noncommutative convexity arises from linear matrix inequalities*, J. Funct. Anal. **240** (2006) no. 1, 105–191

[55] O. Iftume, M.A. Kaashoek, and A. Sasane, *A Grassmannian band method approach to the Nehari-Takagi problem*, J. Math. Anal. Appl. **310** (2005), 97–115.

[56] V. Katsnelson, A. Kheifets, and P. Yuditskii, *An abstract interpolation problem and extension theory of isometric operators*, in: Operators in Spaces of Functions and Problems in Function Theory (ed. V.A. Marchenko), pp. 83–96, **146** Naukova Dumka, Kiev, 1987; English translation in: Topics in Interpolation Theory (eds. H. Dym, B. Fritzsche, V. Katsnelson, and B. Kirstein), pp. 283–298, **OT 95**, Birkhäuser, Basel, 1997.

[57] I.V. Kovalishina and V.P. Potapov, *Seven Papers Translated from the Russian*, Amer. Math. Soc. Transl. (2) **138**, Providence, RI, 1988.

[58] M. Möller, *Isometric and contractive operators in Kreĭn spaces*, St. Petersburg Mathematics J. **3** (1992) no. 3, 595–611.

[59] S. McCullough and T.T. Trent, *Invariant subspaces and Nevanlinna-Pick kernels*, J. Funct. Anal. 178 (2000), no. 1, 226–249.

[60] G. Meinsma, *Polynomial solutions to H_∞ problems*, Int. J. Robust and Nonlinear Control **4** (1994), 323–351.

[61] V. Olshevsky and V. Pan, *A unified superfast algorithm for boundary rational tangential interpolation problems and for inversion and factorization of dense structure*

matrices, in: Proc. 39th IEEE Symposium on Foundations of Computer Science", pp. 192–201, 1998.

[62] V. Olshevsky and A. Shokrollahi, *A superfast algorithm for confluent rational tangential interpolation problem via matrix-vector multiplication for confluent Cauchy-like matrices*, in: Structured Matrices in Mathematics, Computer Science, and Engineering, I (Boulder, CO, 1999), pp. 31–45, Contemp. Math. **280**, Amer. Math. Soc., Providence, RI, 2001.

[63] G. Popescu, *Characteristic functions for infinite sequences of noncommuting operators*, J. Operator Theory **22** (1989), 51–71.

[64] G. Popescu, *Isometric dilations for infinite sequences of noncommuting operators*, Trans. Amer. Math. Soc. **316** (1989) no. 2, 523–536.

[65] G. Popescu, *Multi-analytic operators on Fock spaces*, Math. Ann. **303** (1995), 31–46.

[66] G. Popescu, *Multivariable Nehari problem and interpolation*, J. Funct. Anal. **200** (2003) no. 2, 536–581.

[67] G. Popescu, *Operator theory on noncommutative varieties*, Indiana Univ. Math. J. **55** (2006), no. 2, 389–442.

[68] R. Nevanlinna, *Über beschränkte analytische Funktionen*, Ann. Acad. Sci. Fenn. **A32** no. 7 (1929).

[69] A.A.. Nudelman, *On a new problem of moment type*, Soviet Math. Doklady **18** (1977), 507–510.

[70] M. Rosenblum and J. Rovnyak, *Hardy Classes and Operator Theory*, Oxford University PRess, New York, 1985.

[71] W. Rudin, *Function Theory on Polydisks*, Benjamin, New York, 1969.

[72] D. Sarason, *Generalized interpolation in H^∞*, Trans. Amer. Math. Soc. **127** (1967), 179–203.

[73] S. Treil and A. Volberg, *A fixed point approach to Nehari's problem and its applications*, in: Toeplitz Operators and Related Topics (ed. E.L. Basor and I. Gohberg) **OT 71**, Birkhäuser, Basel, 1994, pp. 165–186.

[74] H.L. Trentelman and J.C. Willems, *H_∞ control in a behavioral context: the full information case*, IEEE Transactions on Automatic Control **44** (1999) no. 3, 521–536.

Joseph A. Ball
Department of Mathematics
Virginia Tech
Blacksburg, VA 24061-0123, USA
e-mail: joball@math.vt.edu

Quanlei Fang
Department of Mathematics & Computer Science
CUNY-BCC
Bronx, NY 10453, USA
e-mail: fangquanlei@gmail.com

Operator Theory:
Advances and Applications, Vol. 222, 73–83

A Brief History of the
Strong Szegő Limit Theorem

Estelle Basor

Abstract. The strong Szegő limit theorem describes the asymptotic behavior of determinants of finite Toeplitz matrices. This article is a survey that describes a simple proof of the strong Szegő limit theorem using some observations and results of Bill Helton. A proof of an exact identity for the determinants is also given along with some applications of the theorem and generalizations to other classes of operators.

Mathematics Subject Classification. 47B35, 82B44.

Keywords. Toeplitz operator, Hankel operator, determinant asymptotics, random matrix theory.

1. Introduction

Toeplitz matrices, that is, matrices whose entries are of the form $(a_{j-k})_{j,k=0}^{n}$, and their properties have been studied for over a century. Most always the coefficients, a_k, are the Fourier coefficients of a function a in L^1 of the unit circle,

$$a_k = \frac{1}{2\pi} \int_0^{2\pi} a(e^{i\theta})e^{-ik\theta}d\theta,$$

and the matrices are denoted by $T_n(a)$. One of the fundamental questions concerns the distribution of the eigenvalues of such matrices as $n \to \infty$ and this question is in turn related to the asymptotics of the determinants.

The first result for determinants was done by Szegő and goes back to 1915 [14] where he proved for positive a that the quotient $\det T_n(a)/\det T_{n-1}(a)$ converges to a limit given by

$$G(a) = \exp\left(\frac{1}{2\pi} \int_0^{2\pi} \log a(e^{i\theta})d\theta\right).$$

Later this theorem was improved by Szegő who showed that, under certain conditions,

$$\lim_{n\to\infty} \det T_n(a)/G(a)^n = E(a)$$

where $E(a) = \exp \sum_{k=1}^{\infty} k(\log a)_k (\log a)_{-k}$ and $(\log a)_k$ denotes the kth Fourier coefficient of any continuous logarithm of a. This second-order result, called the strong Szegö limit theorem, was in response to a question posed to Szegö by Lars Onsager and motivated by a formula for the spontaneous magnetization of the two-dimensional related to the Ising Model. The original positively assumption of Szegö was relaxed by many mathematicians, the result was extended to the block case, and the constant $E(a)$ was reformulated as a determinant of a certain operator.

One of the purposes of this paper is to illustrate two beautiful ideas of Bill Helton that fundamentally made the computing the asymptotics of determinants of finite Toeplitz matrices quite simple. These ideas are also the basis for a simple proof of an explicit identity for Toeplitz determinants. This will also be presented below and also some generalizations of the formulas to other settings.

We begin with a precise form of the classical Strong Szegö Limit Theorem.

2. The precise version and sketch of the proof

For the following we consider the operator $T(a)$ defined on the Hardy space ℓ^2 wtih matrix representation $(a_{j-k})_{j,k=0}^{\infty}$ and the operator $H(a)$ with representation by $(a_{j+k+1})_{j,k=0}^{\infty}$. The finite Toeplitz matrix is a truncation of the Toeplitz operator $T(a)$ and the operator $H(a)$ is called a Hankel operator. The function a is sometimes referred to as the symbol of the operator.

Theorem 1. *Suppose that the bounded function a defined on the unit circle satisfies the condition*

$$\sum_{k=-\infty}^{\infty} |a_k| + \sum_{k=-\infty}^{\infty} |k|a_k|^2 < \infty, \tag{1}$$

and suppose also that a has no zeros on the unit circle and has zero winding number. Then

$$\lim_{n \to \infty} \det T_n(a)/G(a)^n = E(a).$$

The conditions of the theorem assure that the symbol a is continuous and that the operator $T(a)$ is invertible. The constant $E(a)$ can also be written as $\det \left(T(a)T(a^{-1}) \right)$ and this expression makes sense because the conditions of the theorem guarantee that the operator

$$T(a)T(a^{-1}) - I$$

is trace class and thus the infinite determinant of $T(a)T(a^{-1})$ exists.

We also note here that the above conditions can be relaxed to require that the operator $T(a)$ is bounded and invertible and that $\sum_{k=-\infty}^{\infty} |k|a_k|^2 < \infty$. A proof can be found in [15]. However, for our purposes, the slightly less general version of the theorem is more appropriate for this survey. (For more information on these determinants and all other matters concerning Toeplitz operators see [10].)

Here is a sketch of the proof.

The finite Toeplitz matrix $T_n(a)$ can be thought of as the upper left-hand corner of the matrix representation of the operator $T(a)$. We can think of it then as

$$P_n T(a) P_n$$

where

$$P_n : \{x_k\}_{n=0}^{\infty} \in \ell^2 \mapsto \{y_k\}_{k=0}^{\infty} \in \ell^2, \qquad y_k = \begin{cases} x_k & \text{if } k < n \\ 0 & \text{if } k \geq n \end{cases}.$$

The first crucial observation of Helton was that if U is an operator whose matrix representation has an upper triangular form. Then

$$P_n U P_n = U P_n.$$

If L is an operator whose matrix representation has a lower triangular form. Then

$$P_n L P_n = P_n L.$$

So if we had an operator of the form LU, then

$$P_n L U P_n = P_n L P_n U P_n$$

and the corresponding determinants would be easy to compute.

What happens for Toeplitz operators is the opposite. In fact, condition (1) guarantees that the function a has a Wiener-Hopf factorization, that is, $a = a_- a_+$ where a_+ is in H^{∞} (a_+ has only non-zero Fourier coefficients with non-negative indices) and a_- is in $\overline{H^{\infty}}$ (a_- has only non-zero Fourier coefficients with non-positive indices) and also guarantees that the factors also satisfy condition (1). However, despite the fact that such a factorization is extremely useful, it implies that $T(a) = T(a_-)T(a_+)$ and the reader with a moments thought can verify that this is an upper triangular form times a lower triangular form.

Thus we have $T(a) = UL$ where $U = T(a_-)$ and $L = T(a_+)$, and we need to compensate for the factors in the wrong order. The next Helton idea is to use a commutator to fix the problem. We write

$$P_n T(a) P_n = P_n U L P_n = P_n L L^{-1} U L U^{-1} U P_n = P_n L P_n L^{-1} U L U^{-1} P_n U P_n.$$

If we take determinant of this expression, always with respect to the image of the operators, we have

$$\det T_n(a) = \det P_n L P_n \times \det P_n U P_n \times \det P_n L^{-1} U L U^{-1} P_n.$$

A straightforward computation shows that

$$\det P_n L P_n \times \det P_n U P_n = G(a)^n$$

and we are left with the task of computing $\det P_n L^{-1} U L U^{-1} P_n$.

Now we notice that the operator $L^{-1} U L U^{-1}$ is actually

$$T(a_+^{-1}) T(a) T(a^{-1}) T(a_+)$$

and this is I plus a trace class operator and thus has a well-defined infinite determinant. The trace class condition follows from the fact that trace class operators form an ideal and from (1) by using the identity

$$T(a)T(a^{-1}) = I + H(a)H(\tilde{a}^{-1})$$

where $\tilde{a}(e^{i\theta}) = a(e^{-i\theta})$ and noting that condition (1) implies that the operators $H(a)$ and $H(\tilde{a}^{-1})$ are both Hilbert-Schmidt operators. (The hypotheses of the theorem ensure that a^{-1} also satisfies (1).)

The final step of the proof uses the fact that if an operator A is I plus trace class, then

$$\lim_{n\to\infty} \det P_n A P_n = \det A.$$

So putting this all together we have that

$$\det P_n T_n(a) P_n = G(a)^n \times \det P_n L^{-1} U L U^{-1} P_n \qquad (2)$$

and thus

$$\lim_{n\to\infty} \det P_n T_n(a) P_n / G(a)^n = \lim_{n\to\infty} \det P_n L^{-1} U L U^{-1} P_n$$

and since $T(a_+^{-1}) = T(a_+)^{-1}$, this is the same as

$$\lim_{n\to\infty} \det P_n T(a_+^{-1}) T(a) T(a^{-1}) T(a_+) P_n$$

$$= \det T(a_+^{-1}) T(a) T(a^{-1}) T(a_+) = \det T(a) T(a^{-1}).$$

What remains is to show why the last quantity is the same as in Szegö's original constant and in this regard once again the influence of the early work of Helton is critical. The first appearance of this constant is due to Widom [15] who extended the limit theorem to the block case. For the block case the constant, except for some special cases, can only be written as an operator determinant. The extension of the theorem to the block case was a major advance by Widom and nowadays the limit theorem is also known as the Szegö-Widom limit theorem. It should be noted here that the proof that uses the projection properties outlined above appears in [6] and can be extended to the block case. The proof appeared following the pioneering block case results of Widom [15], who first used operator theoretic methods of proof the limit theorem.

To show that the constants agree in the scalar case Widom used the remarkable identity

$$\det(e^A e^B e^{-A} e^{-B}) = e^{\text{trace}(AB - BA)}$$

established by Helton and Howe [13] and independently by Pincus. A simple proof was given by Ehrhardt much later in [12]. In the above, if we let $A = T(\log a_-)$ and $B = T(\log a_+)$ and compute the trace term, the result is the original Szegö constant.

3. A Toeplitz determinant identity

With very little work one can amend the previous section to formulate an exact identity for the determinants. In fact we could have started with the identity and then proved the limit theorem as a corollary.

If we return to (2) we have the identity

$$\det P_n T_n(a) P_n = G(a)^n \times \det P_n L^{-1} U L U^{-1} P_n.$$

Let us analyze the last term $\det P_n L^{-1} U L U^{-1} P_n$. We use a very basic identity (due to Jacobi) that if A is an invertible operator on Hilbert space of the form identity + trace class then for projections P and $Q = I - P$ we have

$$\det PAP = (\det A) \times (\det QA^{-1}Q).$$

We apply this to the above with $P = P_n$, $Q_n = I - P_n$ to obtain

$$\det P_n L^{-1} U L U^{-1} P_n = \det L^{-1} U L U^{-1} \times \det Q_n U L^{-1} U^{-1} L Q_n.$$

Using some operator algebra computations one can show that

$$\det Q_n U L^{-1} U^{-1} L Q_n = \det(I - Q_n H(a_- a_+^{-1}) H(\tilde{a}_-^{-1} \tilde{a}_+) Q_n),$$

and thus we have proved the identity

$$\det P_n T_n(a) P_n = G(a)^n \times E(a) \times \det(I - Q_n H(a_- a_+^{-1}) H(\tilde{a}_-^{-1} \tilde{a}_+) Q_n).$$

Since Q_n tends strongly to zero, and $H(a_- a_+^{-1}) H(\tilde{a}_-^{-1} \tilde{a}_+)$ is trace class, this last determinant tends to one. This approach gives an alternative proof of the strong Szegö limit theorem and can also be used to give a more refined expansion of the limit theorem.

The identity has an interesting history. In 2001 it was established by Borodin and Okounkov [9] in response to some questions posed by Deift and Its that arose in random matrix theory, but the proof was complicated. Simple operator theory proofs were given by the author and Widom in [7] and by Böttcher in [8]. Later it was discovered that a version of the identity had been proved much earlier by Case and Geronimo in 1979 [11].

There are extensions of Szegö's theorem to functions to symbols that are not so nicely behaved and that do not satisfy condition (1) but we will not present those results in this survey. The most important case of these are the functions that are of the so-called Fisher-Hartwig type which include the case of jump discontinuities and/or zeros. Results for those functions and references can be found also in [10].

There are many other operators that are in some sense analogous to Toeplitz operators. For example, Wiener-Hopf operators, Bessel operators, or perturbations of Toeplitz operators. For many of these there are strong Szegö limit type results that are similar to the results presented here and many of the proofs use the same ideas. One class of these is described in the next section.

4. An application of Szegő's theorem

Toeplitz matrices arise in many applications including statistical mechanics, engineering and other areas of mathematical physics. One particular interesting application comes from a fundamental connection between determinants of Toeplitz matrices and random matrix ensembles. This example not only illustrates the usefulness of the limit theorem, but also motivates some additional questions.

For example, one can consider the Circular Unitary Ensemble (CUE). This is an ensemble of random $n \times n$ unitary matrices whose eigenvalues $\{e^{i\theta_1}, \ldots, e^{i\theta_n}\}$ have joint probability density a constant times

$$\prod_{j<k} |e^{i\theta_j} - e^{i\theta_k}|^2.$$

A linear statistic for this ensemble is a random variable of the form

$$S_n = \sum_{j=1}^{n} f(e^{i\theta_j}),$$

and it is this quantity which is connected to a Toeplitz determinant.

More precisely, if we define $g(\lambda)$ to be

$$\frac{1}{(2\pi)^n n!} \int_{-\pi}^{\pi} \cdots \int_{-\pi}^{\pi} \prod_{j=1}^{n} e^{i\lambda f(e^{i\theta_j})} \prod_{j<k} |e^{i\theta_j} - e^{i\theta_k}|^2 d\theta_1 \ldots d\theta_n$$

then $g(\lambda)$ is identically equal to

$$\det\left(\frac{1}{2\pi} \int_{-\pi}^{\pi} e^{i\lambda f(\theta)} e^{-i(j-k)\theta} d\theta\right)_{j,k=0,\ldots,n-1}.$$

The last determinant is a Toeplitz determinant with symbol

$$\phi(\theta) = e^{i\lambda f(e^{i\theta})}.$$

The identity holds because a very old result due to Andréief (1883) says that

$$\frac{1}{n!} \int \cdots \int \det(f_j(x_k)) \det(g_j(x_k)) dx_1 \cdots dx_n$$

$$= \det\left(\int f_j(x) g_k(x) dx\right)_{j,k=1,\cdots,N}.$$

One is interested in g because it is the inverse Fourier transform of the density of the linear statistic. In the opposite sense, the Toeplitz determinant can be thought of as an average or expectation with respect to CUE.

Asymptotics of the determinant gives us information about the linear statistic. This is especially useful when the function f is smooth enough, because we may appeal to the strong Szegő limit theorem to tell us asymptotically the behavior of the density function.

Applying the theorem we have

$$g(\lambda) \sim G(\phi)^n E(\phi), \quad \phi(e^{i\theta}) = e^{i\lambda f(e^{i\theta})}$$

where

$$G(\phi)^n = \exp\left(i\lambda\frac{n}{2\pi}\int_{-\pi}^{\pi} f(e^{i\theta})d\theta\right)$$

and

$$E(\phi) = \exp\left(-\lambda^2\sum_{k=1}^{\infty} kf_kf_{-k}\right).$$

We see that we can interpret the last formula as saying that asymptotically as $n \to \infty$: For a smooth function f the distribution of

$$S_n - n\mu$$

where

$$S_n = \sum_{j=1}^{n} f(e^{i\theta_j}), \quad \mu = \frac{1}{2\pi}\int_{-\pi}^{\pi} f(e^{i\theta})d\theta$$

converges to a Gaussian distribution with mean zero and variance given by

$$\sigma^2 = \sum_{1}^{\infty} kf_kf_{-k} = \sum_{1}^{\infty} k|f_k|^2.$$

(The last equality holds if f is real valued.)

It has also known that if one considers averages for $O^+(2n)$, orthogonal matrices of size $2n$ with determinant 1, then the corresponding determinant is of a finite Toeplitz plus Hankel matrix and is of the form

$$\det\left(a_{j-k} + a_{j+k}\right)_{j,k=0,\ldots,n-1}$$

where subscripts denote Fourier coefficients and the function a is assumed to be even.

Hence we are interested in the determinants of a sum of a finite Toeplitz plus a "certain type" of Hankel matrix, or in perturbations of Toeplitz matrices.

To be a little more general we are going to consider a set of operators and associated spaces we will call compatible pairs. These will include certain Toeplitz plus Hankel operators as a special case.

Let $\mathcal{S}$ stand for a unital Banach algebra of functions on the unit circle continuously embedded into L^∞ of the unit circle with the property that $a \in \mathcal{S}$ implies that $\tilde{a} \in \mathcal{S}$ and $Pa \in \mathcal{S}$ and let $\mathcal{C}_1(\ell^2)$ denote the set of trace class operators. Here $\tilde{a}(e^{i\theta}) = a(e^{-i\theta})$, and P is the Riesz projection defined by

$$P: \sum_{k=-\infty}^{\infty} a_ke^{ik\theta} \to \sum_{k=0}^{\infty} a_ke^{ik\theta}.$$

In addition, define

$$\mathcal{S}_- = \left\{a \in \mathcal{S} : a_n = 0 \text{ for all } n > 0\right\},$$
$$\mathcal{S}_0 = \left\{a \in \mathcal{S} : a = \tilde{a}\right\}.$$

Assume that $M : a \in L^\infty \mapsto M(a) \in \mathcal{L}(\ell^2)$ is a continuous linear map such that:

(a) If $a \in \mathcal{S}$, then $M(a) - T(a) \in \mathcal{C}_1(\ell^2)$ and $\|M(a) - T(a)\|_{\mathcal{C}_1(\ell^2)} \le C\,\|a\|_{\mathcal{S}}$.

(b) If $a \in \mathcal{S}_-$, $b \in \mathcal{S}$, $c \in \mathcal{S}_0$, then $M(abc) = T(a)M(b)M(c)$.

(c) $M(1) = I$.

Then we say M and $\mathcal{S}$ are compatible pairs.

All of the following can be realized as compatible pairs with an appropriate Banach algebra. Recall we define the Hankel operator $H(a)$ with symbol a by its with matrix representation

$$H(a) = (a_{j+k+1}), \quad 0 \le j,k < \infty.$$

(I) $M(a) = T(a) + H(a)$,

(II) $M(a) = T(a) - H(a)$,

(III) $M(a) = T(a) - H(t^{-1}a)$ with $t = e^{i\theta}$,

(IV) $M(a) = (T(a) + H(ta))R$ with $R = \mathrm{diag}(1/2, 1, 1, \dots)$.

The matrix representations of the operators are of the form

$$a_{j-k} \pm a_{j+k-\kappa+1}$$

with $\kappa = 0, 1, -1$.

For each of the previous four examples we can take the Banach algebra to be the Besov class. This is the class of all functions a defined on the unit circle for which

$$\int_{-\pi}^{\pi} \frac{1}{y^2} \int_{-\pi}^{\pi} |a(e^{ix+iy}) + a(e^{ix-iy}) - 2a(e^{ix})|\,dx\,dy < \infty.$$

A function a is in this class if and only if the Hankel operators $H(a)$ and $H(\tilde{a})$ are both trace class.

Moreover the Riesz projection is bounded on this class and an equivalent norm is given by

$$|a_0| + \|H(a)\|_{\mathcal{C}_1} + \|H(\tilde{a})\|_{\mathcal{C}_1}.$$

We are interested in the determinants (where the matrices or operators are always thought of as acting on the image of the projection of the appropriate space) of

$$P_n M(a) P_n.$$

Using the same trick as in the first section to pull out the $G(a)$ term, but with a different approach to computing the infinite determinants one can show the following. For details see [1] and for similar computations see [2, 3, 4, 5]

Theorem 2. *Let M and $\mathcal{S}$ be a compatible pair, and let $b \in \mathcal{S}$ and $a = \exp(b)$. Then*

$$\det P_n M(a) P_n \sim G[a]^n \hat{E}[a] \quad \text{as } n \to \infty,$$

where

$$\hat{E}[a] = \exp\left(\operatorname{trace}(M(b) - T(b)) - \frac{1}{2} \operatorname{trace} H(b)^2 + \operatorname{trace} H(b)H(\tilde{b}) \right).$$

One can also produce an analogue of our exact identity for such matrices. The following is for even functions, but can be made more general.

Let M and $\mathcal{S}$ be a compatible pair, and let $b_+ \in \mathcal{S}_+$. Put $a = a_+ \tilde{a}_+ = \exp(b)$ with $a_+ = \exp(b_+)$, $b = b_+ + \tilde{b}_+$. Then

$$\det P_N M(a) P_N = G[a]^N \hat{E}[a] \det(I + Q_N K Q_N),$$

where

$$\hat{E}[a] = \exp\left(\operatorname{trace}(M(b) - T(b)) + \frac{1}{2}\operatorname{trace} H(b)^2\right),$$

and $K = M(a_+^{-1}) T(a_+) - I$.

An application of the above asymptotics yields an expansion for determinants of finite sections of operators of the form

$$T(a) \pm H(at^\kappa),$$

where κ is an integer. This means that we have asymptotics for finite Toeplitz matrices plus or minus "shifted" finite Hankel matrices.

We again make use of the basic Jacobi identity

$$\det PAP = (\det A) \cdot (\det QA^{-1}Q),$$

where $Q = I - P$. This allows us to reduce the "shifted" Hankel cases to the previous cases and compute them asymptotically. We are not including the proof here, but the above identity combined with the fact that for an even function non-vanishing function,

$$(T(a) + H(a))^{-1} = T(a^{-1}) + H(a^{-1})$$

are the main ingredients.

Theorem 3. *Suppose that* $a = a_- a_0$, *where* a_0 *is even and* $a_- \in \overline{H^\infty}$. *Then*

(a) *Suppose* $\kappa = -2l, l \geq 1$. *Then*

$$\det P_n(T(a) \pm H(at^\kappa)) P_n \sim G[a]^{n+\ell} E_{1,\pm}[a] \det P_\ell(T(a_0^{-1}) \pm H(a_0^{-1})) P_\ell$$

as $n \to \infty$, *where* $E_{1,\pm}[a]$ *is given by*

$$\exp\left(\pm \sum_{k=1}^\infty \log a_{2k+1} - \frac{1}{2} \sum_{k=1}^\infty k[\log a]_k^2 + \sum_{k=1}^\infty k[\log a]_{-k}[\log a]_k\right).$$

(b) *Suppose* $\kappa = -1 - 2l, l \geq 1$. *Then*

$$\det P_k(T(a) - H(at^\kappa)) P_k \sim G[a]^{N+\ell} E_2[a] \det P_\ell(T(a_0^{-1}) - H(a_0^{-1}t^{-1})) P_\ell$$

as $k \to \infty$, *where* $E_2[a]$ *is given by*

$$\exp\left(- \sum_{k=1}^\infty \log a_{2k} - \frac{1}{2} \sum_{k=1}^\infty k[\log a]_k^2 + \sum_{k=1}^\infty k[\log a]_{-k}[\log a]_k\right).$$

(c) *Suppose* $\kappa = 1 - 2l, l \geq 1$. *Then*

$$\det P_n(T(a) + H(at^\kappa)) P_n \sim G[a]^{n+\ell} E_3[a] \det P_\ell(T(a_0^{-1}) + H(a_0^{-1}t)) P_\ell$$

as $N \to \infty$, where $E_3[a]$ is given by

$$\exp\left(- \log 2 + \sum_{n=1}^{\infty} \log a_{2n} - \frac{1}{2} \sum_{k=1}^{\infty} k[\log a]_k^2 + \sum_{k=1}^{\infty} k[\log a]_{-k}[\log a]_k \right).$$

(d) *We have*

$$\det P_n(T(a) + H(at^\kappa))P_k = 0 \qquad \text{if } n \geq \kappa \geq 2,$$
$$\det P_n(T(a) - H(at^\kappa))P_n = 0 \qquad \text{if } n \geq \kappa \geq 1.$$

One might ask how general $M(a)$ can be? Are the four cases mentioned above the only possible cases? Here are some remarks about this. Let us write

$$K(a) = M(a) - T(a).$$

The main properties for compatible pairs implies the following:
Since $M(ab) = M(a)M(b)$ for b even, then

$$K(ab) = K(a)K(b) + T(a)K(b) + K(a)T(b) - H(a)H(b)$$

whenever b is even.

Also, $T(a)M(b) = M(ab)$ for $a \in \mathcal{S}_-$, implies that for $a \in \mathcal{S}_-$ we have $K(a) = 0$ and $T(a)K(b) = K(ab)$ for any b.

Using these algebraic facts one can show that the structure of M is determined by $K(t)$.

In fact

$$K(t) = e_0 x^T$$

where $e_0 x^T$ with $x \in \ell^2$ stand for the rank one operator

$$y \in \ell^2 \mapsto e_0 \langle y, x \rangle \in \ell^2$$

and $e_0 = (1, 0, 0, \dots)$. The question of which x then generate an operator with the proper conditions is still not completely solved. A sufficient condition is given in [1] where all the results and proofs that are contained in this last section can be found.

5. Concluding remarks

The author dedicates this paper to Bill Helton. She wrote her first joint paper with Bill [6] where some of the ideas presented here are described in detail. The author is grateful for the wonderful insight of Bill that contributed to the subject and influenced her work for many years, but is even more grateful for the example he gave her of a fearless mathematician who knows when to push ideas forward and has a wonderful, contagious enthusiasm for everything he does.

References

[1] Basor, E.L., Ehrhardt, T.: Determinant computations for some classes of Toeplitz-Hankel matrices, Oper. Matrices 3 (2009), no. 2, 167–186.

[2] Basor, E.L., Ehrhardt, T.: Asymptotic formulas for the determinants of a sum of finite Toeplitz and Hankel matrices, Math. Nachr. **228** (2001), 5–45.

[3] Basor, E.L., Ehrhardt, T.: Asymptotic formulas for the determinants of symmetric Toeplitz plus Hankel matrices, In: Oper. Theory: Adv. Appl., Vol. 135, Birkhäuser, Basel 2002, 61–90.

[4] Basor, E.L., Ehrhardt, T: Asymptotics of determinants of Bessel operators, Commun. Math. Phys. **234** (2003), 491–516.

[5] Basor, E.L., Ehrhardt, T., Widom, H.: On the determinant of a certain Wiener-Hopf + Hankel operator, Integral Equations Operator Theory **47**, no. 3 (2003), 257–288.

[6] Basor, E.L., Helton, J.W.: A new proof of the Szegö limit theorem and new results for Toeplitz operators with discontinuous symbol, J. Operator Th. **3** (1980) 23–39.

[7] Basor, E.L., Widom, H.: On a Toeplitz determinant identity of Borodin and Okounkov, Integral Equations Operator Theory **37**, no. 4, 397–401 (2000).

[8] Böttcher, A.: On the determinant formulas by Borodin, Okounkov, Baik, Deift and Rains, In: Oper. Theory Adv. Appl., Vol. 135, Birkhäuser, Basel, 2002, 91–99.

[9] Borodin, A., Okounkov, A.: A Fredholm determinant formula for Toeplitz determinants, Integral Equations Operator Theory **37**, no. 4 (2000), 386–396.

[10] Böttcher, A., Silbermann, B.: Analysis of Toeplitz operators, Springer, Berlin 1990.

[11] Case, K.M., Geronimo, J.S.: Scattering theory and polynomials orthogonal on the unit circle. J. Math. Phys. **20**, (1979), no. 2, 299–310.

[12] Ehrhardt, T.: A new algebraic approach to the Szegö-Widom limit theorem. Acta Math. Hungar. 99 (2003), no. 3, 233–261.

[13] Helton, J.W., Howe, R.E.: Integral operators: traces, index, and homology, proceedings of the conference on operator theory, Lecture Notes in Math., **345** Springer-Verlag, Berlin, (1973) 141–209.

[14] Szego, G.: Ein Grenzwertsatz über die Toeplitzschen Determinanten einer reellen positiven Funktion, Math. Ann., **76**, (1915) 490–503.

[15] Widom, H.: Asymptotic Behavior of Block Toeplitz Matarices and Determinants. II, Adv. in Math. **21**, No. 1, (1976), 1–29.

Estelle Basor
American Institute of Mathematics
Palo Alto, CA 94306, USA
e-mail: ebasor@aimath.org

Operator Theory:
Advances and Applications, Vol. 222, 85–112

Riccati or Square Root Equation?
The Semi-separable Case

Patrick Dewilde

Dedicated to Bill Helton on his 65th birthday

Abstract. The quadratic matrix Riccati equation has been celebrated as the main ingredient in many problems of system inversion, estimation and control. However, it is an indirect equation, many of these problems can be viewed as special cases of some type of inner-outer factorization, which then can be solved through a linear equation called a 'square root equation' – as it involves a matrix square root of the unknown in the Riccati equation. The paper reviews the major cases systematically in the setting of semi separable, discrete time systems. When possible, the Riccati equation should be avoided in favor of the direct square root equation. The main reason (aside from simplicity and unicity) is numerical: the numerical condition of the squared equation is also the square of the original. The paper parallels, from the "square root equation" point of view, the great insights, which J. William (Bill) Helton has developed over the years, and which are extensively documented in the two books [11, 10].

Mathematics Subject Classification. 15A06, 15A23, 15A29, 65F05, 65F20, 65F99, 93B10, 93B50.

Keywords. Square-root algorithms, Riccati equations, semi-separable systems, time-varying systems, inner-outer factorization, generalized interpolation, H_∞-control.

1. Introduction

In [11] Bill (J. William) Helton with co-author Matthew Merino gives a systematic account of how system control for optimal global performance, so-called H_∞-control, can be developed from basic concepts relating system theory to operator theory. Already very early on in his career, Bill knew how to connect the classical Beurling-Lax theory of Inner-Outer Factorization, the famous Nagy-Foias Lifting Theorem and the Characteristic Function Theory of Krein with applied problems such as Broadband Matching and H_∞-control [12, 13]. A very readable account for this viewpoint, extending it to nonlinear systems, is given in the book of Helton and James [10]. To be sure, many researchers have contributed to H_∞ control theory. Dante Youla [20] is credited for the first problem statement and solution, known as "Youla Parametrization". Maybe the first truly "H_∞" formulation of the problem is due to George Zames [22], who sadly deceased at a relatively young age. Youla and Zames' viewpoint is purely input-output. The state space approach, which leads to the Riccati equations mentioned before, was pioneered in the seminal paper [14]. This approach was further greatly expanded and put into an Operator Theory context by Helton and coauthors, thereby providing a theory that brings many topics together: Broadband Matching, System Inversion, Interpolation, H_∞-control and Model Reduction Theory.

Given the wealth of theory, it would be a great challenge to write an ultimate treatise on the subject, and, actually Bill already did it in the books cited. My point of view will therefore be very modest. I concentrate here on discrete time, time varying systems, and just on the issue of sound numerical algorithms. The approach has the advantage that I can be very precise numerically, while touching on fundamental principles in a direct matrix framework. My presentation parallels the development in the book of Helton and James, but purely from the point of view of deriving the square root equations that solve the problem at hand. The choice for treating time-varying systems is motivated by the fact that they play a crucial role in non-linear system theory. The differential of the non-linear system equations produces a time varying systems along trajectories, and hence is well suited for a time-varying treatment – in fact, this may be almost the only realistic approach to non-linear control. In matrix algebra, a time-varying system is called a "Semi-separable" system (sometimes a "Quasi-separable" system, but I think that is a confusing term that should be avoided). The notion of semi-separable system goes probably back to Fredholm. A seminal treatment appeared in the paper of Gohberg, Kailath and Koltracht [9] where some of its remarkable numerical properties were recognized. Kailath observed the connection between semi-separable systems and Kalman filtering theory, and contributed a square-root algorithm to solve the Kalman filter (I believe he coined the term "square root algorithm", for a survey, see [15]). From that start, the connection between semi-separable systems and time-varying systems described via time-varying state space models was clear. It took a while to develop the background system theory, which found a systematic treatment, e.g., in [4], and produced the connection between inner-outer factorization, interpolation theory, embedding theory and realization theory in a setting that largely extends the usual classical Linear Time Invariant setting, and produces a wealth of new results even in matrix algebra!

A square-root algorithm is symptomatic for an underlying Inner-Outer factorization. In the semi-separable (or time-varying) theory, this connection is immediate. This is the gist of the present paper. Its most primitive form is the celebrated QR-factorization in Numerical Linear Algebra. The approach of systematically using QR factorizations has been called "array computing". It is an age old method going back to Jacobi and Gauss, has been reinvented many times and has been somewhat confusing to mathematicians who like closed formulas for the quantities they compute. Indeed, when one factorizes a matrix $T = QR$ into a unitary matrix Q and a matrix R in upper echelon form, then one produces a whole set of data (the entries of Q and R) from T, without a precise formula, but with a precise algorithm. That is also what the MATLAB notation conveys: $[Q, R] = \text{QR}(T)$ – the QR algorithm is a function that maps T on its decomposing factors Q and R. In Electrical Engineering, the Inner-Outer decomposition is known as a factorization into a lossless factor and a minimal phase factor, also an essential operation there. It is interesting to see that semi-separable theory is capable of bringing all these cases (and all their applications) together in a single framework.

We restrict ourselves here to a uniformly spaced, discrete time setting. It is possible to translate the major elements to the continuous time case. There are some surmountable technical difficulties to do this, but they are way beyond the scope of the present paper. One of the reasons why Riccati equations have been so popular, is that in the Linear Time Invariant case, the square root equation that normally turns out to be a QR recursion, becomes a QR fixed point equation (see [5] for more information), which technically seems more difficult to solve than the eigenvalue problem one builds on the so-called Hamiltonian derived from the Riccati data. Again, and for numerical reasons, the square root problem is much better conditioned and contains only half the eigenvalues (and the correct ones in addition), at the cost of some superficial algebraic complexity. The question becomes very interesting when discrete and continuous are mixed up, as happens when a distributed system is considered, whereby the subsystems are connected with each other in a linear chain, as is considered in [18]. One then gets a Riccati equation build on semi-separable matrices, a truly more complex case that we shall not discuss here. Another case that we shall not consider in this paper, is how one can compute eigenvalues of semi-separable matrices. Many people have worked on this problem, a good survey is given in [17]. Here also, QR steps are used, and efficient algorithms are derived making use of the special semi-separable structure.

2. Preliminaries

Semi-separable systems are (possibly) time-varying systems that are described via state space equations. In this paper we concentrate on the discrete time version, or, in other terms, the matrix version of such a type of system (there is also a continuous time version, which we do not consider here). The matrix represents the input/output behavior of the system, which we shall typically denote by the symbol T, a (block) matrix whose elements are themselves matrices $T = [T_{i,j}]$. Clearly all $T_{i,j}$ on the same row i should have the same number of rows, called m_i and similarly, all block matrices on the same column j shall have the same column dimension n_j. We shall allow these dimensions to become zero, in which case they become mere "place holders" – a zero dimension matrix is not a zero matrix, but an empty matrix, with the convention that the product of an $m \times 0$

matrix (that is a matrix with no columns) with a $0 \times n$ matrix (a matrix with no rows) is an $m \times n$ zero matrix. All other conventions of matrix calculus remain valid with this very useful extension. We allow i and j to range from $-\infty$ to $+\infty$ and embed finite sequences of indices into infinite sequences, just by padding the non-zero indices with place holders (dimensions zero). This convention allows us to work in a uniform framework without worrying about starting and ending values. We collect the sequences of indices into a row sequence $\mathcal{M} = [m_i]$ and a column sequence $\mathcal{N} = [n_j]$. All matrices that we shall handle shall normally be bounded operators between ℓ_2 sequences. We denote by $\ell_2^{\mathcal{N}}$ all (column) vectors of type $[u_j]$ where the dimension of each u_j is n_j, $\|u_j\|_2$ is the usual Euclidean norm of u_j and

$$\|u\|_2 = \sqrt{\sum_{j=-\infty}^{\infty} \|u_j\|_2^2} \tag{1}$$

(we shall normally drop the index 2 when the norm is clear from the context). Similarly for $y \in \ell_2^{\mathcal{M}}$, and when $y = Tu$ is a bounded map $\ell_2^{\mathcal{N}} \to \ell_2^{\mathcal{M}}$, its norm $\|T\|$ would be the normal operator norm, corresponding to the L_∞ norm of a Linear Time Invariant (LTI) system, whose system matrix would be a doubly infinite (block) Toeplitz matrix with Fourier transform belonging to the L_∞ space of the unit circle in the complex plane (we shall not treat LTI systems in this paper, but it is good to remark that they form a special case of semi-separable systems).

A special role is played by the main diagonal of T, the $T_{i,i}$ blocks. They have the meaning of 'instantaneous operators' mapping a input u_i at index point i to a (partial) output y_i of the same index. They also divide the matrix in a lower part and an upper part. The lower part (including the main diagonal) is seen to be a 'causal' operator, it maps any input series u with support on $[i, \dots, \infty)$ to a series $y = Tu$ with support on the same bearer. We now first introduce a *state space realization* for a lower matrix (strictly upper part zero). It is an indexed collection of four matrices $\{A_i, B_i, C_i, D_i\}$ such that $T_{i,i} = D_i$ and for $i > j$ each $T_{i,j} = C_i A_{i-1} \cdots A_{j+1} B_j$ (where the A-entry disappears of course when $i = j + 1$). Corresponding to a state space realization of a strictly lower operator T there is a computational realization for it, via an intermediate "hidden" variable x_i representing the memory or "state" of the computer at time point i, and the *transition map*

$$\begin{bmatrix} x_{i+1} \\ y_i \end{bmatrix} = \begin{bmatrix} A_i & B_i \\ C_i & D_i \end{bmatrix} \begin{bmatrix} x_i \\ u_i \end{bmatrix} \tag{2}$$

which maps present state and input to next state and output at time point i. In many practical cases, the semi-separable representation, alias the state space realization, is of low dimension, and derived from the physical properties of the system that is being represented. For example, when T is banded of bandwidth M, then the minimal dimension of the state space is also M. The interest of semi-separable representations then becomes immediately obvious, since the inverse operator of the banded matrix will be a full matrix, but it will have a derived state space representation of the same dimension as the original. This property generalizes: the inverse (even Moore-Penrose inverse) of a low dimension semi separable system is again a semi separable system of the same low dimension at each time point i! State space realizations are not unique. At each time point i we may introduce a 'state space transformation', which is an invertible matrix R_i defining a new local state $x_i' = R_i^{-1}x_i$. In terms of the prime matrix representation the new realization becomes

$$\begin{bmatrix} x_{i+1}' \\ y_i \end{bmatrix} = \begin{bmatrix} A_i' & B_i' \\ C_i' & D_i' \end{bmatrix} \begin{bmatrix} x_i' \\ u_i \end{bmatrix} = \begin{bmatrix} R_{i+1}^{-1}A_iR_i & R_{i+1}^{-1}B_i \\ C_iR_i & D_i \end{bmatrix} \begin{bmatrix} x_i' \\ u_i \end{bmatrix}. \tag{3}$$

On sequences of the type $u \in \ell_2^N$ we define a unitary causal shift operator Z by $[Zu]_i = u_{i-1}$ – i.e., the series is shifted forward in time. The inverse of Z is also its Hermitian transpose Z^H and is the anticausal shift that shifts sequences one unit backward in time. The matrix sequences A_i, B_i etc. may be viewed as defining instantaneous or local operators themselves (they all act exclusively at time point i), and may be collected into diagonal operators $A = \text{diag}(A_i)$, $B = \text{diag}(B_i)$ etc., and collecting the state vectors also into a single global vector x, we obtain the global state space equations

$$\begin{cases} Z^Hx &= Ax + Bu \\ y &= Cx + Du. \end{cases} \tag{4}$$

Let us now assume that the state space representation has the property of being "uniformly exponentially stable (u.e.s)", meaning that the spectral radius of the operator ZA, denoted as $\sigma(ZA)$ is strictly less than 1. Then $(I - ZA)$ is invertible (due to the Von Neumann series theorem), and the representation

$$T = D + C(I - ZA)^{-1}ZB \tag{5}$$

for the input/output matrix follows, equivalent to the defining representation $T_{i,j} = C_iA_{i-1}\cdots A_{j+1}B_j$. Methods to derive state space realizations from the entries of T are known as "realization theory" and are well documented in the literature, see [4] for a full account, including conditions on the existence of u.e.s. realizations. In the sequel, we shall just assume realizations to have the u.e.s. property. In the case of finite matrices, this invertibility condition is trivially satisfied.

A realization is called *minimal* when the state dimension at each point is the lowest possible. Realization theory shows that the minimum can be achieved at each point in the sequence, and that the minimal dimension is given by the rank of the reachability operator at that point, or, equally, by the rank of the observability operator at that point. Given an index point i, the reachability operator $\mathcal{R}_i$ at that point maps past inputs to the state, i.e.,

$$\mathcal{R}_i = \begin{bmatrix} \cdots & A_{i-1}A_{i-2}B_{i-3} & A_{i-1}B_{i-2} & B_{i-1} \end{bmatrix}.$$

Dually, the observability operator maps the state to future outputs when present and future inputs are all zero, $\mathcal{O}_i = \begin{bmatrix} C_i^H & A_i^HC_{i+1}^H & A_i^HA_{i+1}^HC_{i+2}^H & \cdots \end{bmatrix}^H$. Together, these operators factor the *Hankel operator* at point i, $H_i = \mathcal{O}_i\mathcal{R}_i$, and minimal realizations are obtained by choosing bases for the range of the reachability operator or, alternatively, the co-range of the observability operator. The

Hankel operator H_i coincides with the lower left corner matrix just left of the diagonal element $T_{i,i}$:

$$H_i = \begin{bmatrix} \cdots & T_{i,i-2} & T_{i,i-1} \\ \cdots & T_{i+1,i-2} & T_{i+1,i-1} \\ \ddots & \vdots & \vdots \end{bmatrix}. \tag{6}$$

When an orthonormal basis is chosen for the rows of the reachability operator, then a realization is obtained for which $\begin{bmatrix} A & B \end{bmatrix}$ is co-isometric (i.e., $AA^H + BB^H = I$, or, equivalently, for each i, $A_i A_i^H + B_i B_i^H = I$), dually, when an orthonormal basis is chosen for the columns of the observability operators, then the resulting $\begin{bmatrix} A \\ C \end{bmatrix}$ will be isometric. In the first case the realization is said to be in *input normal form*, while in the second case it will be in *output normal form*. A system is said to be *reachable* (respect. *observable*) if the reachability (respect. observability) operator has dense co-range (respect. range).

Let us introduce a bit more notation at this point. Besides the shifts Z and $Z^H = Z^{-1}$, we shall need a *diagonal shift*, which we indicate by a triangle-bracketed exponent: $A^{\langle 1 \rangle} := ZAZ^{-1}$ shifts the A operator one notch down the diagonal in the South-East direction, while $A^{\langle -1 \rangle} = Z^{-1}AZ$ does the opposite (taking the indexing scheme along! So A and $A^{\langle 1 \rangle}$ are not necessarily compatible, if A is a causal state transition matrix then $AA^{\langle 1 \rangle}$ is meaningful). By definition $A^{-\langle -1 \rangle} = (A^{\langle -1 \rangle})^{-1} = (A^{-1})^{\langle -1 \rangle}$ – maybe a bit strange. Also convenient is a notation to indicate a realization, we write

$$T \approx \begin{bmatrix} A & B \\ C & D \end{bmatrix} \tag{7}$$

for $T = D + C(I - ZA)^{-1}ZB$, the matrix being known as the *transition matrix* mapping {state,input} to {next state, output}, with A the *state transition matrix* which maps a state to a next state (assuming the input to be zero).

A further mathematical construction puts the semi-separable theory in the framework of a nested algebra (the content of this paragraph can be skipped if the reader is only interested in numerical results). Let us denote the dimension of the ith state space with b_i, let $\mathcal{B} = [\dots, b_i, \dots]$ tally the dimensions of the evolving state space, and stack the individual state spaces into a global state space $\ell_2^\mathcal{B}$. Next we stack the global input, output and state space vectors row-wise into a global matrix $\mathcal{X}^\mathcal{N} = \oplus_{-\infty}^\infty \ell_2^\mathcal{N}$ and likewise $\mathcal{X}^\mathcal{B}$ and $\mathcal{X}^\mathcal{M}$, for each time point i gets one column in the stacked matrix (there is a physical motivation for this construct: when one has to identify a time-varying system from its input-output behavior, one has to foresee test time-series at each time point). A comfortable norm on these stacked spaces is obtained when they are endowed with a global Hilbert space structure (they become Hilbert-Schmidt spaces), we denote them as $\mathcal{X}_2^\mathcal{N}$, $\mathcal{X}_2^\mathcal{B}$ and $\mathcal{X}_2^\mathcal{M}$. $\mathcal{X}_2^\mathcal{N}$ can be further decomposed in a lower space $\mathcal{L}_2^\mathcal{N}$ consisting at level i of series with support starting at index point i, an upper space $\mathcal{U}_2^\mathcal{N}$ of series with support from $-\infty$ to i, and a diagonal space $\mathcal{D}_2^{\mathcal{N}} = \mathcal{L}_2^{\mathcal{N}} \cap \mathcal{U}_2^{\mathcal{N}}$. The operator T induces a bounded linear map $\mathcal{X}_2^{\mathcal{N}} \mapsto \mathcal{X}_2^{\mathcal{M}}$, just by stacking individual maps as in

$$ T \left[\; \ldots, \quad u_k, \quad u_{k+1}, \quad \ldots \; \right] \mapsto \left[\; \ldots, \quad Tu_k, \quad Tu_{k+1}, \quad \ldots \; \right] $$

– here each u_k is a full time series. It shall be causal iff it maps $\mathcal{L}_2^{\mathcal{N}}$ to $\mathcal{L}_2^{\mathcal{M}}$. The (overall) Hankel map is then a map from the "strict past" $Z^H \mathcal{U}_2^{\mathcal{N}}$ to the future $\mathcal{L}_2^{\mathcal{M}}$, $HU = (TU)|_{\mathcal{L}_2}$ for $U \in Z^H \mathcal{U}_2^{\mathcal{N}}$. The *observability space* is the (closed) range of the Hankel operator, namely $\mathbf{O} := \mathcal{L}_2^{\mathcal{N}} \ominus H Z^H \mathcal{U}_2$. This space is *restricted shift invariant* for the anti-causal shift Z^H: $\mathbf{O}(Z^H \cdot) \in \mathbf{O}(\cdot)$, because similarly $Z^H \mathcal{U}_2^{\mathcal{N}} \subset \mathcal{U}_2^{\mathcal{N}}$. A basis for this space is provided by a minimal realization: the columns of $C(I - AZ)^{-1}$, actually $\mathbf{O} = C(I - AZ)^{-1} \mathcal{D}_2^{\mathcal{B}}$ (an expression that requires the realization to be u.e.s.) In a dual vein, the *reachability space* is the co-range of H (i.e., the range of H^H), it is a shift invariant space for the restricted shift Z, and a basis for it are the columns of $B^H Z^H (I - A^H Z^H)^{-1}$.

A causal unitary system is called *inner*. It has a unitary realization as well, i.e., the transition matrix $\begin{bmatrix} A & B \\ C & D \end{bmatrix}$ can be chosen unitary (the converse need not be true in general, but is true in case of a u.e.s. system [4]). Such a unitary realization is at the same time in input and in output normal form. Similarly, a causal isometric system V is such that $V^H V = I$ and it has an isometric realization, i.e., a transition matrix $\mathbf{T} = \begin{bmatrix} A & B \\ C & D \end{bmatrix}$, which is such that $\mathbf{T}^H \mathbf{T} = I$. A causal semi-separable system T is called *right-outer* when $T\mathcal{L}_2^{\mathcal{N}}$ is dense in $\mathcal{L}_2^{\mathcal{M}}$ (Arveson [3] uses a slightly more general definition: dense in the space $T\mathcal{X}_2^{\mathcal{N}} \cap \mathcal{L}_2^{\mathcal{M}}$ with the further assumption that the projection operator on the range of T is diagonal). This means that a right-outer T can be approximately causally right inverted (meaning: has a causal right inverse). If the range of the outer T is actually closed, then T is boundedly right-invertible, and it can be shown that the inverse will be u.e.s. as well [4]. A causal operator is said to be *outer*, if it is at the same time right- and left-outer. The situation where the range of an u.e.s. outer T is not closed is very common, a prime example is $T = I - Z$, an approximate causal and bounded inverse for which $T^{-1} \approx \sum_{i=0}^{\infty} \alpha^i Z^i$ with $\alpha < 1$ and $\alpha \approx 1$. The central theorem of inner-outer theory states that *any causal* $T \in \mathcal{L}(\mathcal{N}, \mathcal{M})$ *has a factorization* $T = V T_o$ *in which* $T_o \in \mathcal{L}(\mathcal{N}, \mathcal{M}_V)$ *for some index sequence* $\mathcal{M}_V$ *with* $\mathcal{M}_V \leq \mathcal{M}$ *is right-outer and* V *is a causal isometry in* $\mathcal{L}(\mathcal{M}_V, \mathcal{M})$. V *shall be inner iff* $\ker(T^H) = \{0\}$ [4]. "Lower" is not essential in these definitions, the same notions of inner and outer exist in a dual way for upper systems as well (one must, of course, agree about which of the two notions one uses, in the literature the two cases occur almost equally often.) In the next sections we describe how V and T_o are computed by a square root algorithm. Semi-separable systems do not exhibit a module structure (except a trivial one), the proper environment that fits them are (a special kind of) Nested Algebras, of which we gave concrete examples in the previous paragraph. Luckily, the inner-outer factorization is valid in such environments (as has been shown in general by Arveson in o.c.) and provides the mathematical structure needed to solve most basic problems of time-varying system theory.

More general matrices (with an upper and a lower part) will have separate semi separable realizations for the upper and for the lower part. The theory for the upper (anti-causal) part parallels the theory for the lower part, now with the shift Z^H replacing the causal shift Z. A full semi-separable matrix then has a representation

$$T = C_\ell(I - ZA_\ell)^{-1}ZB_\ell + D + C_u(I - Z^H A_u)^{-1}Z^H B_u \qquad (8)$$

in which $\{A_\ell, B_\ell, C_\ell\}$ gives a realization for the (strictly) lower system, D is the main diagonal and $\{A_u, B_u, C_u\}$ realizes the strictly upper system.

3. System inversion: the Moore-Penrose case

The traditional Moore-Penrose theory factors a general operator $T = UT_oV$ in which U is upper isometric, V lower co-isometric and T_o is upper and (fully) outer. The *Moore-Penrose inverse* is then given by $T^\dagger = V^H T_o^{-1} U^H$ in which R^{-1} is the possibly unbounded but densely defined causal inverse of the outer T_o (in practice to be replaced by an approximate inverse if T_o happens not to be boundedly invertible, a situation that can easily occur, maybe unfortunately – many systems, especially dissipative ones, just do not have very precise (pseudo-)inverses, even though lack of causality is taken into account by the procedure). In the case of a semi-separable system this factorization can be obtained in three steps. Looking at the desired factorization, we see that V^H has to push T to anti-causality, this will be the first step (but it might not determine V fully). Let V_1 be a minimal upper inner operator such that $T_u := TV_1^H$ is upper. In the two next steps, the matrix TV_1^H must be subjected to inner-outer, upper factorizations, one from the left and one from the right, to determine residual kernels. The right factorization produces $T_u = UR$, as in the regular URV matrix factorization, and can be executed in ascending index order as we shall see. However, there may be a residual kernel in R that may still have to be removed (this is peculiar to the infinitely indexed case), and which will finally produce $T_o = RV_2^H$ in a final outer-inner factorization that will reveal a remaining kernel. Neither U nor V_2 have to be unitary, U will be isometric, and V_2 co-isometric, allowing the Moore-Penrose inverse to be properly computed.

Step 1: external factorization

The URV recursion starts with orthogonal operations on (block) columns, transforming the mixed matrix T to the upper form – actually one alternates (block) column with (block) row operations to achieve a one pass solution (in the case of a finite matrix starting at the left upper corner). However, the block column operations are independent from the row operations, hence we can treat them first and then complete with row operations. The (first) column phase of the

URV factorization consists in getting rid of the lower or causal part of T by post-multiplication with a unitary matrix, working on the semi-separable representation instead of on the original data. If one takes the lower part in input normal form, i.e., $\hat{C}_\ell Z (I - \hat{A}_\ell Z)^{-1} \hat{B}_\ell = C_\ell Z (I - A_\ell Z)^{-1} B_\ell$ such that $\hat{A}_\ell \hat{A}_\ell^H + \hat{B}_\ell \hat{B}_\ell^H = I$, then the realization for (upper) V is given by

$$V \approx \begin{bmatrix} \hat{A}_\ell & \hat{B}_\ell \\ C_V & D_V \end{bmatrix} \tag{9}$$

where C_V and D_V are formed by unitary completion of the isometric $\begin{bmatrix} \hat{A}_\ell & \hat{B}_\ell \end{bmatrix}$ (for an approach familiar to numerical analysts see [19]). V^H is a minimal upper unitary operator, which pushes T to upper from the right, and in addition, takes care of a possible (partial) kernel (the full kernel will follow from step 3): $\begin{bmatrix} T_u & 0 \end{bmatrix} := TV^H$ can be checked to be upper and a realization for T_u follows directly as

$$T_u \approx \left[\begin{array}{cc|c} \hat{A}_\ell^H & 0 & C_V^H \\ B_u \hat{B}_\ell^H & A_u & B_u D_V^H \\ \hline \hat{C}_\ell \hat{A}_\ell^H + D\hat{B}_\ell^H & C_u & \hat{C}_\ell C_V^H + DD_V^H \end{array} \right]. \tag{10}$$

As expected, the new transition matrix combines lower and upper parts and has become larger, but T_u is now upper. Numerically, this step is executed as an LQ factorization as follows (for an introduction to QR and LQ factorizations, see the appendix). Let $x_k = R_k \hat{x}_k$ and let us assume we know R_k at step k, then

$$\begin{bmatrix} A_{\ell,k} R_k & B_{\ell,k} \\ C_{\ell,k} R_k & D_k \end{bmatrix} = \begin{bmatrix} R_{k+1} & 0 & 0 \\ \hat{C}_{\ell,k} \hat{A}_{\ell,k}^H + D_k \hat{B}_{\ell,k}^H & \hat{C}_{\ell,k} \hat{C}_{V,k}^H + D_k D_{V,k}^H & 0 \end{bmatrix} \begin{bmatrix} \hat{A}_{\ell,k} & \hat{B}_{\ell,k} \\ C_{V,k} & D_{V,k} \end{bmatrix}. \tag{11}$$

The LQ factorization of the left-hand matrix computes all the data of the right-hand site, namely the realization for V, the data for the upper factor T_u and the new state transformation matrix R_{k+1}, allowing the recursion to move on to the next index point. Because we have not assumed T to be invertible, we have to allow for an LQ factorization that produces an echelon form rather than a strictly square lower triangular form, hence allowing for a kernel as well, represented by a block column of zeros. The state transformation R_k is the (generalized) square root of a Gramian matrix $M_k = R_k R_k^H$, which satisfies (is the positive definite solution of) a (forward recursive) Lyapunov-Stein equation:

$$M_{k+1} = A_{\ell,k} M_k A_{\ell,k}^H + B_{\ell,k} B_{\ell,k}^H. \tag{12}$$

This (linear) equation could be solved recursively, of course, but the square root, LQ factorization has a much better numerical conditioning (square root of the condition number), and is also more economical in numerical computations.

Step 2: left inner-outer factorization

The next step is an inner/outer factorization of the upper operator T_u to produce an upper and upper invertible operator T_o and an upper unitary operator $\begin{bmatrix} U & W \end{bmatrix}$ such that $T_u = \begin{bmatrix} U & W \end{bmatrix} \begin{bmatrix} T_o \\ 0 \end{bmatrix}$ (allowing for a potential kernel as well.) U is an as large as possible upper and isometric operator U such that $U^H T_u$ is still upper – U^H tries to push T_u back to lower, but it should not destroy its 'upperness'. When it does so, an upper and upper right-invertible factor R should result. Writing out the factorization in terms of the realization, and redefining for brevity $T_u := D + CZ^H(I - AZ^H)^{-1}B$ we obtain

$$
\begin{aligned}
U^H T_u &= \left[D_U^H + B_U^H(I - ZA_U^H)^{-1}ZC_U^H\right]\left[D + CZ^H(I - AZ^H)^{-1}B\right] \\
&= D_U^H D + B_U^H(I - ZA_U^H)^{-1}ZC_U^H D + D_U^H CZ^H(I - AZ^H)^{-1}B \qquad (13) \\
&\quad + B_U^H\{(I - ZA_U^H)^{-1}ZC_U^H CZ^H(I - AZ^H)^{-1}\}B.
\end{aligned}
$$

This expression has the form: 'direct term' + 'strictly lower term' + 'strictly upper term' + 'mixed product'. The last term has what is called 'dichotomy', what stands between $\{\cdot\}$ can again be split in three terms:

$$
\begin{aligned}
&(I - ZA_U^H)^{-1}ZC_U^H CZ^H(I - AZ^H)^{-1} \\
&= (I - ZA_U^H)^{-1}ZA_U^H Y + Y + YAZ^H(I - AZ^H)^{-1}
\end{aligned} \qquad (14)
$$

with Y satisfying the 'Lyapunov-Stein equation'

$$
Z^H Y Z = C_U^H C + A_U^H Y A \qquad (15)
$$

or, with indices: $Y_{k+1} = C_{U,k}^H C_k + A_{U,k}^H Y_k A_k$ (in this equation, not only Y is unknown, but also A_U and C_U, so this equation cannot be solved directly, in contrast with the previous case). The resulting strictly lower term has to be annihilated, hence we should require $C_U^H D + A_U^H Y B = 0$, in fact U should be chosen maximal with respect to this property (beware: Y depends on U!) Once these two equations are satisfied, the realization for R results as $R = (D_U^H D + B_U^H Y B) + (D_U^H C + B_U^H Y A)Z^H(I - AZ^H)^{-1}B$ – we see that R inherits A and B from T and gets new values for the other constituents C_R and D_R. Putting these operations together in one matrix equation (in a somewhat special order) and allowing for a kernel, we obtain

$$
\begin{bmatrix} YB & YA \\ D & C \end{bmatrix} = \begin{bmatrix} B_U & A_U & B_W \\ D_U & C_U & D_W \end{bmatrix} \begin{bmatrix} D_R & C_R \\ 0 & Z^H Y Z \\ 0 & 0 \end{bmatrix}. \qquad (16)
$$

Let us interpret this result without going into the motivating theory (as in done in [4, 19]). We have a QR factorization of the left-hand side. At stage k one must assume knowledge of Y_k, and then perform a QR factorization of $\begin{bmatrix} C_k Y_k & D_k \\ A_k Y_k & B_k \end{bmatrix}$. $D_{R,k}$ will be a right-invertible, upper triangular matrix, so its dimensions are fixed by the row dimension of Y_k. The remainder of the factorization produces $C_{R,k}$ and Y_{k+1}, and, of course, the Q factor that gives a complete realization of $\begin{bmatrix} U & W \end{bmatrix}$ at stage k:

$$
\begin{bmatrix} Y_k B_k & Y_k A_k \\ D_k & C_k \end{bmatrix} = \begin{bmatrix} B_{U,k} & A_{U,k} & B_{W,k} \\ D_{U,k} & C_{U,k} & D_{W,k} \end{bmatrix} \begin{bmatrix} D_{R,k} & C_{R,k} \\ 0 & Y_{k+1} \\ 0 & 0 \end{bmatrix}, \qquad (17)
$$

in which the extra columns represented by B_W and D_W define the isometric operator $W = D_W + C_W Z^H (I - A_W Z^H)^{-1} B_W$, which represents the co-kernel of T.

Y satisfies a (rather general) recursive Riccati equation, obtained by eliminating $\begin{bmatrix} U & W \end{bmatrix}$:

$$M_{k+1} = A_k^H M_k A_k + C_k^H C_k \qquad (18)$$
$$- (C_k^H D_k + A_k^H M_k B_k)(B_k^H M_k B_k + D_k^H D_k)^{-1}(D_k M_k A_k + D_k^H C_k),$$

in which the inverse is guaranteed to exist and the equation also has a guaranteed (recursive) positive definite solution under the current assumptions on the entries (the guarantee follows from the existence of the square root equation). Again, it is not advisable to solve this equation, although sometimes advocated, not only because the square root equation is linear rather than quadratic and has only one solution (while the Riccati equation has many) but, at least as importantly, the numerical conditioning of the Riccati equation is (much) worse.

Step 3: right outer-inner factorization

Although the classical URV factorization on finite matrices ends at this point with an invertible R, this is not the case for systems that run from $-\infty$ to $+\infty$ – the case that is common in System Theory and is also important in Numerical Analysis as it relates to issues of numerical stability of the inverses. What is needed next is a right outer-inner factorization of R to produce $R = T_o V_2$. This step is dual to the previous one and therefore I shall not give further details, let me suffice by remarking that here also, V_2 may not be unitary as a global operator, it may just be co-isometric, i.e., $V_2 V_2^H = I$. When the square root equations for this step are written out, then it becomes obvious that they require a recursion in descending index order. This is unavoidable as U determines singularities at $+\infty$, while V_2 does the same, but then at $-\infty$, and singularities can indeed appear at both ends.

Putting everything together we obtain $T = U T_o V_2 V_1$ and the Moore-Penrose inverse $T^{-1} = V_1^H V_2^H T_o^{-1} U^H$ as expected, with now T_o outer as it should be.

That the third step is necessary, is already exemplified by the simple, half-infinite band matrix

$$R = \begin{bmatrix} 1 & -2 & & \\ & 1 & -2 & \\ & & \ddots & \ddots \end{bmatrix} \qquad (19)$$

for which we easily find

$$R = T_o V_2 = \begin{bmatrix} 2 & -1 & & \\ & 2 & -1 & \\ & & 2 & \ddots \\ & & & \ddots \end{bmatrix} \begin{bmatrix} 1/2 & -3/4 & -3/8 & \cdots \\ & 1/2 & -3/4 & \ddots \\ & & 1/2 & \ddots \\ & & & \ddots \end{bmatrix} \qquad (20)$$

in which V_2 is co-isometric but not unitary (the kernel has dimension 1!).

Remarkably, the operations in steps one and two work on the rows and columns of T in ascending order. That means that the URV algorithm can be executed completely in ascending index order (but then R is potentially not yet outer). The reader may wonder at this point (1) how to start the recursions and (2) whether the proposed recursive algorithms are numerically stable. Assuming the matrix to be half-infinite and to start at index point 0, there is no problem starting out the downward recursion at the upper left corner of the matrix, both A_1 and Y_0 are just empty, the first QR is done on $\begin{bmatrix} D_1 & C_1 \end{bmatrix}$. In case the original system does not start at index 1, but has a system part that runs from $-\infty$ onwards, one must introduce knowledge of the initial condition on Y. This is provided, e.g., by an analysis of the LTI system running from $-\infty$ to 0 if that is indeed the case, see [5] for more details. For the upward recursion the same holds mutatis mutandis. On the matter of numerical stability of the square root equation, we offer two remarks. First, propagating Y_k in step 2 is numerically stable, one can show that a perturbation on any Y_k will die out exponentially if the propagating system is assumed exponentially stable. Second, one can show that the transition matrix Δ of the inverse of the outer part will be exponentially stable as well, when certain conditions on the original system are satisfied [4].

4. Constrained interpolation problems

A more complex case is when constrained interpolation problems are considered, the semi-separable or matrix equivalents of Nevanlinna-Pick, Schur, Hermite-Fejer or Schur-Takagi interpolation. All these interpolation types have in common that (1) interpolation data of some type is prescribed and (2) the solution has to be constrained to norm less or equal to one. The classical Nevanlinna-Pick interpolation problem in the open unit disk of the complex plane specifies a single interpolation condition at, say, n different points ν_i, $i = 1, \ldots, n$. It searches for a contractive function S, which is analytic in the open unit disk and takes the values s_i at the points ν_i, $S(\nu_i) = s_i$. It is known that a solution exists iff the so-called Pick matrix based on the given data is positive definite. When the Pick matrix is strictly positive definite, then all solutions can be expressed as a bilinear expression in terms of the entries of a so-called J-inner matrix and a contractive analytic but otherwise arbitrary "load" S_L. In the case of semi-separable systems, the first question that arises is how to specify the interpolation data. In addition, one looses the notion of analyticity, which one has to replace with causality (lower matrices, e.g.), motivated by the fact that analyticity in the unit disk corresponds to causality of the back Fourier transform. These problems can be solved, we quickly describe the procedures directly in terms of semi-separable systems. As we shall see in a further section, constrained interpolation theory of this kind plays a key role in optimal control and in model reduction theory.

The 'point evaluation' for lower (causal) semi-separable systems that properly generalizes interpolation conditions, was originally introduced in [1] and [6], where also a Nevanlinna-Pick type interpolation theory for semi-separable systems was formulated. In [7] the generalization to Schur-Takagi theory was established. I quickly summarize the evaluation concept here. Let $T = T_0 + T_1 Z + T_2 Z^2 + \cdots$ be a causal (lower) and bounded operator with the given expansion in terms of shifted main diagonals notated as T_i, and let W be a (dimensionally compatible with Z) block diagonal operator such that $\sigma(ZW) < 1$. We define the value of T at W to be a diagonal operator, denoted $T(W)$ (in the notation of the original paper it was denoted in a somewhat cumbersome way by $T^\wedge(W)$) which is such that $T = T(W) + T_r(Z - W)$ for some bounded, causal (lower) remainder T_r. $T(W)$ is the W-transform of T, so called because of the resulting reproducing kernel, see [2], where it is also shown that $T(W)$ is given by the strongly convergent series

$$T(W) = T_o + T_1 W + T_2 W^{\langle 1 \rangle} W + T_3 W^{\langle 2 \rangle} W^{\langle 1 \rangle} W + \cdots . \tag{21}$$

The notion clearly generalizes the evaluation of a complex-valued matrix function $T(z)$ at a point $a \in \mathbf{C}$ as $T(a)$. Because of the non-commutativity of the shift operator Z, it does not have all the properties of the evaluation in the complex plane. We do have the following properties.

1. $T(W)$ is the first anticausal diagonal in the expansion of $T(Z - W)^{-1}$:

$$T(Z - W)^{-1} = (\cdots)Z^{-2} + T(W)Z^{-1} + T_r \tag{22}$$

   in which the $\cdots$ is anticausal.

2. (Chain rule) For P and Q anticausal we have $[PQ](W) = [PQ(W)](W)$. If $Q(W)$ is invertible, we have in addition $[PQ](W) = P(W_1)Q(W)$ where $W_1 = Q(W)^{\langle 1 \rangle} WQ(W)^{-1}$.

   Proof. Writing shorthand $Q_W := Q(W)$, we have $Q_W(Z - W)^{-1} = (ZQ_W^{-1} - WQ_W^{-1})^{-1} = (Q_W^{-\langle 1 \rangle} Z - WQ_W^{-1})^{-1} = (Z - W_1)^{-1} Q_W^{\langle 1 \rangle}$, and hence $PQ(Z - W)^{-1} = PQ_W(Z - W)^{-1} + PQ_r = P_{W_1}(Z - W_1)^{-1} Q_W^{\langle 1 \rangle} + P_r Q_W^{\langle 1 \rangle} + PQ_r$, the last being equal again to $P_{W_1} Q_W(Z - W)^{-1} +$ causal.

3. (Constants) Let D be a compatible diagonal operator, then $[DT](W) = DT(W)$. If D is invertible and compatible, then $[TD](W) = T(W_1)D^{\langle 1 \rangle}$, in which $W_1 = D^{\langle 1 \rangle} WD^{-1}$. For addition we simply have $[T + D](W) = T(W) + D$.

4. (State space formulas) Let $T = D + CZ(I - AZ)^{-1}B$ be a realization for T, assumed to be causal and such that $\sigma(AZ) < 1$. Then $T(W) = D + CMW$ where M solves the Lyapunov-Stein equation

$$M^{\langle 1 \rangle} = B + AMW. \tag{23}$$

   In fact,

$$M = [(I - ZA)^{-1}B](W)^{\langle 1 \rangle} = [B + AB^{\langle 1 \rangle} W + AA^{\langle 1 \rangle} B^{\langle 2 \rangle} W^{\langle 1 \rangle} W + \cdots]^{\langle 1 \rangle} \tag{24}$$

   and hence also

$$T(W) = D + C[Z(I - AZ)^{-1}B](W) \tag{25}$$

   in accordance with the previous rules.

The semi-separable version of the Nevanlinna-Pick problem then specifies the evaluation of the sought after operator S at, say, n diagonals ν_i:

$$S(\nu_i) = s_i \tag{26}$$

and asks for solutions with S causal and (strictly) contractive. As in the classical case, the solution will be given in terms of a so-called J-inner operator, in which J is a signature matrix of the type $J = \begin{bmatrix} I & \\ & -I \end{bmatrix}$ (for simplicity of writing we suppress the dimensions, they have to be compatible). An operator Θ is said to be J-unitary iff $\Theta^H J \Theta = J, \Theta J \Theta^H = J$ (all the J's again with appropriate dimensions). Corresponding to a J-unitary Θ there is always a so-called "scattering matrix" defined as

$$\Sigma = \begin{bmatrix} \Theta_{11} - \Theta_{12}\Theta_{22}^{-1}\Theta_{21} & \Theta_{12}\Theta_{22}^{-1} \\ -\Theta_{22}^{-1}\Theta_{21} & \Theta_{22}^{-1} \end{bmatrix}. \tag{27}$$

Σ exists as a bounded operator because the J-unitarity condition forces the existence of Θ_{22}^{-1} as a bounded operator. Σ is unitary, but it need not be causal. A J-unitary Θ is said to be *J-inner* iff the corresponding Σ is lower (causal). In general, a J-unitary or even J-inner operator need not be causal and not even bounded, we shall soon see various important cases. The theory of J-inner operators is quite involved, they play a central role in scattering theory, for a detailed account in the semi-separable context with motivations, see [4].

Let us now assemble the interpolation data in three (dimensionally compatible) block diagonal matrices

$$V := \begin{bmatrix} \nu_1 & & \\ & \ddots & \\ & & \nu_3 \end{bmatrix}, \quad \xi := \begin{bmatrix} I & \cdots & I \end{bmatrix}, \quad \eta = \begin{bmatrix} s_1 & \cdots & s_n \end{bmatrix} \tag{28}$$

then the Nevanlinna-Pick constrained interpolation problem will have a solution iff there exists a J-inner operator with controllability pair

$$\begin{bmatrix} A \mid B_1 & B_2 \end{bmatrix} = \begin{bmatrix} V^H \mid \xi^H & -\eta^H \end{bmatrix}. \tag{29}$$

This will be the case iff the (recursive) Lyapunov-Stein equation

$$M^{<-1>} = V^H M V + \xi^H \xi - \eta^H \eta \tag{30}$$

has a strictly positive definite solution M. M is the "Pick operator" for the present case. This then means that there is a realization for Θ

$$\Theta \approx \left[\begin{array}{c|cc} R^{-\langle -1 \rangle} A R & R^{-\langle -1 \rangle} B_1 & R^{-\langle -1 \rangle)} B_2 \\ \hline C_1 & D_{11} & 0 \\ C_2 & D_{21} & D_{22} \end{array} \right] \tag{31}$$

which is J-unitary for the extended (compatible)

$$J_1 = \left[\begin{array}{c|c} I_n & \\ \hline & \begin{array}{cc} I & \\ & -I \end{array} \end{array} \right],$$

with the square root $M = R^H R$ and appropriate B and D entries (D_{12} can handily always be chosen zero). Then

$$\Theta = \begin{bmatrix} \Theta_{11} & \Theta_{12} \\ \Theta_{21} & \Theta_{22} \end{bmatrix} \tag{32}$$

is J-inner and all solutions of the interpolation problem are given by the bilinear expression

$$S = (S_L \Theta_{12} - \Theta_{22})^{-1}(\Theta_{21} - S_L \Theta_{11}) \tag{33}$$

in which S_L is any contractive causal operator of compatible dimensions. In particular, one can choose $S_L = 0$, and then $S = -\Theta_{22}^{-1}\Theta_{21}$ is a minimal solution (all solutions with S_L constant will be minimal – the converse is not true, there may be cancellations).

The interpolation problem hence reduces to the solution of either a Lyapunov-Stein equation or a square root equation. The latter has to be preferred, of course. The interpolation problem will have a solution iff the forward recursion

$$\begin{bmatrix} A_k R_k & | & B_{1k} & B_{2k} \end{bmatrix} = \begin{bmatrix} R_{k+1} & | & 0 & 0 \end{bmatrix} \Theta_k^H \tag{34}$$

has a J_1-strictly-positive-definite solution, which will be the case iff all subsequent

$$\begin{bmatrix} \dfrac{R_k^H V_k}{\xi_k} \\ -\eta_k \end{bmatrix} = \Theta_k \begin{bmatrix} R_{k+1}^H \\ 0 \\ 0 \end{bmatrix} \tag{35}$$

are J_1-strictly-positive-definite, the solutions being given by J_1-embedding, to be briefly described next. This latter expression is a numerical J_1-unitary-upper factorization, akin to a QR-factorization but the first block columns of Θ_k form a J_1-unitary sub-basis for the J_1-positive part of the global range space of Θ_k.

Θ_k and R_{k+1}^H can be obtained through a modified QR-algorithm, given the interpolation data and the recursive data contained in R_k. This works as follows. First one uses orthogonal (or unitary) transformations to align the first column of $\begin{bmatrix} R_k^H V_k \\ \xi_k \end{bmatrix}$ respect. $-\eta_k$ with the first natural base vector (i.e., only first entry non-zero). Then one performs a single hyperbolic rotation on the two top values, which will be possible due to the positivity condition on the Pick matrix. This aligns the full first column on the first natural base vector. Next one uses again unitary rotations to align the next column on the already transformed second columns of the two sub-matrices, aligning the first one on the second natural basis vector and the second one on the first natural base vector again (which is possible as this will not create fill-ins on the previous transformation). The procedure is illustrated in the next typical sequence (done on a 7×3 case, with 'u' indicating unitary transformations and 'h' an elementary hyperbolic rotation):

$$
\begin{bmatrix}
* & * & * \\
* & * & * \\
* & * & * \\
\hline
* & * & * \\
* & * & * \\
\hline
* & * & * \\
* & * & *
\end{bmatrix}
\rightarrow^u
\begin{bmatrix}
* & * & * \\
0 & * & * \\
0 & * & * \\
\hline
0 & * & * \\
0 & * & * \\
\hline
* & * & * \\
0 & * & *
\end{bmatrix}
\rightarrow^h
\begin{bmatrix}
* & * & * \\
0 & * & * \\
0 & * & * \\
\hline
0 & * & * \\
0 & * & * \\
\hline
0 & * & * \\
0 & * & *
\end{bmatrix}
\rightarrow^u
\begin{bmatrix}
* & * & * \\
0 & * & * \\
0 & 0 & * \\
\hline
0 & 0 & * \\
0 & 0 & * \\
\hline
0 & * & * \\
0 & 0 & *
\end{bmatrix}
\rightarrow^h
\begin{bmatrix}
* & * & * \\
0 & * & * \\
0 & 0 & * \\
\hline
0 & 0 & * \\
0 & 0 & * \\
\hline
0 & 0 & * \\
0 & 0 & *
\end{bmatrix}
$$

$$
\rightarrow^u
\begin{bmatrix}
* & * & * \\
0 & * & * \\
0 & 0 & * \\
\hline
0 & 0 & 0 \\
0 & 0 & 0 \\
\hline
0 & 0 & * \\
0 & 0 & 0
\end{bmatrix}
\rightarrow^h
\begin{bmatrix}
* & * & * \\
0 & * & * \\
0 & 0 & * \\
\hline
0 & 0 & 0 \\
0 & 0 & 0 \\
\hline
0 & 0 & 0 \\
0 & 0 & 0
\end{bmatrix} . \tag{36}
$$

This method of eliminating entries, using a combination of unitary and hyperbolic transformations is as numerically stable as can possibly be, the unitary rotations "concentrate positive or negative energy" as much as possible, and the hyperbolic rotation then eliminates the presumably smaller negative energy against the positive entry. Under the given conditions, all the hyperbolic rotations will be of the positive type, leaving the original signature intact (in the Schur-Takagi case to be briefly discussed further, mixed types occur). This sequence of events also automatically produces a zero at the D_{12} location – which has some significance for the cascading properties of the result obtained.

It should immediately be clear that the Nevanlinna-Pick problem can be generalized to arbitrary collections of non-scalar, multiple and directional interpolation data simply represented by compatible but otherwise arbitrary matrices $\{V, \xi, \eta\}$. In this fashion one obtains Schur type interpolation (when V is nilpotent) or Hermite-Fejér type (when Schur and Nevanlinna-Pick types are mixed). It is also possible to solve mixed left-write interpolation problems (of the so-called Nudel'man type) by an extension of the machinery. This then leads to a Sylvester equation, known to be notoriously numerically unstable. Under some pretty general stabilizing conditions, the mixed interpolation problem can be transformed into a regular right interpolation problem as treated above. Extensive details can be found in [4].

An interesting generalization is provided by the Schur-Takagi interpolation case, which is also equivalent to exact model order reduction for semi separable systems. We briefly summarize the results (for extensive treatment see [4]). The Schur-Takagi interpolation problem is still a constrained problem, requesting the resulting interpolating function to be contractive (we require strictly contractive), but it now allows it to have an upper (non-causal) part, which it tries to keep minimal in the state equation sense (semi-separable with minimal state dimension).

This problem gets solved with the same square root machinery as developed earlier, but now the J_1-signature is allowed to be mixed, keeping its negative part as small as possible. The solution to this problem was, to the best of our knowledge, presented for the first time in [8]. In that paper the connection to model order reduction for semi-separable systems is also explained.

The interpolation data is again of the type $\{V, \xi, -\eta\}$ as before, and one attempts to build a causal (upper) J-unitary operator (not any more J-inner) based on the reachability data

$$\{A, B_1, B_2\} = \{V^H, \xi^H, -\eta^H\}.$$

This shall be possible iff the Pick operator M defined by the forward recursion

$$M^{\langle -1 \rangle} = AMA^H + B_1 B_1^H - B_2 B_2^H \qquad (37)$$

is strictly non-singular (while before it had to be strictly positive). M then has, at each index point i, a signature matrix $\begin{bmatrix} I_{p_i} & \\ & -I_{q_i} \end{bmatrix}$ and a non-singular state transformation matrix such that

$$M_i = R_i \begin{bmatrix} I_{p_i} & \\ & -I_{q_i} \end{bmatrix} R_i^H. \qquad (38)$$

As a consequence, the state transformation $R_{i+1}^{-1} \cdots R_i$ produces a J_1-isometric set of reachability pairs $\{\hat{A}_i, \hat{B}_{1,i}, \hat{B}_{2,i}\} = \{R_{i+1}^{-1} A_i R_i, R_{i+1}^{-1} B_{1,i}, R_{i+1}^{-1} B_{2,i}\}$ for Θ_i:

$$J_{1,i+1} = \begin{bmatrix} \hat{A}_i & | & \hat{B}_{1,i} & \hat{B}_{2,i} \end{bmatrix} J_{1,i} \begin{bmatrix} \hat{A}_i^H \\ \hat{B}_{1,i} \\ \hat{B}_{2,i} \end{bmatrix} \qquad (39)$$

in which J_1 has the form

$$J_1 = \left[\begin{array}{cc|c} I_p & & \\ & -I_q & \\ \hline & & J \end{array} \right] \qquad (40)$$

(which of course changes in dimensions from index point to index point).

The square-root algorithm in this case is then a little more complicated than explained higher, as we not only have to use hyperbolic rotations of mixed type, but also have to determine a new signature matrix recursively. This works as follows. Instead of working on the $\{A, B\}$ data, let us work as before on the interpolation data directly. Rather than describing the algorithm theoretically (as in [4]), let us work on an example. Let us assume that the signature of $J_{1,i}$ is $[+, +, -|+, +|-, -]$, then a possible sequence of operations, mixing separate unitary transformations on the $+$, respect. $-$ data, concentrating energy in single entries, and then followed by single hyperbolic rotations, could be as follows:

$$
\begin{bmatrix} * & * & * \\ * & * & * \\ * & * & * \\ * & * & * \\ * & * & * \\ * & * & * \\ * & * & * \end{bmatrix}
\to^u
\begin{bmatrix} * & * & * \\ 0 & * & * \\ * & * & * \\ 0 & * & * \\ 0 & * & * \\ 0 & * & * \\ 0 & * & * \end{bmatrix}
\to^h
\begin{bmatrix} * & * & * \\ 0 & * & * \\ 0 & * & * \\ 0 & * & * \\ 0 & * & * \\ 0 & * & * \\ 0 & * & * \end{bmatrix}
\to^u
\begin{bmatrix} * & * & * \\ 0 & * & * \\ 0 & * & * \\ 0 & 0 & * \\ 0 & 0 & * \\ 0 & 0 & * \\ 0 & 0 & * \end{bmatrix}
\to^h
\begin{bmatrix} * & * & * \\ 0 & * & * \\ 0 & 0 & * \\ 0 & 0 & * \\ 0 & 0 & * \\ 0 & 0 & * \\ 0 & 0 & * \end{bmatrix}
$$

$$
\to^u
\begin{bmatrix} * & * & * \\ 0 & * & * \\ 0 & 0 & * \\ 0 & 0 & * \\ 0 & 0 & 0 \\ 0 & 0 & 0 \\ 0 & 0 & 0 \end{bmatrix}
\to^h
\begin{bmatrix} * & * & * \\ 0 & * & * \\ 0 & 0 & 0 \\ 0 & 0 & * \\ 0 & 0 & 0 \\ 0 & 0 & 0 \\ 0 & 0 & 0 \end{bmatrix}
\to^t
\begin{bmatrix} * & * & * \\ 0 & * & * \\ 0 & 0 & * \\ 0 & 0 & 0 \\ 0 & 0 & 0 \\ 0 & 0 & 0 \\ 0 & 0 & 0 \end{bmatrix},
$$

$$(41)$$

where the last step consists of a reordering of the rows to $[1, 2, 4|5|3, 6, 7]$ and signature sequence $[+, +, +| + |-, -, -]$ (same number of pluses and minuses but different division of the signatures). At each hyperbolic rotation the evolution can proceed in two ways, depending whether the plus entry has larger magnitude than the minus entry or vice versa (equal magnitude may not occur since that would result in a singular case, which we ruled out). The square root algorithm is indeed capable of producing all the information *directly on the data*, without the need to compute the Gramian M, which would greatly deteriorate the numerical condition. The conditioning of the square root algorithm is as best as is can be, because the constrained interpolation problem of Schur-Takagi type is equivalent to an inverse scattering problem, whose conditioning is determined by the inverse scattering function, linking input data to reflection coefficients. This relationship is "of hyperbolic type" as can be seen already from elementary examples.

Scattering theory in this context also provides the link to determine the interpolation function from the Θ operator. This function will have a causal (lower) and anti-causal (upper) part, it has so-called "dichotomy", the anticausal part having minimal state dimensions. The pluses and minuses in the state propagation convert into downward (causal) and upward (anticausal) propagation in the general, unitary scattering operator Σ_k derived from Θ_k at each stage k. In other words: the dichotomy is visible in the state signature at each stage, thanks to the contractivity of the partial maps (in engineering terminology called "passivity"). This works as follows (for a full discussion and proofs, see [4], p. 276 ff.). Globally, the operator Θ is J-unitary (with specified and generally different J's at the input and output ports), and hence the resulting Σ is unitary, but it will be a mixed causal/anticausal operator, its state space structure down/up. Locally, Θ_k (which is J_1-unitary – where again the J_1's at the input and output side may be different but shall have the same total number of plusses and minuses) may be further divided, according to the division of the signature of the states:

$$\Theta_k \left[\begin{array}{c} x_{+,k} \\ x_{-,k} \\ \hline a_{1,k} \\ b_{1,k} \end{array} \right] = \left[\begin{array}{c} x_{+,k+1} \\ x_{-,k+1} \\ \hline a_{2,k} \\ b_{2,k} \end{array} \right] \tag{42}$$

in which $a_{1,k}$ and $b_{2,k}$ have positive signature in the inner product and $b_{1,k}$, $a_{2,k}$ negative (in the scattering context the positive signatures correspond to "incoming" waves and the negative signatures to outgoing). A local Σ_k map can now be defined, mapping signals with positive signature into signals with negative signature:

$$\Sigma_k \left[\begin{array}{c} x_{+,k} \\ x_{-,k+1} \\ \hline a_{1,k} \\ b_{2,k} \end{array} \right] = \left[\begin{array}{c} x_{+,k+1} \\ x_{-,k} \\ \hline a_{2,k} \\ b_{1,k} \end{array} \right] \tag{43}$$

(one should realize that, e.g., $x_{+,k+1}$ has positive signature for stage $k + 1$, but negative in stage k as it is "outgoing" there – the connection between stages is "energy conserving"). Because the algebraic relations in Σ_k are the same as for Θ_k, it is a proper state space realization for the global Σ, and we see that the realization has become mixed with a causal state propagating downwards and an anticausal state propagating upwards (the two get mixed up in each stage, of course). We know already, from the J-unitarity of Θ, that Θ_{22} must be boundedly invertible to produce Σ_{22}, it is now a semi-separable operator of mixed lower/upper type, and it can be seen that its anti-causal state space structure corresponds to the $x_{-,k}$ (which were kept as small as possible in the construction of the Θ_k). As a final result (going far beyond the present paper devoted to the square root algorithm), one shows that, indeed, all interpolating functions with minimal anti-causal part are given by eq. (33), and that the dimension of their state space at stage k is the dimension of $x_{-,k}$.

Schur-Takagi interpolation provides a direct and impressively complete solution to the *model reduction problem* for semi-separable systems. The conversion from model reduction to interpolation goes as follows. We suppose that we are given a "high-order" lower (strictly causal and bounded) semi-separable operator T, which has to be model order reduced, i.e., approximated by a semi-separable operator with low state space dimensions, which we shall call T_a. To be sure, we have to decide on a level of precision and a norm (these are critical decisions for the theory to work!). For the desired precision of the approximation, we allow for a block-diagonal invertible and hermitian operator Γ, which tallies the precision at each stage (we could suffice with just a constant $\Gamma = \epsilon$, but that would limit flexibility, as we might wish different precisions in different portions of the operator, which the theory perfectly allows). More delicate is the norm to be used. The straight operator norm appears to be too strong, and just a quadratic norm is too weak, this we know from experience with the LTI case. The norm that works is the Hankel norm. For comfort, we assume our original operator to be strictly causal (i.e., main diagonal zero), one can always shift the operator one notch left (or even more than one to increase precision – or else one may chose to keep the main diagonal intact). The Hankel norm is just the norm of the Hankel operator associated with the operator, denoted as $\|T\|_H$, it is only defined on strictly causal systems. Hence, one looks for a T_a such that $\|(T-T_a)\Gamma^{-1}\|_H \ll 1$ (we require strict inequality, so that the resulting interpolation problem will be strict as well). For the high-order model T one often takes a sufficiently high-order series expansion: with $T = T_1 Z + T_2 Z^2 + \cdots + T_n Z^n$ we could take

$$
A = \begin{bmatrix} 0 & I & & \\ & \ddots & \ddots & \\ & & \ddots & I \\ & & & 0 \end{bmatrix}, \quad B = \begin{bmatrix} T_1^{\langle -1 \rangle} \\ T_2^{\langle -2 \rangle} \\ \vdots \\ T_n^{\langle -n \rangle} \end{bmatrix}, \tag{44}
$$
$$
C = \begin{bmatrix} I & 0 & \cdots & 0 \end{bmatrix}, \quad D = 0
$$

which puts the realization for T directly in output normal form (as we shall assume now, allowing for any such realization – warning: the global block decomposition given in the previous equation consists of diagonal matrices, at each stage k there is an entry in each block).

Hence, we assume $T = C(I - ZA)^{-1}ZB$ in output normal form, i.e., $A^H A + C^H C = I$. The model reduction algorithm consists in executing the following steps:

1. Determine a minimal external factorization for $T = U\Delta^H$ in which U is inner and Δ anti-causal. A semi separable realization for U borrows $\{A, C\}$ from T and determines two new entries B_U and D_U that make U causal unitary: $U = D_U + C(I - ZA)^{-1}ZB_U$ (in the case of the series expansion exemplified above, U is trivially equal to Z^n).
2. Find a causal Θ matrix with $\begin{bmatrix} A & | & B_U & -B\Gamma^{-1} \end{bmatrix}$ as reachability data.
3. A minimal model reduced system is obtained from $T'^H = T^H - \Gamma \Sigma_{12} U^H$ and is given by
$$
T_a = \mathbf{P}_{sc} T' \tag{45}
$$
   in which $\mathbf{P}_{sc}$ projects on the strictly lower (causal) part.

It turns out that the causal part of T' has the same state dimensions as Σ_{22} – and hence will be minimal for the given degree of accuracy. This may seem like hocus-pocus, but it is easy to see why it works. From the formula for T' it is immediately clear that $\|(T - T_a)\Gamma^{-1}\|_H \le \|\mathbf{P}_{sc} U\Sigma_{12}^H\|_H \ll 1$, because $U\Sigma_{12}^H$ is strictly contractive, and its Hankel norm certainly less than 1 as it is the operator norm of a subsystem of an already contractive system. The important point is that T_a meets the state complexity requirement. Consider the following expression:

$$
\Theta \begin{bmatrix} U^H \\ -\Gamma^{-1} T^H \end{bmatrix} := \begin{bmatrix} X \\ -Y \end{bmatrix} \tag{46}
$$

As Θ borrows the reachability data of the right factor in the left member of this equation, it actually pushes it to causality, both the resulting X and Y are causal

operators (one could also say: $\begin{bmatrix} U & -T\Gamma^{-1} \end{bmatrix} = \begin{bmatrix} X^H & -Y^H \end{bmatrix} \Theta$ is a right external factorization with a causal J-unitary factor, similar to the normal right external
factorization with an inner factor). Now, $T' = Y^H \Sigma_{22}^H \Gamma$, and as $T_a = \mathbf{P}_{sc} T'$ and
Y^H is anti-causal, its state space dimensions are at most equal to the state space
dimensions of $\mathbf{P}_{sc} \Sigma_{22}^H$, i.e., of the anti-causal part of Σ_{22}, in turn exactly equal to
the number of negative signs q_k at stage k in the realization of Θ_k.

The Hankel norm model reduction theory shows that the problem reduces
to (is equivalent to) the solution of a Schur-Takagi interpolation problem on the
data given by $\begin{bmatrix} A & | & B_u & -B\Gamma^{-1} \end{bmatrix}$. Unfortunately, the actual computation of the
reduced model turns out to be a bit messy, although straightforward – some matrix
multiplications. Be that as it may, the essential ingredients are obtained through
a square root algorithm just like in the Schur-Takagi interpolation case.

5. H_∞-control for the semi-separable case

Also in this section, we shall show that we can give a solution to the H_∞ control
problem for semi-separable systems, just by using a sequence of two square root
algorithms, very much as is the case for general Moore-Penrose inversion, but now
using J-inner factors rather than the inner factors used there. I follow Kimura's
[16] approach to H_∞ control (Kimura uses Riccati equations and treats the LTI
case). I am indebted to my student Xiaode Yu for working out the main ideas in
her thesis ([21], this work has not been published so far).

The H_∞-control is intended to keep a plant's error of operation measured
as a "sensitivity operator" within a specified worst case limit (the norm of the
operator). It assumes a known (linear) plant model, described by four operators:

$$\begin{bmatrix} P_{11} & P_{12} \\ P_{21} & P_{22} \end{bmatrix} \begin{bmatrix} u \\ w \end{bmatrix} = \begin{bmatrix} z \\ y \end{bmatrix} \tag{47}$$

in which u is the control input, y the input data available to the controller, w
summarizes external disturbances and z describes measurement data available
to gauge the performance of the system. The plant model P describes all the
particular relations between these quantities. We are asked to design a controller
K such that when $u = Ky$ the influence of the disturbances on the measurements
z is a strictly contractive map Φ (we can incorporate a precision measure Γ in
scaling z as desired). This influence shall be well defined if the operator $(I - P_{21}K)$
is boundedly invertible and is given by

$$\Phi = P_{12} + P_{11}K(I - P_{21}K)^{-1} P_{22} \tag{48}$$

and the control condition becomes $\|\Phi\| \ll 1$. To be able to solve the problem
meaningfully, we have to assume that the available gauging measurements are rich
enough, we just require P_{22} to be invertible. In that case, a chain operator G exists,
that connects $\{u, y\}$ to test and measurement data $\{z, w\}$:

$$G = \begin{bmatrix} P_{11} - P_{12}P_{22}^{-1}P_{21} & P_{12}P_{22}^{-1} \\ -P_{22}^{-1}P_{21} & P_{22}^{-1} \end{bmatrix} \tag{49}$$

The Kimura method now prescribes to find a J-inner-outer factorization of $G = \Theta G_o$, in which Θ is J-inner and G_o is causal outer. Following Kimura, we define the Homographic transformation on a chain operator G as

$$\text{HM}(G, K) = (G_{11}K + G_{12})(G_{21}K + G_{22})^{-1} \qquad (50)$$

(in scattering terms that is G loaded by K). Then any control that will guarantee Φ to be contractive is given by $K = \text{HM}(G_o, S_L)$ where S_L is an arbitrary causal contractive operator (e.g., one can choose $S_L = 0$) and then it turns out that $\Phi = \text{HM}(G, K)$. That this works is indeed directly seen from the resulting relation $\Phi = \text{HM}(\Theta, S_L)$, which will automatically be causal contractive, as this property is conserved by a homographic transformation by a J-inner operator. There exists also a dual case, where the assumption is that P_{11} is boundedly invertible, I describe this case briefly at the end of this section. So far, the semi-separable theory is formally precisely equal to the classical (Kimura) theory. We saw already in the previous section that J-inner functions are characteristic for constrained interpolation problems. In the present case, the J-inner function needed is more general than previously considered and involves two types of J-inner operators, namely causal ones and anti-causal ones, much like in the Moore-Penrose inversion problem, but now with J-inner instead of inner (remark that in the case of the Schur-Takagi problem a causal J-unitary function was involved, not a J-inner one, that makes the Model Reduction essentially different from the H_∞ Control problem!).

Hence we assume G semi-separable of mixed causal/anti-causal type and given by

$$G = C_c(I - ZA_c)^{-1}ZB_c + D + C_a(I - Z^H A_a)^{-1}Z^H B_a, \qquad (51)$$

in which we assume the causal (lower) and anti-causal (upper) parts given by their semi-separable realizations. Finding Θ and G_o parallels the strategy we followed for the Moore-Penrose case (Section 2), but now using J-inner functions. However, there is a major difference. In the Moore-Penrose case, we were able to devise a solution procedure that recurses in one direction (which we chose forwards), with an inner function acting on the right to convert the mixed system to anti-causal (upper), followed by an inner-outer factorization on the left of the anti-causal system, progressing downwards in the matrix. This resulted in an URV-type factorization, with inner factors at both sides. However, in the present case, the J-inner factorization must produce a right J-inner factor, meaning that both J-inner factors must act at the same side (in this case left). This should produce first a $G = \Theta_- G_1$ in which Θ_- is both J-inner and anti-causal (a type of object we have not encountered so far) and next, a J-inner-outer decomposition on $G_1 = \Theta_+ G_o$ should yield another, this time causal J-inner factor Θ_+ as in the constrained interpolation case treated before. The final result is then $G = \Theta G_o$ with global $\Theta = \Theta_- \Theta_+$. If both Θ_- and Θ_+ are J-inner, then the total product $\Theta = \Theta_- \Theta_+$ will be J-inner as well – a property that follows directly from the scattering properties of a J-inner function.

The first factorization $G = \Theta_- G_1$ requires an anti-causal, J-inner Θ_- that provides for the anti-causal part in G. To achieve this, it must borrow the observability pair $\{A_a, C_a\}$ of G. We need:

Lemma 1 ([21]). *There exists an anti-causal, u.e.s. J-inner operator Θ_- with observability data $\{A_a, C_a\}$ iff the Lyapunov-Stein equation*

$$Q^{\langle -1 \rangle} = A_a^H Q A_a + C_a^H J C_a \tag{52}$$

has a strictly negative definite solution $Q = -R^H R$. In that case, a realization for Θ_- is obtained by J-unitary completion of the basis $\begin{bmatrix} R A_a R^{-\langle -1 \rangle} \\ C_a R^{-\langle -1 \rangle} \end{bmatrix}$, which is J_1-unitary for $J_1 = \begin{bmatrix} -I & \\ & J \end{bmatrix}$.

This is a forward recursion, based on the observability space of the anti-causal part of G (as before with the Moore-Penrose inversion), which has to be J_1-positive, to be executed by a QR factorization producing the square root R, just as before, but now with a J-unitary factor as explained in the previous section. The Θ-upper factorization (like QR but now with a J-unitary factor as in the case of the constrained interpolation problem) proceeds as follows. Let us define (as also before in the Moore-Penrose case) the transformed quantities $\hat{A}_a := R A_a R^{-\langle -1 \rangle}$ and $\hat{C}_a = C_a R^{-\langle -1 \rangle}$ (also the observability data for Θ_-), then the factorization gives

$$\begin{bmatrix} R_k A_{a,k} & R_k B_{a,k} \\ C_{a,k} & D \end{bmatrix} = \begin{bmatrix} A_{\Theta,k} & B_{\Theta,k} \\ C_{\Theta,k} & D_{\Theta,k} \end{bmatrix} \begin{bmatrix} R_{k+1} & \hat{A}_{a,k}^H - \hat{C}_{a,k} J D \\ 0 & -J B_{\Theta,k}^H \hat{B}_k + J D_{\Theta,k} J D \end{bmatrix}, \tag{53}$$

assuming of course that the existence condition is indeed satisfied, for which the algorithm provides discretion. The expression for the right most factor is found by inverting the Θ-factor and multiplying out, indeed

$$\begin{bmatrix} A_{\Theta,k} & B_{\Theta,k} \\ C_{\Theta,k} & D_{\Theta,k} \end{bmatrix}^{-1} = \begin{bmatrix} A_{\Theta,k}^H & -C_{\Theta,k}^H \\ -B_{\Theta,k}^H & D_{\Theta,k}^H \end{bmatrix}$$

because of the J_1-unitarity. One also computes the realization for $G_1 = \Theta_-^{-1} G$ directly and one obtains as realization

$$G_1 \approx \begin{bmatrix} A_{11} & A_{12} & B_1 \\ A_{21} & A_{22} & B_2 \\ C_1 & C_2 & D_n \end{bmatrix} := \begin{bmatrix} \hat{A}_a^H & \hat{C}_a^H J C_c & -\hat{A}_a^H B_a + \hat{C}_a^H J D \\ 0 & A_c & B_c \\ J B_{\Theta}^H & J D_{\Theta}^H J C_c & -J B_{\Theta}^H \hat{B}_a + J D_{\Theta}^H J D \end{bmatrix} \tag{54}$$

where G_1 is now causal and its dimensionality has picked up that of the anticausal part of G. Reverting to the Θ-upper factorization, we see that it contains all the necessary new data:

$$\begin{bmatrix} R_k A_{a,k} & R_k B_{a,k} \\ C_{a,k} & D \end{bmatrix} = \begin{bmatrix} \hat{A}_{a,k} & B_{\Theta,k} \\ \hat{C}_{a,k} & D_{\Theta,k} \end{bmatrix} \begin{bmatrix} R_{k+1} & -B_{1,k} \\ 0 & D_{n,k} \end{bmatrix}. \tag{55}$$

The next step is the J-inner-outer decomposition of G_1 with a left J-inner factor. It turns out to produce a backward recursion. It is now based, not on observability or reachability spaces of G_1 but on its zero space as a restricted causal operator (this is actually the reachability space of the anticausal part of the inverse operator). This is the crucial step that results in a Riccati equation in the classical theory, but from the previous case we can already guess that, again, a square root algorithm will suffice, as is true for any "inner-outer" type equation – the fact that we need a J-inner factor only puts an extra positivity condition on the data, making the case equivalent to a constrained interpolation problem of the type treated in the previous section.

We need $G_1 = \Theta_+ G_o$ in which Θ_+ has to be causal J-inner (as in the section on constrained interpolation), and G_o outer. The square-root algorithm that produces the factorization is a $\Theta - L$-factorization, which is easily derived from the state space data for G_1, by working out $G_o = \Theta_+^{-1} G_1$ and the fact that $\Theta_+^{-1} = J\Theta_+^H J$ from the J-unitarity of Θ_+ and it produces, just as in the Moore-Penrose section, mutatis mutandis:

$$\begin{bmatrix} Y_k A_k & Y_k B_k \\ C_k & D_{n,k} \end{bmatrix} = \begin{bmatrix} A_{\Theta_+,k} & B_{\Theta_+,k} \\ C_{\Theta_+,k} & D_{\Theta_+,k} \end{bmatrix} \begin{bmatrix} Y_{k-1} & 0 \\ C_{o,k} & D_{o,k} \end{bmatrix}. \tag{56}$$

The square root recursion involves the intermediate recursive quantity Y_k. The problem of obtaining the J-inner Θ_+ and outer G_o will have a solution iff all the subsequent $\begin{bmatrix} A_{\Theta_+,k} \\ C_{\Theta_+,k} \end{bmatrix}$ that result from the algorithm (which is executed from right to left at each step) turn out to be J_1-positive, with $J_1 = \begin{bmatrix} I & \\ & J \end{bmatrix}$. This is equivalent to the condition that the backward recursive algebraic Riccati equation

$$\begin{aligned} M_{k-1} = {} & A_k^H M_k A_k + C_k^H J C_k \\ & - (A_k^H M_k B_k + C_k^H J D_{n,k})(B_k^H M_k B_k + D_{n,k}^H J D_{n,k})^{-1} \\ & \times (B_k^H M_k A_k + D_{n,k}^H J C_k) \end{aligned} \tag{57}$$

has a positive definite solution M_k for all k. We then have $M_k = Y_k^H Y_k$ and Y_k solves the square root equation with a J-inner Θ. Evidently, the condition on the square root equation is much easier to verify algorithmically than the positivity condition on all M_k, although at first glance it may seem more complicated (although it is not!) Again, we find that the square root algorithm does a much better job with much improved numerical conditioning.

In the beginning of this section we assumed the plant, disturbance and measurement data to be such that P_{22} is invertible. There is a dual approach possible, also proposed by Kimura, whereby it is assumed that P_{11} is invertible instead. In that case, another chain operator exists, namely the one that connects test and measurement data $\{z, w\}$ to the plant input and output data $\{u, y\}$. That chain operator than has the expression

$$G_d = \begin{bmatrix} P_{11}^{-1} & -P_{11}^{-1}P_{12} \\ P_{21}P_{11}^{-1} & P_{22} - P_{21}P_{11}^{-1}P_{12} \end{bmatrix}. \qquad (58)$$

The procedure then follows the same steps as before in a dual way, now an outer-J-inner factorization is sought of the type $G_d = G_o\Theta_+\Theta_-$ resulting again in two square-root equations of the same type as before, but now with all orders reversed.

6. Appendix: QR/QL/LQ/RQ factorization

Let $u = \begin{bmatrix} u_1 \\ u_2 \end{bmatrix}$ be a unit norm vector with first component u_1, the remaining of the vector being denoted as u_2 (in the real case one usually chooses u_1 positive). Then

$$\rho_u := \begin{bmatrix} u_1^* & u_2^H \\ -u_2\frac{u_1^*}{|u_1|} & I - \frac{1}{1+|u_1|}u_2u_2^H \end{bmatrix} \qquad (59)$$

is a unitary (rotation) matrix that rotates u to the first natural basis vector: $\rho_u u = \begin{bmatrix} 1 \\ 0 \end{bmatrix}$ (if $u_1 = 0$ one puts $u_1^*/|u_1| = 1$. Numerical analysts traditionally use so-called Householder reflections to achieve the same feat. However, this is not advisable as the Householder transformation unpleasantly changes the sign of the determinant in the real case. The transformation given here is just as computationally efficient as the Householder transformation and has no negative side effect). Suppose that now, somewhere in an algorithm, a non-zero (sub-)vector $a = \begin{bmatrix} a_1 \\ a_2 \end{bmatrix}$ with first component a_1 has to be rotated with a unitary transformation so that only its first position is non-zero and equal to $|a|$. Then one can just apply $\rho_{a/|a|}$ to annihilate all but the top position. This procedure can be applied in subsequent steps, starting in the upper left corner and moving from column to column, gradually decreasing the size of the vectors to avoid creating fill ins. Let, e.g., $A = \begin{bmatrix} a_1 & a_2 & \cdots & a_n \end{bmatrix}$ be a matrix consisting of the columns $a_i : i = 1, \ldots, n$. Then either $a_1 = 0$ and the algorithm skips the first column to move to the next, or it is non-zero, and $\rho_{a_1/|a_1|}$ is applied, transforming a_1 to $\begin{bmatrix} |a_1| \\ 0 \end{bmatrix}$ and, of course, transforming the other columns to $\rho_{a_1/|a_1|}a_k, k = 2, \ldots, n$ as well. Then the algorithm moves to the second column. If the first was zero, it will apply the original procedure again, but now on the second column, which may also turn out be zero, in which case the algorithm moves directly to the third column. Otherwise, the algorithm deflates with one unit, leaves the first row henceforth unchanged, and starts applying the procedure on the submatrix of column vectors with positions starting at 2. Transformations applied to these vectors will leave the crossed out rows unchanged. The result of these transformations is a so-called *echelon matrix*, which looks as follows:

$$\begin{bmatrix} 0 & X & * & * & * & \cdots & * \\ 0 & 0 & 0 & X & * & \cdots & * \\ 0 & 0 & 0 & 0 & X & \cdots & * \\ 0 & & & \cdots & & & 0 \end{bmatrix}. \tag{60}$$

Here the 'X' indicates entries that are strictly positive. Writing $\rho_{i,j}$ for a rotation matrix affecting the entries $i,\ldots,j$, we see that in this case the total rotation matrix will be a composition of the type $\rho = \rho_{3,m}\rho_{2,m}\rho_{1,m}$. Taking $Q = \rho^H$ and writing R for the echelon matrix, we get $A = QR$. The non-zero rows of R form a basis for the rows of A, of a special form. Splitting $Q = \begin{bmatrix} Q_1 & Q_2 \end{bmatrix}$ accordingly, we see that the columns of Q_2 forms a basis for the co-kernel of A, while the columns of Q_1 form a basis for the range.

Instead of starting the algorithm in the top left corner and working in the South-East direction, one could just as well start in the bottom right corner and work in the North-West direction. This would then produce a QL-factorization, where the L-factor again contains a basis for the rows of A, but now favoring the last entries.

A different result is obtained when one compresses rows rather than columns. This would produce an LQ-factorization (when one starts with the first row and compresses to the left) or an RQ-factorization (starting on the last row and concentrating to the right). Rotations are now applied to the right side of the matrix, compressing entries in the South-West corner as in:

$$\begin{bmatrix} 0 & 0 & \cdots & 0 & 0 & \cdots & 0 \\ X & 0 & \cdots & 0 & 0 & \cdots & 0 \\ * & 0 & \cdots & 0 & 0 & \cdots & 0 \\ * & X & \cdots & 0 & 0 & \cdots & 0 \\ \vdots & \vdots & & \vdots & \vdots & & \vdots \\ * & * & \cdots & X & 0 & \cdots & 0 \\ * & * & \cdots & * & 0 & \cdots & 0 \end{bmatrix} \tag{61}$$

and in the RQ case the echelon form would look like

$$\begin{bmatrix} 0 & \cdots & 0 & * & * & * \\ 0 & \cdots & 0 & X & * & * \\ 0 & \cdots & 0 & 0 & X & * \\ 0 & \cdots & 0 & 0 & 0 & * \\ \vdots & & \vdots & \vdots & \vdots & \vdots \\ 0 & \cdots & 0 & 0 & 0 & X \\ 0 & \cdots & 0 & 0 & 0 & 0 \end{bmatrix}. \tag{62}$$

With $A = RQ$, the columns of R now form a basis for the columns of A. Splitting $Q = \begin{bmatrix} Q_1 \\ Q_2 \end{bmatrix}$, we see that the columns of Q_1^H form a basis for the kernel of A in this last case.

The numerical accuracy of these procedures is "good", in the sense that they are backward stable (meaning that the accumulation of numerical errors can be mapped back to an error matrix on the original data that is in magnitude of the same order). When the original matrix has small singular values (its conditioning is poor), then the procedures may lead to erratic results in the determination of the rank of the matrix given a certain tolerance. In that case a singular value decomposition (SVD) is advisable, in which a decision can be made about which singular values to neglect. The QR or other form can then be recovered if necessary.

References

[1] D. Alpay and P. Dewilde. Time-varying signal approximation and estimation. In A.C.M. Ran M.A. Kaashoek, J.H. van Schuppen, editor, *Signal Processing, Scattering and Operator Theory, and Numerical Methods, vol. III*, pages 1–22, 1990.

[2] D. Alpay, P. Dewilde, and H. Dym. Lossless Inverse Scattering and reproducing kernels for upper triangular operators. In I. Gohberg, editor, *Extension and Interpolation of Linear Operators and Matrix Functions*, volume 47 of *Operator Theory, Advances and Applications*, pages 61–135. Birkhäuser Verlag, 1990.

[3] W. Arveson. Interpolation problems in nest algebras. *J. Functional Anal.*, 20:208–233, 1975.

[4] P. Dewilde and A.-J. van der Veen. *Time-varying Systems and Computations*. Kluwer, 1998.

[5] P. Dewilde and A.-J. van der Veen. Inner-outer factorization and the inversion of locally finite systems of equations. *Linear Algebra and its Applications*, 313:53–100, 2000.

[6] P.M. Dewilde. A course on the algebraic Schur and Nevanlinna-Pick interpolation problems. In Ed. F. Deprettere and A.J. van der Veen, editors, *Algorithms and Parallel VLSI Architectures*, volume A, pages 13–69. Elsevier, 1991.

[7] P.M. Dewilde and A.J. van der Veen. On the Hankel-norm approximation of upper-triangular operators and matrices. *Integral Eq. Operator Th.*, 17(1):1–45, 1993.

[8] P.M. Dewilde and A.J. van der Veen. On the Hankel-norm approximation of upper-triangular operators and matrices. *Integral Eq. Operator Th.*, 17(1):1–45, 1993.

[9] I. Gohberg, T. Kailath, and I. Koltracht. Linear complexity algorithms for semiseparable matrices. *Integral Equations and Operator Theory*, 8:780–804, 1985.

[10] J. William Helton and Matthew R. James. *Extending H^∞ Control to Nonlinear Systems*. SIAM, 1999.

[11] J. William Helton and Orlando Merino. *Classical Control using H^∞ Mehtods*. SIAM, 1998.

[12] J.W. Helton. The characteristic functions of operator theory and electrical network realization. *Indiana Univ. Math. J.*, 22(5):403–414, 1972.

[13] J.W. Helton. Passive network realization using abstract operator theory. *IEEE Trans. Circuit Th.*, 1972.

[14] P.P. Khargonekar J.C. Doyle, K. Glover and B.A. Francis. State space solutions to standard H_2 and H_∞ control problems. *IEEE Trans. on Automatic Control*, 34:831–846, 1989.

[15] T. Kailath. *Lectures on Wiener and Kalman Filtering*. Springer Verlag, CISM Courses and Lectures No. 140, Wien, New York, 1981.

[16] H. Kimura. *Chain-Scattering Approach to H^∞-Control*. Birkhäuser, Boston, 1997.

[17] M. Van Barel R. Vandebril and N. Mastronardi. *Matrix Computations and Semiseparable Matrices*. The John Hopkins University Press, 2008.

[18] J.K. Rice and M. Verhaegen. Distributed control: A sequentially semi-separable approach for spatially heterogeneous linear systems. *IEEE Transactions on Automatic Control*, 54:1270–1284, 2009.

[19] S. Chandrasekaran, P. Dewilde, M. Gu, T. Pals, A.-J. van der Veen and J. Xia. *A fast backward stable solver for sequentially semi-separable matrices*, volume HiPC202 of *Lecture Notes in Computer Science*, pages 545–554. Springer Verlag, Berlin, 2002.

[20] D.C. Youla, H.A. Jabr, and J.J. Bongiorno. Modern Wiener-Hopf design of optimal controllers: Part II. *IEEE Trans. Automat. Contr.*, 21:319, 1976.

[21] Xiaode Yu. *Time-varying System Identification, J-lossless Factorization, and H_∞ Control*. PhD thesis, Delft Univ. of Technology, Delft, The Netherlands, May 1996.

[22] G. Zames. Feedback and optimal sensitivity: Model reference transformations, multiplicative seminorms, and approximate inverses. *IEEE Trans. on Automatic Control*, AC-26:301–320, 1981.

Patrick Dewilde
TU München
Institute for Advanced Study
e-mail: `p.dewilde@me.com`

Operator Theory:
Advances and Applications, Vol. 222, 113–128
© 2012 Springer Basel

Spectral Inclusion Theorems

Ronald G. Douglas and Jörg Eschmeier

Dedicated to Professor J. William Helton on the occasion of his 65th birthday

Abstract. In this note we use Gelfand theory to show that the validity of a spectral mapping theorem for a given representation $\Phi : \mathcal{M} \to L(X)$ of a Banach function algebra $\mathcal{M}$ on a bounded open set in $\mathbb{C}^n$ implies the validity of one-sided spectral mapping theorems for all subspectra. In particular, a spectral mapping theorem for the Taylor spectrum yields at least a one-sided spectral mapping theorem for the essential Taylor spectrum. In our main example $\mathcal{M}$ is the multiplier algebra of a Banach space X of analytic functions and Φ is the canonical representation of $\mathcal{M}$ on X. In this case, we show that interpolating sequences for $\mathcal{M}$, or suitably defined Berezin transforms, can sometimes be used to obtain the missing inclusion for the essential Taylor spectrum or its parts.

Mathematics Subject Classification. Primary 47A10, 47A13;
secondary 47A60, 47B32.

Keywords. Spectral inclusion, functional calculus, Banach spaces of analytic functions, Taylor spectrum, semi-Fredholmness.

1. Introduction

Originally, operator theory arose from the study of integral and differential equations. As a consequence, and due in part to the relation with matrix theory, the basic question was that of invertibility and hence determining the spectrum of an operator. Subsequently, one considered how to construct operators out of other operators and, in particular, obtaining operators as functions of simpler operators.

The first step in this direction came from algebra and represented one operator as a polynomial in another one. Given the focus on the spectrum and the

---

Research on this paper was begun during a visit of the first-named author at the University of the Saarland.

algebraic context, one easily proved that $\sigma(p(T)) = p(\sigma(T))$. Next, F. Riesz defined a functional calculus for an operator T for analytic functions f with a domain sufficiently large containing $\sigma(T)$ and obtained the same result. Other more refined functional calculi were introduced such as that of Sz.-Nagy and Foiaş [16] for a bounded function ϕ of a contraction operator T, in which only an inclusion in one direction holds and the meaning of $\phi(\sigma(T))$ requires interpretation [10]. This note continues this topic for n-tuples of operators in which the Taylor spectrum is used.

The basic operators considered are multiplication operators on Banach or Hilbert spaces of holomorphic functions defined on bounded domains in $\mathbb{C}^n$. Examples are the Hardy and Bergman spaces on the unit ball or polydisk or the Drury-Arveson space. The basic data are the functions defined on the domain and the corresponding Koszul complex defined by them. Following the development of Taylor [17], one relates the operator-valued Koszul complex to that of the function-valued one. Gelfand theory is used to show that the validity of a spectral mapping formula for a representation $\Phi : \mathcal{M} \to L(X)$ of a Banach algebra $\mathcal{M}$ that consists of complex-valued functions on a bounded open set Ω in $\mathbb{C}^n$ and contains the polynomials implies the validity of one-sided spectral mapping theorems for all subspectra. Thus, in particular, a spectral mapping formula for the Taylor spectrum or its parts implies at least a one-sided spectral mapping formula for the corresponding parts of the essential Taylor spectrum. In the case where $\mathcal{M}$ is the multiplier algebra of a Banach space X of analytic functions and Φ is the representation associating with each multiplier the induced multiplication operator on X, we show how the existence of sufficiently many interpolating sequences for $\mathcal{M}$, or a suitably defined Berezin transform, can be used to obtain the missing inclusion for the essential Taylor spectrum. In the final part of the paper we give applications to concrete functional Banach and Hilbert spaces and their multiplier algebras.

For the definitions, and the necessary background on the Taylor spectrum, the essential Taylor spectrum and its parts, we refer the reader to the monograph [9].

2. Gelfand theory

Let $\mathcal{M}$ be a unital Banach algebra of complex-valued functions on a bounded open subset Ω of $\mathbb{C}^n$ such that the coordinate functions z_i ($1 \le i \le n$) belong to $\mathcal{M}$ and let X be a complex Banach space. Consider a unital algebra homomorphism $\Phi : \mathcal{M} \to L(X)$. Then $B = \overline{\Phi(\mathcal{M})} \subset L(X)$ is a unital commutative Banach subalgebra. Denote by $c(B)$ the set of all tuples $a = (a_1, \dots, a_r) \in B^r$ of arbitrary finite length r, and write $c(\mathcal{M})$ for the set of all finite tuples in $\mathcal{M}$. By a spectral system on B we mean a rule which assigns to each element $a = (a_1, \dots, a_r) \in c(B)$ a closed subset $\sigma(a) \subset \mathbb{C}^r$ in such a way that for all $a, b \in c(B)$

1. $p\,\sigma(a, b) = \sigma(a),\ q\,\sigma(a, b) = \sigma(b)$,
2. $\sigma(a) \subset \sigma_B(a)$.

Here, for given $a = (a_1, \ldots, a_r), b = (b_1, \ldots, b_s) \in c(B)$, we denote by $p : \mathbb{C}^{r+s} \to \mathbb{C}^r$ and $q : \mathbb{C}^{r+s} \to \mathbb{C}^s$ the projections of $\mathbb{C}^{r+s}$ onto its first r and last s coordinates, respectively, and

$$\sigma_B(a) = \left\{ z \in \mathbb{C}^r; \ 1 \notin \sum_{i=1}^{r} (z_i - a_i)B \right\}$$

is the usual joint spectrum of a in the commutative Banach algebra B.

Standard results going back to J.L. Taylor (see, e.g., Proposition 2.6.1 in [9]), show that for any spectral system σ on B, the set

$$\Delta_\sigma = \{\lambda \in \Delta_\mathcal{M}; \ \widehat{f}(\lambda) \in \sigma(\Phi(f)) \text{ for all } f \in c(\mathcal{M})\}$$

is the unique closed subset of the maximal ideal space $\Delta_\mathcal{M}$ of $\mathcal{M}$ with the property that $\widehat{f}(\Delta_\sigma) = \sigma(\Phi(f))$ for all $f \in c(\mathcal{M})$. Here $\Phi(f) = (\Phi(f_1), \ldots, \Phi(f_r))$ and $\widehat{f} = (\widehat{f}_1, \ldots, \widehat{f}_r) : \Delta_\mathcal{M} \to \mathbb{C}^r$ is the tuple consisting of the Gelfand transforms $\widehat{f}_i$ of f_i. In the following we shall denote by $\mathcal{U}(\zeta)$ the system of all open neighbourhoods of a given point ζ in $\mathbb{C}^n$, and we write $\pi = (z_1, \ldots, z_n) \in c(\mathcal{M})$ for the tuple consisting of the coordinate functions.

Theorem 1. *Let σ be a spectral system on B and let $\Delta = \Delta_\sigma \subset \Delta_\mathcal{M}$ be the closed subset of the maximal ideal space of $\mathcal{M}$ associated with σ. Define $a = (a_1, \ldots, a_n) = (\Phi(z_1), \ldots, \Phi(z_n))$. If Φ satisfies*

(1) $\sigma(\Phi(f)) = \bigcap \left(\overline{f(U \cap \Omega)}; \ U \supset \sigma(a) \text{ open} \right) \quad (f \in c(\mathcal{M}))$,

then Φ satisfies

(2) $\widehat{f}(\lambda) \in \bigcap \left(\overline{f(U \cap \Omega)}; \ U \in \mathcal{U}(\widehat{\pi}(\lambda)) \right) \quad (\lambda \in \Delta, f \in c(\mathcal{M}))$.

Conversely, if Φ satisfies (2), then at least the inclusions

(3) $\sigma(\Phi(f)) \subset \bigcap \left(\overline{f(U \cap \Omega)}; \ U \supset \sigma(a) \text{ open} \right) \quad (f \in c(\mathcal{M}))$

hold.

Proof. Suppose that condition (1) holds. Note that the set Δ_0, consisting of all $\lambda \in \Delta$ which satisfy condition (2) for every tuple $f \in c(\mathcal{M})$, is a closed subset of Δ. Indeed, if $(\lambda_i)_{i \in I}$ is a net in Δ_0 which converges to a functional $\lambda \in \Delta$ and U is an open neighbourhood of $\widehat{\pi}(\lambda)$, then there is an index $i_0 \in I$ such that $U \in \mathcal{U}(\widehat{\pi}(\lambda_i))$ for all $i \geq i_0$. But then it follows that

$$\widehat{f}(\lambda) = \lim_i \widehat{f}(\lambda_i) \in \overline{f(U \cap \Omega)}$$

for all $f \in c(\mathcal{M})$.

To prove (2) it suffices to check that

$$\sigma(\Phi(f)) \subset \widehat{f}(\Delta_0)$$

for every tuple $f \in c(\mathcal{M})$. To this end, let us fix a tuple $f \in c(\mathcal{M})$ and a point $w \in \sigma(\Phi(f))$. Condition (1) implies that there is a sequence of points $z_k \in \Omega$ with $\mathrm{dist}(z_k, \sigma(a)) < \frac{1}{k}$ and $\|w - f(z_k)\| < \frac{1}{k}$ for all $k \geq 1$. By passing to a subsequence we may suppose that the limit $z = \lim_{k \to \infty} z_k \in \sigma(a)$ exists. Since the maximal ideal space $\Delta_{\mathcal{M}}$ of $\mathcal{M}$ is compact, we can choose a subnet $(z_{k_i})_{i \in I}$ of the sequence $(z_k)_{k \geq 1}$ such that $\lim_i z_{k_i} = \chi$ in $\Delta_{\mathcal{M}}$ for a suitable character χ on $\mathcal{M}$. Because

$$w = \lim_i f(z_{k_i}) = \chi(f)$$

it suffices to show that $\chi \in \Delta_0$. Note that

$$\widehat{\pi}(\chi) = \lim_i z_{k_i} = z \in \sigma(a).$$

Hence for $g \in c(\mathcal{M})$ and every open neighbourhood U of $\widehat{\pi}(\chi)$, we have that

$$\widehat{g}(\chi) = \lim_i g(z_{k_i}) \in \overline{g(U \cap \Omega)}.$$

Using condition (1) we find in particular that $\widehat{g}(\chi) \in \sigma(\Phi(g))$ for all $g \in c(\mathcal{M})$. Therefore $\chi \in \Delta_0$ and the proof of the implication (1) $\Rightarrow$ (2) is complete.

Suppose that condition (2) holds and fix a tuple $f \in c(\mathcal{M})$. Since $\sigma(\Phi(f)) = \widehat{f}(\Delta)$, we find that

$$\sigma(\Phi(f)) \subset \bigcup_{\lambda \in \Delta} \bigcap \left(\overline{f(U \cap \Omega)}; \; U \in \mathcal{U}(\widehat{\pi}(\lambda)) \right) \subset \bigcap \left(\overline{f(U \cap \Omega)}; \; U \supset \sigma(a) \text{ open} \right).$$

Hence condition (3) holds. $\qquad\qquad\qquad\qquad\qquad\qquad\qquad\qquad\qquad\qquad\square$

Remark 2. (a) For $z \in \Omega$, let $\delta_z : \Omega \to \mathbb{C}, f \mapsto f(z)$, be the point evaluation at z. We shall say that $\mathcal{M}$ satisfies the corona property if $\{\delta_z; \; z \in \Omega\} \subset \Delta_{\mathcal{M}}$ is dense, or equivalently, if

$$\sum_{i=1}^{r} f_i \mathcal{M} = \mathcal{M}$$

for every tuple $f = (f_1, \ldots, f_r) \in c(\mathcal{M})$ with $0 \notin \overline{f(\Omega)}$. It is well known that in the setting of Theorem 1 the validity of the corona property for $\mathcal{M}$ implies the spectral inclusion

$$\sigma(\Phi(f)) \subset \bigcap \left(\overline{f(U \cap \Omega)}; \; U \supset \sigma(a) \text{ open} \right)$$

for every spectral system σ on $B = \overline{\Phi(\mathcal{M})}$ and every tuple $f \in c(\mathcal{M})$ (Theorem 2.6 in [15]). To recall the elementary argument, fix a point $w \in \sigma(\Phi(f))$ and a character $\lambda \in \Delta_\sigma$ with $w = \widehat{f}(\lambda)$. The corona property allows us to choose a net (z_i) in Ω such that $\lambda = \lim_i \delta_{z_i}$. Then $\lim_i z_i = \widehat{\pi}(\lambda) \in \sigma(a)$ and, for every open set $U \supset \sigma(a)$, we find that

$$w = \lim_i f(z_i) \in \overline{f(U \cap \Omega)}.$$

(b) On the other hand, even if $\mathcal{M}$ satisfies the corona property, the reverse inclusion need not be true. Examples in [10] show that there are completely non-unitary contractions T on a Hilbert space H and functions $f \in H^\infty(\mathbb{D})$ such that

$$\sigma(\Phi(f)) = \{0\} \subset \overline{\mathbb{D}} = \bigcap \left(\overline{f(U \cap \mathbb{D})}; \; U \supset \sigma(T) \text{ open} \right),$$

where $\sigma(T)$ is the usual spectrum of T and $\Phi : H^\infty(\mathbb{D}) \to L(H)$ denotes the H^∞-functional calculus of T.

Since condition (2) in Theorem 1 is preserved when passing from a given spectral system σ to a smaller one, the validity of a spectral mapping theorem for σ yields at least a one-sided spectral mapping theorem for all subspectra.

Corollary 3. *Let σ and σ_0 be spectral systems on $B = \overline{\Phi(\mathcal{M})}$ such that the inclusion $\sigma_0(b) \subset \sigma(b)$ holds for every tuple $b \in c(B)$. Suppose that $a = (\Phi(z_1), \ldots, \Phi(z_n))$ satisfies the spectral mapping formula*

$$\sigma(\Phi(f)) = \bigcap \left(\overline{f(U \cap \Omega)};\ U \supset \sigma(a) \text{ open} \right) \quad (f \in c(\mathcal{M})).$$

Then σ_0 has the property that

$$\sigma_0(\Phi(f)) \subset \bigcap \left(\overline{f(U \cap \Omega)};\ U \supset \sigma_0(a) \text{ open} \right) \quad (f \in c(\mathcal{M})).$$

Proof. Denote by $\Delta = \Delta_\sigma$ and $\Delta_0 = \Delta_{\sigma_0}$ the subsets of $\Delta_\mathcal{M}$ carrying σ and σ_0. Then $\Delta_0 \subset \Delta$ and hence condition (2) in Theorem 1 is preserved when passing from Δ to Δ_0. But then Theorem 1 shows that σ_0 satisfies at least the spectral inclusion formula (3). $\qquad\qquad \square$

In the next section, we shall apply the above results to the case where $\mathcal{M}$ is the multiplier algebra of a given Banach space of functions and Φ is the representation associating with each multiplier the induced multiplication operator.

3. Interpolating sequences and Berezin transforms

A functional Banach (Hilbert) space on an arbitrary set Ω is a Banach (Hilbert) space X of complex-valued functions on Ω such that all point evaluations $\delta_z : X \to \mathbb{C}, f \mapsto f(z)$ $(z \in \Omega)$, are continuous. Suppose that X is a functional Banach space on Ω with $1 \in X$. Let $\mathcal{M}(X) = \{h : \Omega \to \mathbb{C}; hX \subset X\}$ be the multiplier space of X. By the closed graph theorem, for every $h \in \mathcal{M}(X)$, the induced multiplication operator $M_h : X \to X, f \mapsto hf$, is continuous. An elementary and well-known argument shows that the estimates

$$\|h\|_{\infty,\Omega} \le r(M_h) \le \|M_h\|$$

hold for every multiplier $h \in \mathcal{M}(X)$. Here $r(M_h)$ denotes the spectral radius of M_h. Indeed, $M_h'(\delta_z) = h(z)\delta_z$ for all $z \in \Omega$. Hence $|h(z)| \le \|M_h\|$ for all $z \in \Omega$ and

$$\|h\|_{\infty,\Omega} \le \inf_{n\ge 1} \|M_{h^n}\|^{\frac{1}{n}} = r(M_h).$$

The induced algebra homomorphism $\Phi : \mathcal{M}(X) \to L(X), h \mapsto M_h$, is injective with WOT-closed range. It is not hard to see that the range of Φ is even a reflexive subalgebra of $L(X)$. Indeed, if $C \in L(X)$ leaves invariant every closed subspace of X which is invariant under ran Φ, then

$$0 = \delta_z(C(f - f(z))) = (Cf - f(z)C1)(z)$$

for $f \in X$ and $z \in \Omega$. Hence $Cf = (C1)f$ for all $f \in X$.

In particular, $\mathcal{M}(X)$ becomes a unital commutative Banach algebra with respect to the multiplier norm $\|h\|_{\mathcal{M}} = \|M_h\|$. Let Ω be a bounded open subset of $\mathbb{C}^n$. We shall say that X satisfies the ℓ^∞-interpolation property if each sequence (z_k) in Ω converging to a boundary point $z \in \Omega$ contains a subsequence (w_k) with the property that

$$\{(h(w_k)); h \in \mathcal{M}(X)\} = \ell^\infty.$$

For a tuple $f = (f_1, \ldots, f_m) \in \mathcal{M}(X)^m$ of multipliers on X, we denote by $M_f = (M_{f_1}, \ldots, M_{f_m}) \in L(X)^m$ the induced tuple of multiplication operators on X.

Theorem 4. *Let X be a functional Banach space on a bounded open set $\Omega \subset \mathbb{C}^n$ with $1 \in X$ such that X satisfies the ℓ^∞-interpolation property. Then for $m \geq 1$ and $f \in \mathcal{M}(X)^m$, the spectral inclusion*

$$\bigcap \overline{(f(U \cap \Omega)}; U \supset \partial\Omega \text{ open}) \subset \sigma_{re}(M_f)$$

holds.

Proof. An elementary argument shows that the set on the left-hand side consists precisely of all points $w \in \mathbb{C}^m$ for which there is a sequence $(z_k)_{k \geq 1}$ in Ω converging to a boundary point $\lambda \in \partial\Omega$ such that $w = \lim_{k \to \infty} f(z_k)$. Fix such a point w. Clearly we may suppose that $w = 0$. By passing to a subsequence, we can achieve that

$$r : \mathcal{M}(X) \longrightarrow \ell^\infty, g \longmapsto (g(z_k))_{k \geq 1}$$

is a surjective continuous linear operator between Banach spaces. Note that the continuity of r follows from the closed graph theorem. By the open mapping principle there is a zero sequence $(f_N)_{N \geq 1}$ in $\mathcal{M}(X)^m$ such that

$$f_N(z_k) = \begin{cases} 0 & ; \ k < N \\ f(z_k) & ; \ k \geq N \end{cases}$$

for every $N \geq 1$. Then the sequence $(g_N)_{N \geq 1} = (f - f_N)_{N \geq 1}$ converges to f in $\mathcal{M}(X)^m$. Denote by $\delta_k : X \to \mathbb{C}$, $h \mapsto h(z_k)$, the evaluation maps at the points z_k. Since $\mathcal{M}(X) \subset X$ and $(z_k)_{k \geq 1}$ is interpolating for $\mathcal{M}(X)$, the family $(\delta_k)_{k \geq 1}$ is linearly independent. But then the observation that for each $N \geq 1$ the composition

$$X^m \xrightarrow{M_{g_N}} X \xrightarrow{\delta_k} \mathbb{C}$$

is zero for $k \geq N$, implies that the quotients $X/M_{g_N}X^m$ are infinite dimensional for every $N \geq 1$. Hence $0 \in \sigma_{re}(M_{g_N})$ for every $N \geq 1$ and hence $0 \in \sigma_{re}(M_f)$. $\square$

In Section 4 we shall give a few applications of Theorem 4. But before this, we indicate an alternative method which works in the case of functional Hilbert spaces and uses generalized Berezin transforms instead of interpolating sequences.

Let H be a functional Hilbert space on an arbitrary set Ω. Then for each point $w \in \Omega$, there is a unique function $K_w \in H$ such that

$$\langle f, K_w \rangle = f(w) \quad (f \in H).$$

The resulting map $K : \Omega \times \Omega \to \mathbb{C}, K(z, w) = K_w(z)$, is called the reproducing kernel of H. In the following we denote by $k_w = K_w / \|K_w\|$ the normalized reproducing kernel vectors.

The next lemma will show that, in the case of functional Hilbert spaces, the Berezin method is more general than the method using interpolating sequences.

Lemma 5. *Let H be a functional Hilbert space on a bounded open set $\Omega \subset \mathbb{C}^n$ with reproducing kernel $K : \Omega \times \Omega \to \mathbb{C}$. Suppose that $1 \in H$ and that H has the ℓ^∞-interpolation property. Then weak-$\lim_{z \to \partial\Omega} k_z = 0$ and $\lim_{z \to \partial\Omega} \|K_z\| = \infty$.*

Proof. Because of $1/\|K_z\| = \langle 1, k_z \rangle$ it suffices to prove the first assertion. Arguing by contradiction, let us assume that there are a weak zero neighbourhood $U \subset H$ and a sequence (z_j) in Ω with $\text{dist}(z_j, \partial\Omega) \xrightarrow{j} 0$, but $k_{z_j} \notin U$ for all j. By passing to a subsequence we can achieve that in addition (z_j) is interpolating for $\mathcal{M}(H)$. By a result going back to Marshall and Sundberg (see, e.g., Theorem 9.19 in [1]), it follows that

$$\{(\langle f, k_{z_j} \rangle)_{j \geq 0}; f \in H\} = \ell^2.$$

Hence (k_{z_j}) is a weak zero sequence. This contradiction completes the proof. $\square$

We define the generalized Berezin transform in a slightly more general setting (cf. [6]). Let $\mathcal{E}$ be an arbitrary (possibly infinite-dimensional) Hilbert space. By an $\mathcal{E}$-valued functional Hilbert space on a given set Ω we mean a Hilbert space $H(\mathcal{E})$ of $\mathcal{E}$-valued functions on Ω such that all point evaluations

$$\delta_z : H(\mathcal{E}) \to \mathcal{E}, f \mapsto f(z) \quad (z \in \Omega)$$

are continuous. The positive definite map $K : \Omega \times \Omega \to L(\mathcal{E}), K(z, w) = \delta_z \delta_w^*$, called the reproducing kernel of $H(\mathcal{E})$, is the unique function such that $K(\cdot, z)x \in H(\mathcal{E})$ and

$$\langle f, K(\cdot, z)x \rangle = \langle f(z), x \rangle$$

holds for all $f \in H(\mathcal{E}), z \in \Omega$ and $x \in \mathcal{E}$. Conversely, for every positive definite function $K : \Omega \times \Omega \to L(\mathcal{E})$, there is a unique $\mathcal{E}$-valued functional Hilbert space $H(\mathcal{E})$ on Ω with reproducing kernel K.

In the following let $H(\mathcal{E})$ be an $\mathcal{E}$-valued functional Hilbert space on a bounded open set $\Omega \subset \mathbb{C}^n$ such that all point evaluations $\delta_z : H(\mathcal{E}) \to \mathcal{E}$ are surjective. Let $\delta_z^* = V_z Q_z$ be the polar decomposition of δ_z^*. Then $Q_z \in L(\mathcal{E})$ is a positive invertible operator and $V_z \in L(\mathcal{E}, H(\mathcal{E}))$ is an isometry. For a given operator $X \in L(H(\mathcal{E}))$, we define its (generalized) Berezin transform $\Gamma(X) : \Omega \to L(\mathcal{E})$

by

$$\Gamma(X)(z) = V_z^* X V_z.$$

Under suitable conditions the Berezin transform of every compact operator vanishes on the boundary of Ω. To describe a typical case, let us observe that

$$\|Q_z^{-1}\| = m(Q_z)^{-1} = m(\delta_z^*)^{-1} \qquad (z \in \Omega),$$

where $m(T) = \inf_{\|x\|=1} \|Tx\|$ denotes the minimum modulus of a given bounded operator T.

Proposition 6. *Let $H(\mathcal{E})$ be an $\mathcal{E}$-valued functional Hilbert space on a bounded open set $\Omega \subset \mathbb{C}^n$ such that all point evaluations $\delta_z : H(\mathcal{E}) \to \mathcal{E}$ $(z \in \Omega)$ are surjective. Then we have:*

(a) *if $\lim_{z \to \partial\Omega} m(\delta_z^*) = \infty$ and if the bounded functions are dense in $H(\mathcal{E})$, then SOT-$\lim_{z \to \partial\Omega} V_z^* = 0$;*

(b) *WOT-$\lim_{z \to \partial\Omega} V_z = 0$ if and only if SOT-$\lim_{z \to \partial\Omega} \Gamma(C)(z) = 0$ for every compact operator C on $H(\mathcal{E})$;*

(c) *SOT-$\lim_{z \to \partial\Omega} V_z^* = 0$ if and only if $\lim_{z \to \partial\Omega} \|\Gamma(C)(z)\| = 0$ for every compact operator C on $H(\mathcal{E})$.*

Proof. Let $z \in \Omega, f \in H(\mathcal{E})$ and $x \in \mathcal{E}$ be given. Since

$$\langle f, K(\cdot, z)x \rangle = \langle f(z), x \rangle = \langle f, \delta_z^* x \rangle,$$

it follows that $V_z x = \delta_z^* Q_z^{-1} x = K(\cdot, z) Q_z^{-1} x$. Similarly, from

$$\langle V_z^* f, x \rangle = \langle f, K(\cdot, z) Q_z^{-1} x \rangle = \langle f(z), Q_z^{-1} x \rangle = \langle Q_z^{-1}(f(z)), x \rangle$$

we deduce that $V_z^* f = Q_z^{-1}(f(z))$. But then

$$\|V_z^* f\| \leq \|Q_z^{-1}\| \|f(z)\| = m(\delta_z^*)^{-1} \|f(z)\| \xrightarrow{z \to \partial\Omega} 0$$

for each bounded function $f \in H(\mathcal{E})$. This implies part (a).

For $f, g \in H(\mathcal{E})$, let $g \otimes f \in L(H(\mathcal{E}))$ be the rank-one operator acting as $g \otimes f(h) = \langle h, f \rangle g$. Then $\Gamma(g \otimes f)(z) = \langle \cdot, V_z^* f \rangle V_z^* g$. To prove part (b), let us first recall that a compact operator C on $H(\mathcal{E})$ maps each weak zero sequence to a sequence converging to zero in norm. On the other hand, if SOT-$\lim_{z \to \partial\Omega} \Gamma(C)(z) = 0$ for each compact operator C, then it suffices to choose a function $g \in H(\mathcal{E})$ with $\|V_z^* g\| = 1$ and to observe that

$$|\langle V_z x, f \rangle| = \|(\Gamma(g \otimes f)(z))(x)\| \xrightarrow{z \to \partial\Omega} 0 \quad (x \in \mathcal{E}, f \in H(\mathcal{E})).$$

To prove (c), suppose first that SOT-$\lim_{z \to \partial\Omega} V_z^* = 0$. Then for $f, g \in H(\mathcal{E})$,

$$\|\Gamma(g \otimes f)(z)\| = \|V_z^* f\| \|V_z^* g\| \xrightarrow{z \to \partial\Omega} 0.$$

Since Γ is a contractive linear map from $L(H(\mathcal{E}))$ into the Banach space of all bounded $L(\mathcal{E})$-valued functions on Ω (equipped with the uniform norm) and since every compact operator on $H(\mathcal{E})$ is the norm limit of a sequence of finite-rank operators, it follows that $\lim_{z \to \partial\Omega} \|\Gamma(C)(z)\| = 0$ for every compact operator C on $H(\mathcal{E})$. Conversely, if the latter condition holds, then the above argument with $f = g$ yields that SOT-$\lim_{z \to \partial\Omega} V_z^* = 0$. $\qquad\square$

Note that the condition SOT- $\lim_{z \to \partial \Omega} V_z^* = 0$ implies that WOT- $\lim_{z \to \partial \Omega} V_z = 0$.
If $\dim \mathcal{E} < \infty$, then these two conditions are easily seen to be equivalent.

Let $H = H(K)$ be a scalar-valued functional Hilbert space given by a repro-
ducing kernel $K : \Omega \times \Omega \to \mathbb{C}$. Then for any Hilbert space $\mathcal{E}$, the $\mathcal{E}$-valued functional Hilbert space $H(\mathcal{E})$ determined by the reproducing kernel $K_\mathcal{E} : \Omega \times \Omega \longrightarrow$
$L(\mathcal{E})$, $(z, w) \mapsto K(z, w) 1_\mathcal{E}$, can be identified with the Hilbertian tensor product
$H(K) \otimes \mathcal{E}$. The hypothesis that all point evaluations $\delta_z : H(\mathcal{E}) \to \mathcal{E}$ $(z \in \Omega)$
are surjective is equivalent to the condition that $H(K)$ has no common zeros. An
elementary exercise shows that in this case the operators Q_z and V_z are given by

$$Q_z = \|K(\cdot, z)\| 1_\mathcal{E} \quad \text{and} \quad V_z : \mathcal{E} \to H(\mathcal{E}), x \mapsto k_z \otimes x,$$

where as before $k_z = K(\cdot, z)/\|K(\cdot, z)\|$ are the normalized reproducing kernel
vectors. In particular, WOT-$\lim_{z \to \partial \Omega} V_z = 0$ if and only if weak-$\lim_{z \to \partial \Omega} k_z = 0$.
By Lemma 5 both conditions are satisfied when $1 \in H(K)$ and $H(K)$ satisfies the
ℓ^∞-interpolation property.

Let us return to the case of general vector-valued functional Hilbert spaces.
The following result shows that, in many cases, the Berezin transform can be used
to recover the symbol from a given Toeplitz operator.

Lemma 7. *Let $H(\mathcal{E})$ be an $\mathcal{E}$-valued functional Hilbert space with reproducing kernel
$K : \Omega \times \Omega \to L(\mathcal{E})$ such that all point evaluations $\delta_z : H(\mathcal{E}) \to \mathcal{E}$ $(z \in \Omega)$ are
surjective. Then the identity*

$$f(z) K(z, z) g(z)^* = Q_z \Gamma(M_f M_g^*)(z) Q_z \qquad (z \in \Omega)$$

holds for every pair of multipliers $f, g \in \mathcal{M}(H(\mathcal{E}))$.

Proof. Let $f, g \in \mathcal{M}(H(\mathcal{E}))$ be multipliers. Then the identities

$$f(z) K(z, z) g(z)^* = f(z) \delta_z \delta_z^* g(z)^* = \delta_z M_f M_g^* \delta_z^*$$
$$= Q_z (V_z^* M_f M_g^* V_z) Q_z = Q_z \Gamma(M_f M_g^*)(z) Q_z$$

hold for every $z \in \Omega$. $\qquad\qquad\qquad\qquad\qquad\qquad\qquad\qquad\qquad\qquad\qquad$ □

Proposition 6 can be used to deduce a lower bound for the semi-Fredholm
spectrum of a multiplier tuple. A corresponding result for contractive quasi-free
Hilbert modules over the ball algebra $A(\mathbb{B}_n)$ is contained in [7].

Theorem 8. *Let $H(\mathcal{E})$ be an $\mathcal{E}$-valued functional Hilbert space with reproducing
kernel $K : \Omega \times \Omega \to L(\mathcal{E})$. Suppose that all point evaluations $\delta_z : H(\mathcal{E}) \to \mathcal{E}$
$(z \in \Omega)$ are surjective and that*

$$\text{SOT-} \lim_{z \to \partial \Omega} V_z^* = 0.$$

*Then, for a given multiplier tuple $f = (f_1, \ldots, f_m) \in \mathcal{M}(H(\mathcal{E}))^m$ with the property
that the row multiplication*

$$M_f : H(\mathcal{E})^m \to H(\mathcal{E}), (g_i) \mapsto \sum_{i=1}^m f_i g_i$$

has finite-codimensional range, there exist a positive real number $c > 0$ and an open neighbourhood U of $\partial\Omega$ such that the inequality

$$\sum_{i=1}^{m} f_i(z)K(z,z)f_i(z)^* \geq cK(z,z)$$

holds for all $z \in U \cap \Omega$.

Proof. The hypothesis implies that there is a finite-rank projection $F \in L(H(\mathcal{E}))$ such that the operator

$$M_f M_f^* + F : \ H(\mathcal{E}) \to H(\mathcal{E})$$

is bounded below. Hence there is a constant $c > 0$ such that

$$\Gamma(M_f M_f^* + F)(z) \geq \Gamma(2c\, 1_{H(\mathcal{E})})(z) = 2c\, 1_{\mathcal{E}} \quad (z \in \Omega).$$

By Proposition 6 there is an open neighbourhood U of $\partial\Omega$ such that $\Gamma(M_f M_f^*)(z) \geq c1_{\mathcal{E}}$ for $z \in U \cap \Omega$. Using Lemma 7 we find that

$$\sum_{i=1}^{m} f_i(z)K(z,z)f_i(z)^* = Q_z\Gamma(M_f M_f^*)(z)Q_z \geq c\, Q_z^2$$

$$= cQ_z V_z^* V_z Q_z = c\delta_z \delta_z^* = cK(z,z)$$

for all $z \in U \cap \Omega$. □

Let $\mathcal{E}$ be an arbitrary Hilbert space and let $H(\mathcal{E}) = H \otimes \mathcal{E}$ be given by a scalar-valued reproducing kernel $K : \Omega \times \Omega \to \mathbb{C}$. Suppose that $H(K)$ has no common zeros and that

$$\text{weak-} \lim_{z \to \partial\Omega} k_z = 0.$$

The proof of part (a) of Proposition 6 shows that

$$\|V_z^*(f \otimes x)\| = \|Q_z^{-1}(f(z)x)\| = \frac{|f(z)|\|x\|}{\|K(\cdot,z)\|} = |\langle f, k_z\rangle|\, \|x\|$$

converges to zero as $z \to \partial\Omega$ for any given $f \in H(K)$ and $x \in \mathcal{E}$. Hence the hypothesis of Theorem 8, that is, the condition that

$$\text{SOT-} \lim_{z \to \partial\Omega} V_z^* = 0$$

is satisfied in this case.

Consequently, for a given multiplier tuple $f = (f_1, \ldots, f_m) \in \mathcal{M}(H(\mathcal{E}))^m$ with the property that the operator $M_f : H(\mathcal{E})^m \to H(\mathcal{E})$ has finite-codimensional range, there exist a positive real number $c > 0$ and an open neighbourhood U of $\partial\Omega$ such that

$$\sum_{i=1}^{m} f_i(z)f_i(z)^* \geq c\, 1_{\mathcal{E}}$$

for $z \in U \cap \Omega$. In the scalar case $\mathcal{E} = \mathbb{C}$, this estimate reduces to

$$\sum_{i=1}^{m} |f_i(z)|^2 \geq c$$

for $z \in U \cap \Omega$. Hence, in the particular case of scalar-valued functional Hilbert spaces, we obtain the following result.

Corollary 9. *Let $H = H(K)$ be a functional Hilbert space with reproducing kernel $K : \Omega \times \Omega \to \mathbb{C}$ on a bounded open set $\Omega \subset \mathbb{C}^n$ such that H has no common zeros and*

$$\text{weak-} \lim_{z \to \partial\Omega} k_z = 0.$$

Then for $m \geq 1$ and $f \in \mathcal{M}(H)^m$, the spectral inclusion

$$\bigcap \left(\overline{f(U \cap \Omega)}; U \supset \partial\Omega \,\text{open} \right) \subset \sigma_{re}(M_f)$$

holds.

Another possible proof of Corollary 9, which is formulated under more restrictive conditions, but also works in the setting of the above corollary, is given by Lin and Salinas in [14] (Proposition 2.2).

4. Applications

Let $(H_\alpha)_{\alpha \geq -(n+1)}$ be the family of functional Hilbert spaces defined on the open Euclidean unit ball $\mathbb{B}_n \subset \mathbb{C}^n$ by the reproducing kernels

$$K_{-(n+1)}(z, w) = 1 + \log \frac{1}{1 - \langle z, w \rangle}$$

and

$$K_\alpha(z, w) = \frac{1}{(1 - \langle z, w \rangle)^{n+1+\alpha}} \quad (\alpha > -(n+1)).$$

In the terminology of Zhao and Zhu [19], the spaces H_α are the generalized Bergman spaces A_α^2. In the notation of Costea-Sawyer-Wick [5], the space H_α is the analytic Besov-Sobolev space $B^2_{(\alpha+n+1)/2}$. Since the polynomials are dense in H_α and $\lim_{z \to \partial\Omega} \|K_\alpha(\cdot, z)\| = \infty$, Corollary 9 applies to each of these spaces H_α.

Theorem 10. *For $\alpha \geq -(n+1)$ and $H_\alpha = H(K_\alpha)$ as above, the right Taylor spectrum of a multiplier tuple $f \in \mathcal{M}(H_\alpha)^m$ is given by*

$$\sigma_r^T(M_f) = \overline{f(\mathbb{B}_n)}$$

and the right essential Taylor spectrum of M_f is given by

$$\sigma_{re}^T(M_f) = \bigcap \left(\overline{f(U \cap \mathbb{B}_n)}; U \supset \partial\mathbb{B}_n \,\text{open} \right).$$

For $-n \geq \alpha \geq -(n+1)$ and f as above, we obtain in addition the equalities

$$\sigma^T(M_f) = \sigma_r^T(M_f) \quad \text{and} \quad \sigma_e^T(M_f) = \sigma_{re}^T(M_f).$$

Proof. The multiplier algebra $\mathcal{M} = M(H_\alpha)$ equipped with its multiplier norm becomes a unital commutative Banach algebra. Denote by

$$\Phi : \mathcal{M}(H_\alpha) \to L(H_\alpha),\ f \mapsto M_f$$

the representation associating with each multiplier $f \in \mathcal{M}$ the induced multiplication operator M_f on H_α. Since H_α contains the constant functions, the inclusion $\overline{f(\mathbb{B}_n)} \subset \sigma_r^T(M_f)$ holds for every tuple $f \in \mathcal{M}^m$. The reverse inclusion

$$\sigma_r^T(M_f) \subset \overline{f(\mathbb{B}_n)}$$

follows from Theorem 2 in [5]. It is well known that the coordinate functions are multipliers of H_α with $\sigma^T(M_z) = \overline{\mathbb{B}}_n = \sigma_r^T(M_z)$ and $\sigma_e^T(M_z) = \partial \mathbb{B}_n = \sigma_{re}^T(M_z)$. As an application of Corollary 3 and Corollary 9 we obtain that

$$\sigma_{re}^T(M_f) = \bigcap \left(\overline{f(U \cap \mathbb{B}_n)}\, ; U \supset \partial \mathbb{B}_n \text{ open} \right)$$

for every tuple $f \in \mathcal{M}^m$. In the cited paper [5] of Costea-Sawyer-Wick it is shown that $\mathcal{M} = \mathcal{M}(H_\alpha)$ satisfies the corona property for $-n \geq \alpha \geq -(n+1)$. Hence, in these cases, the missing inclusions follow directly from part (a) of Remark 2. $\qquad \square$

Let $\mathcal{M} = H^\infty(\mathbb{D}^n)$ be the multiplier space of the Hardy space $H^2(\mathbb{D}^n)$ on the unit polydisc $\mathbb{D}^n$ and let $\Phi : \mathcal{M} \to L(H^2(\mathbb{D}^n)),\ f \mapsto M_f$, be the induced representation. By a recent result of T. Trent [18], which improves corresponding H^p-corona theorems of S.-Y. Li [12] and K.-C. Lin [13], the spectral mapping formula

$$\sigma^T(\Phi(f)) = \overline{f(\mathbb{D}^n)}$$

holds for all $f \in \mathcal{M}^m, m \geq 1$. Standard tensor product arguments [8] show that $\sigma_e^T(M_z) = \partial \mathbb{D}^n$. Since the polynomials are dense in $H^2(\mathbb{D}^n)$ and since the reproducing kernel

$$K : \mathbb{D}^n \times \mathbb{D}^n \to \mathbb{C},\ K(z,w) = \prod_{i=1}^{n}(1 - z_i \overline{w_i})^{-1}$$

of $H^2(\mathbb{D}^n)$ has the property that $\lim_{z \to \partial \mathbb{D}^n} \|K(\cdot, z)\| = \infty$, an application of Corollary 3 and Corollary 9 yields the following result.

Theorem 11. *For $f \in H^\infty(\mathbb{D}^n)^m$, the essential Taylor spectrum of M_f on $H^2(\mathbb{D}^n)$ is given by*

$$\sigma_e^T(M_f) = \sigma_{re}^T(M_f) = \bigcap(\overline{f(U \cap \mathbb{D}^n)}; U \supset \partial \mathbb{D}^n \text{ open}).$$

A classical result of Carleson [4] implies that the Hardy space $H^2(\mathbb{D})$ on the open unit disc $\mathbb{D}$ in $\mathbb{C}$ satisfies the ℓ^∞-interpolation property. To conclude we indicate an elementary way to reduce some multidimensional cases to this classical situation.

Let X be a functional Banach space on a bounded open set $\Omega \subset \mathbb{C}^n$ such that $1 \in X$.

Lemma 12. *Suppose that, for every boundary point $\lambda \in \partial\Omega$, there is a multiplier $h \in \mathcal{M}(X)$ with $h(\Omega) \subset \mathbb{D}$, $\lim_{z \to \lambda} h(z) = 1$ and*

$$\{f \circ h; \ f \in H^\infty(\mathbb{D})\} \subset \mathcal{M}(X).$$

Then X satisfies the ℓ^∞-interpolation property.

Proof. Let (z_k) be a sequence in Ω converging to a boundary point $\lambda \in \partial\Omega$. Choose a multiplier $h \in \mathcal{M}(X)$ for λ as described in the hypothesis of the lemma. By the classical result of Carleson referred to above, there is a subsequence (w_k) of (z_k) such that

$$\{(f(h(w_k)))_k; \ f \in H^\infty(\mathbb{D})\} = \ell^\infty.$$

Since by hypothesis $f \circ h \in \mathcal{M}(X)$ for every function $f \in H^\infty(\mathbb{D})$, the assertion follows. $\qquad\square$

If the underlying set Ω has suitable additional properties, then for every boundary point $\lambda \in \partial\Omega$, there are multipliers $h \in \mathcal{M}(X)$ satisfying the first two properties required in Lemma 12

Proposition 13. *Let X be a functional Banach space on a bounded open set $\Omega \subset \mathbb{C}^n$ with $1 \in X$. Suppose that Ω satisfies one of the following conditions:*

(i) Ω *is convex and $z_1, \ldots, z_n \in \mathcal{M}(X)$;*
(ii) Ω *is an analytic polyhedron and $\mathcal{O}(\overline{\Omega}) \subset \mathcal{M}(X)$;*
(iii) Ω *is strictly pseudoconvex and $\mathcal{O}(\overline{\Omega}) \subset \mathcal{M}(X)$.*

Then for each boundary point $\lambda \in \partial\Omega$, there exists a multiplier $h \in \mathcal{M}(X)$ such that $h(\Omega) \subset \mathbb{D}$ and $\lim_{z \to \lambda} h(z) = 1$.

Proof. Suppose that Ω is convex and that $\lambda \in \partial\Omega$. Then by a standard separation theorem, there is a homogeneous polynomial $p(z) = \sum_{i=1}^n a_i z_i$ of degree one such that

$$\operatorname{Re} p(z) < \operatorname{Re} p(\lambda) \quad (z \in \Omega).$$

If the coordinate functions are multipliers, then also $e^p = \sum_{k=0}^\infty \frac{p^k}{k!}$ is a multiplier of X. Since $\left|e^{p(z)}\right| < \left|e^{p(\lambda)}\right|$ for $z \in \Omega$, the function $h = e^p / e^{p(\lambda)} \in \mathcal{M}(X)$ has the required properties.

Suppose that Ω is an analytic polyhedron and that $\lambda \in \partial\Omega$. Then by definition there are analytic functions $f_1, \ldots, f_r \in \mathcal{O}(U)$ on an open neighbourhood U of $\overline{\Omega}$ such that

$$\Omega = \{z \in U; \ \max_{1 \le j \le r} |f_j(z)| < 1\} \subset \overline{\Omega} \subset U.$$

Then $|f_j(\lambda)| = 1$ for some $j = 1, \ldots, r$ and $h = f_j$ has the required properties.

Finally, if Ω is strictly pseudoconvex, then by a well-known result from several complex variables, for every boundary point $\lambda \in \partial\Omega$, there is an analytic function $h \in \mathcal{O}(\overline{\Omega})$ such that $h(\lambda) = 1$ and $|h(z)| < 1$ for $z \in \overline{\Omega} \setminus \{\lambda\}$. This completes the proof in the third case. $\qquad\square$

The question whether h can be chosen in such a way that $M_h \in L(X)$ possesses an $H^\infty(\mathbb{D})$-functional calculus is more delicate. There are two obvious cases in which this is true. First, if $\mathcal{M}(X) = H^\infty(\Omega)$, then M_h admits an $H^\infty(\mathbb{D})$-functional calculus by trivial reasons. Secondly, if X is a Hilbert space and $\|h\|_{\mathcal{M}} \leq 1$, then M_h can easily be shown to be a $C_{.0}$-contraction. Hence also in this case, M_h possesses an $H^\infty(\mathbb{D})$-functional calculus.

Corollary 14. *For each $\alpha > -(n+1)$, the functional Hilbert space H_α considered in Theorem 10 possesses the ℓ^∞-interpolation property.*

Proof. Fix a point $\lambda \in \partial\mathbb{B}_n$ and choose a unitary operator $U : \mathbb{C}^n \to \mathbb{C}^n$ such that $U\lambda = (1, 0, \ldots, 0)$. Denote by $\pi_1 : \mathbb{C}^n \to \mathbb{C}$ the projection onto the first coordinate and consider the function $h = \pi_1 \circ (U|_{\mathbb{B}_n})$. Clearly, $h(\mathbb{B}_n) \subset \mathbb{D}$ and $\lim_{z \to \lambda} h(z) = 1$. It is well known that $z_i \in \mathcal{M}(H_\alpha)$ with $\|z_i\|_{\mathcal{M}} = 1$ for $i = 1, \ldots, n$. Since

$$\frac{1 - h(z)\overline{h(w)}}{(1 - \langle z, w \rangle)^{n+1+\alpha}} = \frac{1 - \pi_1(U(z))\overline{\pi_1(U(w))}}{(1 - \langle Uz, Uw \rangle)^{n+1+\alpha}}$$

is positive definite as a function of $(z, w) \in \mathbb{B}_n \times \mathbb{B}_n$, it follows that $h \in \mathcal{M}(H_\alpha)$ with $\|h\|_{\mathcal{M}} \leq 1$. Now the assertion follows from Lemma 12 and the remarks preceding the corollary. $\qquad\square$

We conclude this note by giving some Banach space examples. For a bounded open set Ω in $\mathbb{C}^n$, let $L_a^p(\Omega)$ be the Bergman space consisting of all analytic functions on Ω that are p-integrable with respect to the $2n$-dimensional Lebesgue measure on Ω.

Corollary 15. *Let $1 \leq p < \infty$ be a real number and let $\Omega \subset \mathbb{C}^n$ be a bounded open set. In each of the following cases the Bergman space $L_a^p(\Omega)$ satisfies the ℓ^∞-interpolation property:*

  (i) *Ω is convex;*
  (ii) *Ω is an analytic polyhedron;*
  (iii) *Ω is strictly pseudoconvex.*
In each of these cases, we have that

$$\bigcap \left(\overline{f(U \cap \Omega)}; U \supset \partial\Omega \text{ open} \right) \subset \sigma_{re}^T(M_f) \quad (f \in H^\infty(\Omega)^m).$$

If $p = 2$, then in each of these cases we obtain that

$$\sigma_e^T(M_f) = \sigma_{re}^T(M_f) = \bigcap \left(\overline{f(U \cap \Omega)}; U \supset \partial\Omega \text{ open} \right) \quad (f \in H^\infty(\Omega)^m).$$

Proof. Suppose that Ω satisfies one of the conditions (i), (ii) or (iii).

Since $\mathcal{M}(L_a^p(\Omega)) = H^\infty(\Omega)$, Proposition 13 and the remarks following it, show that the hypotheses of Lemma 12 are satisfied. Hence $L_a^p(\Omega)$ satisfies the ℓ^∞-interpolation property. Then Theorem 4 implies that the spectral inclusion

$$\bigcap \left(\overline{f(U \cap \Omega)}; U \supset \partial\Omega \text{ open} \right) \subset \sigma_{re}^T(M_f)$$

holds for every $m \geq 1$ and $f \in H^\infty(\Omega)^m$.

Let us specialize to the case $p = 2$. Then for every bounded pseudoconvex domain $\Omega \subset \mathbb{C}^n$ and every tuple $f \in H^\infty(\Omega)^m)$, the formula

$$\sigma^T(M_f) = \sigma_r^T(M_f) = \overline{f(\Omega)}$$

holds for the multiplication tuple $M_f \in L(L_a^2(\Omega))^m$ (Corollary 8.2.3 in [9]). By Hörmander's exactness results for the $\bar\partial$-sequence with L^2-bounds, it follows that (Theorem 2.2.3 in [11] and Theorem 8.1.1 in [9])

$$\sigma_e^T(M_z) = \sigma_{re}^T(M_z) \subset \partial\Omega,$$

where $M_z = (M_{z_1}, \ldots, M_{z_n}) \in L(L_a^2(\Omega))^n$. As an application of Corollary 3, we find that

$$\sigma_e^T(M_f) \subset \bigcap \left(\overline{f(U \cap \Omega)}; \, U \supset \sigma_e^T(M_z) \text{ open} \right).$$

Hence, in each of the three cases, the first part of the proof yields the missing inclusions. □

Let $\Omega \subset \mathbb{C}^n$ be a bounded strictly pseudoconvex domain with C^2-boundary. For $p = 2$, the above formula for the essential Taylor spectrum of M_f on the Bergman space $L_a^2(\Omega)$ can also be found in [9]. In the proof given here we have replaced some non-trivial, and more specific, arguments from several variable local spectral theory by general Gelfand theory.

In [2], Andersson and Carlsson show that the spectral mapping formula

$$\sigma^T(M_f) = \sigma_r^T(M_f) = \overline{f(\Omega)}$$

also holds for analytic Toeplitz tuples M_f ($f \in H^\infty(\Omega)^m$) on Hardy spaces $H^p(\Omega)$ ($1 \leq p < \infty$) over strictly pseudoconvex domains $\Omega \subset \mathbb{C}^n$ with C^3-boundary. In [3], Andersson and Sandberg use $\bar\partial$-techniques to obtain the corresponding formulas

$$\sigma_e^T(M_f) = \sigma_{re}^T(M_f) = \bigcap \left(\overline{f(U \cap \Omega)}; \, U \supset \partial\Omega \text{ open} \right)$$

for the essential spectra. For $p = 2$, well-known C^*-algebra methods based on the essential normality of M_z on $H^2(\Omega)$ show that $\sigma_e^T(M_z) = \partial\Omega$. This observation together with Theorem 1 can be used to reduce the essential spectral mapping formula of Andersson and Sandberg [3] directly to the corresponding formula for the full Taylor spectrum from [2]. The same idea should be applicable to some other spaces occurring in [3].

References

[1] J. Agler, J.E. McCarthy, Pick interpolation and Hilbert function spaces, Graduate Studies in Mathematics, 44, AMS, Providence, RI, 2002.

[2] M. Andersson, H. Carlsson, Estimates of solutions of the H^p and $BMOA$ corona problem, Math. Ann. 316 (2000), 83–102.

[3] M. Andersson, S. Sandberg, The essential spectrum of holomorphic Toeplitz operators on H^p spaces, Studia Math. 154 (2003), 223–231.

[4] L. Carleson, An interpolation problem for bounded analytic functions, Amer. J. Math. 80 (1958), 921–930.

[5] S. Costea, E. Sawyer, B.D. Wick, The Corona Theorem for the Drury-Arveson Hardy space and other holomorphic Besov Sobolev spaces on the unit ball in $\mathbb{C}^n$, Anal. PDE, to appear.

[6] K.R. Davidson, R.G. Douglas, The generalized Berezin transform and commutator ideals, Pacific J. Math. 222 (2005), 29–56.

[7] R.G. Douglas, J. Sarkar, A note on semi-Fredholm Hilbert modules, Operator Theory: Advances and Appl. 202, 143–150, Birkhäuser, Basel, 2010.

[8] J. Eschmeier, Tensor products and elementary operators, J. reine angew. Math. 390 (1988), 47–66.

[9] J. Eschmeier, M. Putinar, Spectral decompositions and analytic sheaves, LMS Monograph Series, Vol. 10, Clarendon Press, Oxford 1996.

[10] C. Foiaş, W. Mlak, The extended spectrum of completely non-unitary contractions and the spectral mapping theorem, Studia Math. 26 (1966), 239–245.

[11] L. Hörmander, L^2-estimates and existence theorems for the $\bar{\partial}$-operator, Acta Math. 113(1965), 89–152.

[12] S.-Y. Li, Corona problems of several complex variables, Madison symposium of complex analysis, Contemporary Mathematics, vol. 137, Amer. Math. Soc., Providence, RI, 1991.

[13] K.-C. Lin, The H^p-corona theorem for the polydisc, Trans. Amer. Math. Soc. 341 (1994), 371–351.

[14] Q. Lin, N. Salinas, Proper holomorphic maps and analytic Toeplitz n-tuples, Indiana Univ. Math. J. 39 (1990), 547–562.

[15] K. Rudol, Spectral mapping theorems for analytic functional calculi, Operator Theory: Advances and Appl. 17 (1986), 331–340.

[16] B. Sz.-Nagy, C. Foiaş, Harmonic analysis of operators on Hilbert spaces, North-Holland, Amsterdam, 1970.

[17] J.L. Taylor, The analytic functional calculus for several commuting operators, Acta Math. 125 (1970), 1–48.

[18] T. Trent, On the Taylor spectrum of n-tuples of Toeplitz operators on the polydisc, multivariable operator theory workshop, Fields Institute, August 2009.

[19] R. Zhao, K. Zhu, Theory of Bergman spaces on the unit ball of $\mathbb{C}^n$, Mém. Soc. Math. Fr. (NS), 115 (2008).

Ronald G. Douglas
Department of Mathematics
Texas A&M University
College Station, TX, USA
e-mail: `rdouglas@math.tamu.edu`

Jörg Eschmeier
Fachrichtung Mathematik
Universität des Saarlandes
Saarbrücken, Germany
e-mail: `eschmei@math.uni-sb.de`

Operator Theory:
Advances and Applications, Vol. 222, 129–149

Tutorial on a Nehari Problem and Related Reproducing Kernel Spaces

Harry Dym

To Bill on his 65th, who was one of the first to realize the
importance of the AAK papers and their manifold applications

Abstract. This is an expository paper on a Nehari problem that is formulated
for $p \times q$ rational matrix-valued functions. The symbol for the associated
Hankel operator is holomorphic in the open left half-plane. The interplay
with a number of related finite-dimensional reproducing kernel Hilbert spaces
and Krein spaces is emphasized.

Mathematics Subject Classification. 47B32, 47B35, 47B50, 93B28.

Keywords. Nehari problem, realization theory, reproducing kernel spaces.

1. Introduction

This paper is devoted to an expository account of a Nehari problem for rational mvf's (matrix-valued functions) and a number of related finite-dimensional
RKHS's (reproducing kernel Hilbert spaces) and finite-dimensional RKKS's (reproducing kernel Krein spaces). The only prerequisites are some familiarity with
the theory of rational mvf's with entries in the Hardy spaces H_2 and H_∞ of the
right half-plane (most of which is easily verified by Cauchy's integral formula) and
a little bit of realization theory. A short introduction to the latter may be found
in [D07]; for more extensive discussion and many applications see [ZDG96]. The
monograph [F87] and the tutorial papers [FD87] and [Gl89] partially overlap the
subject matter of this paper and extend it in many other directions connected
with control theory.

The given data for the problem of interest is a rational $p \times q$ mvf $R(\lambda)$ with
minimal realization

$$R(\lambda) = C(\lambda I_n - A)^{-1}B \quad \text{and} \quad \sigma(A) \subset \Omega_+, \tag{1}$$

where $C \in \mathbb{C}^{p \times n}$, $A \in \mathbb{C}^{n \times n}$, $B \in \mathbb{C}^{n \times q}$,

Ω_+ (resp., Ω_-) denotes the open right (resp., left) half-plane and the minimality assumption will be explained in the next section.

Problem: Given R, describe the set

$$\mathcal{N}_R = \{f \in \mathcal{R} \cap L_\infty^{p \times q} : f - R \in H_\infty^{p \times q} \quad \text{and} \quad \|f\|_\infty \le 1\},$$

where

$$L_\infty^{p \times q} = \{p \times q \text{ mvf's with entries in } L_\infty(i\mathbb{R})\},$$
$$H_\infty^{p \times q} = \{p \times q \text{ mvf's with entries in } H_\infty(\Omega_+)\},$$
$$\mathcal{R} = \text{the space of rational mvf's.}$$

We shall also need the notation

$$H_2^{p \times q} = H_2^{p \times q}(\Omega_+) = \{p \times q \text{ mvf's with entries in } H_2(\Omega_+)\},$$
$$(H_2^\perp)^{p \times q} = H_2^{p \times q}(\Omega_-) = \{p \times q \text{ mvf's with entries in } H_2(\Omega_-)\},$$
$$H_2^p = H_2^{p \times 1} \quad \text{and} \quad (H_2^\perp)^q = (H_2^\perp)^{q \times 1},$$
$$\|A\| = \text{the maximum singular value of a matrix } A,$$
$$A^H \quad \text{the Hermitian transpose of } A,$$
$$\mathcal{S}^{p \times q} = \{f \in H_\infty^{p \times q} : \|f(\lambda)\| \le 1 \quad \text{for every point } \lambda \in \Omega_+\},$$
$$\Pi_+ = \text{the orthogonal projection of } L_2^p \text{ onto } H_2^p(\Omega_+),$$
$$\Pi_- = \text{the orthogonal projection of } L_2^p \text{ onto } H_2^p(\Omega_-),$$
$$f^\#(\lambda) = f(-\bar{\lambda})^H,$$
$$\langle f, g \rangle_{st} = \int_{-\infty}^\infty \text{trace}\,\{g(i\nu)^H f(i\nu)\} d\nu \quad \text{for } f, g \in L_2^{p \times q}(i\mathbb{R}) \text{ and}$$
$$\langle f, g \rangle_{nst} = \frac{1}{2\pi} \langle f, g \rangle_{st}$$

The notation $(H_2^{p \times q})^\perp$ stems from the fact that

$$f \in H_2^{p \times q} \text{ and } g \in (H_2^{p \times q})^\perp \implies \langle f, g \rangle_{st} = 0.$$

The Poisson formula for the right half-plane

$$f(a + ib) = \frac{a}{\pi} \int_{-\infty}^\infty \frac{f(i\nu)}{a^2 + (\nu - b)^2} d\nu, \quad \text{for } a > 0 \text{ and } b \in \mathbb{R} \tag{2}$$

holds for $f \in \mathcal{R} \cap H_\infty^{p \times q}$ and, as

$$\frac{a}{\pi} \int_{-\infty}^\infty \frac{1}{a^2 + (\nu - b)^2} d\nu = 1 \quad \text{for } a > 0 \text{ and } b \in \mathbb{R},$$

leads easily to the maximum principle

$$\|f(a + ib)\| \le \max_{\nu \in \mathbb{R}} \|f(i\nu)\| \quad \text{for } a > 0 \text{ and } b \in \mathbb{R}. \tag{3}$$

The operator Γ acting from H_2^q into $(H_2^\perp)^p$ that is defined by the formula

$$\Gamma g = \Pi_- R g. \tag{4}$$

is called the **Hankel** operator (with symbol R). It is easily seen that if $f \in \mathcal{N}_R$, then

$$\Pi_-(f - R)g = 0 \quad \text{for every } g \in H_2^q \tag{5}$$

and hence, upon setting $\Gamma_f = \Pi_- f|_{H_2^q}$, that

$$\mathcal{N}_R \subseteq \mathcal{N}_\Gamma \overset{\text{def}}{=} \{ f \in \mathcal{R} \cap L_\infty^{p \times q} : \Gamma_f = \Gamma \quad \text{and} \quad \|f\|_\infty \le 1 \}. \tag{6}$$

In fact it is not hard to show that equality holds in (6):

$$f \in \mathcal{N}_\Gamma \implies (f - R)\frac{I_q}{\lambda + 1} \in \mathcal{R} \cap H_2^{p \times q}.$$

Therefore, $f - R$ is holomorphic in Ω_+ with boundary values in $L_\infty^{p \times q}$. Consequently, by the maximum principle (3), $f - R \in \mathcal{R} \cap H_\infty^{p \times q}$. Thus,

$$\mathcal{N}_R = \mathcal{N}_\Gamma. \tag{7}$$

2. Two reproducing kernel Hilbert spaces

Let

$$F_o(\lambda) = C(\lambda I_n - A)^{-1} \quad \text{and} \quad F_c(\lambda) = B^H(\lambda I_n + A^H)^{-1} \tag{8}$$

be defined in terms of the matrices in the realization (1), let

$$P_o = \frac{1}{2\pi} \int_{-\infty}^\infty F_o(i\nu)^H F_o(i\nu) d\nu, \quad P_c = \frac{1}{2\pi} \int_{-\infty}^\infty F_c(i\nu)^H F_c(i\nu) d\nu \tag{9}$$

and let

$$\mathcal{M}_o = \{ F_o(\lambda)u : u \in \mathbb{C}^n \} \quad (\text{resp.}, \ \mathcal{M}_c = \{ F_c(\lambda)u : u \in \mathbb{C}^n \}) \tag{10}$$

endowed with the inner product

$$\langle F_o u, F_o v \rangle_{\mathcal{M}_o} = v^H P_o u \quad (\text{resp.}, \ \langle F_c u, F_c v \rangle_{\mathcal{M}_c} = v^H P_c u) \tag{11}$$

for every choice of $u, v \in \mathbb{C}^n$.

The assumption that the realization (1) is minimal means that the pair (C, A) is observable and the pair (A, B) is controllable, i.e.,

$$\bigcap_{k=0}^{n-1} \ker C A^k = \{0\} \quad \text{and} \quad \bigcap_{k=0}^{n-1} \ker B^H (A^H)^k = \{0\}$$

and hence that if

$$F_o(\lambda)u = 0 \quad \text{for every } \lambda \in \mathbb{C} \setminus \sigma(A)$$

$$(\text{resp.}, F_c(\lambda)u = 0 \quad \text{for every } \lambda \in \mathbb{C} \setminus \sigma(-A^H)), \text{ then } u = 0.$$

Thus the n columns of $F_o(\lambda)$ (resp., $F_c(\lambda)$) form a basis for the vector space $\mathcal{M}_o$ (resp., $\mathcal{M}_c$). Moreover, the Hermitian matrices P_o and P_c defined in (9) are both positive definite.

Lemma 1. *The spaces $\mathcal{M}_o$ and $\mathcal{M}_c$ are both n-dimensional RKHS's with respect to the inner products defined in* (11) *with RK's (reproducing kernels)*

$$K_\omega^o(\lambda) = F_o(\lambda)P_o^{-1}F_o(\omega)^H \quad \textit{for } \lambda, \omega \in \mathbb{C} \setminus \sigma(A) \tag{12}$$

and

$$K_\omega^c(\lambda) = F_c(\lambda)P_c^{-1}F_c(\omega)^H \quad \textit{for } \lambda, \omega \in \mathbb{C} \setminus \sigma(-A^H), \tag{13}$$

respectively. Moreover, the inner products defined in (11) *coincide with the normalized standard inner product.*

Proof. To verify the assertion for $\mathcal{M}_o$, it suffices to show that

(1) $K_\omega^o v \in \mathcal{M}_o$ for every $v \in \mathbb{C}^p$ and $\omega \in \mathbb{C} \setminus \sigma(A)$; and

(2) $\langle f, K_\omega^o v\rangle_{\mathcal{M}_o} = v^H f(\omega)$ for every $f \in \mathcal{M}_o$, $v \in \mathbb{C}^p$ and $\omega \in \mathbb{C} \setminus \sigma(A)$.

But this is easy, since $f \in \mathcal{M}_o$ means that $f = F_o u$ for some $u \in \mathbb{C}^n$. The verification for $\mathcal{M}_c$ is similar. □

Lemma 2. *The matrix P_o (resp., P_c) is the only solution of the Lyapunov equation*

$$A^H P_o + P_o A - C^H C = 0 \quad (\textit{resp.,} \quad AP_c + P_c A^H - BB^H = 0). \tag{14}$$

Proof. Let $E \in \mathbb{C}^{n\times n}$, $Q \in \mathbb{C}^{n\times n}$, and assuming that $\sigma(E) \cap i\mathbb{R} = \emptyset$, let

$$X_R = -\frac{1}{2\pi}\int_{-R}^{R} (i\nu I_n + E^H)^{-1} Q (i\nu I_n - E)^{-1} d\nu \quad \text{for } R > 0.$$

Then, in view of (8) and (9),

$$\lim_{R\uparrow\infty} X_R = \begin{cases} P_o & \text{if } E = A \quad \text{and} \quad Q = C^H C \\ P_c & \text{if } E = -A^H \quad \text{and} \quad Q = BB^H \end{cases}.$$

The rest of the proof amounts to showing that X_R tends to a solution of a Lyapunov equation as $R \uparrow \infty$. The argument is broken into steps.

(a) Show that

$$E^H X_R + X_R E = -\frac{1}{2\pi}\int_{-R}^{R} Q(i\nu I_n - E)^{-1} d\nu + \frac{1}{2\pi}\int_{-R}^{R} (i\nu I_n + E^H)^{-1} Q d\nu. \tag{15}$$

(b) Show that if $\sigma(E) \subset \Omega_+$ and R is large enough, then the right-hand side of (15) is equal to

$$Q + \frac{1}{2\pi}\int_{-\pi/2}^{\pi/2} \left\{(Re^{i\theta} I_n + E^H)^{-1} Q - Q(Re^{i\theta} I_n - E)^{-1}\right\} Re^{i\theta} d\theta \tag{16}$$

(c) Show that the last expression tends to Q as $R \uparrow \infty$.

(d) Show that if $P = \lim_{R\uparrow\infty} X_R$, then

$$E^H P + PE = \begin{cases} Q & \text{if } \ \sigma(E) \subset \Omega_+ \\ -Q & \text{if } \ \sigma(E) \subset \Omega_- \end{cases}.$$

Finally, the uniqueness follows from the fact that $\sigma(A) \cap \sigma(-A^H) = \emptyset$; see, e.g., Section 18.2 of [D07]. □

Remark 3. Since $\sigma(A) \subset \Omega_+$, the matrices P_o and P_c are also given by the formulas

$$P_o = \int_0^\infty e^{-tA^H} C^H C e^{-tA} dt \quad \text{and} \quad P_c = \int_0^\infty e^{-tA} BB^H e^{-tA^H} dt. \quad (17)$$

This may be verified by invoking Parseval's formula; another way is to show that the integrals are solutions of the Lyapunov equations (14).

The mvf

$$\theta_c(\lambda) = I_q - F_c(\lambda) P_c^{-1} B = I_q - B^H (\lambda I_n + A^H)^{-1} P_c^{-1} B \quad (18)$$

is inner with respect to Ω_+. This follows easily from the identity

$$I_q - \theta_c(\lambda)\theta_c(\omega)^H = (\lambda + \overline{\omega}) F_c(\lambda) P_c^{-1} F_c(\omega)^H, \quad (19)$$

which is verified by straightforward calculation with the help of the Lyapunov equation for P_c.

Similarly, the mvf

$$\theta_o(\lambda) = I_p + F_o(\lambda) P_o^{-1} C^H = I_p + C(\lambda I_n - A)^{-1} P_o^{-1} C^H \quad (20)$$

is inner with respect to Ω_-. This follows easily from the identity

$$I_p - \theta_o(\lambda)\theta_o(\omega)^H = -(\lambda + \overline{\omega}) F_o(\lambda) P_o^{-1} F_o(\omega)^H, \quad (21)$$

which is verified by straightforward calculation with the help of the Lyapunov equation for P_o.

Lemma 4. *The RKHS's $\mathcal{M}_o$ and $\mathcal{M}_c$ can be identified as*

$$\mathcal{M}_o = (H_2^p)^\perp \ominus \theta_o (H_2^p)^\perp \quad \text{and} \quad \mathcal{M}_c = H_2^q \ominus \theta_c H_2^q,$$

respectively, with RK's

$$K_\omega^o(\lambda) = -\frac{I_p - \theta_o(\lambda)\theta_o(\omega)^H}{\lambda + \overline{\omega}} \quad \text{for } \lambda, \, \omega \in \mathbb{C} \setminus \sigma(A)$$

and

$$K_\omega^c(\lambda) = \frac{I_q - \theta_c(\lambda)\theta_c(\omega)^H}{\lambda + \overline{\omega}} \quad \text{for } \lambda, \, \omega \in \mathbb{C} \setminus \sigma(-A^H).$$

Proof. The formulas for the RK's are immediate from (19) and (21) and the formulas in Lemma 1; the identification of the spaces will follow from the evaluations in the next section. □

3. Some evaluations

Lemma 5. *If Γ and $R(\lambda)$ are defined by formulas (4) and (1), respectively, then*

$$(\Gamma f)(\lambda) = F_o(\lambda) \left\{ \frac{1}{2\pi} \int_{-\infty}^\infty (A - i\nu I_n)^{-1} B f(i\nu) d\nu \right\}$$

$$= F_o(\lambda) \left\{ \frac{1}{2\pi} \int_{-\infty}^\infty F_c(i\nu)^H f(i\nu) d\nu \right\} \quad (22)$$

for every $f \in H_2^q$ when $\lambda \notin \sigma(A)$.

134 H. Dym

Proof. To ease the calculation, suppose first that $A = \text{diag}\{\omega_1, \ldots, \omega_n\}$ for some distinct set of points $\omega_1, \ldots, \omega_n$ in Ω_+, let $f \in H_2^q$ and let e_j denote the jth column of I_n. Then, since

$$C(\lambda I_n - A)^{-1}B = \sum_{j=1}^n C \frac{e_j e_j^H}{\lambda - \omega_j} B, \qquad \frac{f(\lambda) - f(\omega_j)}{\lambda - \omega_j} \text{ belongs to } H_2^q$$

and

$$\frac{f(\omega_j)}{\lambda - \omega_j} \text{ belongs to } (H_2^\perp)^q,$$

it is readily seen that

$$(\Pi_- Rf)(\lambda) = \sum_{j=1}^n C \left\{ \frac{e_j e_j^H}{\lambda - \omega_j} \right\} Bf(\omega_j)$$

$$= C \left\{ \sum_{j=1}^n \frac{e_j e_j^H}{\lambda - \omega_j} \right\} \sum_{k=1}^n e_k e_k^H Bf(\omega_k)$$

$$= C(\lambda I_n - A)^{-1} \left\{ \sum_{k=1}^n \frac{e_k e_k^H}{\lambda - \omega_k} \right\} Bf(\omega_k).$$

By Cauchy's formula for H_2^q,

$$f(\omega_k) = -\frac{1}{2\pi} \int_{-\infty}^\infty \frac{f(i\nu)}{i\nu - \omega_k} d\nu \quad \text{when } \omega_k \in \Omega_+.$$

Thus,

$$(\Pi_- Rf)(\lambda) = -F_o(\lambda) \frac{1}{2\pi} \int_{-\infty}^\infty \sum_{k=1}^n \frac{e_k e_k^H}{i\nu - \omega_k} Bf(i\nu) d\nu$$

$$= F_o(\lambda) \frac{1}{2\pi} \int_{-\infty}^\infty (A - i\nu I_n)^{-1} Bf(i\nu) d\nu,$$

which coincides with (22). This completes the proof of formula (22) when A is a diagonal matrix. The same conclusion holds if A is diagonalizable. Therefore, it also holds for general $A \in \mathbb{C}^{n \times n}$ with $\sigma(A) \subset \Omega_+$, since every such matrix can be approximated arbitrarily well by a diagonalizable matrix. $\square$

Lemma 6. *If Γ and $R(\lambda)$ are defined by formulas (4) and (1), respectively, then*

(1) *The adjoint Γ^* of Γ with respect to the standard inner product (normalized or not) maps $(H_2^\perp)^p$ into H_2^q via the formula*

$$\Gamma^* g = \Pi_+ R^\# g. \tag{23}$$

(2) *If $g \in (H_2^\perp)^p$, then*

$$(\Gamma^* g)(\lambda) = F_c(\lambda) \frac{1}{2\pi} \int_{-\infty}^\infty F_o(i\nu)^H g(i\nu) d\nu. \tag{24}$$

(3) *The formulas*

$$\Gamma F_c u = F_o P_c u \quad and \quad \Gamma^* F_o v = F_c P_o v \tag{25}$$

hold for every choice of u, $v \in \mathbb{C}^n$.

(4) *If* $f \in H_2^q$ *is orthogonal to* $F_c u$ *for every* $u \in \mathbb{C}^n$, *then* $\Gamma f = 0$.

(5) *If* $g \in (H_2^\perp)^p$ *is orthogonal to* $F_o v$ *for every* $v \in \mathbb{C}^n$, *then* $\Gamma^* g = 0$.

Proof. The first assertion follows easily from the observation that if $f \in H_2^q(\Omega_+)$ and $g \in H_2^p(\Omega_-)$, then

$$\begin{aligned}
\langle \Gamma f, g \rangle_{nst} &= \langle \Pi_- R f, g \rangle_{nst} = \langle f, R^\# g \rangle_{nst} \\
&= \langle f, \Pi_+ R^\# g \rangle_{nst}.
\end{aligned}$$

The justification of (24) is similar to the verification of (22). Again it is easiest to first verify it first for $A = \text{diag}\{\omega_1, \ldots, \omega_n\}$ and then to approximate general $A \in \mathbb{C}^{n\times n}$ by diagonalizable matrices.

Finally, (3), (4) and (5) are easy consequences of formulas (22) and (24). □

Theorem 7. *The Hankel operator* Γ *maps* $\mathcal{M}_c$ *injectively onto* $\mathcal{M}_o$ *and* Γ^* *maps* $\mathcal{M}_o$ *injectively onto* $\mathcal{M}_c$.

Proof. This is an easy consequence of the definition of the spaces $\mathcal{M}_c$ and $\mathcal{M}_o$ in (10), the formulas in (25) and the fact that (C, A) and (B^H, A^H) are observable pairs. Thus, for example, if $\Gamma f = 0$ for some $f \in \mathcal{M}_c$, then, since $f = F_c u$ for some vector $u \in \mathbb{C}^n$, $\Gamma f = \Gamma F_c u = F_o P_c u = 0$. Therefore, since (C, A) is an observable pair and P_c is invertible, $u = 0$. This proves that Γ maps the n-dimensional space $\mathcal{M}_c$ injectively into the n-dimensional space $\mathcal{M}_o$. Therefore, the mapping is also automatically onto. The asserted properties of Γ^* may be verified in much the same way. □

Lemma 8. *If* $P_c^{1/2} P_o P_c^{1/2} = UDU^H$, *where* $UU^H = U^H U = I_n$, $u_1 \ldots, u_n$ *denote the columns of* U, $D = \text{diag}\{s_1^2, \ldots, s_n^2\}$, $s_1 \geq \cdots \geq s_n > 0$ *and if*

$$f_j = F_c P_c^{-1/2} u_j \quad and \quad g_j = \frac{1}{s_j} F_o P_c^{1/2} u_j \quad for \ j = 1, \ldots, n,$$

then

$$\Gamma f_j = s_j g_j \quad and \quad \Gamma^* g_j = s_j f_j \quad for \ j = 1, \ldots, n. \tag{26}$$

Moreover,

$$\langle f_j, f_k \rangle_{nst} = \langle g_j, g_k \rangle_{nst} = \begin{cases} 0 & if \ j \neq k \\ 1 & if \ j = k \end{cases} \tag{27}$$

and

$$\langle f_j, g_k \rangle_{nst} = 0 \quad for \ j, k = 1, \ldots, n. \tag{28}$$

Proof. This is an easy consequence of the formulas in (25). □

4. J spectral factorization

This section is adapted from [F87], who cites [BR86] as a source; for additional insight on this approach, see also [GGLD88] and [Gr92]. This section is included because it leads to an interesting factorization of the mvf Θ that plays a key role in Theorem 14. It is possible to skip Sections 4 and 5 and to rely instead on the definition of Θ given in Section 7.

$$G(\lambda) = \begin{bmatrix} I_p & R(\lambda) \\ 0 & I_q \end{bmatrix} \quad \text{and} \quad J = \begin{bmatrix} I_p & 0 \\ 0 & -I_q \end{bmatrix}. \tag{29}$$

Theorem 9. *If $\|\Gamma\| < 1$, then there exists exactly one mvf $G_+ \in \mathcal{R} \cap H_\infty^{m \times m}(\Omega_+)$ such that $G_+^{-1} \in \mathcal{R} \cap H_\infty^{m \times m}(\Omega_+)$, $G_+(\infty) = I_m$ and*

$$G^{\#}(\lambda) J G(\lambda) = G_+^{\#}(\lambda) J G_+(\lambda). \tag{30}$$

It is given by the formula

$$G_+(\lambda) = I_m + \begin{bmatrix} C P_c N \\ B^H N \end{bmatrix} (\lambda I_n + A^H)^{-1} \begin{bmatrix} C^H & -P_o B \end{bmatrix}, \tag{31}$$

where

$$N = (I_n - P_o P_c)^{-1} \tag{32}$$

and

$$G_+(\lambda)^{-1} = I_m - \begin{bmatrix} C P_c \\ B^H \end{bmatrix} (\lambda I_n + A^H)^{-1} \begin{bmatrix} N C^H & -N P_o B \end{bmatrix}. \tag{33}$$

Proof. Suppose first that there exists a mvf $G_+ \in H_\infty^{m \times m}(\Omega_+)$ that meets the stated conditions and let

$$G_+(\lambda) = \begin{bmatrix} g_{11}(\lambda) & g_{12}(\lambda) \\ g_{21}(\lambda) & g_{22}(\lambda) \end{bmatrix} \quad \text{and} \quad G_+(\lambda)^{-1} = \begin{bmatrix} h_{11}(\lambda) & h_{12}(\lambda) \\ h_{21}(\lambda) & h_{22}(\lambda) \end{bmatrix}.$$

Then, since

$$G(\lambda) G_+(\lambda)^{-1} = J G^{\#}(\lambda)^{-1} G_+^{\#}(\lambda) J,$$

it follows that

$$\begin{bmatrix} I_p & R \\ 0 & I_q \end{bmatrix} \begin{bmatrix} h_{11} & h_{12} \\ h_{21} & h_{22} \end{bmatrix} = \begin{bmatrix} I_p & 0 \\ R^{\#} & -I_q \end{bmatrix} \begin{bmatrix} g_{11}^{\#} & -g_{21}^{\#} \\ g_{12}^{\#} & -g_{22}^{\#} \end{bmatrix}, \tag{34}$$

and hence, in particular, that

$$h_{11} - I_p + R h_{21} = g_{11}^{\#} - I_p \tag{35}$$

and

$$h_{21} = R^{\#} g_{11}^{\#} - g_{12}^{\#}. \tag{36}$$

This is a set of two equations with four unknown mvf's. Nevertheless it is uniquely solvable because of the constraints $G_+^{\pm 1} \in H_\infty^{m \times m}$ and $G_+(\infty) = I_m$. An application of the orthogonal projection Π_- to both sides of (35) and Π_+ to both sides of (36) column by column (in a self-evident abuse of notation) serves to eliminate $h_{11} - I_p$ in (35) and $g_{12}^{\#}$ in (36), leaving

$$\Gamma h_{21} = g_{11}^{\#} - I_p \quad \text{and} \quad h_{21} = \Gamma^*(g_{11}^{\#} - I_p) + R^{\#},$$

and implies further that the columns of $g_{11}^{\#} - I_p$ belong to the range of Γ while the columns of $h_{21} - R^{\#}$ belong to the range of Γ^*. Thus, in view of (3)–(5) of Lemma 6,

$$g_{11}^{\#}(\lambda) - I_p = F_o(\lambda) X_{11} \quad \text{for some matrix} \quad X_{11} \in \mathbb{C}^{p \times p},$$

and hence

$$h_{21} = \Gamma^* F_o X_{11} + R^{\#} = F_c P_o X_{11} - F_c C^H$$

and

$$F_o X_{11} = \Gamma h_{21} = \Gamma \{ F_c P_o X_{11} - F_c C^H \} = F_o P_c \{ P_o X_{11} - C^H \}.$$

Therefore,

$$X_{11} = P_c \{ P_o X_{11} - C^H \},$$

since (C, A) is observable. Moreover, as the assumption $\|\Gamma\| < 1$ guarantees that $I_n - P_o P_c$ and $I_n - P_c P_o$ are invertible, this implies that

$$X_{11} = -(I_n - P_c P_o)^{-1} P_c C^H.$$

Consequently,

$$g_{11}^{\#} = I_p - F_o (I_n - P_c P_o)^{-1} P_c C^H \quad \text{and} \quad h_{21} = -F_c (I_n - P_o P_c)^{-1} C^H.$$

Thus, upon substituting these last two formulas into (35) and (36), it is readily checked (with the aid of (14)) that

$$h_{11} = I_p - C P_c (\lambda I_n + A^H)^{-1} (I_n - P_o P_c)^{-1} C^H$$

and

$$g_{12}^{\#} = B^H P_o (\lambda I_n - A)^{-1} P_c N C^H.$$

In much the same way the second block columns in formula (34) may be used to compute h_{12}, h_{22}, g_{21} and g_{22} (still assuming that there exists a mvf $G_+(\lambda)$ with the stated properties) and thus to fill in the remaining entries in formulas (31) and (33). The details are left to the reader.

Conversely, it is readily checked that if G_+ is given by formula (31), then G_+^{-1} is given by formula (33) and that both of these mvf's belong to $\mathcal{R} \cap H_{\infty}^{m \times m}$, that $G_+(\lambda) G_+(\lambda)^{-1} = I_m$ and that (30) holds.

Finally, although uniqueness is really a consequence of the formulas obtained for G_+ in the first part of the proof, it is instructive to verify uniqueness a second way: suppose that there is a second mvf $\widetilde{G}_+ \in H_{\infty}^{m \times m}$ such that $(\widetilde{G}_+)^{-1} \in H_{\infty}^{m \times m}$ and (30) holds with $\widetilde{G}_+$ in place of G_+. Then, in view of (30), each entry in the left-hand side of the formula

$$J(G_+^{\#})^{-1} \widetilde{G}_+^{\#} J = G_+ (\widetilde{G}_+)^{-1}$$

is bounded in $\overline{\Omega_-}$, whereas each entry in the right-hand side is bounded in $\overline{\Omega_+}$. Thus, each entry in $G_+(\widetilde{G}_+)^{-1}$ is bounded in the whole complex plane and by Liouville's theorem must be constant. Since $G_+(\infty) = I_m$, this implies that

$$\widetilde{G}_+(\lambda) = \widetilde{G}_+(\infty) G_+(\lambda) \tag{37}$$

for all points $\lambda \in \mathbb{C}$.   □

5. The mvf $\Theta = GG_+^{-1}$

The blocks of the mvf $\Theta = GG_+^{-1}$ that are considered in this section will be the coefficients of a linear fractional description of the set $\mathcal{N}_R$ that will be developed in Section 8. This definition of Θ is taken from [F87]. However, the analysis in Section 8 is quite different from that in [F87]. As noted earlier, it is possible to skip this section and to rely instead on the characterization of Θ that will be presented in Section 7. The formulas for the blocks of Θ given below in (39) correspond to those in [F87] and [BR87].

The block entries in the mvf

$$\Theta = GG_+^{-1} = \begin{bmatrix} \theta_{11} & \theta_{12} \\ \theta_{21} & \theta_{22} \end{bmatrix} = \begin{bmatrix} h_{11} + Rh_{21} & h_{12} + Rh_{22} \\ h_{21} & h_{22} \end{bmatrix} \tag{38}$$

are given by the formulas

$$\theta_{11} = I_p - F_o(\lambda)P_cNC^H, \quad \theta_{12} = F_o(\lambda)N^H B$$
$$\theta_{21} = -F_c(\lambda)NC^H \quad \text{and} \quad \theta_{22} = I_q + F_c(\lambda)P_oN^H B. \tag{39}$$

Thus,

$$\Theta(\lambda) = I_m - \begin{bmatrix} C & 0 \\ 0 & B^H \end{bmatrix} \begin{bmatrix} \lambda I_n - A & 0 \\ 0 & \lambda I_n + A^H \end{bmatrix}^{-1} \begin{bmatrix} P_cN & N^H \\ N & P_oN^H \end{bmatrix} \begin{bmatrix} C^H & 0 \\ 0 & B \end{bmatrix} J,$$

with J as in (29). Moreover, as

$$\begin{bmatrix} P_cN & N^H \\ N & P_oN^H \end{bmatrix} = \begin{bmatrix} -P_o & I_n \\ I_n & -P_c \end{bmatrix}^{-1} \tag{40}$$

the formula for $\Theta(\lambda)$ can be rewritten as

$$\Theta(\lambda) = I_m - \tilde{C}(\lambda I_{2n} - \tilde{A})^{-1}\tilde{P}^{-1}\tilde{C}^H J, \tag{41}$$

where

$$\tilde{C} = \begin{bmatrix} C & 0 \\ 0 & B^H \end{bmatrix}, \quad \tilde{A} = \begin{bmatrix} A & 0 \\ 0 & -A^H \end{bmatrix} \quad \text{and} \quad \tilde{P} = \begin{bmatrix} -P_o & I_n \\ I_n & -P_c \end{bmatrix}. \tag{42}$$

Also

$$\theta_{11}^{-1} = I_p + C(\lambda I_n - A_1)^{-1}P_cNC^H, \quad \text{where } A_1 = A + P_cNC^HC, \tag{43}$$

and

$$\theta_{22}^{-1} = I_q - B^H(\lambda I_n + A_2)^{-1}P_oN^H B, \quad \text{where } A_2 = A^H + NP_oBB^H. \tag{44}$$

Lemma 10. $\sigma(A_1) \subset \Omega_+, \sigma(A_2) \subset \Omega_+$ and $A_1^H = N^{-1}A_2N$.

Proof. Let $P = NP_o$ and $Q = BB^H$, so that

$$A_2 = A^H + PQ \quad \text{and} \quad A_2^H = A + QP.$$

Then, since

$$A^H P_o + P_o A = C^H C \quad \text{and} \quad AP_c + P_c A^H = BB^H \tag{45}$$

it is readily checked that

$$P^{-1}A_2 + A_2^H P^{-1} = P^{-1}A^H + AP^{-1} + 2Q$$
$$= P_o^{-1}(I_n - P_oP_c)A^H + AP_o^{-1}(I_n - P_oP_c) + 2Q$$
$$= P_o^{-1}A^H + AP_o^{-1} + Q = P_o^{-1}\{A^H P_o + P_oA\}P_o^{-1} + BB^H$$
$$= P_o^{-1}C^H CP_o^{-1} + BB^H.$$

Therefore, if $A_2 x = \lambda x$ for some nonzero vector $x \in \mathbb{C}^n$ and $B^H x \neq 0$, then

$$(\lambda + \bar{\lambda})\langle P^{-1}x, x\rangle = \langle P^{-1}A_2 x, x\rangle + \langle A_2^H P^{-1}x, x\rangle$$
$$= x^H \{P_o^{-1}C^H CP_o^{-1} + BB^H\}x > 0.$$

Thus, as P is positive definite, $\lambda + \bar{\lambda} > 0$.

On the other hand, if $A_2 x = \lambda x$ for some nonzero vector $x \in \mathbb{C}^n$ and $B^H x = 0$, then

$$\lambda x = A_2 x = A^H x + PBB^H x = A^H x,$$

which exhibits λ as an eigenvalue of A^H, and $\sigma(A^H) \subset \Omega_+$, by assumption. This completes the proof that $\sigma(A_2) \subset \Omega_+$.

Next, the formulas in (45) imply that

$$A^H P_oP_c + P_oAP_c = C^H CP_c \quad \text{and} \quad P_oAP_c + P_oP_cA^H = P_oBB^H.$$

Therefore, upon subtracting the second equation from the first,

$$A^H(P_oP_c - I_n) + (I_n - P_oP_c)A^H = C^H CP_c - P_oBB^H,$$

or, equivalently,

$$-A^H N^{-1} + N^{-1}A^H = C^H CP_c - P_oBB^H.$$

Thus,

$$A_1^H = A^H + C^H CP_cN = N^{-1}(A^H + NP_oBB^H)N = N^{-1}A_2N,$$

which serves to show that $\sigma(A_1^H) = \sigma(A_2)$. Therefore, since $\sigma(A_2) \subset \Omega_+$, it follows that $\sigma(A_1) \subset \Omega_+$ also. □

Remark 11. The verification of the fact that $\sigma(A_1) \subset \Omega_+$ can also be carried out much as in Step 1. To do this, let

$$\Pi_1 = P_cN \quad \text{and} \quad Q_1 = C^H C$$

for short. Then

$$A_1^H \Pi_1^{-1} + \Pi_1^{-1}A_1 = (A^H + Q_1\Pi_1)\Pi_1^{-1} + \Pi_1^{-1}(A + \Pi_1Q_1)$$
$$= A^H(I_n - P_oP_c)P_c^{-1} + (I_n - P_oP_c)P_c^{-1}A + 2Q_1$$
$$= P_c^{-1}\{P_cA^H + AP_c\}P_c^{-1} + C^H C = P_c^{-1}BB^H P_c^{-1} + C^H C.$$

Thus, if $A_1 x = \lambda x$ for some nonzerovector $x \in \mathbb{C}^n$, then

$$(\lambda + \bar{\lambda})\langle \Pi_1^{-1}x, x\rangle = \langle \Pi_1^{-1}A_1 x, x\rangle + \langle A_1^H \Pi_1^{-1}x, x\rangle = \|B^H P_c^{-1}x\|^2 + \|Cx\|^2.$$

Consequently, as Π_1 is positive definite, $\lambda + \overline{\lambda} > 0$ if $Cx \neq 0$. However, if $Cx = 0$, then $A_1 x = Ax$, i.e., $\lambda \in \sigma(A) \subset \Omega_+$.

Theorem 12. *The blocks in the mvf Θ enjoy the following properties*

$$(\theta_{11}^\#)^{\pm 1} \in \mathcal{R} \cap H_\infty^{p \times p}, \qquad \theta_{12}^\# \in \mathcal{R} \cap H_\infty^{q \times p},$$

$$\theta_{21} \in \mathcal{R} \cap H_\infty^{q \times p} \quad and \quad \theta_{22}^{\pm 1} \in \mathcal{R} \cap H_\infty^{q \times q}.$$

Proof. This is immediate from Lemma 10 and the exhibited formulas for the blocks of Θ. $\qquad\square$

Remark 13. If $f \in \mathcal{R} \cap H_\infty^{q \times p}$ and $\lim_{|\lambda| \uparrow \infty} f(\lambda) = \gamma$, then f belongs to the Wiener plus space (with respect to Ω_+)

$$\mathcal{W}_+^{q \times p}(\gamma) = \{f : f(\lambda) = \gamma + \int_0^\infty e^{-\lambda t} h(t) dt \quad and \quad h \in L_1^{q \times p}(\mathbb{R}_+)\}.$$

Thus, in view of Theorem 12 and formulas (39), (43) and (44),

$$(\theta_{11}^\#)^{\pm 1} \in \mathcal{R} \cap \mathcal{W}_+^{p \times p}(I_p), \qquad \theta_{12}^\# \in \mathcal{R} \cap \mathcal{W}_+^{q \times p}(0)$$

$$\theta_{21} \in \mathcal{R} \cap \mathcal{W}_+^{q \times p}(0) \quad and \quad \theta_{22}^{\pm 1} \in \mathcal{R} \cap \mathcal{W}_+^{q \times q}(I_q).$$

Thus, Θ belongs to the class called $\mathcal{W}_r^{q \times p}(j_{pq})$ in [ArD12] (but with respect to Ω_+ not $\mathbb{C}_+$).

6. Detour on kernel formulas

Let

$$F(\lambda) = \begin{bmatrix} F_o(\lambda) & 0 \\ 0 & F_c(\lambda) \end{bmatrix} \quad and \quad \mathfrak{h}_F = \mathbb{C} \setminus \{\sigma(A) \cup \sigma(-A^H)\}. \tag{46}$$

The space

$$\mathcal{M} = \begin{matrix} \mathcal{M}_o \\ \oplus \\ \mathcal{M}_c \end{matrix} = \{F(\lambda)x : x \in \mathbb{C}^{2n}\}$$

endowed with the normalized standard inner product is an RKHS with RK

$$K_\omega(\lambda) = \begin{bmatrix} F_o(\lambda) P_o^{-1} F_o(\omega)^H & 0 \\ 0 & F_c(\lambda) P_c^{-1} F_c(\omega)^H \end{bmatrix} \quad \text{on } \mathfrak{h}_F \times \mathfrak{h}_F.$$

Thus,

$$K_\omega x \in \mathcal{M} \quad and \quad x^H F(\omega) y = \langle Fy, K_\omega x \rangle_{nst}$$

for every choice of $\omega \in \mathfrak{h}_F$ and $x, y \in \mathbb{C}^{2n}$.

The restriction of the operator

$$\begin{bmatrix} I & -\Gamma \\ -\Gamma^* & I \end{bmatrix}$$

to $\mathcal{M}$ can also be expressed in terms of a kernel: If

$$G_\omega(\lambda) = F(\lambda) \begin{bmatrix} P_o^{-1} & -I_n \\ -I_n & P_c^{-1} \end{bmatrix} F(\omega)^H \quad \text{on } \mathfrak{h}_F \times \mathfrak{h}_F,$$

then

$$x^H \left(\begin{bmatrix} I & -\Gamma \\ -\Gamma^* & I \end{bmatrix} \begin{bmatrix} F_o u \\ F_c v \end{bmatrix} \right) (\omega) = x^H \begin{bmatrix} F_o(\omega)(u - P_c v) \\ F_c(\omega)(v - P_o u) \end{bmatrix} = \left\langle F \begin{bmatrix} u \\ v \end{bmatrix}, G_\omega x \right\rangle_{nst}$$

for every choice of $\omega \in \mathfrak{h}_F$, $u, v \in \mathbb{C}^n$ and $x \in \mathbb{C}^{2n}$.

The two kernels are simply related:

$$G_\omega(\lambda) = K_\omega(\lambda) + F(\lambda) \begin{bmatrix} 0 & I_n \\ I_n & 0 \end{bmatrix} F(\omega)^H \quad \text{on } \mathfrak{h}_F \times \mathfrak{h}_F. \tag{47}$$

7. An alternate characterization of Θ

Since the columns of

$$\Theta - I_m \quad \text{and} \quad \begin{bmatrix} 0 & R \\ R^\# & 0 \end{bmatrix}.$$

belong to $\mathcal{M}$, formulas (22) and (23) imply that if Γ and Γ^* act on mvf's column by column, then $\Theta - I_m$ can be characterized as the unique solution of the operator equation

$$\begin{bmatrix} I & -\Gamma \\ -\Gamma^* & I \end{bmatrix} \begin{bmatrix} \theta_{11} - I_p & \theta_{12} \\ \theta_{21} & \theta_{22} - I_q \end{bmatrix} = \begin{bmatrix} 0 & R \\ R^\# & 0 \end{bmatrix} \tag{48}$$

with columns in $\mathcal{M}$ when $\|\Gamma\| < 1$.

It is easy to derive the formulas for the blocks of Θ given in (39) directly from (48) when $\|\Gamma\| < 1$ by invoking the characterizations of the spaces $\mathcal{M}_o$ and $\mathcal{M}_c$ given in (10) and the evaluations in (25). Thus, for example, in view of (10), $\theta_{22} - I_q = F_c X_{22}$ and $\theta_{12} = F_o X_{12}$ for some choice of $X_{22} \in \mathbb{C}^{q \times q}$ and $X_{12} \in \mathbb{C}^{p \times q}$. Consequently, the formula

$$(I - \Gamma^* \Gamma)(\theta_{22} - I_q) = \Gamma^* R,$$

that is obtained from (48), translates to

$$(I - \Gamma^* \Gamma) F_c X_{22} = \Gamma^* F_o B,$$

which, in view of (25), leads to

$$F_c X_{22} - F_c P_o P_c X_{22} = F_c P_o B.$$

Therefore, since (B^H, A) is an observable pair,

$$\theta_{22} - I_q = F_c X_{22} = F_c (I_n - P_c P_o)^{-1} P_o B,$$

which coincides with the formula for θ_{22} in (39). The other blocks of Θ can be obtained in much the same way.

Characterizations of *resolvent matrices* for the Nehari problems analogous to (48) are valid in much more general settings; see, e.g., [KMA86], [Dy89], [ArD12] and the references cited therein; and, for a comprehensive treatise on the discrete Nehari problem, [Pe03].

8. Linear fractional transformations based on Θ

The main objective of this section is to parametrize the set $\mathcal{N}_R$. Analogous descriptions were obtained earlier by other means in [BH83].

Theorem 14. *If $\|\Gamma\| < 1$ and Θ is given by (41) and (42) (or by (48)), then*

$$\mathcal{N}_R = \{T_\Theta[\varepsilon] : \varepsilon \in \mathcal{R} \cap \mathcal{S}^{p \times q}\} \qquad (49)$$

where

$$T_\Theta[\varepsilon] = (\theta_{11}\varepsilon + \theta_{12})(\theta_{21}\varepsilon + \theta_{22})^{-1}. \qquad (50)$$

Proof. The claim follows from Lemmas 16 and 18, which will be established below.
$\qquad\qquad\qquad\qquad\qquad\qquad\qquad\qquad\qquad\qquad\qquad\qquad\qquad\qquad\quad\square$

It is convenient, however, to first introduce the Potapov-Ginzburg transform Σ of Θ:

$$\Sigma = \begin{bmatrix} \sigma_{11} & \sigma_{12} \\ \sigma_{21} & \sigma_{22} \end{bmatrix} = \begin{bmatrix} \theta_{11} & \theta_{12} \\ 0 & I_q \end{bmatrix} \begin{bmatrix} I_p & 0 \\ \theta_{21} & \theta_{22} \end{bmatrix}^{-1}$$

$$= \begin{bmatrix} \theta_{11} - \theta_{12}\theta_{22}^{-1}\theta_{21} & \theta_{12}\theta_{22}^{-1} \\ -\theta_{22}^{-1}\theta_{21} & \theta_{22}^{-1} \end{bmatrix},$$

which is unitary on $i\mathbb{R}$. The blocks of Σ are given by the formulas

$$\sigma_{11} = I_p - CP_c(\lambda I_n + A_2)^{-1}NC^H,$$

$$\sigma_{12} = C(\lambda I_n - A)^{-1}P_o^{-1}(\lambda I_n + A^H)(\lambda I_n + A_2)^{-1}P_oN^HB, \qquad (51)$$

$$\sigma_{21} = B^H(\lambda I_n + A_2)^{-1}NC^H \quad \text{and} \quad \sigma_{22} = I_q - B^H(\lambda I_n + A_2)^{-1}P_oN^HB,$$

whereas,

$$\sigma_{11}^{-1} = I_p + CP_cN(\lambda I_n + A^H)^{-1}C^H, \quad \sigma_{22}^{-1} = I_q + B^H(\lambda I_n + A^H)^{-1}P_oN^HB,$$

and

$$\sigma_{12} - R = CP_c(\lambda I_n + A_2)^{-1}P_oN^HB.$$

Lemma 15. $\sigma_{11}^{\pm 1} \in \mathcal{R} \cap H_\infty^{p \times p}$, $\sigma_{21} \in \mathcal{R} \cap H_\infty^{p \times q}$, $\sigma_{22}^{\pm 1} \in \mathcal{R} \cap H_\infty^{q \times q}$ and $\|\sigma_{21}(\omega)\| \le \delta < 1$ *for every point* $\omega \in \overline{\Omega_+}$.

Proof. The stated inclusions are clear from Lemma 10 and the formulas for $\sigma_{11}^{\pm 1}$, σ_{21} and $\sigma_{22}^{\pm 1}$ that are displayed just above. Let $\mathfrak{s}_1(i\nu) \ge \cdots \ge \mathfrak{s}_q(i\nu)$ denote the singular values of $\sigma_{22}(i\nu)$. Then, since Σ is unitary on $i\mathbb{R}$,

$$\sigma_{21}(i\nu)\sigma_{21}(i\nu)^H = I_q - \sigma_{22}(i\nu)\sigma_{22}(i\nu)^H$$

and hence

$$\|\sigma_{21}(i\nu)\|^2 = 1 - \mathfrak{s}_q(i\nu)^2.$$

Therefore, since

$$\mathfrak{s}_q(i\nu)^{-1} = \|\sigma_{22}(i\nu)^{-1}\|$$

and there exists a constant $\rho > 1$ such that

$$\|\sigma_{22}(i\nu)^{-1}\| \le \rho \quad \text{for all } \nu \in \mathbb{R},$$

it follows that

$$\|\sigma_{21}(i\nu)\|^2 \le 1 - \rho^{-2},$$

which serves to establish the advertised bound with $\delta = (1 - \rho^{-2})^{1/2}$.   □

Lemma 16. *If $\|\Gamma\| < 1$, $Y \in \mathcal{R} \cap \mathcal{S}^{p \times q}$ and $\Theta(\lambda)$ is defined by formula (38), then*

$$T_\Theta[Y] \in \mathcal{N}_R. \tag{52}$$

Proof. The formula

$$T_\Theta[Y] - R = (h_{11}Y + h_{12})(h_{21}Y + h_{22})^{-1}$$

follows easily from (38) and (50). Moreover, as Θ is J-unitary on $i\mathbb{R}$, it is readily checked that

$$Y \in \mathcal{R} \cap \mathcal{S}^{p \times q} \implies \|(T_\Theta[Y])(i\nu)\| \le 1 \quad \text{for } \nu \in \mathbb{R}.$$

Thus, it remains to show that

$$T_\Theta[Y] - R \in H_\infty^{p \times q}. \tag{53}$$

Since $(h_{11}Y + h_{12}) \in \mathcal{R} \cap H_\infty^{p \times q}$ and

$$(h_{21}Y + h_{22})^{-1} = (I_q - \sigma_{21}Y)^{-1}\sigma_{22},$$

this reduces to checking that

$$(I_q - \sigma_{21}Y)^{-1}\sigma_{22} \in \mathcal{R} \cap H_\infty^{q \times q}.$$

But, as $\sigma_{22} \in \mathcal{R} \cap H_\infty^{q \times q}$, this follows easily from the bounds established in Lemma 15:

$$\|(I_q - (\sigma_{21}Y)(\lambda))^{-1}\| \le \sum_{k=0}^{\infty} \|(\sigma_{21}Y)(\lambda)\|^k \le \sum_{k=0}^{\infty} \delta^k = (1 - \delta)^{-1}$$

for every point $\lambda \in \Omega_+$, which serves to complete the proof of the theorem.   □

At this point it is convenient to pause from the main development in order to establish some facts from linear algebra that will be needed to establish a converse to Lemma 16.

Lemma 17. *If $T \in \mathbb{C}^{q \times q}$ and $T + T^H \ge \delta I_q$ for some $\delta > 0$, then T is invertible and*

$$\|T^{-1}\| \le \frac{2}{\delta}. \tag{54}$$

If $U \in \mathbb{C}^{q \times q}$ and $\|U\| < 1$, then $I_q + U$ is invertible and

$$(I_q + U)^{-1} + (I_q + U^H)^{-1} \ge I_q. \tag{55}$$

Proof. If $T + T^H \ge \delta I_q$ and $\langle Tx, x \rangle = a + ib$ (with $a, b \in \mathbb{R}$), then

$$2a = \langle (T + T^H)x, x \rangle \ge \delta \langle I_q x, x \rangle = \delta \|x\|^2.$$

Therefore, since

$$|a| \le \sqrt{a^2 + b^2} = |\langle Tx, x \rangle| \le \|Tx\| \, \|x\|,$$

it follows that

$$\|Tx\| \geq \frac{\delta}{2}\|x\| \quad \text{for every } x \in \mathbb{C}^q, \tag{56}$$

which proves that T is invertible and, upon setting $x = T^{-1}y$, that

$$\|T^{-1}y\| \leq \frac{2}{\delta}\|y\| \quad \text{for every } y \in \mathbb{C}^q,$$

which justifies (54).

Next, observe that if $\|U\| < 1$, then $I_q + U$ is invertible and the real part of the Cayley transform

$$V = (I_q - U)(I_q + U)^{-1} = (2I_q - (I_q + U))(I_q + U)^{-1} = 2(I_q + U)^{-1} - I_q$$

of U is positive semidefinite, i.e.,

$$V + V^H \geq 0.$$

Thus, as $(I_q + U)^{-1} = (V + I_q)/2$,

$$(I_q + U)^{-1} + (I_q + U^H)^{-1} = I_q + (V + V^H)/2 \geq I_q,$$

which justifies (55). $\square$

Lemma 18. *If $\|\Gamma\| < 1$ and $\Theta(\lambda)$ is defined by formulas (41) and (42), then*

$$\mathcal{N}_R \subseteq \{T_\Theta[\varepsilon] : \varepsilon \in \mathcal{R} \cap \mathcal{S}^{p \times q}\}. \tag{57}$$

Proof. Let $S \in \mathcal{N}_R$ and set

$$Y = T_{\Theta^{-1}}[S] \quad \text{on } i\mathbb{R}.$$

Then

$$S \in \mathcal{R} \cap L_\infty^{p \times q}, \quad S - R \in \mathcal{R} \cap H_\infty^{p \times q}, \quad \|S(i\nu)\| \leq 1 \text{ for } \nu \in \mathbb{R}$$

and, since $\Theta(i\nu)^H J\Theta(i\nu) = J$ for $\nu \in \mathbb{R}$, $\|Y(i\nu)\| \leq 1$ for $\nu \in \mathbb{R}$.

It remains to show that $Y \in \mathcal{S}^{p \times q}$. Towards this end, it is convenient to first reexpress $S = T_\Theta[Y]$ in terms of the entries in the Potapov-Ginzburg transform Σ of Θ as

$$S = T_\Theta[Y] = \sigma_{12} + \sigma_{11}Y(I_q - \sigma_{21}Y)^{-1}\sigma_{22}$$

and then to proceed in steps.

1. $Y(I_q - \sigma_{21}Y)^{-1} \in \mathcal{R} \cap H_\infty^{p \times q}$:

Since $\sigma_{12} = T_\Theta[0]$, Lemma 16 guarantees that $\sigma_{12} \in \mathcal{N}_R$ and hence that $\sigma_{12} - R \in \mathcal{R} \cap H_\infty^{p \times q}$. Thus,

$$S - \sigma_{12} = (S - R) - (\sigma_{12} - R) \in \mathcal{R} \cap H_\infty^{p \times q}.$$

Therefore, as $\sigma_{11}^{-1} \in \mathcal{R} \cap H_\infty^{p \times p}$ and $\sigma_{22}^{-1} \in \mathcal{R} \cap H_\infty^{q \times q}$, it follows that

$$Y(I_q - \sigma_{21}Y)^{-1} = \sigma_{11}^{-1}(S - \sigma_{12})\sigma_{22}^{-1} \quad \text{belongs to } \mathcal{R} \cap H_\infty^{p \times q}.$$

2. $(I_q - \sigma_{21}Y) \in \mathcal{R} \cap H_\infty^{p \times q}$:

Since $\sigma_{21} \in \mathcal{R} \cap H_\infty^{q \times p}$, Step 1 and the identity

$$I_q + \sigma_{21}Y(I_q - \sigma_{21}Y)^{-1} = (I_q - \sigma_{21}Y)^{-1}$$

imply that the mvf

$$f = (I_q - \sigma_{21}Y)^{-1}$$

belongs to $\mathcal{R} \cap H_\infty^{q \times q}$. The bound on σ_{21} in Lemma 15 implies that

$$\|\sigma_{21}(i\nu)Y(i\nu)\| \le \|\sigma_{21}(i\nu)\| \le \delta < 1 \quad \text{for every } \nu \in \mathbb{R},$$

Lemma 17 implies that

$$f(i\nu) + f(i\nu)^H \ge I_q,$$

the Poisson formula for the right half-plane (2) implies that

$$f(a+ib) + f(a+ib)^H = \frac{a}{\pi} \int_{-\infty}^{\infty} \frac{f(i\nu) + f(i\nu)^H}{a^2 + (\nu - b)^2} d\nu$$

$$\ge \frac{a}{\pi} \int_{-\infty}^{\infty} \frac{I_q}{a^2 + (\nu - b)^2} d\nu = I_q \quad \text{for } a > 0 \text{ and } b \in \mathbb{R}.$$

Therefore, by another application of Lemma 17, f is invertible in Ω_+ and $\|f(a+ib)^{-1}\| \le 2$ for $a > 0$. Thus,

$$f^{-1} = I_q - \sigma_{21}Y = (g_{21}\varepsilon + g_{22})^{-1} \quad \text{belongs to } \mathcal{R} \cap H_\infty^{q \times q}.$$

3. $Y \in \mathcal{R} \cap \mathcal{S}^{p \times q}$:

The preceding steps imply that

$$Y(I_q - \sigma_{21}Y)^{-1} \in \mathcal{R} \cap H_\infty^{p \times q} \quad \text{and} \quad (I_q - \sigma_{21}Y) \in \mathcal{R} \cap H_\infty^{q \times q}.$$

Therefore, since Y is the product of these two mvf's, it belongs to $\mathcal{R} \cap H_\infty^{p \times q}$. Moreover, since $\|Y(i\nu)\| \le 1$, the maximum principle (3) guarantees that $Y \in \mathcal{R} \cap \mathcal{S}^{p \times q}$. □

Remark 19. The proof of Lemma 18 may also be based on the formula

$$\Theta^{-1} = G_+ G^{-1} = \begin{bmatrix} g_{11} & -g_{11}R + g_{12} \\ g_{21} & -g_{21}R + g_{22} \end{bmatrix}.$$

Then

$$Y = T_{\Theta^{-1}}[S] = T_{G_+}[S - R] = (g_{11}\varepsilon + g_{12})(g_{21}\varepsilon + g_{22})^{-1},$$

with $\varepsilon = S - R \in \mathcal{R} \cap H_\infty^{p \times q}$, and

$$I_q - \sigma_{21}Y = (g_{21}\varepsilon + g_{22})^{-1} \quad \text{and} \quad Y(I_q - \sigma_{21}Y)^{-1} = (g_{11}\varepsilon + g_{12}).$$

In this formalism, it is easy to see that $(g_{11}\varepsilon + g_{12}) \in \mathcal{R} \cap H_\infty^{p \times q}$ and $(g_{21}\varepsilon + g_{22}) \in \mathcal{R} \cap H_\infty^{q \times q}$, but not so easy to verify that $(g_{21}\varepsilon + g_{22})^{-1} \in \mathcal{R} \cap H_\infty^{q \times q}$.

Remark 20. (An example connected with study of g_{22}^{-1})
   If $R(\lambda) = c(\lambda - a)^{-1}b$ with $a > 0$ and $b = \bar{c}$, then

$$P_o = P_c = \frac{|c|^2}{2a}, \quad N = \frac{4a^2}{4a^2 - |c|^4}$$

and

$$A - N^H BB^H P_o = a \left\{ 1 - \frac{2|c|^4}{4a^2 - |c|^4} \right\}.$$

There is no guarantee that this number is positive, since the only constraint is that $N > 0$, which is not strong enough (it only guarantees that $4a^2 - |c|^4 > 0$). However, if is also assumed that $\|R(i\nu)\| \leq 1$ for all $\nu \in \mathbb{R}$, then there is an additional constraint:

$$\frac{|c|^4}{\nu^2 + a^2} \leq 1 \quad \text{for all } \nu \in \mathbb{R}$$

and hence that

$$\frac{|c|^4}{a^2} \leq 1,$$

in which case $\sigma(A - N^H BB^H P_o) \subset \Omega_+$.

9. The finite-dimensional Krein space $\mathcal{K}(\Theta)$

If $\widetilde{A}$, $\widetilde{C}$ and $\widetilde{P}$ are defined by formula (42), then it is readily checked (with the aid of the Lyapunov equations (14) that

$$\widetilde{A}^H \widetilde{P} + \widetilde{P}\widetilde{A} + \widetilde{C}^H J \widetilde{C} = 0 \tag{58}$$

and hence that the mvf Θ defined by formula (41) satisfies the identity

$$J - \Theta(\lambda)J\Theta(\omega)^H = (\lambda + \overline{\omega})\widetilde{C}(\lambda I_{2n} - \widetilde{A})^{-1}\widetilde{P}^{-1}(\overline{\omega}I_{2n} - \widetilde{A}^H)^{-1}\widetilde{C}^H \tag{59}$$

for $\lambda, \omega \in \mathbb{C} \setminus \sigma(\widetilde{A})$.

Lemma 21. *The matrix $\widetilde{P}$ defined in (42) has n positive eigenvalues and n negative eigenvalues (counting multiplicities).*

Proof. By Schur complements,

$$\widetilde{P} = \begin{bmatrix} I_n & -P_c^{-1} \\ 0 & I_n \end{bmatrix} \begin{bmatrix} P_c^{-1} - P_o & 0 \\ 0 & -P_c \end{bmatrix} \begin{bmatrix} I_n & 0 \\ -P_c^{-1} & I_n \end{bmatrix}.$$

Therefore, the signature of $\widetilde{P}$ is equal to the signature of the matrix

$$\begin{bmatrix} P_c^{-1} - P_o & 0 \\ 0 & -P_c \end{bmatrix}.$$

The asserted claim now follows easily from the fact that

$$P_c^{-1} - P_o = P_c^{-1/2}(I_n - P_c^{1/2}P_oP_c^{1/2})P_c^{1/2}$$

is positive definite and $-P_c$ is negative definite. □

Thus, as the pair $(\widetilde{C}, \widetilde{A})$ is observable, the preceding analysis implies that:

Theorem 22. *If $F(\lambda)$ is defined as in (46), then the space*

$$\mathcal{M} = \{F(\lambda)x : x \in \mathbb{C}^{2n}\},$$

endowed with the inner product

$$\langle Fx, Fy \rangle = y^H \widetilde{P}x \quad \text{for every choice of } x, y \in \mathbb{C}^{2n}$$

is a finite-dimensional RKKS with n negative squares and RK

$$K_\omega(\lambda) = F(\lambda)\widetilde{P}^{-1}F(\omega)^H = \frac{J - \Theta(\lambda)J\Theta(\omega)^H}{\lambda + \overline{\omega}}$$

for $\lambda, \omega \in \mathbb{C} \setminus \sigma(\widetilde{A})$.

Thus, the characterization of the set $\{T_\Theta[s] : s \in \mathcal{S}^{p \times q}]\}$ that was developed in [DD09] and [DD10] is applicable. However, because of lack of space and time, we shall not pursue this here. Another route to information on this set will be considered in the next section.

10. Detour on J-inner mvf's

A lengthy but straightforward calculation leads to the identity

$$W(\lambda) = \begin{bmatrix} \theta_o(\lambda) & 0 \\ 0 & \theta_c(\lambda) \end{bmatrix}^{-1} \Theta(\lambda) = I_m - V(\lambda I_{2n} - M)^{-1}Q^{-1}V^H J, \qquad (60)$$

where

$$V = \begin{bmatrix} 0 & CP_o^{-1} \\ B^H P_c^{-1} & 0 \end{bmatrix}, \quad M = \begin{bmatrix} A & 0 \\ 0 & -A^H \end{bmatrix} \quad \text{and} \quad Q = \begin{bmatrix} P_c^{-1} & -I_n \\ -I_n & P_o^{-1} \end{bmatrix}$$

Moreover, since

$$J - W(\lambda)JW(\omega)^H = (\lambda + \overline{\omega})V(\lambda I_{2n} - M)^{-1}Q^{-1}(\overline{\omega}I_{2n} - M^H)^{-1}V^H$$

and Q is positive definite, it follows that W is J-inner with respect to Ω_+ (i.e., $J - W(\omega)JW(\omega)^* \geq 0$ for $\omega \in \Omega_+$ with equality on $i\mathbb{R}$). Therefore,

$$T_W[\varepsilon] \in \mathcal{S}^{p \times q} \quad \text{for every } \varepsilon \in \mathcal{S}^{p \times q}$$

and, in view of (60),

$$T_\Theta[\varepsilon] = \theta_o \, T_W[\varepsilon] \, \theta_c^{-1}.$$

Since both of the multipliers, θ_o and θ_c^{-1} are contributing poles in Ω_+, it seems at first glance that $T_\Theta[\varepsilon]$ may have up to $2n$ poles in Ω_+. However, this is not the case because $s = T_W[\varepsilon]$ has compensating zeros. In fact the characterization of the set $\{T_W[\varepsilon] : \varepsilon \in \mathcal{S}^{p \times q}\}$ in [Dy03] implies that if mvf $F(\lambda) = V(\lambda I_{2n} - M)^{-1}$, then

$$[I_p \quad - s]Fx \in H_2^p \qquad \text{for every } x \in \mathbb{C}^{2n} \qquad (61)$$

and

$$[-s^\# \quad I_q]Fx \in (H_2^q)^\perp \quad \text{for every } x \in \mathbb{C}^{2n}. \qquad (62)$$

In the present setting the constraint (61) implies that

$$sB^H P_c^{-1}(\lambda I_n - A)^{-1}u \in H_2^p \quad \text{for every } u \in \mathbb{C}^n$$

and hence that

$$s\theta_c^{-1} = s(I_p + B^H P_c^{-1}(\lambda I_n - A)^{-1}B) \quad \text{belongs to } H_\infty^{p \times q}.$$

Similarly, the constraint (62) implies that

$$v^H(\lambda I_n - A)^{-1}P_o^{-1}C^H s \in H_2^{q \times 1} \quad \text{for every } v \in \mathbb{C}^n$$

and hence that

$$\theta_o s = (I_p + C(\lambda I_n - A)^{-1}P_o^{-1}C^H)s \quad \text{belongs to } H_\infty^{p \times q}.$$

Thus, s will have at most n poles in $\mathbb{C}_+$.

Remark 23. Formula (60) exhibits a factorization of the J-inner mvf W into the product of a diagonal matrix based on the *associated pairs* $\{\theta_o^{-1}, \theta_c\}$ of W and a *gamma generating matrix* Θ. The general theory of such factorizations originate in the work of D.Z. Arov in the late eighties; see [Ar89]; and for additional discussion, developments and references, [ArD08].

I thank V. Derkach, B. Francis, M. Porat and M. Putinar for reading and commenting on early versions of this paper.

References

[AAK71a] Vadim M. Adamjan, Damir Z. Arov and Mark G. Krein, Analytic properties of Schmidt pairs for a Hankel operator and the generalized Schur-Takagi problem, Math. USSR sb. **15** (1971), 31–73.

[Ar89] Damir Z. Arov, γ-generating matrices, j-inner matrix-functions and related extrapolation problems. Teor. Funktsii Funktsional. Anal. i Prilozhen, I, **51** (1989), 61–67; II, **52** (1989), 103–109; translation in J. Soviet Math. I, **52** (1990), 3487–3491; III, **52** (1990), 3421–3425.

[ArD08] Damir Z. Arov and Harry Dym, *J-contractive matrix valued functions and related topics*, Encyclopedia of Mathematics and its Applications, **116**, Cambridge University Press, Cambridge, 2008.

[ArD12] Damir Z. Arov and Harry Dym, *Bitangential Direct and Inverse Problems for Systems of Differential Equations*, Cambridge University Press, Cambridge, 2012.

[BH83] Joseph A. Ball and J. William Helton, A Beurling-Lax theorem for the Lie group $U(m,n)$ which contains most classical interpolation theory, J. Operator Theory, **9** (1983), 107–142.

[BR86] Joseph A. Ball and Andre C.M. Ran, Hankel norm approximation of a rational matrix function in terms of its realization, in: C.I. Bymes and A. Lindquist, eds., *Modeling, Identification and Robust Control*, North-Holland, Amsterdam, 1986, pp. 285–296.

[BR87] Joseph A. Ball and Andre C.M. Ran, Optimal Hankel norm model reduction and Wiener-Hopf factorization, I: The canonical case, SIAM J. Control Optim., 25 (1987), 362–383.

[DD10] Vladimir Derkach and Harry Dym, Bitangential interpolation in generalized Schur classes. Complex Anal. Oper. Theory 4 (2010), no. 4, 701–765,

[DD09]   Vladimir Derkach and Harry Dym, On linear fractional transformations associated with generalized J-inner matrix functions. Integral Equations Operator Theory 65 (2009), no. 1, 1–50

[Dy89]   Harry Dym, *J contractive matrix functions, reproducing kernel Hilbert spaces and interpolation*, CBMS Regional Conference Series in Mathematics, **71**, American Mathematical Society, Providence, RI, 1989.

[D07]   Harry Dym, *Linear Algebra in Action*, American Mathematical Society, Providence, R.I., 2007.

[Dy03]   Harry Dym, Linear fractional transformations, Riccati equations and bitangential interpolation, revisited, in: *Reproducing kernel spaces and applications*, Oper. Theory Adv. Appl., **143**, Birkhäuser, Basel, 2003, pp. 171–212.

[F87]   Bruce A. Francis, *A Course in H_∞ Control Theory*, Lecture Notes in Control and Information Sciences, **88**, Springer-Verlag, Berlin, 1987.

[FD87]   Bruce A. Francis and John C. Doyle, Linear control theory with an $H\infty$ optimality criterion, SIAM J. Control Optim. **25** (1987), no. 4, 815–844,

[Gl89]   Keith Glover, A tutorial on Hankel norm approximation, in: *Data to Model*, J.C. Willems, ed., Springer Verlag, New York, 1989

[Gr92]   Michael Green, H_∞ controller synthesis by J-lossless coprime factorization, SIAM J. Control and Optimization, **30**, no. 3 (1992), 522–547.

[GGLD88] Michael Green, Keith Glover, David J.N. Limebeer and John C. Doyle, A J-spectral factorization approach to H_∞ control, SIAM J. Control and Optimization, **28** (1988), 1350–1371.

[KMA86] Mark G. Krein and Felix E. Melik-Adamjan, Matrix-continuous analogues of the Schur and the Carathéodory-Toeplitz problem, (Russian) Izv. Akad. Nauk Armyan, SSR Ser. Mat. **21** (1986), no. 2, 10–141, 207.

[Pe03]   Vladimir V. Peller, *Hankel Operators and their Applications*, Springer-Verlag, New York 2003.

[ZDG96] Kemin Zhou, John C. Doyle and Keith Glover, *Robust and Optimal Control*, Prentice Hall, New Jersey, 1996.

Harry Dym
Department of Mathematics
The Weizmann Institute of Science
Rehovot 76100, Israel
e-mail: harry.dym@weizmann.ac.il

Operator Theory:
Advances and Applications, Vol. 222, 151–172
© 2012 Springer Basel

Optimal Solutions to Matrix-valued Nehari Problems and Related Limit Theorems

A.E. Frazho, S. ter Horst and M.A. Kaashoek

*Dedicated to J. William Helton, on the occasion of his
65th birthday, with admiration and friendship.*

Abstract. In a 1990 paper Helton and Young showed that under certain conditions the optimal solution of the Nehari problem corresponding to a finite rank Hankel operator with scalar entries can be efficiently approximated by certain functions defined in terms of finite-dimensional restrictions of the Hankel operator. In this paper it is shown that these approximations appear as optimal solutions to restricted Nehari problems. The latter problems can be solved using relaxed commutant lifting theory. This observation is used to extent the Helton and Young approximation result to a matrix-valued setting. As in the Helton and Young paper the rate of convergence depends on the choice of the initial space in the approximation scheme.

Mathematics Subject Classification. Primary 47A57, 47B35;
secondary 93B15, 93B36.

Keywords. Nehari problem, Hankel operators, H-infinity theory, relaxed commutant lifting, approximation.

1. Introduction

Since the 1980s, the Nehari problem played an important role in system and control theory, in particular, in the H^∞-control solutions to sensitivity minimization and robust stabilization, cf., [9]. In system and control theory the Nehari problem appears mostly as a distance problem: Given G in L^∞, determine the distance of G to H^∞, that is, find the quantity $d := \inf\{\|G - F\|_\infty \mid F \in H^\infty\}$ and, if possible, find an $F \in H^\infty$ for which this infimum is attained. Here all functions are complex-valued functions on the unit circle $\mathbb{T}$. It is well known that the solution to

The research of the first author was partially supported by a visitors grant from NWO (Netherlands Organisation for Scientific Research).

this problem is determined by the Hankel operator H which maps H^2 into $K^2 = L^2 \ominus H^2$ according to the rule $Hf = P_-(Gf)$, where P_- is the orthogonal projection of L^2 onto K^2. Note that H is uniquely determined by the Fourier coefficients of G with negative index. Its operator norm determines the minimal distance. In fact, $d = \|H\|$ and the infimum is attained. Furthermore, if H has a maximizing vector φ, that is, if φ is a non-zero function in H^2 such that $\|H\varphi\| = \|H\| \|\varphi\|$, then the AAK theory [1, 2] (see also [18]) tells us that the best approximation $\widehat{G}$ of G in H^∞ is unique and is given by

$$\widehat{G}(e^{it}) = G(e^{it}) - \frac{(H\varphi)(e^{it})}{\varphi(e^{it})} \quad \text{a.e.} \tag{1.1}$$

By now the connection between the Nehari problem and Hankel operators is well established, also for matrix-valued and operator-valued functions, and has been put into the larger setting of metric constrained interpolation problems, see, for example, the books [6, Chapter IX], [13, Chapter XXXV], [7, Chapter I], [17, Chapter 5] and [3, Chapter 7], and the references therein.

The present paper is inspired by Helton-Young [14]. Note that formula (1.1) and the maximizing vector φ, may be hard to compute, especially if H has large or infinite rank. Therefore, to approximate the optimal solution (1.1), Helton-Young [14] replaces H by the restriction $\dot{H} = H|_{H^2 \ominus z^n q H^2}$ to arrive at

$$\widetilde{G}(e^{it}) = G(e^{it}) - \frac{(\dot{H}\widetilde{\varphi})(e^{it})}{\widetilde{\varphi}(e^{it})}, \quad \text{a.e.} \tag{1.2}$$

as an approximant of $\widehat{G}$. Here n is a positive integer, q is a polynomial and $\widetilde{\varphi}$ is a maximizing vector of $\dot{H}$. Note that a maximizing vector $\widetilde{\varphi}$ of $\dot{H}$ always exists, since rank $\dot{H} \leq n + \deg q < \infty$, irrespectively of the rank of H being finite, or not.

In [14] it is shown that $\widetilde{G}$ is a computationally efficient approximation of the optimal solution $\widehat{G}$ when the zeros of the polynomial q are close to the poles of G in the open unit disk $\mathbb{D}$ that are close to the unit circle $\mathbb{T}$. To be more precise, it is shown that if G is rational, i.e., rank $H < \infty$, and $\|H\|$ is a simple singular value of H, then $\|\widehat{G} - \widetilde{G}\|_\infty$ converges to 0 as $n \to \infty$. This convergence is proportional to r^n if the poles of G in $\mathbb{D}$ are within the disc $\mathbb{D}_r = \{z \in \mathbb{C} \mid |z| < r\}$, and the rate of convergence can be improved by an appropriate choice of the polynomial q.

It is well known that the Nehari problem fits in the commutant lifting framework, and that the solution formula (1.1) follows as a corollary of the commutant lifting theorem. We shall see that the same holds true for formula (1.2) provided one uses the relaxed commutant lifting framework of [8]; cf., Corollary 2.5 in [8].

To make the connection with relaxed commutant lifting more precise, define R_n to be the orthogonal projection of H^2 onto $H^2 \ominus z^{n-1} q H^2$, and put $Q_n = SR_n$, where S is the forward shift on H^2. Then the operators R_n and Q_n both map H^2 into $H^2 \ominus z^n q H^2$, and the restriction operator $H_n := H|_{H^2 \ominus z^n q H^2}$ satisfies the intertwining relation $V_- H_n R_n = H_n Q_n$. Here V_- is the compression of the forward shift V on L^2 to K^2. Given this intertwining relation, the relaxed commutant lifting theorem [8, Theorem 1.1] tells us that there exists an operator B_n from $H^2 \ominus z^n q H^2$ into L^2 such that

$$P_- B_n = H_n, \quad V B_n R_n = B Q_n, \quad \|B_n\| = \|H_n\|. \tag{1.3}$$

The second identity in (1.3) implies (see Lemma 2.2 below) that for a solution B_n to (1.3) there exists a unique function $\Phi_n \in L^2$ such that the action of B_n is given by

$$(B_n h)(e^{it}) = \Phi_n(e^{it}) h(e^{it}) \quad a.e. \quad (h \in H^2 \ominus z^n q H^2). \tag{1.4}$$

Furthermore, since H_n has finite rank, there exists only one solution B_n to (1.3) (see Proposition 2.3 below), and if $\psi_n = \widetilde{\varphi}$ is a maximizing vector of H_n, then this unique solution is given by (1.4) with Φ_n equal to

$$\Phi_n(e^{it}) = \frac{(H_n \psi_n)(e^{it})}{\psi_n(e^{it})} = \frac{(\dot{H}\widetilde{\varphi})(e^{it})}{\widetilde{\varphi}(e^{it})}, \quad a.e.. \tag{1.5}$$

Thus $G - \widetilde{G}$ in (1.2) appears as an optimal solution to a relaxed commutant lifting problem.

This observation together with the relaxed commutant lifting theory developed in the last decade, enabled us to extent the Helton-Young convergence result for optimal solutions in [14] to a matrix-valued setting, that is, to derive an analogous convergence result for optimal solutions to matrix-valued Nehari problems; see Theorem 3.1 below. A complication in this endeavor is that formula (1.1) generalizes to the vector-valued case, but not to the matrix-valued case. Furthermore, in the matrix-valued case there is in general no unique solution. We overcome the latter complication by only considering the central solutions, which satisfy an additional maximum entropy-like condition. On the way we also derive explicit state space formulas for optimal solutions to the classical and restricted Nehari problem assuming that the Hankel operator is of finite rank and satisfies an appropriate additional condition on its maximizing vectors. These state space formulas play an essential role in the proof of the convergence theorem.

This paper consists of 6 sections including the present introduction. In Section 2, which has a preliminary character, we introduce a restricted version of the matrix-valued Nehari problem, and use relaxed commutant lifting theory to show that it always has an optimal solution. Furthermore, again using relaxed commutant lifting theory, we derive a formula for the (unique) central optimal solution. In Section 3 we state our main convergence result. In Section 4 the formula for the (unique) central optimal solution derived in Section 2 is developed further, and in Section 5 this formula is specified for the classical Nehari problem. Using these formulas Section 6 presents the proof of the main convergence theorem.

Notation and terminology. We conclude this introduction with a few words about notation and terminology. Given p, q in $\mathbb{N}$, the set of positive integers, we write $L^2_{q \times p}$ for the space of all $q \times p$-matrices with entries in L^2, the Lebesgue space of square integrable functions on the unit circle. Analogously, we write $H^2_{q \times p}$ for the space of all $q \times p$-matrices with entries in the classical Hardy space H^2, and $K^2_{q \times p}$ stands for the space of all $q \times p$-matrices with entries in the space $K^2 = L^2 \ominus H^2$, the orthogonal compliment of H^2 in L^2. Note that each $F \in L^2_{q \times p}$ can be written uniquely as a sum $F = F_+ + F_-$ with $F_+ \in H^2_{q \times p}$ and $F_- \in K^2_{q \times p}$. We shall refer to F_+ as the *analytic part* of F and to F_- as its *co-analytic part*. When there is only one column we simply write L^2_p, H^2_p and K^2_p instead of $L^2_{p \times 1}$, $H^2_{p \times 1}$ and $K^2_{p \times 1}$. Note that L^2_p, H^2_p and K^2_p are Hilbert spaces and $K^2_p = L^2_p \ominus H^2_p$. Finally, $L^\infty_{q \times p}$ stands for the space of all $q \times p$-matrices whose entries are essentially bounded on the unit circle with respect to the Lebesque measure, and $H^\infty_{q \times p}$ stands for the space of all $q \times p$-matrices whose entries are analytic and uniformly bounded on the open unit disc $\mathbb{D}$. Note that each $F \in L^\infty_{q \times p}$ belongs to $L^2_{q \times p}$ and hence the analytic part F_+ and the co-analytic part F_- of F are well defined. These functions belong to $L^2_{q \times p}$ and it may happen that neither F_+ nor F_- belong to $L^\infty_{q \times p}$. In the sequel we shall need the following embedding and projection operators:

$$E : \mathbb{C}^p \to H^2_p, \quad Eu(\lambda) = u \quad (\lambda \in \mathbb{D}); \qquad (1.6)$$

$$\Pi : K^2_q \to \mathbb{C}^q, \quad \Pi f = \frac{1}{2\pi} \int_0^{2\pi} e^{it} f(e^{it}) \, dt. \qquad (1.7)$$

Throughout $G \in L^\infty_{q \times p}$, and $H : H^2_p \to K^2_q$ is the Hankel operator defined by the co-analytic part of G, that is, $Hf = P_-(Gf)$ for each $f \in H^2_p$. Here P_- is the orthogonal projection of L^2_q onto K^2_q. Note that $V_- H = HS$, where S is the forward shift on H^2_p and V_- is the compression to K^2_q of the forward shift V on L^2_q.

Finally, we associate with the Hankel operator H two auxiliary operators involving the closure of its range, i.e., the space $\mathcal{X} = \overline{\operatorname{Im} H}$, as follows:

$$Z : \mathcal{X} \to \mathcal{X}, \quad Z = V_-|_{\mathcal{X}}, \qquad (1.8)$$

$$W : H^2_p \to \mathcal{X}, \quad Wf = Hf \quad (f \in H^2_p). \qquad (1.9)$$

Note that $\mathcal{X} := \overline{\operatorname{Im} H}$ is a V_--invariant subspace of K^2_q. Hence Z is a well-defined contraction. Furthermore, if $\operatorname{rank} H$ is finite, then the spectral radius $r_{\operatorname{spec}}(Z)$ is strictly less than one and the co-analytic part G_- of G is the rational matrix function given by

$$G_-(\lambda) = (\Pi|_{\mathcal{X}})(\lambda I - Z)^{-1} W E.$$

In system theory the right-hand side of the above identity is known as the restricted backward shift realization of G_-; see, for example, [5, Section 7.1]. This realization is minimal, and hence the eigenvalues of Z coincide with the poles of G_- in $\mathbb{D}$. In particular, $r_{\operatorname{spec}}(Z) < 1$. Since $V_- H = HS$, we have $ZW = WS$. Furthermore, $\operatorname{Ker} H^* = K^2_q \ominus \mathcal{X}$.

2. Restricted Nehari problems and relaxed commutant lifting

In this section we introduce a restricted version of the Nehari problem, and we prove that it is equivalent to a certain relaxed commutant lifting problem. Throughout $\mathcal{M}$ is a subspace of H^2_p such that

$$S^* \mathcal{M} \subset \mathcal{M}, \quad \operatorname{Ker} S^* \subset \mathcal{M}. \qquad (2.1)$$

With $\mathcal{M}$ we associate operators $R_{\mathcal{M}}$ and $Q_{\mathcal{M}}$ acting on H_p^2, both mapping H_p^2 into $\mathcal{M}$. By definition $R_{\mathcal{M}}$ is the orthogonal projection of H_p^2 onto $S^*\mathcal{M}$ and $Q_{\mathcal{M}} = SR_{\mathcal{M}}$.

We begin by introducing the notion of an $\mathcal{M}$-norm. We say that $\Phi \in L_{q\times p}^2$ has a *finite $\mathcal{M}$-norm* if $\Phi h \in L_q^2$ for each $h \in \mathcal{M}$ and the map $h \mapsto \Phi h$ is a bounded linear operator, and in that case we define

$$\|\Phi\|_{\mathcal{M}} = \sup\{\|\Phi h\|_{L_q^2} \mid h \in \mathcal{M}, \quad \|h\|_{H_p^2} \le 1\}.$$

If $\mathcal{M}$ is finite dimensional, then each $\Phi \in L_{q\times p}^2$ has a finite $\mathcal{M}$-norm. Furthermore, $\Phi \in L_{q\times p}^\infty$ has a finite $\mathcal{M}$-norm for every choice of $\mathcal{M}$, and in this case $\|\Phi\|_{\mathcal{M}} \le \|\Phi\|_\infty$, with equality if $\mathcal{M} = H_p^2$. Note that $\Phi \in L_{q\times p}^2$ has a finite $\mathcal{M}$-norm and $G \in L_{q\times p}^\infty$ imply $G - \Phi$ has a finite $\mathcal{M}$-norm.

We are now ready to formulate the $\mathcal{M}$-restricted Nehari problem. Given $G \in L_{q\times p}^\infty$ and a subspace $\mathcal{M}$ of H_p^2, we define the *optimal $\mathcal{M}$-restricted Nehari problem* to be the problem of determining the quantity

$$d_{\mathcal{M}} := \inf\{\|G - F\|_{\mathcal{M}} \mid F \in H_{q\times p}^2 \text{ and } F \text{ has a finite } \mathcal{M}\text{-norm}\}, \qquad (2.2)$$

and, if possible, to find a function $F \in H_{q\times p}^2$ of finite $\mathcal{M}$-norm at which the infimum is attained. In this case, a function F attaining the infimum is called an *optimal solution*. The suboptimal variant of the problem allows the norm $\|G - F\|_{\mathcal{M}}$ to be larger than the infimum. When $\mathcal{M} = H_p^2$, the problem coincides with the classical matrix-valued Nehari problem in $L_{q\times p}^\infty$. In [15, 16] the case where $\mathcal{M} = H_p^2 \ominus S^k H_p^2$, with $k \in \mathbb{N}$, was considered.

Proposition 2.1. *Let $G \in L_{q\times p}^\infty$, and let $\mathcal{M}$ be a subspace of H_p^2 satisfying the conditions in (2.1). Then the $\mathcal{M}$-restricted Nehari problem has an optimal solution and the quantity $d_{\mathcal{M}}$ in (2.2) is equal to $\gamma_{\mathcal{M}} := \|H|_{\mathcal{M}}\|$, where $H : H_p^2 \to K_q^2$ is the Hankel operator defined by the co-analytic part of G.*

We shall derive the above result as a corollary to the relaxed commutant lifting theorem [8, Theorem 1.1], in a way similar to the way one proves the Nehari theorem using the classical commutant lifting theorem (see, for example, [6, Section II.3]). For this purpose we need the following notion. We say that an operator B from $\mathcal{M}$ into L_q^2 is *defined by* a $\Phi \in L_{q\times p}^2$ if the action of B is given by

$$(Bh)(e^{it}) = \Phi(e^{it})h(e^{it}) \quad \text{a.e.} \quad (h \in \mathcal{M}). \qquad (2.3)$$

In that case, Φ has a finite $\mathcal{M}$-norm, and $\|\Phi\|_{\mathcal{M}} = \|B\|$. When (2.3) holds we refer to Φ as the *defining function* of B. The following lemma characterizes operators B from $\mathcal{M}$ into L_q^2 defined by a function $\Phi \in L_{q\times p}^2$ in terms of an intertwining relation.

Lemma 2.2. *Let $\mathcal{M}$ be a subspace of H_p^2 satisfying (2.1), and let B be a bounded operator from $\mathcal{M}$ into L_q^2. Then B is defined by a $\Phi \in L_{q\times p}^2$ if and only if B satisfies the intertwining relation $VBR_{\mathcal{M}} = BQ_{\mathcal{M}}$. In that case, $\Phi(\cdot)u = BEu(\cdot)$ for any $u \in \mathbb{C}^p$ and $\|B\| = \|\Phi\|_{\mathcal{M}}$.*

Proof. This result follows by a modification of the proof of Lemma 3.2 in [11]. We omit the details. □

Proof of Proposition 2.1. Put $\gamma_{\mathcal{M}} = \|H|_{\mathcal{M}}\|$. Recall that the Hankel operator H satisfies the intertwining relation $V_- H = HS$. This implies $V_- H|_{\mathcal{M}} R_{\mathcal{M}} = H|_{\mathcal{M}} Q_{\mathcal{M}}$. Here $R_{\mathcal{M}}$ and $Q_{\mathcal{M}}$ are the operators defined in the first paragraph of the present section. Since $Q_{\mathcal{M}}^* Q_{\mathcal{M}} = R_{\mathcal{M}}^* R_{\mathcal{M}}$ and V is an isometric lifting of V_-, the quintet

$$\{H|_{\mathcal{M}}, V_-, V, R_{\mathcal{M}}, Q_{\mathcal{M}}, \gamma_{\mathcal{M}}\} \tag{2.4}$$

is a lifting data set in the sense of Section 1 in [8]. Thus Theorem 1.1 in [8] guarantees the existence of an operator B from $\mathcal{M}$ into L_q^2 with the properties

$$P_- B = H|_{\mathcal{M}}, \quad VBR_{\mathcal{M}} = BQ_{\mathcal{M}}, \quad \|B\| = \gamma_{\mathcal{M}}. \tag{2.5}$$

By Lemma 2.2 the second equality in (2.5) tells us there exists a $\Phi \in L_{q \times p}^2$ defining B, that is, the action of B is given by (2.3). As $\Phi(\cdot)u = BEu(\cdot)$, the first identity in (2.5) shows that $G_- = \Phi_-$, and hence $F := G - \Phi \in H_{q \times p}^2$. Furthermore,

$$\|G - F\|_{\mathcal{M}} = \|\Phi\|_{\mathcal{M}} = \|B\| = \gamma_{\mathcal{M}},$$

because of the third identity in (2.5). Thus the quantity $d_{\mathcal{M}}$ in (2.2) is less than or equal to $\gamma_{\mathcal{M}}$.

It remains to prove that $d_{\mathcal{M}} \geq \gamma_{\mathcal{M}}$. In order to do this, let $\tilde{F} \in H_{q \times p}^2$ and have a finite $\mathcal{M}$-norm. Put $\tilde{\Phi} = G - \tilde{F}$. Then $\tilde{\Phi}$ has a finite $\mathcal{M}$-norm. Let $\tilde{B}$ be the operator from $\mathcal{M}$ into L_q^2 defined by $\tilde{\Phi}$. Since $\tilde{F} \in H_{q \times p}^2$, we have $G_- = \tilde{\Phi}_-$, and hence the first identity in (2.5) holds with $\tilde{B}$ in place of B. It follows that

$$\|G - \tilde{F}\|_{\mathcal{M}} = \|\tilde{\Phi}\|_{\mathcal{M}} = \|\tilde{B}\| \geq \|H|_{\mathcal{M}}\| = \gamma_{\mathcal{M}}.$$

This completes the proof. □

In the scalar case, or more generally in the case when $p = 1$, the optimal solution is unique. Moreover this unique solution is given by a formula analogous to (1.2); cf., [1]. This is the contents of the next proposition which is proved in much the same way as the corresponding result for the Nehari problem. We omit the details.

Proposition 2.3. *Assume $p = 1$, that is, $G \in L_q^\infty$ and $\mathcal{M}$ a subspace of H^2 satisfying (2.1). Assume that $H|_{\mathcal{M}}$ has a maximizing vector $\psi \in \mathcal{M}$. Then there exists only one optimal solution F to the $\mathcal{M}$-restricted Nehari problem (2.5), and this solution is given by*

$$F(e^{it}) = G(e^{it}) - \frac{(H\psi)(e^{it})}{\psi(e^{it})} \quad a.e. \tag{2.6}$$

In general, if $p > 1$ the optimal solution is not unique. To deal with this non-uniqueness, we shall single out a particular optimal solution.

First note that the proof of Proposition 2.1 shows that there is a one-to-one correspondence between the optimal solutions of the $\mathcal{M}$-restricted Nehari problem of G and all interpolants for $H|_\mathcal{M}$ with respect to the lifting data set (2.4), that is, all operators B from $\mathcal{M}$ into L_q^2 satisfying (2.5). This correspondence is given by

$$B \mapsto F = G - \Phi, \text{ where } \Phi \text{ is the defining function of } B. \qquad (2.7)$$

Next we use that the relaxed commutant lifting theory tells us that among all interpolants for $H|_\mathcal{M}$ with respect to the lifting data set (2.4) there is a particular one, which is called the central interpolant for $H|_\mathcal{M}$ with respect to the lifting data set (2.4); see [8, Section 4]. Since V is a minimal isometric lifting of V_-, this central interpolant is uniquely determined by a maximum entropy principle (see [8, Section 8]) and given by an explicit formula using the operators appearing in the lifting data set.

Using the correspondence (2.7) we say that an optimal solution F of the $\mathcal{M}$-restricted Nehari problem of G is the *central optimal solution* whenever $\Phi := G - F$ is the defining function of the central interpolant B for $H|_\mathcal{M}$ with respect to the lifting data set (2.4). Furthermore, using the formula given in [8, Section 4] for the central interpolant the correspondence (2.7) allows us to derive a formula for the central optimal solution. To state this formula we need to make some preparations.

As before $\gamma_\mathcal{M} = \|H|_\mathcal{M}\|$. Note that $\|HP_\mathcal{M}S\| \leq \|HP_\mathcal{M}\| = \|H|_\mathcal{M}\|$, where $P_\mathcal{M}$ is the orthogonal projection of $H^2(\mathbb{C}^p)$ on $\mathcal{M}$. This allows us to define the following defect operators acting on $H^2(\mathbb{C}^p)$

$$D_\mathcal{M} = (\gamma_\mathcal{M}^2 I - P_\mathcal{M}H^*HP_\mathcal{M})^{1/2} \text{ on } H^2(\mathbb{C}^p), \qquad (2.8)$$

$$D_\mathcal{M}^\circ = (\gamma_\mathcal{M}^2 I - S^*P_\mathcal{M}H^*HP_\mathcal{M}S)^{1/2} \text{ on } H^2(\mathbb{C}^p). \qquad (2.9)$$

For later purposes we note that $S^*D_\mathcal{M}^2 S = D_\mathcal{M}^{\circ 2}$. Next define

$$\omega = \begin{bmatrix} \omega_1 \\ \omega_2 \end{bmatrix} : H_p^2 \to \begin{bmatrix} \mathbb{C}^q \\ H_p^2 \end{bmatrix}, \qquad (2.10)$$

$$\omega(D_\mathcal{M}Q_\mathcal{M}) = \begin{bmatrix} \Pi H R_\mathcal{M} \\ D_\mathcal{M}R_\mathcal{M} \end{bmatrix} \quad \text{and} \quad \omega|_{\mathrm{Ker}\, Q_\mathcal{M}^* D_\mathcal{M}} = 0. \qquad (2.11)$$

From the relaxed commutant lifting theory we know that ω is a well-defined partial isometry with initial space $\mathcal{F} = \overline{\mathrm{Im}\, D_\mathcal{M}Q_\mathcal{M}}$. Furthermore, the forward shift operator V on L_q^2 is the Sz.-Nagy-Schäffer isometric lifting of V_-. Then as a consequence of [8, Theorem 4.3] and the above analysis we obtain the following result.

Proposition 2.4. *Let $G \in L_{q \times p}^\infty$, and let $\mathcal{M}$ be a subspace of H_p^2 satisfying the conditions in (2.1). Then the central optimal solution $F_\mathcal{M}$ to the $\mathcal{M}$-restricted Nehari problem is given by $F_\mathcal{M} = G - \Phi_\mathcal{M}$, where $\Phi_\mathcal{M} \in L_{q \times p}^2$ has finite $\mathcal{M}$-norm, the co-analytic part of $\Phi_\mathcal{M}$ is equal to G_-, and the analytic part $\Phi_{\mathcal{M},+}$ of $\Phi_\mathcal{M}$ is given by*

$$\Phi_{\mathcal{M},+}(\lambda) = \omega_1(I - \lambda\omega_2)^{-1}D_\mathcal{M}E. \qquad (2.12)$$

Here E is defined by (1.6), and ω_1 and ω_2 are defined by (2.10) and (2.11).

It is this central optimal solution $F_\mathcal{M}$ we shall be working with.

3. Statement of the main convergence result

Let $G \in L^\infty_{q \times p}$, and let H be the Hankel operator defined by the co-analytic part of G. In our main approximation result we shall assume that the following two conditions are satisfied:

(C1) H has finite rank,
(C2)   none of the maximizing vectors of H belongs SH^2_p.

Note that (C1) is equivalent to G being the sum of a rational matrix function with all its poles in $\mathbb{D}$ and a matrix-valued H^∞ function.

In the scalar case conditions (C1) and (C2) are equivalent to the conditions assumed in the Helton-Young paper [14]. To see that this is the case, assume (C1) holds and $p = q = 1$. It suffices to show (C2) is equivalent to $\|H\|$ being a simple singular value, in other words, that the span of maximizing vectors of H is a one-dimensional subspace. Fist assume $\|H\|$ is simple, but (C2) does not holds. Let Sv be a maximizing vector of H. Then $v \in H^2$ is non-zero and

$$\|H\|\|v\| = \|H\|\|Sv\| = \|HSv\| = \|V_-Hv\| \leq \|Hv\| \leq \|H\|\|v\|.$$

Thus the inequalities are equalities, and v is a maximizing vector of H. As the space spanned by the maximizing vectors of H is assumed to be one dimensional, v must be a scalar multiple of Sv, which can only happen when $v = 0$, which contradicts $v \neq 0$. Thus $\|H\|$ being simple implies (C2). Conversely, assume (C2) holds and that $v_1, v_2 \in H^2$ are maximizing vectors. Then $w = v_2(0)v_1 - v_1(0)v_2$ is in the span of maximizing vectors and has $w(0) = 0$. Hence $w \in SH^2$. By (C2), w is not maximizing, and thus necessarily $w = 0$. Hence v_1 is a scalar multiple of v_2, or conversely. We conclude that the span of maximizing vectors has dimension 1, hence that $\|H\|$ is simple.

For our approximation scheme we fix a finite-dimensional subspace $\mathcal{M}_0$ of H^2_p invariant under S^*, and we define recursively

$$\mathcal{M}_k = \operatorname{Ker} S^* \oplus S\mathcal{M}_{k-1}, \quad k \in \mathbb{N}. \tag{3.1}$$

Since $\mathcal{M}_0$ is invariant under S^*, the space $\mathcal{M}_0^\perp$ is invariant under S, and the Beurling-Lax theorem tells us that $\mathcal{M}_0^\perp = \Theta H^2_\ell$, where $\Theta \in H^\infty_{p \times \ell}$ and can be taken to be inner. Using this representation one checks that $\mathcal{M}_k = H^2_p \ominus z^k \Theta H^2_\ell$ for each $k \in \mathbb{N}$. It follows that $\mathcal{M}_0 \subset \mathcal{M}_1 \subset \mathcal{M}_2 \subset \cdots$ and $\bigvee_{k \geq 0} \mathcal{M}_k = H^2_p$. Furthermore,

$$S^*\mathcal{M}_k \subset \mathcal{M}_k \quad \text{and} \quad \operatorname{Ker} S^* \subset \mathcal{M}_k, \quad k \in \mathbb{N}. \tag{3.2}$$

Note that the spaces $\mathcal{M}_k = H^2 \ominus z^k q H^2$, $k = 1, 2, \ldots$, appearing in [14] satisfy (3.1) with $\mathcal{M}_0 = H^2 \ominus qH^2$.

Theorem 3.1. *Let $G \in L^\infty_{q \times p}$. Assume that conditions (C1) and (C2) are satisfied, and let the sequence of subspaces $\{\mathcal{M}_k\}_{k \in \mathbb{N}}$ be defined by (3.1) with $\mathcal{M}_0$ a finite-dimensional S^*-invariant subspace of H^2_p. Let F be the central optimal solution to the Nehari problem for G, and for each $k \in \mathbb{N}$ let F_k be the central optimal solution to the $\mathcal{M}_k$-restricted Nehari problem. Then $G - F$ is a rational function in $L^\infty_{q \times p}$,* and for $k \in \mathbb{N}$ *sufficiently large, the same holds true for* $G - F_k$. *Furthermore,* $\|F_k - F\|_\infty \to 0$ *for* $k \to \infty$. *More precisely, if all the poles of* G *inside* $\mathbb{D}$ *are within the disk* $\mathbb{D}_r = \{\lambda \mid |\lambda| < r\}$, *for* $r < 1$, *then there exists a number* $L > 0$ *such that* $\|F_k - F\|_\infty < Lr^k$ *for* k *large enough.*

Improving the rate of convergence is one of the main issues in [14], where it is shown that for the case when the poles of G inside $\mathbb{D}$ are close to the unit circle, that is, r close to 1, convergence with $\mathcal{M}_0 = \{0\}$ may occur at a slow rate. In [14] it is also shown how to choose (in the scalar case) a scalar polynomial q so that the choice $\mathcal{M}_0 = H^2 \ominus qH^2$ increases the rate of convergence. In fact, if the roots of q coincide with the poles of G in $\mathbb{D}_r \backslash \mathbb{D}_0$, then starting with $\mathcal{M}_0 = H^2 \ominus qH^2$ the convergence is of order $O(r_0^k)$ rather than $O(r^k)$. In Section 6 we shall see that Theorem 3.1 remains true if $r < 1$ is larger than the spectral radius of the operator $V_-|_{H\mathcal{M}_0^\perp}$, and thus again the convergence rate can be improved by an appropriate choice of $\mathcal{M}_0$. To give a trivial example: when $\mathcal{M}_0$ is chosen in such a way that it includes $\operatorname{Im} H^*$, all the central optimal solutions F_k in Theorem 3.1 coincide with the central optimal solution solution F to the Nehari problem.

4. The central optimal solution revisited

As before $G \in L_{q\times p}^\infty$ and H is the Hankel operator defined by the co-analytic part of G. Furthermore, $\mathcal{M}$ is a subspace of H_p^2 satisfying (2.1). In this section we assume that $\|HP_\mathcal{M}S\| < \gamma_\mathcal{M} = \|HP_\mathcal{M}\|$. In other words, we assume that the defect operator $D_\mathcal{M}^\circ$ defined by (2.9) is invertible. This additional condition allows us to simplify the formula for the central optimal solution to the $\mathcal{M}$-restricted Nehari problem presented in Proposition 2.4. We shall prove the following theorem.

Theorem 4.1. *Let* $G \in L_{q\times p}^\infty$, *and let* $\mathcal{M}$ *be a subspace of* H_p^2 *satisfying* (2.1). *Assume the defect operator* $D_\mathcal{M}^\circ$ *defined by* (2.9) *is invertible, and put*

$$\Lambda_\mathcal{M} = D_\mathcal{M}^{\circ -2} S^* D_\mathcal{M}^2. \tag{4.1}$$

Then $r_{\mathrm{spec}}(\Lambda_\mathcal{M}) \leq 1$, *and the central optimal solution* $F_\mathcal{M}$ *to the* $\mathcal{M}$-*restricted Nehari problem is given by* $F_\mathcal{M} = G - \Phi_\mathcal{M}$, *where* $\Phi_\mathcal{M} \in L_{q\times p}^2$ *has finite* $\mathcal{M}$-*norm, the co-analytic part of* $\Phi_\mathcal{M}$ *is equal to* G_-, *and the analytic part of* $\Phi_\mathcal{M}$ *is given by*

$$\Phi_{\mathcal{M},+}(\lambda) = \Pi H(I - \lambda\Lambda_\mathcal{M})^{-1}\Lambda_\mathcal{M}E = N_\mathcal{M}(\lambda)M_\mathcal{M}(\lambda)^{-1} \quad (\lambda \in \mathbb{D}), \tag{4.2}$$

where

$$N_\mathcal{M}(\lambda) = \Pi H(I - \lambda S^*)^{-1}\Lambda_\mathcal{M}E, \quad M_\mathcal{M}(\lambda) = I - \lambda E^*(I - \lambda S^*)^{-1}\Lambda_\mathcal{M}E. \tag{4.3}$$

In particular, $M(\lambda)$ *is invertible for each* $\lambda \in \mathbb{D}$.

The formulas in the above theorem for the central optimal solution are inspired by the formulas for the central suboptimal solution in Sections IV.3 and IV.4 of [7].

We first prove two lemmas. In what follows $P_\mathcal{M}$ and $R_\mathcal{M}$ are the orthogonal projections of H_p^2 onto $\mathcal{M}$ and $S^*\mathcal{M}$, respectively, and $Q_\mathcal{M} = SR_\mathcal{M}$.

Lemma 4.2. *Let* $\mathcal{M}$ *be a subspace of* H_p^2 *satisfying* (2.1). *Then*

$$R_{\mathcal{M}} = S^* P_{\mathcal{M}} S, \quad R_{\mathcal{M}} S^* = S^* P_{\mathcal{M}}, \quad Q_{\mathcal{M}} = P_{\mathcal{M}} S. \tag{4.4}$$

Proof. Note that

$$(S^* P_{\mathcal{M}} S)^2 = S^* P_{\mathcal{M}} S S^* P_{\mathcal{M}} S = S^* P_{\mathcal{M}} S - S^* P_{\mathcal{M}} (I - SS^*) P_{\mathcal{M}} S.$$

Since $I - SS^*$ is the orthogonal projection onto $\operatorname{Ker} S^*$, the second part of (2.1) implies that $P_{\mathcal{M}}(I - SS^*) = I - SS^*$. Thus $(S^* P_{\mathcal{M}} S)^2 = S^* P_{\mathcal{M}} S$, and hence $S^* P_{\mathcal{M}} S$ is an orthogonal projection. The range of this orthogonal projection is $S^* \mathcal{M}$, and therefore the first identity in (4.4) is proved.

Using this first identity and $P_{\mathcal{M}}(I - SS^*) = I - SS^*$ we see that

$$R_{\mathcal{M}} S^* = S^* P_{\mathcal{M}} S S^* = S^* P_{\mathcal{M}} - S^* P_{\mathcal{M}} (I - SS^*) = S^* P_{\mathcal{M}}.$$

Thus the second identity in (4.4) also holds. Finally,

$$Q_{\mathcal{M}} = S R_{\mathcal{M}} = (R_{\mathcal{M}} S^*)^* = (S^* P_{\mathcal{M}})^* = P_{\mathcal{M}} S.$$

Thus (4.4) is proved. □

Lemma 4.3. *Let* $G \in L^\infty_{q \times p}$, *and let* $\mathcal{M}$ *be a subspace of* H_p^2 *satisfying* (2.1). *Assume the defect operator* $D_{\mathcal{M}}^\circ$ *defined by* (2.9) *is invertible. Then the range* $\mathcal{F}$ *of the operator* $D_{\mathcal{M}} Q_{\mathcal{M}}$ *is closed and the orthogonal projection of* H_p^2 *onto* $\mathcal{F}$ *is given by*

$$P_{\mathcal{F}} = D_{\mathcal{M}} Q_{\mathcal{M}} D_{\mathcal{M}}^{\circ -2} Q_{\mathcal{M}}^* D_{\mathcal{M}}. \tag{4.5}$$

Proof. We begin with two identities:

$$D_{\mathcal{M}} P_{\mathcal{M}} = P_{\mathcal{M}} D_{\mathcal{M}}, \quad D_{\mathcal{M}}^\circ R_{\mathcal{M}} = R_{\mathcal{M}} D_{\mathcal{M}}^\circ. \tag{4.6}$$

Since $P_{\mathcal{M}}$ is an orthogonal projection, the first equality in (4.6) follows directly from the definition of $D_{\mathcal{M}}$ in (2.8). To prove the second, we use the second identity in (4.4). Taking adjoints and using the fact that $R_{\mathcal{M}}$ and $P_{\mathcal{M}}$ are orthogonal projections, we see that $P_{\mathcal{M}} S = S R_{\mathcal{M}}$. It follows that $D_{\mathcal{M}}^\circ$ is also given by

$$D_{\mathcal{M}}^\circ = (\gamma_{\mathcal{M}}^2 I - R_{\mathcal{M}} S^* H^* H S R_{\mathcal{M}})^{1/2}. \tag{4.7}$$

From this formula for $D_{\mathcal{M}}^\circ$ the second identity in (4.6) is clear.

Now assume that $D_{\mathcal{M}}^\circ$ is invertible, and let P be the operator defined by the right-hand side of (4.5). Clearly, P is selfadjoint. Let us prove that P is a projection. Using the second equality in (4.6) we have

$$P^2 = D_{\mathcal{M}} Q_{\mathcal{M}} D_{\mathcal{M}}^{\circ -2} Q_{\mathcal{M}}^* D_{\mathcal{M}}^2 Q_{\mathcal{M}} D_{\mathcal{M}}^{\circ -2} Q_{\mathcal{M}}^* D_{\mathcal{M}}$$

$$= D_{\mathcal{M}} Q_{\mathcal{M}} D_{\mathcal{M}}^{\circ -2} (R_{\mathcal{M}} S^* D_{\mathcal{M}}^2 S R_{\mathcal{M}}) D_{\mathcal{M}}^{\circ -2} Q_{\mathcal{M}}^* D_{\mathcal{M}}$$

$$= D_{\mathcal{M}} Q_{\mathcal{M}} R_{\mathcal{M}} D_{\mathcal{M}}^{\circ -2} (S^* D_{\mathcal{M}}^2 S) D_{\mathcal{M}}^{\circ -2} R_{\mathcal{M}} Q_{\mathcal{M}}^* D_{\mathcal{M}}.$$

Observe that $Q_{\mathcal{M}} R_{\mathcal{M}} = S R_{\mathcal{M}}^2 = S R_{\mathcal{M}} = Q_{\mathcal{M}}$. Since $D_{\mathcal{M}}^{\circ 2} = S^* D_{\mathcal{M}}^2 S$, it follows that

$$P^2 = D_{\mathcal{M}} Q_{\mathcal{M}} D_{\mathcal{M}}^{\circ -2} Q_{\mathcal{M}}^* D_{\mathcal{M}} = P.$$

Thus P is an orthogonal projection. This implies that $D_{\mathcal{M}} Q_{\mathcal{M}}$ has a closed range, and $P_{\mathcal{F}} = P$. □

Proof of Theorem 4.1. Our starting point is formula (2.12). Recall that ω_1 and ω_2 are zero on $\operatorname{Ker} Q_{\mathcal{M}}^* D_{\mathcal{M}}$. From Lemma 4.3 we know that $D_{\mathcal{M}} Q_{\mathcal{M}}$ has a closed range. It follows that $\omega_1 = \omega_1 P_{\mathcal{F}}$ and $\omega_2 = \omega_2 P_{\mathcal{F}}$, where $P_{\mathcal{F}}$ is the orthogonal projection of H_p^2 onto $\mathcal{F} = \operatorname{Im} D_{\mathcal{M}} Q_{\mathcal{M}}$. Using the formula for $P_{\mathcal{F}}$ given by (4.5), the second intertwining relation in (4.6), the identities in (4.4) and the definition of ω in (2.10), (2.10) we compute

$$\omega_1 D_{\mathcal{M}} = \omega_1 P_{\mathcal{F}} D_{\mathcal{M}} = \omega_1 D_{\mathcal{M}} Q_{\mathcal{M}} D_{\mathcal{M}}^{\circ -2} Q_{\mathcal{M}}^* D_{\mathcal{M}}^2$$
$$= \Pi H R_{\mathcal{M}} D_{\mathcal{M}}^{\circ -2} R_{\mathcal{M}} S^* D_{\mathcal{M}}^2 = \Pi H D_{\mathcal{M}}^{\circ -2} R_{\mathcal{M}} S^* D_{\mathcal{M}}^2$$
$$= \Pi H D_{\mathcal{M}}^{\circ -2} S^* R_{\mathcal{M}} D_{\mathcal{M}}^2 = \Pi H \Lambda_{\mathcal{M}} P_{\mathcal{M}},$$

and

$$\omega_2 D_{\mathcal{M}} = \omega_2 P_{\mathcal{F}} D_{\mathcal{M}} = \omega_2 D_{\mathcal{M}} Q_{\mathcal{M}} D_{\mathcal{M}}^{\circ -2} Q_{\mathcal{M}}^* D_{\mathcal{M}}^2$$
$$= D_{\mathcal{M}} R_{\mathcal{M}} D_{\mathcal{M}}^{\circ -2} Q_{\mathcal{M}}^* D_{\mathcal{M}}^2 = D_{\mathcal{M}} \Lambda_{\mathcal{M}} P_{\mathcal{M}}.$$

Furthermore, using the intertwining relations in (4.6) and the second identity in (4.4) we see that $R_{\mathcal{M}} \Lambda_{\mathcal{M}} = \Lambda_{\mathcal{M}} P_{\mathcal{M}}$. In particular, $\Lambda_{\mathcal{M}}$ leaves $\mathcal{M}$ invariant.

Let us now prove that $r_{\mathrm{spec}}(\Lambda_{\mathcal{M}}) \leq 1$. Note that

$$r_{\mathrm{spec}}(\omega_2) = r_{\mathrm{spec}}(\omega_2 P_{\mathcal{F}}) = r_{\mathrm{spec}}(D_{\mathcal{M}} R_{\mathcal{M}} D_{\mathcal{M}}^{\circ -2} S^* D_{\mathcal{M}})$$
$$= r_{\mathrm{spec}}(R_{\mathcal{M}} D_{\mathcal{M}}^{\circ -2} S^* D_{\mathcal{M}}^2) = r_{\mathrm{spec}}(R_{\mathcal{M}} \Lambda_{\mathcal{M}}) = r_{\mathrm{spec}}(\Lambda_{\mathcal{M}} P_{\mathcal{M}}).$$

Thus $r_{\mathrm{spec}}(\Lambda_{\mathcal{M}} P_{\mathcal{M}}) \leq 1$, because ω_2 is contractive. Since $\Lambda_{\mathcal{M}}$ leaves $\mathcal{M}$ invariant, we see that relative to the orthogonal decomposition $H_p^2 = \mathcal{M} \oplus \mathcal{M}^\perp$ the operator $\Lambda_{\mathcal{M}}$ decomposes as

$$\Lambda_{\mathcal{M}} = \begin{bmatrix} P_{\mathcal{M}} \Lambda_{\mathcal{M}} P_{\mathcal{M}} & \star \\ 0 & (I - P_{\mathcal{M}}) \Lambda_{\mathcal{M}} (I - P_{\mathcal{M}}) \end{bmatrix}. \tag{4.8}$$

Note that $(I - P_{\mathcal{M}})(I - R_{\mathcal{M}}) = (I - P_{\mathcal{M}})$. Using the latter identity, the formulas (2.8) and (4.7), and the intertwining relations in (4.6), we obtain

$$(I - P_{\mathcal{M}}) \Lambda_{\mathcal{M}} (I - P_{\mathcal{M}}) = (I - P_{\mathcal{M}})(I - R_{\mathcal{M}}) \Lambda_{\mathcal{M}} (I - P_{\mathcal{M}})$$
$$= (I - P_{\mathcal{M}})(I - R_{\mathcal{M}}) S^* (I - P_{\mathcal{M}})$$
$$= (I - P_{\mathcal{M}}) S^* (I - P_{\mathcal{M}}).$$

Thus $(I - P_{\mathcal{M}}) \Lambda_{\mathcal{M}} (I - P_{\mathcal{M}})$ is a contraction. Hence $r_{\mathrm{spec}}((I - P_{\mathcal{M}}) \Lambda_{\mathcal{M}} (I - P_{\mathcal{M}}) \leq 1$. But then (4.8) shows that $r_{\mathrm{spec}}(\Lambda_{\mathcal{M}}) \leq 1$.

Next, using that $\Lambda_{\mathcal{M}} \mathcal{M} \subset S^* \mathcal{M} \subset \mathcal{M}$ and $\operatorname{Im} E = \operatorname{Ker} S^* \subset \mathcal{M}$, we obtain for each $\lambda \in \mathbb{D}$ that

$$\Phi_{\mathcal{M},+}(\lambda) = \omega_1 (I - \lambda \omega_2)^{-1} D_{\mathcal{M}} E = \omega_1 D_{\mathcal{M}} (I - \lambda \Lambda_{\mathcal{M}} P_{\mathcal{M}})^{-1} E$$
$$= \Pi H \Lambda_{\mathcal{M}} P_{\mathcal{M}} (I - \lambda \Lambda_{\mathcal{M}} P_{\mathcal{M}})^{-1} E = \Pi H (I - \lambda \Lambda_{\mathcal{M}} P_{\mathcal{M}})^{-1} \Lambda_{\mathcal{M}} P_{\mathcal{M}} E$$
$$= \Pi H (I - \lambda \Lambda_{\mathcal{M}})^{-1} \Lambda_{\mathcal{M}} E,$$

which gives formula (4.2).

Finally, to see that (4.3) holds, note that $\Lambda_{\mathcal{M}} S = I$. Hence $\Lambda_{\mathcal{M}}$ is a left inverse of S. Since E is an isometry with $\operatorname{Im} E = \operatorname{Ker} S^*$, we have $\Lambda_{\mathcal{M}} = S^* + \Lambda_{\mathcal{M}} E E^*$. Therefore, for each $\lambda \in \mathbb{D}$,

$$
\begin{aligned}
\Phi_{\mathcal{M},+}(\lambda) &= \Pi H (I - \lambda \Lambda_{\mathcal{M}})^{-1} \Lambda_{\mathcal{M}} E = \Pi H (I - \lambda S^* - \lambda \Lambda_{\mathcal{M}} E E^*)^{-1} \Lambda_{\mathcal{M}} E \\
&= \Pi H (I - \lambda (I - \lambda S^*)^{-1} \Lambda_{\mathcal{M}} E E^*)^{-1} (I - \lambda S^*)^{-1} \Lambda_{\mathcal{M}} E \\
&= \Pi H (I - \lambda S^*)^{-1} \Lambda_{\mathcal{M}} E (I - \lambda E^* (I - \lambda S^*)^{-1} \Lambda_{\mathcal{M}} E)^{-1} \\
&= N(\lambda) M(\lambda)^{-1}.
\end{aligned}
$$

In particular, $M(\lambda)$ is invertible. □

Remark. From $R_{\mathcal{M}} \Lambda_{\mathcal{M}} = \Lambda_{\mathcal{M}} P_{\mathcal{M}}$ we see that $\Lambda_{\mathcal{M}}$ leaves $\mathcal{M}$ invariant. Thus, if $\mathcal{M}$ in Theorem 4.1 is finite dimensional, then $\Phi_{\mathcal{M},+}$ in (4.2) is a rational function in $H_{p \times q}^2$, and hence $\Phi_{\mathcal{M},+}$ is a rational $p \times q$ matrix function which has no pole in the closed unit disk.

Next we present a criterion in terms of maximizing vectors under which Theorem 4.1 applies.

Proposition 4.4. *Assume* $\operatorname{rank} H P_{\mathcal{M}}$ *is finite. Then* $D_{\mathcal{M}}^{\circ}$ *is invertible if and only if none of the maximizing vectors of* $H P_{\mathcal{M}}$ *belongs to* $S H_p^2$.

Proof. A vector $h \in H_p^2$ is a maximizing vector of $H P_{\mathcal{M}}$ if and only if $0 \neq h \in \mathcal{D}_{\mathcal{M}}^{\perp}$. Thus we have to show that invertibility of $D_{\mathcal{M}}^{\circ}$ is equivalent to $\mathcal{D}_{\mathcal{M}}^{\perp} \cap S H_p^2 = \{0\}$.

Assume $\mathcal{D}_{\mathcal{M}}^{\perp} \cap S H_p^2 \neq \{0\}$. Thus, using the definition of a maximizing vector, there exists $S v$ with $v \neq 0$ such that $\|H P_{\mathcal{M}} S v\| = \gamma_{\mathcal{M}} \|S v\|$. Since S is an isometry we see that $\|H P_{\mathcal{M}} S v\| = \gamma_{\mathcal{M}} \|v\|$. It follows that v is in the kernel of $D_{\mathcal{M}}^{\circ}$, and hence $D_{\mathcal{M}}^{\circ}$ is not invertible.

Conversely, assume that $\mathcal{D}_{\mathcal{M}}^{\perp} \cap S H_p^2 = \{0\}$. Note that $\operatorname{rank}(H P_{\mathcal{M}} S)$ is also finite. Hence $H P_{\mathcal{M}} S$ has a maximizing vector, say v. We may assume that $\|v\| = 1$. By our assumption the vector $S v$ is not a maximizing vector of $H P_{\mathcal{M}}$. Hence

$$
\|H P_{\mathcal{M}} S\| = \|H P_{\mathcal{M}} S\| \|v\| = \|H P_{\mathcal{M}} S v\| < \|H P_{\mathcal{M}}\| \|S v\| = \gamma_{\mathcal{M}} \|S v\| = \gamma_{\mathcal{M}}.
$$

Therefore $D_{\mathcal{M}}^{\circ 2} = \gamma_{\mathcal{M}}^2 I - S^* P_{\mathcal{M}} H^* H P_{\mathcal{M}} S$ is positive definite, and thus invertible. Consequently, $D_{\mathcal{M}}^{\circ}$ is invertible. □

It is straightforward to prove that $D_{\mathcal{M}}^{\circ}$ is invertible if and only if the operator $\gamma_{\mathcal{M}}^2 I - H P_{\mathcal{M}} S S^* P_{\mathcal{M}} H^*$ is invertible, and in that case we have

$$
\Lambda_{\mathcal{M}} P_{\mathcal{M}} H^* = R_{\mathcal{M}} H^* V_-^* (\gamma_{\mathcal{M}}^2 I - H P_{\mathcal{M}} S S^* P_{\mathcal{M}} H^*)^{-1} \times
$$
$$
\times (\gamma_{\mathcal{M}}^2 I - H P_{\mathcal{M}} H^*), \qquad (4.9)
$$
$$
\Lambda_{\mathcal{M}} E = -R_{\mathcal{M}} H^* V_-^* (\gamma_{\mathcal{M}}^2 I - H P_{\mathcal{M}} S S^* P_{\mathcal{M}} H^*)^{-1} H E. \qquad (4.10)
$$

These formulas can be simplified further using the operators Z and W associated to the Hankel operator H which have been introduced at the end of Section 1, see (1.8) and (1.9). Recall that $\mathcal{X} = \overline{\operatorname{Im} H}$. Since $K_q^2 \ominus \mathcal{X} = \operatorname{Ker} H^*$, the space $\mathcal{X}$ is a reducing subspace for the operators $\gamma_{\mathcal{M}}^2 I - HP_{\mathcal{M}}SS^*P_{\mathcal{M}}H^*$ and $\gamma_{\mathcal{M}}^2 I - HP_{\mathcal{M}}H^*$. Furthermore,

$$\Delta_{\mathcal{M}} := (\gamma_{\mathcal{M}}^2 I - HP_{\mathcal{M}}SS^*P_{\mathcal{M}}H^*)|_{\mathcal{X}} = \gamma_{\mathcal{M}}^2 I_{\mathcal{X}} - ZWR_{\mathcal{M}}W^*Z^*, \qquad (4.11)$$

$$\Xi_{\mathcal{M}} := (\gamma_{\mathcal{M}}^2 I - HP_{\mathcal{M}}H^*)|_{\mathcal{X}} = \gamma_{\mathcal{M}}^2 I_{\mathcal{X}} - WP_{\mathcal{M}}W^*. \qquad (4.12)$$

Note that $\Delta_{\mathcal{M}}$ is invertible if and only if $D_{\mathcal{M}}^\circ$ is invertible. Using the above operators, (4.9) and (4.10) can be written as

$$\Lambda_{\mathcal{M}}P_{\mathcal{M}}W^* = R_{\mathcal{M}}W^*Z^*\Delta_{\mathcal{M}}^{-1}\Xi_{\mathcal{M}},$$
$$\Lambda_{\mathcal{M}}E = -R_{\mathcal{M}}W^*Z^*\Delta_{\mathcal{M}}^{-1}WE. \qquad (4.13)$$

Corollary 4.5. *Let $G \in L_{q \times p}^\infty$, and let $\mathcal{M}$ be a subspace of H_p^2 satisfying (2.1). Assume the operator $\Delta_{\mathcal{M}}$ defined by (4.11) is invertible. Then the defect operator $D_{\mathcal{M}}^\circ$ defined by (2.9) is invertible, and the functions $N_{\mathcal{M}}$ and $M_{\mathcal{M}}$ appearing in (4.3) are also given by*

$$N_{\mathcal{M}}(\lambda) = N_{\mathcal{M},1}(\lambda) + N_{\mathcal{M},2}(\lambda), \qquad (4.14)$$

$$N_{\mathcal{M},1}(\lambda) = -\Pi HW^*(I - \lambda Z^*)^{-1}Z^*\Delta_{\mathcal{M}}^{-1}WE \qquad (4.15)$$

$$N_{\mathcal{M},2}(\lambda) = \Pi H(I - \lambda S^*)^{-1}(I - R_{\mathcal{M}})W^*Z^*\Delta_{\mathcal{M}}^{-1}WE. \qquad (4.16)$$

and

$$M_{\mathcal{M}}(\lambda) = M_{\mathcal{M},1}(\lambda) + M_{\mathcal{M},2}(\lambda), \qquad (4.17)$$

$$M_{\mathcal{M},1}(\lambda) = I + \lambda E^*W^*(I - \lambda Z^*)^{-1}Z^*\Delta_{\mathcal{M}}^{-1}WE \qquad (4.18)$$

$$M_{\mathcal{M},2}(\lambda) = -\lambda E^*(I - \lambda S^*)^{-1}(I - R_{\mathcal{M}})W^*Z^*\Delta_{\mathcal{M}}^{-1}WE. \qquad (4.19)$$

Furthermore, if $r_{\text{spec}}(Z^*\Delta_{\mathcal{M}}^{-1}\Xi_{\mathcal{M}}) < 1$, then $M_{\mathcal{M},1}(\lambda)$ is invertible for $|\lambda| \leq 1$ and

$$M_{\mathcal{M},1}(\lambda)^{-1} = I - \lambda E^*W^*(I - \lambda Z^*\Delta_{\mathcal{M}}^{-1}\Xi_{\mathcal{M}})^{-1}Z^*\Delta_{\mathcal{M}}^{-1}WE, \quad |\lambda| \leq 1. \quad (4.20)$$

Proof. For operators A and B the invertibility of $I + AB$ is equivalent to the invertibility of $I + BA$. Using this fact it is clear that the invertibility of $D_{\mathcal{M}}^\circ$ follows form the invertibility of $\Delta_{\mathcal{M}}$. Hence we can apply Theorem 4.1. Writing $R_{\mathcal{M}}$ as $I - (I - R_{\mathcal{M}})$ and using (4.13) we see that (4.14) holds with $N_{\mathcal{M},2}$ being given by (4.16) and with

$$N_{\mathcal{M},1}(\lambda) = -\Pi H(I - \lambda S^*)^{-1}W^*Z^*\Delta_{\mathcal{M}}^{-1}WE. \qquad (4.21)$$

The intertwining relation $WS = ZW$ yields $(I - \lambda S^*)^{-1}W^* = W^*(I - \lambda Z^*)^{-1}$. Using the latter identity in (4.21) yields (4.15). In a similar way one proves the identities (4.17)–(4.19).

To complete the proof assume $r_{\rm spec}(Z^*\Delta_{\mathcal{M}}^{-1}\Xi_{\mathcal{M}}) < 1$. Then the inversion formula for $M_{\mathcal{M},1}(\lambda)$ follows from the standard inversion formula from [4, Theorem 2.2.1], where we note that the state operator in the inversion formula equals

$$
\begin{aligned}
Z^* - Z^*\Delta_{\mathcal{M}}^{-1}WEE^*W^* &= Z^*\Delta_{\mathcal{M}}^{-1}(\gamma_{\mathcal{M}}^2 I - ZWR_{\mathcal{M}}W^*Z^* - WEE^*W^*)\\
&= Z^*\Delta_{\mathcal{M}}^{-1}(\gamma_{\mathcal{M}}^2 I - W(SR_{\mathcal{M}}S^* + EE^*)W^*)\\
&= Z^*\Delta_{\mathcal{M}}^{-1}(\gamma_{\mathcal{M}}^2 I - W(SS^*P_{\mathcal{M}} + EE^*P_{\mathcal{M}})W^*)\\
&= Z^*\Delta_{\mathcal{M}}^{-1}(\gamma_{\mathcal{M}}^2 I - WP_{\mathcal{M}}W^*) = Z^*\Delta_{\mathcal{M}}^{-1}\Xi_{\mathcal{M}},
\end{aligned}
$$

as claimed. Here we used the second identity in (4.4), and the fact that $P_{\mathcal{M}}E = E$, because $\mathrm{Im}\,E = \mathrm{Ker}\,S^* \subset \mathcal{M}$. □

5. The special case where $\mathcal{M} = H_p^2$

Throughout this section $\mathcal{M} = H_p^2$, that is, we are dealing with the H_p^2-restricted Nehari problem, which is just the usual Nehari problem. Since $\mathcal{M} = H_p^2$, we will suppress the index $\mathcal{M}$ in our notation, and just write D, D°, $\mathcal{D}$, $\mathcal{D}^\circ$, Λ, etc. instead of $D_{\mathcal{M}}$, $D_{\mathcal{M}}^\circ$, $\mathcal{D}_{\mathcal{M}}$, $\mathcal{D}_{\mathcal{M}}^\circ$, $\Lambda_{\mathcal{M}}$, etc. In particular,

$$\gamma = \|H\|, \quad D = (\gamma^2 I - H^*H)^{1/2}, \quad D^\circ = (\gamma^2 I - S^*H^*HS)^{1/2}. \tag{5.1}$$

We shall assume (cf., the first paragraph of Section 3) that the following two conditions are satisfied

(C1) H has finite rank,
(C2)   none of the maximizing vectors of H belongs SH_p^2.

Note that the space spanned by the maximizing vectors of H is equal to $\mathrm{Ker}\,D = \mathcal{D}^\perp$, where $\mathcal{D}$ is the closure of the range of D. We see, using Proposition 4.4, that under condition (C1):

$$(\mathrm{C2}) \iff \mathcal{D} \cap SH_p^2 = \{0\} \iff D^\circ \text{ invertible}. \tag{5.2}$$

Let Z and W be the operators defined by (1.8) and (1.9), respectively, and

$$\Delta = \gamma^2 I_{\mathcal{X}} - ZWW^*Z^*, \quad \Xi = \gamma^2 I_{\mathcal{X}} - WW^*. \tag{5.3}$$

We shall prove the following theorem.

Theorem 5.1. *Let $G \in L_{q\times p}^2$, and assume that the Hankel operator H associated with the co-analytic part of G satisfies conditions (C1) and (C2). Then the operator Δ defined by the first identity in (5.3) is invertible and the central optimal solution F to the Nehari problem associated with G is given by $F = G_+ - \Phi_+$, where G_+ is the analytic part of G and Φ_+ is the rational $q\times p$ matrix-valued H^∞ function defined by*

$$
\begin{aligned}
\Phi_+(\lambda) &= N(\lambda)M(\lambda)^{-1}, \quad \text{where}\\
N(\lambda) &= -\Pi HW^*(I_{\mathcal{X}} - \lambda Z^*)^{-1}Z^*\Delta^{-1}WE,\\
M(\lambda) &= I + \lambda E^*W^*(I_{\mathcal{X}} - \lambda Z^*)^{-1}Z^*\Delta^{-1}WE.
\end{aligned}
$$

Furthermore, $r_{\mathrm{spec}}(Z^*\Delta^{-1}\Xi) < 1$, *and the inverse of* $M(\lambda)$ *is given by*

$$M(\lambda)^{-1} = I - \lambda E^* W^* (I_{\mathcal{X}} - \lambda Z^* \Delta^{-1} \Xi)^{-1} Z^* \Delta^{-1} W E.$$

Here Ξ *is the operator defined by the second identity in* (5.3).

It will be convenient first to prove the following lemma.

Lemma 5.2. *Assume conditions* (C1) *and* (C2) *are satisfied. Then the following holds:*

(i) *the subspace* $\mathfrak{M} := \mathrm{Ker}\, \omega_2 D$ *of* H_p^2 *is cyclic for* S,
(ii) *the operators* ω_2 *and* Λ *are strongly stable.*

Proof. We split the proof into three parts. The first part has a preparatory character. The two other parts deal with items (i) and (ii), respectively.

Part 1. We first prove that

$$\omega_2 = D(S^* D^2 S)^{-1} S^* D, \quad \Lambda = (S^* D^2 S)^{-1} S^* D^2. \tag{5.4}$$

Since (C1) and (C2) are satisfied, we can apply Proposition 4.4 with $\mathcal{M} = H_p^2$ to show that $D^{\circ 2} = S^* D^2 S$ is invertible. By definition, see (2.11) with $\mathcal{M} = H_p^2$, the operator ω_2 is zero on $\mathrm{Ker}\, S^* D$. Thus in order to prove the first equality in (5.4), it suffices to show that $\omega_2 f = D(S^* D^2 S)^{-1} S^* Df$ for each $f \in \mathcal{F} = \overline{\mathrm{Im}\, DS} = \overline{\mathrm{Im}\, DS} = (\mathrm{Ker}\, S^* D)^{\perp}$. But the latter follows by specifying (4.5) and (2.11) for $\mathcal{M} = H_p^2$, using $D^{\circ 2} = S^* D^2 S$.

The second equality in (5.4) follows from (4.1), again using $D^{\circ 2} = S^* D^2 S$ and $\mathcal{M} = H_p^2$.

Part 2. We prove (i). Using the result of the previous part we have

$$\mathfrak{M} = \mathrm{Ker}\, \omega_2 D = \mathrm{Ker}\, D(S^* D^2 S)^{-1} S^* D^2.$$

Fix $h \perp \bigvee_{k \geq 0} S^k \mathfrak{M}$. Then $S^{*k} h \perp \mathfrak{M}$, and thus

$$S^{*k} h = D^2 S(S^* D^2 S)^{-1} Dh_k, \quad \text{for some } h_k \in H_p^2.$$

It follows that $S^{*k+1} h = S^* D^2 S(S^* D^2 S)^{-1} Dh_k = Dh_k$, and hence

$$S^{*k} h = D^2 S(S^* D^2 S)^{-1} S^{*k+1} h = \Lambda^* S^{*k+1} h.$$

Taking $k = 0$, we conclude that

$$h = \Lambda^* S^* h = \Lambda^{*2} S^{*2} h = \Lambda^{*k} S^{*k} h \quad (k \in \mathbb{N}).$$

Next observe that $D\Lambda = \omega_2 D$. This yields

$$\Lambda^{k+1} = D^{\circ -2} S^* D^2 \Lambda^k = D^{\circ -2} S^* D\omega_2^k D. \tag{5.5}$$

Since ω_2 is contractive, we conclude $\sup_{k \in \mathbb{N}} \|\Lambda^k\| < \infty$. Therefore, since S^* is strongly stable, we see that $\Lambda^{*k} S^{*k} h \to 0$ if $k \to \infty$. Hence $h = 0$, and we have proved the cyclicity of $\mathrm{Ker}\, \omega_2 D$.

Part 3. We prove (ii). By definition $\omega_2 DS = D$. Thus for each $k \in \mathbb{N}$ we have $\omega_2^k DS^k = D$. Now, fix $n \in \mathbb{N}$, and take $0 \leq j \leq n-1$. Then

$$\omega_2^n DS^j \mathfrak{M} = \omega_2^{n-j} D\mathfrak{M} = \omega_2^{n-j-1}\omega_2 D\mathfrak{M} = \{0\}.$$

In other words, $\operatorname{Ker}\omega_2^n D$ includes $\mathcal{X}_n = \bigvee_{j=0}^{n-1} S^j \mathfrak{M}$.

Let P_n be the orthogonal projection of H_p^2 onto $H_p^2 \ominus \mathcal{X}_n$. By (i) we have $P_n \to 0$ pointwise. Now take $h \in H_p^2$. Then

$$\|\omega_2^n Dh\| = \|\omega_2^n DP_n h\| \leq \|DP_n h\| \to 0 \quad (n \to \infty).$$

From the first identity in (5.4) we see that ω_2 is zero on $\operatorname{Ker} D = (\operatorname{Im} D)^\perp$. Thus $\omega_2^n(\operatorname{Im} D)^\perp = \{0\}$, and we conclude that ω_2 is strongly stable.

Finally, using (5.5), we see that $\Lambda^k h \to 0$, as $k \to \infty$, for any $h \in H_p^2$. Hence Λ is strongly stable too. $\qquad\square$

Proof of Theorem 5.1. We already observed that D° is invertible. Since the invertibility of D° implies the invertibility of Δ, we can apply Theorem 4.1 and Corollary 4.5 with $\mathcal{M} = H_p^2$ to get the desired formula for Φ_+. Note that $R_{H_p^2} = I$, and hence in this case the functions appearing in (4.16) and (4.19) are identically zero.

Put $T = Z^* \Delta^{-1} \Xi$. Next we show that $r_{\text{spec}}(T) < 1$. By specifying the first identity in (4.13) we see that $\Lambda W^* = W^* T$, and thus $\Lambda^k W^* = W^* T^k$ for each $k \in \mathbb{N}$. Since Λ is strongly stable (by Lemma 5.2 (ii)), we arrive at $\lim_{k\to\infty} W^* T^k x = 0$. The fact that H has finite rank, implies that the range of H is closed, and hence W is surjective. But then $(WW^*)^{-1}W$ is a left inverse of W^*, and $T^k x = (WW^*)^{-1}WT^k x \to 0$ if $k \to \infty$. Thus T is strongly stable. Since the underlying space $\mathcal{X}$ is finite dimensional, we conclude that $r_{\text{spec}}(Z^*\Delta\Xi) = r_{\text{spec}}(T) < 1$.

Finally, since $r_{\text{spec}}(Z^*\Delta\Xi) = r_{\text{spec}}(T) < 1$, the invertibility of $M(\lambda)$ for $|\lambda| < 1$ and the formula for its inverse follow by specifying the final part of Corollary 4.5 for the case when $\mathcal{M} = H_p^2$. $\qquad\square$

6. Convergence of central optimal solutions

Throughout $G \in L_{q\times p}^\infty$ and H is the Hankel operator defined by the co-analytic part of G. We assume that conditions (C1) and (C2) formulated in the first paragraph of Section 3 are satisfied. Furthermore, $\mathcal{M}_0$ is a finite-dimensional S^*-invariant subspace of H_p^2, and $\mathcal{M}_0, \mathcal{M}_1, \mathcal{M}_2, \ldots$ is a sequence of subspaces of H_p^2 defined recursively by (3.1). We set $P_k = P_{\mathcal{M}_k}$. From the remarks made in the paragraph preceding Theorem 3.1 one sees that

$$I - P_k = S^k(I - P_0)S^{*k}, \quad S^* P_k = P_{k-1}S^*, \quad P_k E = E \quad (k \in \mathbb{N}). \tag{6.1}$$

Here E is the embedding operator defined by (1.6).

In this section we will prove Theorem 3.1. In fact we will show that with an appropriate choice of the initial space $\mathcal{M}_0$ convergence occurs at an ever faster rate than stated in Theorem 3.1. We start with a lemma that will be of help when proving the increased rate of convergence.

Lemma 6.1. *Let Z and W be the operators defined by* (1.8) *and* (1.9), *respectively. Then the space $\mathcal{X}_0 = W\mathcal{M}_0^\perp$ is a Z-invariant subspace of $\mathcal{X} = \operatorname{Im} W$. Furthermore, let the operators $Z_0 : \mathcal{X}_0 \to \mathcal{X}_0$ and $W_0 : H_p^2 \to \mathcal{X}_0$ be defined by $Z_0 = Z|_{\mathcal{X}_0}$ and $W_0 = \Pi_{\mathcal{X}_0} W$, where $\Pi_{\mathcal{X}_0}$ is the orthogonal projection of $\mathcal{X}$ onto $\mathcal{X}_0$. Then $r_{\mathrm{spec}}(Z_0) \le r_{\mathrm{spec}}(Z)$ and*

$$Z^k W(I - P_0) = \Pi_{\mathcal{X}_0}^* Z_0^k W_0(I - P_0), \quad k = 0, 1, 2, \dots. \tag{6.2}$$

Proof. Since $ZW = WS$ and $\mathcal{M}_0^\perp$ is invariant under S, we see that $\mathcal{X}_0$ is invariant under Z, and thus $r_{\mathrm{spec}}(Z_0) \le r_{\mathrm{spec}}(Z)$. From the definition of Z_0 and W_0 we see that $Z\Pi_{\mathcal{X}_0}^* = \Pi_{\mathcal{X}_0}^* Z_0$ and $\Pi_{\mathcal{X}_0}^* W_0(I - P_0) = W(I - P_0)$. Thus

$$Z^k W(I - P_0) = Z^k \Pi_{\mathcal{X}_0}^* W_0(I - P_0) = \Pi_{\mathcal{X}_0}^* Z_0^k W_0(I - P_0), \quad k = 0, 1, 2, \dots$$

This proves (6.2). $\square$

Assume $0 < r < 1$ such that the poles of G inside $\mathbb{D}$ are in the open disc $\mathbb{D}_r$. As mentioned in the introduction, the poles of G inside $\mathbb{D}$ coincide with the eigenvalues of Z. Thus $r_{\mathrm{spec}}(Z) < r$. By Lemma 6.1, $r_{\mathrm{spec}}(Z_0) \le r_{\mathrm{spec}}(Z) < r$. In what follows we fix $0 < r_0 < 1$ such that $r_{\mathrm{spec}}(Z_0) < r_0 < r$. We will show that the convergence of the central optimal solutions F_k in Theorem 3.1 is proportional to r_0^k.

In the sequel we will use the following notation. For a sequence of operators X_n, $n \in \mathbb{N}$, with $X_n \to_r X$ we mean that there exists a positive number $L_X > 0$ such that $\|X - X_n\| \le L_X r^n$. Hence $X_n \to_r X$ if and only if $\|X - X_n\| \to_r 0$. Note that if $X_n \to_r X$ and if Y_n is another sequence of operators of appropriate size, with $Y_n \to Y$, $n \in \mathbb{N}$, then $X_n Y_n \to_r XY$. Moreover, if $X_n \to_r X$, the operators X, X_n are invertible, and $\sup_{n \in \mathbb{N}} \|X_n^{-1}\| < \infty$, then $X_n^{-1} \to_r X^{-1}$. Finally, if K, K_n, $n \in \mathbb{N}$, are functions defined on $\overline{\mathbb{D}}$, then $K_n \to_r K$ will indicate convergence proportional to r^n, uniform on $\overline{\mathbb{D}}$, that is, there exists a positive number $L_K > 0$, independent of λ, such that $\|K(\lambda) - K_n(\lambda)\| \le L_K r^n$ for each $n \in \mathbb{N}$ and $\lambda \in \overline{\mathbb{D}}$.

For simplicity, we will adapt the notation of Section 5, and write γ, Δ, N and M instead of $\gamma_{H_p^2}$, $\Delta_{H_p^2}$, $N_{H_p^2}$ and $M_{H_p^2}$. Furthermore, we use the abbreviated notation P_k, γ_k, Λ_k, Ξ_k, and Δ_k for the operators $P_{\mathcal{M}_k}$, $\gamma_{\mathcal{M}_k}$, $\Lambda_{\mathcal{M}_k}$, $\Xi_{\mathcal{M}_k}$, and $\Delta_{\mathcal{M}_k}$ appearing in Section 4 for $\mathcal{M} = \mathcal{M}_k$.

As a first step towards the proof of our convergence result we prove the following lemma.

Lemma 6.2. *Assume conditions* (C1) *and* (C2) *are satisfied. Then $\Delta_k \to_{r_0^2} \Delta$, and for $k \in \mathbb{N}$ large enough Δ_k is invertible, and $\Delta_k^{-1} \to_{r_0^2} \Delta^{-1}$.*

Proof. We begin with a few remarks. Recall that for $\mathcal{M}$ in (2.1) the operator $R_{\mathcal{M}}$ is defined to be the orthogonal projection of H_p^2 onto $S^*\mathcal{M}$; see the first paragraph of Section 2. For $\mathcal{M} = \mathcal{M}_k$ we have $S^*\mathcal{M}_k = \mathcal{M}_{k-1}$ by (3.1), and thus $\mathcal{M} = \mathcal{M}_k$ implies $R_{\mathcal{M}_k} = P_{k-1}$. It follows that the operator Δ_k is given by $\Delta_k = \gamma_k^2 I_{\mathcal{X}} - ZW P_{k-1} W^* Z^*$; cf., the second part of (4.11). From the invertibility of $D^\circ_{\mathcal{M}_k}$ we obtain that Δ_k is invertible as well; see the first paragraph of the proof Corollary 4.5. The identities in (4.13) for $\mathcal{M} = \mathcal{M}_k$ now take the form

$$\Lambda_k P_k W^* = P_{k-1} W^* Z^* \Delta_k^{-1} \Xi_k, \quad \Lambda_k E = -P_{k-1} W^* Z^* \Delta_k^{-1} W E. \tag{6.3}$$

Observe that $\gamma_k^2 = \|HP_k\|^2 = \|P_k H^*\|^2 = r_{\mathrm{spec}}(HP_k H^*) = \|HP_k H^*\|$. By a similar computation $\gamma^2 = \|HH^*\|$. Thus, using (6.1) and (6.2),

$$|\gamma^2 - \gamma_k^2| = |\|HH^*\| - \|HP_k H^*\|| \leq \|HH^* - HP_k H^*\|$$
$$= \|HS^k(I - P_0)S^{*k}H^*\| = \|Z_0^k W_0(I - P_0)W_0^* Z_0^{*k}\|$$
$$\leq \|Z_0^k\| \|H\| \|(I - P_0)\| \|H^*\| \|Z_0^{*k}\| = \|H\|^2 \|Z_0^k\|^2.$$

It follows that $\gamma_k^2 \to_{r_0^2} \gamma^2$. Next, again by (6.1) and (6.2), we obtain

$$\Delta_k = \gamma_k^2 I - ZWP_{k-1}W^*Z^*$$
$$= \gamma_k^2 I - ZWW^*Z^* + ZWS^{k-1}(I - P_0)S^{*k-1}W^*Z^*$$
$$= \Delta + (\gamma_k^2 - \gamma^2)I + P_{\mathcal{X}_0} Z_0^{k-1} W_0(I - P_0)W_0^* Z_0^{*k} P_{\mathcal{X}_0}.$$

Clearly the second and third summand converge to zero proportional to r_0^{2k}, and thus we may conclude that $\Delta_k \to_{r_0^2} \Delta$.

Since Δ is invertible by Theorem 5.1. The result of the previous paragraph implies that for k large enough Δ_k is invertible and $\|\Delta_k^{-1}\| < L$ for some $L > 0$ independent of k. Consequently $\Delta_k^{-1} \to_{r_0^2} \Delta^{-1}$.   □

Proof of Theorem 3.1 (*with r_0^k-convergence*). We split the proof into four parts. Throughout $k \in \mathbb{N}$ is assumed to be large enough so that Δ_k is invertible; see Lemma 6.2.

Part 1. Let N and M be as in Theorem 5.1. Put

$$N_{k,1}(\lambda) = -\Pi HW^*(I - \lambda Z^*)^{-1}Z^*\Delta_k^{-1}WE, \tag{6.4}$$
$$M_{k,1}(\lambda) = I + \lambda E^*W^*(I - \lambda Z^*)^{-1}Z^*\Delta_k^{-1}WE. \tag{6.5}$$

Since the only dependence on k in $N_{k,1}$ and $M_{k,1}$ occurs in the form of Δ_k, it follows from Lemma 6.2 that

$$M_{k,1} \to_{r_0^2} M \quad \text{and} \quad N_{k,1} \to_{r_0^2} N. \tag{6.6}$$

Part 2. From Corollary 4.5 we know that

$$N_k(\lambda) = N_{k,1}(\lambda) + N_{k,2}(\lambda), \quad N_{k,2}(\lambda) = \Pi H \gamma_k(\lambda)Z^*\Delta_k^{-1}WE, \tag{6.7}$$
$$M_k(\lambda) = M_{k,1}(\lambda) + M_{k,2}(\lambda), \quad M_{k,2}(\lambda) = -\lambda E^*\gamma_k(\lambda)Z^*\Delta_k^{-1}WE. \tag{6.8}$$

Here $\gamma_k(\lambda) = (I - \lambda S^*)^{-1}(I - P_{k-1})W^*$. In this part we show that $M_{k,2} \to_{r_0} 0$.

Using the first identity in (6.1), the intertwining relation $ZW = WS$, and (6.2) we see that

$$\gamma_k(\lambda) = (I - \lambda S^*)^{-1}S^{k-1}(I - P_0)S^{*k-1}W^*$$
$$= (I - \lambda S^*)^{-1}S^{k-1}(I - P_0)W_0^* Z_0^{*k-1}\Pi_{\mathcal{X}_0}.$$

Next we use that

$$(I - \lambda S^*)^{-1} S^{k-1} = \sum_{j=0}^{k-2} \lambda^j S^{k-1-j} + \lambda^{k-1}(I - \lambda S^*)^{-1}.$$

Thus $\gamma_k(\lambda) = \gamma_{k,1}(\lambda) + \gamma_{k,2}(\lambda)$, where

$$\gamma_{k,1}(\lambda) = \Big(\sum_{j=0}^{k-2} \lambda^j S^{k-1-j} \Big)(I - P_0) W_0^* Z_0^{*k-1} \Pi_{\chi_0},$$

$$\gamma_{k,2}(\lambda) = \lambda^{k-1}(I - \lambda S^*)^{-1}(I - P_0) W_0^* Z_0^{*k-1} \Pi_{\chi_0}.$$

Now recall that $\mathcal{M}_0$ is S^*-invariant, and write $S_0 = P_0 S P_0 = P_0 S$. The fact that $\mathcal{M}_0$ is finite dimensional implies $r_{\mathrm{spec}}(S_0) < 1$. The computation

$$(I - \lambda S^*)^{-1}(I - P_0) W_0^* = (I - \lambda S^*)^{-1} W_0^* - (I - \lambda S^*)^{-1} P_0 W_0^*$$

$$= W_0^*(I - \lambda Z_0^*)^{-1} - (I - \lambda S_0^*)^{-1} P_0 W_0^*,$$

shows that $(I - \lambda S^*)^{-1}(I - P_0) W_0^*$ is uniformly bounded on $\mathbb{D}$. Since $r_{\mathrm{spec}}(Z_0) < r_0 < 1$, we conclude that $\gamma_{k,2} \to_{r_0} 0$.

Next observe that $E^* \big(\sum_{j=0}^{k-2} \lambda^j S^{k-1-j} \big) = 0$, and thus $E^* \gamma_{k,1}(\lambda) = 0$ for each $k \in \mathbb{N}$. We conclude that

$$M_{k,2}(\lambda) = -\lambda E^* \gamma_{k,2}(\lambda) Z^* \Delta_k^{-1} W E.$$

But then $\gamma_{k,2} \to_{r_0} 0$ implies that the same holds true for $M_{k,2}$, that is, $M_{k,2} \to_{r_0} 0$. Indeed, this follows from the above identity and the fact that the sequence Δ_k^{-1} is uniformly bounded.

Part 3. In this part we show that $N_{k,2} \to_{r_0} 0$. To do this we first observe that

$$\Pi H S^{k-1-j} = \Pi V_-^{k-1-j} H = \Pi V_-^{k-1-j} P_\chi W = \Pi P_\chi Z^{k-1-j} W.$$

Post-multiplying this identity with $I - P_0$ and using (6.2) yields

$$\Pi H S^{k-1-j}(I - P_0) = \Pi P_{\chi_0} Z_0^{k-1-j} W_0(I - P_0).$$

It follows that

$$N_{k,2}(\lambda) = \Big(\sum_{j=0}^{k-2} \lambda^j \Pi P_{\chi_0} Z_0^{k-1-j} \Big) W_0(I - P_0) W_0^* Z_0^{*k-1}$$

$$+ \Pi H \gamma_{k,2}(\lambda) Z^* \Delta_k^{-1} W E. \qquad (6.9)$$

From the previous part of the proof we know that $\gamma_{k,2} \to_{r_0} 0$, and by Lemma 6.2 the sequence Δ_k^{-1} is uniformly bounded. It follows that the second term in the right-hand side of (6.9) converges to zero with a rate proportional to r_0^k. Note that for $\lambda \in \mathbb{D}$ we have

$$\Big\| \sum_{j=0}^{k-2} \lambda^j \Pi P_{\chi_0} Z_0^{k-1-j} \Big\| \leq \sum_{j=0}^{k-2} \|Z_0\|^{k-1-j} \leq \sum_{j=1}^{\infty} \|Z_0^j\| \leq \frac{L_0 r_0}{1 - r_0}.$$

Since $r_{\mathrm{spec}}(Z_0) < r_0 < 1$, we also have $\|Z_0^{*k-1}\| \to_{r_0} 0$. It follows that the first term in the right-hand side of (6.9) converges to zero with a rate proportional to r_0^k. We conclude that $N_{k,2} \to_{r_0} 0$.

Part 4. To complete the proof, it remains to show that $M_k^{-1}(\lambda) \to_{r_0} M^{-1}(\lambda)$ uniformly on $\overline{\mathbb{D}}$. By similar computations as in the proof of Lemma 6.2, it follows that $\Xi_k \to_{r_0^2} \Xi$. Hence $Z^* \Delta_k^{-1} \Xi_k \to_{r_0^2} Z^* \Delta^{-1} \Xi$. By Theorem 5.1 we have $r_{\mathrm{spec}}(Z^* \Delta^{-1} \Xi) < 1$. Thus for k large enough also $r_{\mathrm{spec}}(Z^* \Delta_k^{-1} \Xi_k) < 1$, and $M_{k,1}(\lambda)$ is invertible on $\overline{\mathbb{D}}$. From the fact that $M_{k,1} \to_{r_0^2} M$, we see that $M_{k,1}^{-1} \to_{r_0^2} M^{-1}$, with $M_{k,1}^{-1}$ and M^{-1} indicating here the functions on $\overline{\mathbb{D}}$ with values $M_{k,1}(\lambda)^{-1}$ and $M(\lambda)^{-1}$ for each $\lambda \in \overline{\mathbb{D}}$. In particular, the functions $M_{k,1}^{-1}$ are uniformly bounded on $\overline{\mathbb{D}}$ by a constant independent of k, which implies

$$I + M_{k,1}^{-1} M_{k,2} \to_{r_0} I, \quad (I + M_{k,1}^{-1} M_{k,2})^{-1} \to_{r_0} I.$$

As a consequence

$$M_k^{-1} = (M_{k,1} + M_{k,2})^{-1} = (I + M_{k,1}^{-1} M_{k,2})^{-1} M_{k,1}^{-1} \to_{r_0} I \cdot M^{-1} = M^{-1},$$

which completes the proof.   □

Concluding remarks

Note that the functions $M_{k,1}$ and $N_{k,1}$ given by (6.4) and (6.5) converge with a rate proportional to r_0^{2k} rather than r_0^k; cf., (6.6). Consequently the same holds true for $M_{k,1}^{-1}$. Thus a much faster convergence may be achieved when $N_{k,1} M_{k,1}^{-1}$ are used instead of $N_k M_k^{-1}$. However, for the inverse of $M_{k,1}$ to exist on $\overline{\mathbb{D}}$ we need k to be large enough to guarantee $r_{\mathrm{spec}}(Z^* \Delta_k \Xi_k) < 1$, and it is at present not clear how large k should be.

For the scalar case condition (C2) is rather natural. Indeed (see the second paragraph of Section 3) for the scalar case condition (C2) is equivalent to the requirement that the largest singular value of the Hankel operator is simple. The latter condition also appears in model reduction problems. In the matrix-valued case (C2) seems rather special; in this paper it serves a practical purpose, namely that the central optimal solutions admit a more explicit description under this assumption.

If conditions (C1) and (C2) are fulfilled and the space spanned by the maximizing vectors of the Hankel operator H has dimension p, then the central optimal solution to the Nehari problem is the only optimal solution. To see this note that the additional assumption implies that $H_p^2 = \mathrm{Ker}\, S^* \dot{+} \mathcal{D}$. Using the later identity and some elementary Fredholm theory from [12, Chapter XI] one proves that $\mathcal{F} = \overline{\mathrm{Im}\, DS} = \mathrm{Im}\, DS$ is equal to $\mathcal{D}$. But then the optimal solution is unique.

Computational examples show that it may happen that the approximations of the optimal solution to the Nehari problem considered in this paper oscillate to the optimal solution when the initial space $\mathcal{M}_0 = \{0\}$. Although the rate of convergence can be improved considerably by choosing a different initial space $\mathcal{M}_0$, the same examples show that the approximations still oscillate in much the same way as before to the optimal solution. This suggests that approximating the optimal solution may not be practical in some problems. In this case, one may have to adjust these approximating optimal solutions. We plan to return to this phenomenon in a later paper.

Acknowledgement

The authors thank Joe Ball for mentioning the Helton-Young paper [14] to the second author.

References

[1] Adamjan, V.M.; Arov, D.Z.; Kreĭn, M.G., Infinite Hankel matrices and generalized problems of Carathéodory-Fejér and F. Riesz. *Functional Analyis and Applications* **2** (1968), 1–18 [English translation].

[2] V.M. Adamjan, D.Z. Arov, M.G. Krein, Infinite block Hankel matrices and their connection with the interpolation problem. *Amer. Math. Soc. Transl.* (2) **111** (1978), 133–156 [Russian original 1971].

[3] D.Z. Arov and H. Dym, *J-contractive matrix valued functions and related topics*, Cambridge University Press, 2008.

[4] H. Bart, I. Gohberg, M.A. Kaashoek, and A.C.M. Ran, *Factorization of matrix and operator functions: the state space method*, OT **178**, Birkhäuser Verlag, Basel, 2008.

[5] M.J. Corless and A.E. Frazho, *Linear systems and control*, Marcel Dekker, Inc., New York, 2003.

[6] C. Foias and A.E. Frazho, *The Commutant Lifting Approach to Interpolation Problems,* OT **44**, Birkhäuser Verlag, Basel, 1990.

[7] C. Foias, A.E. Frazho, I. Gohberg and M.A. Kaashoek, *Metric Constrained Interpolation, Commutant Lifting and Systems*, OT **100**, Birkhäuser Verlag, Basel, 1998.

[8] C. Foias, A.E. Frazho, and M.A. Kaashoek, Relaxation of metric constrained interpolation and a new lifting theorem, *Integral Equations Operator Theory* **42** (2002), 253–310.

[9] B.A. Francis, *A Course in H_∞ Control Theory*, Lecture Notes in Control and Information Sciences **88**, Springer, Berlin, 1987.

[10] A.E. Frazho, S. ter Horst, and M.A. Kaashoek, All solutions to the relaxed commutant lifting problem, *Acta Sci. Math. (Szeged)* **72** (2006), 299–318.

[11] A.E. Frazho, S. ter Horst, and M.A. Kaashoek, Relaxed commutant lifting: an equivalent version and a new application, in: *Recent Advances in Operator Theory and Applications*, OT **187**, Birkhäuser Verlag, Basel, 2009, pp. 157–168.

[12] I. Gohberg, S. Goldberg and M.A. Kaashoek,*Classes of Linear Operators*, Volume I, OT **49**, Birkhäuser Verlag, Basel, 1990.

[13] I. Gohberg, S. Goldberg and M.A. Kaashoek,*Classes of Linear Operators*, Volume II, OT **63**, Birkhäuser Verlag, Basel, 1993.

[14] J.W. Helton, and N.J. Young, Approximation of Hankel operators: truncation error in an H^∞ design method, in: *Signal Processing, Part II, IMA Vol. Math. Appl.*, 23 Springer, New York, 1990, pp. 115–137.

[15] S. ter Horst, *Relaxed commutant lifting and Nehari interpolation*, Ph.D. dissertation, VU University, Amsterdam, 2007.

[16] S. ter Horst, Relaxed commutant lifting and a relaxed Nehari problem: Redheffer state space formulas, *Math. Nachr.* **282** (2009), 1753–1769.

[17] V.V. Peller, *Hankel Operators and their Applications*, Springer Monographs in Mathematics, Springer 2003.

[18] D. Sarason, Generalized interpolation in H^∞, *Trans. Amer. Math. Soc.* **127** (1967), 179–203.

A.E. Frazho
Department of Aeronautics and Astronautics
Purdue University
West Lafayette, IN 47907, USA
e-mail: `frazho@ecn.purdue.edu`

S. ter Horst
Department of Mathematics
Utrecht University
P.O. Box 80010
NL-3508 TA Utrecht, The Netherlands
e-mail: `s.ter.horst@uu.nl`

M.A. Kaashoek
Department of Mathematics
VU University Amsterdam
De Boelelaan 1081a
NL-1081 HV Amsterdam, The Netherlands
e-mail: `m.a.kaashoek@vu.nl`

Operator Theory:
Advances and Applications, Vol. 222, 173–187

On Theorems of Halmos and Roth

P.A. Fuhrmann and U. Helmke

Dedicated to Bill Helton on the occasion of his 65th birthday

Abstract. This paper was motivated by a result of Halmos [1971] on the characterization of invariant subspaces of finite-dimensional, complex linear operators. It presents a purely algebraic approach, using polynomial and rational models over an arbitrary field, that yields a functional proof of an extension of the result by Halmos. This led to a parallel effort to give a simplified, matrix-oriented, proof. In turn, we explore the connection of Halmos' result with a celebrated Theorem of Roth [1952]. The method presented here has the advantage of generalizing to a class of infinite-dimensional shift operators.

Mathematics Subject Classification. 15A24, 15A54.

Keywords. Invariant subspaces, commutant, functional models, Roth's theorem.

1. Introduction

The present paper re-examines the central result of Halmos [1971] that characterizes invariant subspaces of finite-dimensional complex linear operators as kernels of intertwining maps. It was first brought to our attention by reading Domanov [2010], who presented a short proof based on elementary matrix calculations and a clever choice of coordinates. In this paper we present an even more streamlined, coordinate-free proof, that is valid over an arbitrary field. All these matrix-oriented proofs by Halmos, Domanov and ourselves depend on a theorem of Frobenius [1896] as the essential step. Next, we extend the result to the more general context of polynomial models, introduced in Fuhrmann [1976]. The advantage of this approach is that it does not use Frobenius' result and that it extends to some interesting, infinite-dimensional situations. Realizing that the polynomial proof does not resort

The second author was supported by the DFG under Grant HE 1858/12-1.

to Frobenius' theorem, we looked again for a matrix proof of the same kind. This brought to light some interesting connections to a theorem of Roth [1952], that deserve further investigations.

The paper is structured as follows. In Section 2, we give a short, streamlined proof of the Halmos result. Section 3 is devoted to a brief description of the relevant results from the theory of polynomial models. In Section 4, we present the polynomial model based analog of the Halmos result. Finally, in Section 5, we prove the equivalence of a special case of Halmos' theorem and a Theorem of Roth. This calls for establishing the equivalence of Halmos' theorem and an appropriate generalization of Roth's theorem. This we leave open for a future publication.

Finally, we would like to thank M. Porat for several helpful remarks and for pointing out a gap in the proof of Theorem 7.

2. The Halmos theorem

Halmos [1971] has shown that any invariant subspace $\mathcal{V}$ of an arbitrary complex $n \times n$ matrix A is the image of a complex matrix B, that commutes with A. Similarly, $\mathcal{V} = \operatorname{Ker} C$ for a matrix C commuting with A. Halmos uses the Hilbert space structure of $\mathbb{C}^n$, so his proof does not immediately extend to matrices over arbitrary fields. On the other hand, an essential part of his argument is based on the Frobenius theorem, stating that every square matrix is similar to its transpose $A^\top$. This result holds over any field. His presentation of the main proof idea is convoluted, due to an awkward notation and misleading comments. On the other hand, if one deletes all the unnecessary detours made by Halmos, i.e., using adjoints of complex matrices, allowing matrix multiplication on the right and not only on the left and avoiding basis descriptions, the proof condenses to an extremely short argument that is presented below. The proof holds for an arbitrary field $\mathbb{F}$.

Theorem 1. *Let A denote an arbitrary $n \times n$ matrix over a field $\mathbb{F}$ and $\mathcal{V}$ denote an invariant subspace of A. Then there exist matrices B, C, both commuting with A, such that* Im $B = \mathcal{V}$ *and* Ker $C = \mathcal{V}$.

Proof. Let $\mathcal{V}$ be a subspace invariant under A and let X be a basis matrix for $\mathcal{V}$. By invariance, there exists a matrix Λ for which

$$AX = X\Lambda, \tag{1}$$

and

$$\mathcal{V} = \operatorname{Im} X. \tag{2}$$

By a theorem of Frobenius [1896], see also Fuhrmann [1983], every matrix $A \in \mathbb{F}^{n \times n}$ is similar to its transpose $A^\top$. Let S be such a similarity matrix, i.e., we have $S^{-1}A^\top S = A$. Analogously, there exists a matrix $T \in \mathbb{F}^{p \times p}$ for which $T\Lambda^\top T^{-1} = \Lambda$. Substituting into (1), we get

$$S^{-1}A^\top SX = XT\Lambda^\top T^{-1}, \quad \text{or} \quad A^\top(SXT) = (SXT)\Lambda^\top.$$

Setting $Y = SXT$, we have

$$A^\top Y = Y\Lambda^\top. \tag{3}$$

We define now $B = XY^\top$ and compute

$$\begin{cases} AB = AXY^\top = X\Lambda Y^\top \\ BA = XY^\top A = X\Lambda Y^\top, \end{cases}$$

i.e., we have $AB = BA$. Now we note that both X and Y have full column rank. In particular, $Y^\top$ is surjective which implies

$$\operatorname{Im} B = \operatorname{Im} X. \tag{4}$$

Similarly, there exists a full row rank matrix Z for which we have the kernel representation

$$\mathcal{V} = \operatorname{Ker} Z.$$

This shows the existence of a matrix L for which

$$ZA = LZ. \tag{5}$$

Applying Frobenius' theorem once again, there exists a nonsingular matrix U for which $L = U^{-1}L^\top U$. Substituting in (5), thus $ZS^{-1}A^\top S = U^{-1}L^\top UZ$, or $UZS^{-1}A^\top = L^\top UZS^{-1}$. Setting $W = UZS^{-1}$, we have

$$WA^\top = L^\top W. \tag{6}$$

Defining $C = W^\top Z$ and noting that $W^\top$ is injective. We conclude, that

$$\mathcal{V} = \operatorname{Ker} C. \tag{7}$$

To show that C commutes with A, we note that

$$\begin{cases} AC = AW^\top Z = W^\top LZ \\ CA = W^\top ZA = W^\top LZ, \end{cases}$$

i.e., we have $AC = CA$. $\qquad\square$

In the present paper we extend the result to the context of polynomial models, introduced in Fuhrmann [1976]. The advantage of this approach is that it does not use Frobenius' result and that it extends to some interesting, infinite-dimensional situations.

3. Preliminaries

We begin by giving a brief review of the basic results on polynomial and rational models that will be used in the sequel. We omit proofs which can be found in various papers, e.g., Fuhrmann [1976, 2010b].

3.1. Polynomial models

Let $\mathbb{F}$ denote an arbitrary field. We will denote by $\mathbb{F}^m$ the space of all m-vectors with coordinates in $\mathbb{F}$. By $\mathbb{F}((z^{-1}))$ we denote the field of truncated Laurent series, namely, the space of series of the form $h(z) = \sum_{j=-\infty}^{n(h)} h_j z^j$ with $n(h) \in \mathbb{Z}$. By $z^{-1}\mathbb{F}[[z^{-1}]]$ we denote the subspace of $\mathbb{F}[[z^{-1}]]$ consisting of all power series with vanishing constant term. As $\mathbb{F}$-linear spaces, we have the direct sum representation

$$\mathbb{F}((z^{-1}))^m = \mathbb{F}[z]^m \oplus z^{-1}\mathbb{F}[[z^{-1}]]^m. \qquad (8)$$

We denote by π_+ and π_- the projections of $\mathbb{F}((z^{-1}))^m$ on $\mathbb{F}[z]^m$ and $z^{-1}\mathbb{F}[[z^{-1}]]^m$ respectively, i.e., given by

$$\begin{aligned}
\pi_- \textstyle\sum_{j=-\infty}^{N} h_j z^j &= \sum_{j=-\infty}^{-1} h_j z^j \\
\pi_+ \textstyle\sum_{j=-\infty}^{N} h_j z^j &= \sum_{j=0}^{N} h_j z^j
\end{aligned} \qquad (9)$$

Clearly, π_+ and π_- are complementary projections, i.e., satisfy $\pi_\pm^2 = \pi_\pm$ and $\pi_+ + \pi_- = I$.

Polynomial models are defined as concrete representations of quotient modules of the form $\mathbb{F}[z]^m/\mathcal{M}$, where $\mathcal{M} \subset \mathbb{F}[z]^m$ is a full submodule, i.e., that $\mathbb{F}[z]^m/\mathcal{M}$ is required to be a torsion module. It can be shown that this is equivalent to a representation $\mathcal{M} = D(z)\mathbb{F}[z]^m$ with $D(z) \in \mathbb{F}[z]^{m \times m}$ nonsingular. Defining a projection map $\pi_D : \mathbb{F}[z]^m \longrightarrow \mathbb{F}[z]^m$ by

$$\pi_D f = D\pi_- D^{-1} f \qquad f \in F[z]^m, \qquad (10)$$

we have the isomorphism

$$X_D = \operatorname{Im} \pi_D \simeq \mathbb{F}[z]^m / D(z)\mathbb{F}[z]^m, \qquad (11)$$

which gives concrete, but non canonical, representations for the quotient module. The **shift operator** $S_D : X_D \longrightarrow X_D$ is defined by

$$S_D f = \pi_D z f = z f - D(z)\xi_f, \qquad f \in X_D, \qquad (12)$$

where $\xi_f = (D^{-1} f)_{-1}$.

It is known that $\lambda \in \mathbb{F}$ is an eigenvalue of S_D if and only if $\operatorname{Ker} D(\lambda) \neq 0$. In fact, we have

$$\operatorname{Ker}(\lambda I - S_D) = \left\{ \frac{D(z)\xi}{z - \lambda} \,\middle|\, \xi \in \operatorname{Ker} D(\lambda) \right\} \qquad (13)$$

The next theorem explores the close relationship between factorizations of polynomial matrices and invariant subspaces, thereby it provides a link between geometry and arithmetic. It is one of the principal results which makes the study of polynomial models so useful.

Theorem 2. *Let $D(z) \in \mathbb{F}[z]^{m \times m}$ be nonsingular. A subset $\mathcal{V} \subset X_D$ is a submodule, or equivalently an S_D-invariant subspace, if and only if $\mathcal{V} = D_1 X_{D_2}$ for some factorization $D(z) = D_1(z)D_2(z)$ with $D_i(z) \in \mathbb{F}[z]^{m \times m}$ also nonsingular. Moreover, we have*

$$S_D | D_1 X_{D_2} = D_1 S_{D_2} D_1^{-1}. \qquad (14)$$

3.2. $\mathbb{F}[z]$-Homomorphisms

Polynomial models have two basic structures, that of an $\mathbb{F}$-vector space and that of an $\mathbb{F}[z]$-module. The $\mathbb{F}[z]$-homomorphisms are of particular importance in interpolation and the following theorem gives their characterization.

Theorem 3. *Let $D_1(z) \in \mathbb{F}[z]^{m \times m}$ and $D_2(z) \in \mathbb{F}[z]^{p \times p}$ be nonsingular. An $\mathbb{F}$-linear map $Z : X_{D_1} \longrightarrow X_{D_2}$ is an $\mathbb{F}[z]$-homomorphism, or a map intertwining S_{D_1} and S_{D_2}, i.e., it satisfies*

$$S_{D_2} Z = Z S_{D_1} \tag{15}$$

if and only if there exist $N_1(z), N_2(z) \in \mathbb{F}[z]^{p \times m}$ such that

$$N_2(z)D_1(z) = D_2(z)N_1(z) \tag{16}$$

and

$$Zf = \pi_{D_2} N_2 f. \tag{17}$$

Theorem 4. *Let $Z : X_{D_1} \longrightarrow X_{D_2}$ be the $\mathbb{F}[z]$-module homomorphism defined by*

$$Zf = \pi_{D_2} N_2 f. \tag{18}$$

with

$$N_2(z)D_1(z) = D_2(z)N_1(z) \tag{19}$$

holding. Then

1. *$\operatorname{Ker} Z = E_1 X_{F_1}$, where $D_1(z) = E_1(z)F_1(z)$ and $F_1(z)$ is a g.c.r.d. of $D_1(z)$ and $N_1(z)$.*
2. *$\operatorname{Im} Z = E_2 X_{F_2}$, where $D_2(z) = E_2(z)F_2(z)$ and $E_2(z)$ is a g.c.l.d. of $D_2(z)$ and $N_2(z)$.*
3. *Z is invertible if and only if $D_1(z)$ and $N_1(z)$ are right coprime and $D_2(z)$ and $N_2(z)$ are left coprime.*
4. *$D_1(z)$ and $N_1(z)$ are right coprime and $D_2(z)$ and $N_2(z)$ are left coprime if and only if there exist polynomial matrices $X_1(z), Y_1(z), X_2(z), Y_2(z)$ for which the following doubly coprime factorization holds*

$$\begin{pmatrix} Y_2(z) & -X_2(z) \\ -N_2(z) & D_2(z) \end{pmatrix} \begin{pmatrix} D_1(z) & X_1(z) \\ N_1(z) & Y_1(z) \end{pmatrix} = \begin{pmatrix} I & 0 \\ 0 & I \end{pmatrix}$$
$$\begin{pmatrix} D_1(z) & X_1(z) \\ N_1(z) & Y_1(z) \end{pmatrix} \begin{pmatrix} Y_2(z) & -X_2(z) \\ -N_2(z) & D_2(z) \end{pmatrix} = \begin{pmatrix} I & 0 \\ 0 & I \end{pmatrix}. \tag{20}$$

5. *In terms of the doubly coprime factorizations (20), $Z^{-1} : X_{D_2} \longrightarrow X_{D_1}$ is given by*

$$Z^{-1}g = -\pi_{D_1} X_1 g, \qquad g \in X_{D_2}. \tag{21}$$

4. Kernel and image representations

We have now at hand the necessary machinery to prove the analog of Halmos' result in the context of polynomial models. In preparation, we recall the concept of skew-primeness of polynomial matrices and the principal result.

Definition 5. Let $D_1(z), D_2(z) \in \mathbb{F}[z]^{p \times p}$ be nonsingular polynomial matrices. The ordered pair $(D_1(z), D_2(z))$ is called **left skew prime** if there exist polynomial matrices $\overline{D}_1(z)$ and $\overline{D}_2(z)$ such that

1. $D_1(z)D_2(z) = \overline{D}_2(z)\overline{D}_1(z)$
2. $D_1(z)$ and $\overline{D}_2(z)$ are left coprime
3. $D_2(z)$ and $\overline{D}_1(z)$ are right coprime.

In this case we will say that the pair $(\overline{D}_2(z), \overline{D}_1(z))$ is a **skew complement** of $(D_1(z), D_2(z))$. Note that a sufficient, but not necessary, condition for a pair $(D_1(z), D_2(z))$ to be left skew prime is that $\det D_1(z), \det D_2(z)$ are coprime.

For the following result, which we state without proof, see Fuhrmann [2005]. The geometric interpretation of skew-primeness is due to Khargonekar, Georgiou and Özgüler[1983].

Theorem 6. *Let $D_1(z), D_2(z) \in \mathbb{F}[z]^{p \times p}$ be nonsingular polynomial matrices. Then the following statements are equivalent.*

1. *$D_1(z)$ and $D_2(z)$ are left skew prime.*
2. *The submodule $D_1 X_{D_2} \subset X_{D_1 D_2}$ is an $\mathbb{F}[z]$-direct summand, i.e., it has a complementary submodule.*
3. *The equation*

$$X(z)D_1(z) + D_2(z)Y(z) = I \qquad (22)$$

   has a polynomial solution.
4. *With the factorization $D(z) = D_1(z)D_2(z)$, the invariant subspace $D_1 X_{D_2} \subset X_{D_1 D_2}$ is the kernel of a projection module homomorphism $f : X_D \longrightarrow X_D$.*

The next theorem can be considered a special case of Theorem 1, thus needs no proof. However, in the polynomial approach that we adopt here we aim at more than just a different proof of the Halmos result. What we are after is a full description of the set of all operators Z commuting with a linear operator A that have an invariant subspace $\mathcal{V}$ of A as their kernel. What we describe below are preliminary results.

Theorem 7. *Let $\mathbb{F}$ be a field, $D(z) \in \mathbb{F}[z]^{m \times m}$ be nonsingular and let S_D be the shift operator defined by (12). Then*

1. *A subspace $\mathcal{V} \subset X_D$ is an S_D-invariant subspace if and only if it is the kernel of a map Z that commutes with S_D.*
2. *A subspace $\mathcal{V} \subset X_D$ is an S_D-invariant subspace if and only if it is the image of a map W that commutes with S_D.*

Proof. **1.** The "if" part is trivial.

To prove the "only if" part, we assume that $\mathcal{V} \subset X_D$ is an S_D-invariant subspace, hence, by Theorem 2, there exists a factorization

$$D(z) = D_1(z)D_2(z), \tag{23}$$

for which $\mathcal{V} = D_1 X_{D_2}$.

Case 1: We further assume that $\mathcal{V}$ is reducing, i.e., has a complementary invariant subspace. This, by Theorem 6, implies the existence of a factorization

$$D(z) = D_1(z)D_2(z) = \overline{D}_2(z)\overline{D}_1(z), \tag{24}$$

with $D_1(z), \overline{D}_2(z)$ left coprime and $D_2(z), \overline{D}_1(z)$ right coprime. The coprimeness conditions are equivalent to the existence of polynomial solutions to the Bezout equations

$$D_1(z)X(z) + \overline{D}_2(z)\overline{Y}(z) = I \tag{25}$$

and

$$\overline{X}(z)\overline{D}_1(z) + Y(z)D_2(z) = I. \tag{26}$$

Equations (24), (25) and (26) can be put in matrix form as

$$\begin{pmatrix} \overline{D}_2(z) & D_1(z) \\ \overline{X}(z) & -Y(z) \end{pmatrix} \begin{pmatrix} \overline{Y}(z) & \overline{D}_1(z) \\ X(z) & -D_2(z) \end{pmatrix} = \begin{pmatrix} I & 0 \\ K(z) & I \end{pmatrix}.$$

Multiplying on the left by $\begin{pmatrix} I & 0 \\ -K(z) & I \end{pmatrix}$ and appropriately redefining $\overline{X}(z)$ and $Y(z)$, we obtain the doubly coprime factorization

$$\begin{aligned} \begin{pmatrix} \overline{D}_2(z) & D_1(z) \\ \overline{X}(z) & -Y(z) \end{pmatrix} \begin{pmatrix} \overline{Y}(z) & \overline{D}_1(z) \\ X(z) & -D_2(z) \end{pmatrix} &= \begin{pmatrix} I & 0 \\ 0 & I \end{pmatrix}, \\ \begin{pmatrix} \overline{Y}(z) & \overline{D}_1(z) \\ X(z) & -D_2(z) \end{pmatrix} \begin{pmatrix} \overline{D}_2(z) & D_1(z) \\ \overline{X}(z) & -Y(z) \end{pmatrix} &= \begin{pmatrix} I & 0 \\ 0 & I \end{pmatrix}. \end{aligned} \tag{27}$$

In particular, we have

$$\overline{Y}(z)D_1(z) - \overline{D}_1(z)Y(z) = 0 \tag{28}$$

and the Bezout equation

$$X(z)D_1(z) + D_2(z)Y(z) = I. \tag{29}$$

Next, we define

$$\begin{aligned} M(z) &= Y(z)D_2(z) \\ N(z) &= \overline{D}_2(z)\overline{Y}(z). \end{aligned} \tag{30}$$

We compute

$$N(z)D(z) = \overline{D}_2(z)\overline{Y}(z)D_1(z)D_2(z) = \overline{D}_2(z)\overline{D}_1(z)Y(z)D_2(z) = D(z)M(z).$$

In turn, this implies that the map $Z : X_D \longrightarrow X_D$ defined by

$$Zg = \pi_D N g \tag{31}$$

is intertwining S_D. Comparing the factorizations (23) and (30), we conclude that $D_2(z)$ is a common right divisor of $M(z)$ and $D(z)$. It is a g.c.r.d. as, by the

Bezout equation (29), $Y(z)$ and $D_1(z)$ are right coprime. Applying Theorem 4, we conclude that

$$\text{Ker } Z = D_1 X_{D_2}. \tag{32}$$

Case 2: We assume that there exists an injective homomorphism $W : X_{D_1} \longrightarrow X_{D_2}$. This implies the existence of polynomial matrices $N_2(z)$, $M_1(z)$ satisfying

$$N_2(z)D_1(z) = D_2(z)M_1(z) \tag{33}$$

with $D_1(z), M_1(z)$ right coprime. In turn, this implies the equality

$$(D_1(z)N_2(z))(D_1(z)D_2(z)) = (D_1(z)D_2(z))(M_1(z)D_2(z)). \tag{34}$$

Defining

$$\begin{aligned} N(z) &= D_1(z)N_2(z) \\ M(z) &= M_1(z)D_2(z), \end{aligned} \tag{35}$$

we obtain

$$N(z)D(z) = D(z)M(z). \tag{36}$$

The right coprimeness of $D_1(z), M_1(z)$ shows that a g.c.r.d. of $D(z), M(z)$ is $D_2(z)$. Applying Theorem 4, we get the representation (32).

2. The proof can be derived from Part 1 by duality considerations. However, this is rather intricate and we prefer to give a direct proof.

As before, the "if" part is trivial. To prove the "only if" part, we assume that $\mathcal{V} \subset X_D$ is an S_D-invariant subspace, hence, by Theorem 2, there exists a factorization

$$D(z) = D_1(z)D_2(z), \tag{37}$$

for which $\mathcal{V} = D_1 X_{D_2}$. As we assume that $\mathcal{V}$ is reducing, i.e., has a complementary invariant subspace, Theorem 6, implies the existence of a factorization

$$D(z) = \overline{D}_2(z)\overline{D}_1(z), \tag{38}$$

with $D_1(z), \overline{D}_2(z)$ left coprime and $D_2(z), \overline{D}_1(z)$ right coprime. As in Part 1, we conclude the existence of polynomial matrices $X(z), Y(z), \overline{X}(z), \overline{Y}(z)$, for which the doubly coprime factorization (27) holds. We define now

$$\begin{aligned} N(z) &= D_1(z)X(z) \\ M(z) &= \overline{X}(z)\overline{D}_1(z). \end{aligned} \tag{39}$$

We compute

$$N(z)D(z) = D_1(z)X(z)\overline{D}_2(z)\overline{D}_1(z) = D_1(z)D_2(z)\overline{X}(z)\overline{D}_1(z) = D(z)M(z).$$

In turn, this implies that the map $Z : X_D \longrightarrow X_D$ defined by

$$Zg = \pi_D Ng \tag{40}$$

is intertwining S_D. Comparing the factorizations (37) and (39), we conclude that $D_1(z)$ is a common left divisor of $N(z)$ and $D(z)$. It is a g.c.l.d. as, by the Bezout equation (29), $X(z)$ and $D_2(z)$ are left coprime. Applying Theorems 3 and 4, we conclude that

$$\operatorname{Im} Z = D_1 X_{D_2}. \tag{41}$$

$\square$

Noting that we have the isomorphism $A \simeq S_{zI-A}$, it follows that Theorem 1 is a consequence of Theorem 7

We can prove Theorem 7 in a slightly different way. For this we state and prove the following proposition, a variation on a similar result in Fuhrmann [2010a].

Proposition 8. *Let $D(z) \in \mathbb{F}[z]^{p \times p}$ be nonsingular with proper inverse $D(z)^{-1}$. Choose any $\lambda \in \mathbb{F}$ such that $D(\lambda)$ is nonsingular. Then*

1. *$\lambda I - S_D$ is invertible.*
2. *The Bezout equation*

$$(z - \lambda)\Phi(z) + D(z)\Psi(z) = I \tag{42}$$

   has a unique solution for which $D(z)^{-1}\Phi(z)$ is strictly proper. $\Psi(z) = \Psi$ is necessarily constant and we have

$$\Psi = D(\lambda)^{-1} \tag{43}$$

   and

$$\Phi(z) = \frac{I - D(z)D(\lambda)^{-1}}{z - \lambda}. \tag{44}$$

3. *Defining a map $Z : X_D \longrightarrow X_D$ by*

$$Zg = \pi_D \Phi g \qquad g \in X_D, \tag{45}$$

   then we have

$$(S_D - \lambda I)^{-1} = Z. \tag{46}$$

4. *The following is a doubly coprime factorization*

$$\begin{pmatrix} D(\lambda)^{-1} & \dfrac{I - D(\lambda)^{-1}D(z)}{z - \lambda} \\ -(z - \lambda)I & D(z) \end{pmatrix}$$

$$\times \begin{pmatrix} D(z) & -\dfrac{I - D(z)D(\lambda)^{-1}}{z - \lambda} \\ (z - \lambda)I & D(\lambda)^{-1} \end{pmatrix} = \begin{pmatrix} I & 0 \\ 0 & I \end{pmatrix}$$

$$\begin{pmatrix} D(z) & -\dfrac{I - D(z)D(\lambda)^{-1}}{z - \lambda} \\ (z - \lambda)I & D(\lambda)^{-1} \end{pmatrix}$$

$$\times \begin{pmatrix} D(\lambda)^{-1} & \dfrac{I - D(\lambda)^{-1}D(z)}{z - \lambda} \\ -(z - \lambda)I & D(z) \end{pmatrix} = \begin{pmatrix} I & 0 \\ 0 & I \end{pmatrix} \tag{47}$$

Proof. **1.** By (13), the assumption that $D(\lambda)$ is nonsingular is equivalent to the invertibility of $\lambda I - S_D$.

2. The nonsingularity of $D(\lambda)$ is also equivalent to the left (and right) co-primeness of $(z - \lambda)I$ and $D(z)$. This implies the solvability of the Bezout equation (42). Choosing the unique solution for which $\Phi(z)D(z)^{-1}$ is strictly proper, forces $(z - \lambda)^{-1}\Psi(z)$ to be strictly proper too, i.e., $\Psi(z)$ is necessarily constant. Evaluating (42) at $z = \lambda$, we obtain (43). (44) follows from (42) by extracting $\Phi(z)$.

3. For $g \in X_D$, we compute

$$
\begin{aligned}
g &= \pi_D I g = \pi_D((z - \lambda)\Phi + D\Psi)g \\
  &= \pi_D(z - \lambda)\Phi g = \pi_D(z - \lambda)\pi_D \Phi g = (S_D - \lambda I)Zg,
\end{aligned}
$$

which proves (46).

4. Can be checked by direct computation.   □

We proceed to give an alternative proof of the "only if" part of Theorem 7, using the same notation. We do this as this approach extends to the infinite-dimensional, Hardy space oriented, generalization of the theorem. We do not make the strict properness assumption on $M(z)D(z)^{-1}$ as it suffices to construct any $M_2(z)$ satisfying the required conditions. We note that the nonsingularity of $D_1(z)$ implies the existence of a $\lambda \in \mathbb{F}$ for which $D(\lambda)$ is invertible. We invoke now Proposition 8, but with $D_1(z)$ replacing $D(z)$. We define

$$
M_2(z) = (z - \lambda)I \tag{48}
$$

The coprime factorization (47) implies the right coprimeness of $M_2(z), D_1(z)$. We define the other polynomial matrices by

$$
\begin{aligned}
\overline{M}_2(z) &= (z - \lambda)I \\
\overline{D}_2(z) &= D_2(z) \\
\overline{D}_1(z) &= D_1(z). \\
M(z) &= (z - \lambda)D_2(z) \\
N(z) &= (z - \lambda)D_2(z)
\end{aligned} \tag{49}
$$

We observe that $M_2(z), D_1(z)$ are right coprime because of the following identity

$$
\begin{aligned}
&D_1(\lambda)^{-1}D_1(z) + \frac{I - D_1(\lambda)^{-1}D_1(z)}{z - \lambda}M_2(z) \\
&= D_1(\lambda)^{-1}D_1(z) + \frac{I - D_1(\lambda)^{-1}D_1(z)}{z - \lambda}(z - \lambda)I = I.
\end{aligned} \tag{50}
$$

The rest of the proof is as before.

5. Some interesting connections

We observe that the proof of Theorem 7, which is the polynomial model version of Halmos' result, did not need to use Frobenius' theorem. This leads us to consider the possibility of also finding an elementary, matrix proof which does not use the

Frobenius result. While this remains an open problem, it reveals some interesting connections, especially with Roth [1952], the Sylvester equation and the concept of skew-primeness. In the following we describe some of these connections.

We begin with some heuristic analysis. Let $T : \mathbb{F}^n \longrightarrow \mathbb{F}^n$ be linear and let $\mathcal{V} \subset \mathbb{F}^n$ be a k-dimensional T-invariant subspace. Choose a basis so that $T = \begin{pmatrix} A & 0 \\ C & B \end{pmatrix}$, i.e.,

$$\mathcal{V} = \mathrm{Im} \begin{pmatrix} 0 \\ I \end{pmatrix} = \left\{ \begin{pmatrix} 0 \\ \xi \end{pmatrix} \mid \xi \in \mathbb{F}^k \right\}.$$

Assume:

(a) The matrix $\begin{pmatrix} X & U \\ Z & Y \end{pmatrix}$ commutes with $\begin{pmatrix} A & 0 \\ C & B \end{pmatrix}$.

(b) $\mathrm{Ker} \begin{pmatrix} X & U \\ Z & Y \end{pmatrix} = \mathcal{V}$.

Condition (b) implies that $U = 0$, $Y = 0$ and $\begin{pmatrix} X \\ Z \end{pmatrix}$ is left invertible.

The commutativity condition translates into

$$\begin{cases} XA &= AX \\ ZA &= BZ + CX. \end{cases} \tag{51}$$

Thus we have a connection to the Sylvester equation. Equation (51) can be rewritten as the following matrix equation.

$$\begin{pmatrix} X \\ Z \end{pmatrix} A = \begin{pmatrix} A & 0 \\ C & B \end{pmatrix} \begin{pmatrix} X \\ Z \end{pmatrix}. \tag{52}$$

This indicates that $\mathcal{W} = \mathrm{Im} \begin{pmatrix} X \\ Z \end{pmatrix}$ is an $\begin{pmatrix} A & 0 \\ C & B \end{pmatrix}$-invariant subspace. In turn, we have the isomorphism

$$A \simeq \begin{pmatrix} A & 0 \\ C & B \end{pmatrix} \Big|_{\mathcal{W}}. \tag{53}$$

Clearly, $\mathcal{W}$ is a complementary subspace to $\mathcal{V}$ if and only if X is nonsingular.

In order to prove the Halmos result, we need to construct matrices X, Z for which (51) holds and $\begin{pmatrix} X \\ Z \end{pmatrix}$ is left invertible.

Here are some special cases:

1. The characteristic polynomials of A and B are coprime.
   Under this assumption, the Sylvester equation $ZA - BZ = C$ has a unique solution. Choose $X = I$.
2. The matrices A, B are similar.
   In this case, there exists a nonsingular R for which $RA = BR$. Choose $X = 0$, $Z = R$ and we are done.

3. Assume $\begin{pmatrix} A & 0 \\ C & B \end{pmatrix}, \begin{pmatrix} A & 0 \\ 0 & B \end{pmatrix}$ are similar.

   In this case, by Roth's Theorem [1952], the Sylvester equation $ZA - BZ = C$ is solvable. Choosing $X = I$, we are done.

4. A special case of the previous item is $C = 0$.

   Choose $Z = 0$ and $X = I$.

In all these cases, the constructed matrix $\begin{pmatrix} X \\ Z \end{pmatrix}$ is left invertible.

Lemma 9. *Given* $A \in \mathbb{F}^{(n-k)\times(n-k)}$, $B \in \mathbb{F}^{k\times k}$, $C \in \mathbb{F}^{k\times(n-k)}$, $X \in \mathbb{F}^{(n-k)\times(n-k)}$ *and* $Z \in \mathbb{F}^{k\times(n-k)}$.

1. *We have*

$$\begin{pmatrix} X \\ Z \end{pmatrix} A = \begin{pmatrix} A & 0 \\ C & B \end{pmatrix} \begin{pmatrix} X \\ Z \end{pmatrix}. \tag{54}$$

if and only if

$$\begin{pmatrix} A & 0 \\ C & B \end{pmatrix} \begin{pmatrix} X & 0 \\ Z & I \end{pmatrix} = \begin{pmatrix} X & 0 \\ Z & I \end{pmatrix} \begin{pmatrix} A & 0 \\ 0 & B \end{pmatrix}. \tag{55}$$

2. *If (55) holds with X nonsingular, then we can assume, without loss of generality, that for some Z we have*

$$\begin{pmatrix} A & 0 \\ C & B \end{pmatrix} \begin{pmatrix} I & 0 \\ Z & I \end{pmatrix} = \begin{pmatrix} I & 0 \\ Z & I \end{pmatrix} \begin{pmatrix} A & 0 \\ 0 & B \end{pmatrix}. \tag{56}$$

Proof. **1.** By a simple computation.

2. Multiplying on the right by $\begin{pmatrix} X^{-1} & 0 \\ 0 & I \end{pmatrix}$ and redefining Z. □

If (55) holds and X is invertible, then the matrices $\begin{pmatrix} A & 0 \\ C & B \end{pmatrix}, \begin{pmatrix} A & 0 \\ 0 & B \end{pmatrix}$ are similar. By a theorem of Roth [1952], this is equivalent to the solvability of the Sylvester equation $ZA - BZ = C$. Our next theorem clarifies the connection between this special case of Halmos' theorem, namely the case where (54) is solvable with X invertible, and the theorem of Roth.

Theorem 10. *Given matrices* $A \in \mathbb{F}^{(n-k)\times(n-k)}$, $B \in \mathbb{F}^{k\times k}$ *and* $C \in \mathbb{F}^{k\times(n-k)}$. *The following statements are equivalent:*

1. *We have the following similarity*

$$\begin{pmatrix} A & 0 \\ C & B \end{pmatrix} \simeq \begin{pmatrix} A & 0 \\ 0 & B \end{pmatrix}. \tag{57}$$

2. *The subspace* $\operatorname{Im} \begin{pmatrix} 0 \\ I \end{pmatrix}$ *has a complementary* $\begin{pmatrix} A & 0 \\ C & B \end{pmatrix}$-*invariant subspace.*

3. *There exists a solution of the following Sylvester equation*

$$ZA - BZ = C. \tag{58}$$

4. *There exists a matrix commuting with* $\begin{pmatrix} A & 0 \\ C & B \end{pmatrix}$ *whose kernel is* $\mathrm{Im} \begin{pmatrix} 0 \\ I \end{pmatrix}$

   and whose image is complementary to $\mathrm{Im} \begin{pmatrix} 0 \\ I \end{pmatrix}$.

Proof. $(4) \Rightarrow (2)$

Let $\begin{pmatrix} X & U \\ Z & Y \end{pmatrix}$ be a matrix for which:

(a) The matrix $\begin{pmatrix} X & U \\ Z & Y \end{pmatrix}$ commutes with $\begin{pmatrix} A & 0 \\ C & B \end{pmatrix}$

(b) We have $\mathrm{Ker} \begin{pmatrix} X & U \\ Z & Y \end{pmatrix} = \mathrm{Im} \begin{pmatrix} 0 \\ I \end{pmatrix}$.

Condition (b) implies that $U = 0$, $Y = 0$ and $\begin{pmatrix} X \\ Z \end{pmatrix}$ is left invertible.

The commutativity condition (a) translates into

$$\begin{cases} XA &= AX \\ ZA &= BZ + CX. \end{cases} \tag{59}$$

Equation (59) can be rewritten as the following matrix equation.

$$\begin{pmatrix} X \\ Z \end{pmatrix} A = \begin{pmatrix} A & 0 \\ C & B \end{pmatrix} \begin{pmatrix} X \\ Z \end{pmatrix}. \tag{60}$$

This indicates that $\mathrm{Im} \begin{pmatrix} X \\ Z \end{pmatrix}$ is an $\begin{pmatrix} A & 0 \\ C & B \end{pmatrix}$-invariant subspace. The complementary assumption implies that X is nonsingular. Applying Lemma 9.2, we get the similarity (56).

$(3) \Rightarrow (1)$

Assume Z solves the Sylvester equation (58). This implies the identity (56) and hence the similarity (56).

$(2) \Rightarrow (1)$

Let $\mathrm{Im} \begin{pmatrix} X \\ Z \end{pmatrix}$ be an $\begin{pmatrix} A & 0 \\ C & B \end{pmatrix}$-invariant subspace which is complementary to

$\mathrm{Im} \begin{pmatrix} 0 \\ I \end{pmatrix}$. Without loss of generality, we can assume that $\begin{pmatrix} X \\ Z \end{pmatrix}$ is left invertible.

This implies the existence of a matrix K for which

$$\begin{pmatrix} A & 0 \\ C & B \end{pmatrix} \begin{pmatrix} X \\ Z \end{pmatrix} = \begin{pmatrix} X \\ Z \end{pmatrix} K. \tag{61}$$

The complementarity assumption implies that X is nonsingular. From equation (61) we obtain

$$\begin{pmatrix} A & 0 \\ C & B \end{pmatrix} \begin{pmatrix} I \\ ZX^{-1} \end{pmatrix} = \begin{pmatrix} I \\ ZX^{-1} \end{pmatrix} (XKX^{-1}). \tag{62}$$

This implies $XKX^{-1} = A$. Redefining Z, we have

$$\begin{pmatrix} A & 0 \\ C & B \end{pmatrix} \begin{pmatrix} I \\ Z \end{pmatrix} = \begin{pmatrix} I \\ Z \end{pmatrix} A. \tag{63}$$

Applying Lemma 9.2, we get the similarity (56).

$(2) \Rightarrow (3)$
The Sylvester equation (58) follows from (63).

$(3) \Rightarrow (4)$
Let Z be a solution of the Sylvester equation (58). This implies (63). In turn, we have

$$\begin{pmatrix} A & 0 \\ C & B \end{pmatrix} \begin{pmatrix} I & 0 \\ Z & 0 \end{pmatrix} = \begin{pmatrix} I & 0 \\ Z & 0 \end{pmatrix} \begin{pmatrix} A & 0 \\ C & B \end{pmatrix}.$$

This shows that $\begin{pmatrix} I & 0 \\ Z & 0 \end{pmatrix}$ commutes with $\begin{pmatrix} A & 0 \\ C & B \end{pmatrix}$. Clearly, Ker $\begin{pmatrix} I & 0 \\ Z & 0 \end{pmatrix} =$ Im $\begin{pmatrix} 0 \\ I \end{pmatrix}$.

$(1) \Rightarrow (3)$
This follows from Roth [1952].   □

We wish to point out that Roth's theorem is also about existence of complementary subspaces. In the polynomial model case this connects to the concept of skew-primeness, introduced in Wolovich [1978]. For a geometric interpretation of skew-primeness, see Khargonekar, Georgiou and Özgüler[1983]. Fuhrmann [1994] contains an infinite-dimensional generalization of skew-primeness. This opens up the possibility of establishing the analog of Halmos's theorem in the context of backward shift invariant subspaces.

Because of space constraints, we just state the next result. The context is the n-vectorial H^2 of the unit disk. We use Beurling's representation of invariant subspaces in terms of inner functions, the relation between factorizations of $Q(z)$ and invariant subspaces of the model operator defined by

$$S_Q f = P_{H(Q)} z f, \qquad f(z) \in H(Q), \tag{64}$$

and, finally, the representation of elements of the commutant of S_Q. For most of the background, one can consult Fuhrmann [1994].

Theorem 11. *Let $Q(z) \in H^\infty_{n \times n}$ be an inner function and let $H(Q) = \{QH^2_n\}^\perp = H^2_n \ominus QH^2_n$. Let $S_Q : H(Q) \longrightarrow H(Q)$ be defined by (64). Then*

    1. *A subspace $V \subset H(Q)$ is an S_Q-invariant subspace if and only if it is the kernel of a bounded linear operator W that commutes with S_Q.*

    2. *A subspace $V \subset H(Q)$ is an S_Q-invariant subspace if and only if it is the image of a bounded linear operator W that commutes with S_Q.*

References

[2010] I. Domanov "On invariant subspaces of matrices: A new proof of a theorem of Halmos", *Lin. Alg. Appl.*, 433, 2255–2256.

[1896] F.G. Frobenius, "Über die mit einer Matrix vertauschbaren Matrizen", *Ges. Abh.*, Bd. 3, Springer, pp. 415–427.

[1976] P.A. Fuhrmann, "Algebraic system theory: An analyst's point of view", *J. Franklin Inst.* 301, 521–540.

[1977] P.A. Fuhrmann, "On strict system equivalence and similarity", *Int. J. Contr.*, 25, 5–10.

[1983] P.A. Fuhrmann, "On symmetric rational matrix functions", *Lin. Alg. Appl*, Special issue on Linear Control Theory, (1983), 167–250.

[1994] P.A. Fuhrmann, "On skew primeness of inner functions", *Lin. Alg. Appl.*, vols. 208–209, 539–551.

[1996] P.A. Fuhrmann, *A Polynomial Approach to Linear Algebra*, Springer Verlag, New York.

[2005] P.A. Fuhrmann, "Autonomous subbehaviors and output nulling subspaces", *Int. J. Contr.*, (2005), 78, 1378–1411.

[2006] P.A. Fuhrmann, "On duality in some problems of geometric control", *Acta Applicandae Mathematicae*, 91, 207–251.

[2010a] P.A. Fuhrmann, "A functional approach to the Stein equation", *Lin. Alg. Appl.*, 432, 3031–3071.

[2010b] P.A. Fuhrmann, "On tangential matrix interpolation", *Lin. Alg. Appl.*, 433, 2018–2059.

[1971] P.R. Halmos "Eigenvectors and adjoints", *Lin. Alg. Appl.*, 4, 11–15.

[1983] P.P. Khargonekar, T.T. Georgiou and A.B. Özgüler, "zeros-Prime polynomial matrices, the polynomial model approach", *Lin. Alg. Appl.*, 50, 403–435.

[1952] W.E. Roth, "The equations $AX - YB = C$ and $AX - XB = C$ in matrices", *Proc. Amer. Math. Soc.*, 3, 392–396.

[1970] B. Sz.-Nagy and C. Foias, *Harmonic Analysis of Operators on Hilbert Space*, North-Holland, Amsterdam.

[1974] W.A. Wolovich, *Linear Multivariable Systems*, Springer Verlag, New York.

[1978] W.A. Wolovich, "Skew prime polynomial matrices", *IEEE Trans. Automat. Control*, 23, 880–887.

P.A. Fuhrmann
Department of Mathematics, Ben-Gurion University of the Negev
Beer Sheva, Israel
e-mail: `fuhrmannbgu@gmail.com`

U. Helmke
Universität Würzburg, Institut für Mathematik
Würzburg, Germany
e-mail: `helmke@mathematik.uni-wuerzburg.de`

Operator Theory:
Advances and Applications, Vol. 222, 189–194
© 2012 Springer Basel

Pfister's Theorem Fails in the Free Case

Martin Harrison

Dedicated to Professor Helton, on the occasion of his 65th birthday

Abstract. Artin solved Hilbert's 17th problem by showing that every positive semidefinite polynomial can be realized as a sum of squares of rational functions. Pfister gave a bound on the number of squares of rational functions: if p is a positive semi-definite polynomial in n variables, then there is a polynomial q so that $q^2 p$ is a sum of at most 2^n squares. As shown by D'Angelo and Lebl, the analog of Pfister's theorem fails in the case of Hermitian polynomials. Specifically, it was shown that the rank of any multiple of the polynomial $\|z\|^{2d} \equiv (\sum_j |z_j|^2)^d$ is bounded below by a quantity depending on d. Here we prove that a similar result holds in a free $*$-algebra.

Mathematics Subject Classification. 16S10, 16W10.

Keywords. Pfister's theorem, Hilbert's 17th problem, positive polynomials, free $*$-algebra, sums of squares.

1. Introduction

The aim of this section is to define the main objects and to review some related work. We work in the real free $*$-algebra $\mathbb{R}\langle X, X^* \rangle$ generated by the n noncommuting (NC) variables $X_1, \ldots, X_n$ and their *adjoints* X_j^*. After taking a representation we can think of these variables as real square matrices, and the $*$ function on $\mathbb{R}\langle X, X^* \rangle$ as the transpose operation. In particular, $*$ respects addition and multiplication by scalars and is defined on monomials by $(X_{j_1} \cdots X_{j_k})^* = X_{j_k}^* \cdots X_{j_1}^*$ and $(X_j^*)^* = X_j$. We use multi-indices α, tuples of non-negative integers from 0 to $2n$, to index monomials: $X^\alpha \equiv X_{\alpha_1} X_{\alpha_2} \cdots X_{\alpha_k}$. $X^\emptyset$ is simply the empty word, denoted by 1. For $0 < j \leq n$, we define $X_{j+n} \equiv X_j^*$. We define conjugation and concatenation of multi-indices α and β by the equations $X^{\alpha^*} = (X^\alpha)^*$ and $X^{\alpha \circ \beta} = X^\alpha X^\beta$.

Evaluation of $p \in \mathbb{R}\langle X, X^* \rangle$ at a tuple $(M_1, \ldots, M_n)$ of square matrices of the same size is defined by the substitution of M_j for X_j and M_j^T for X_j^*.

We say that $p \in \mathbb{R}\langle X, X^* \rangle$ is symmetric when $p^* = p$. Such a polynomial p is said to be *matrix positive* if the matrix $p(M)$ is positive semidefinite (or *PSD*) for every tuple M of square matrices. It was shown by Helton in [3] that every matrix positive polynomial is a sum of squares (*SOS*). The minimal number of squares required to express a matrix positive polynomial as a sum of squares is not known in general, although upper bounds are easy to obtain. The question is open in the commutative case as well, and in both cases amounts to a problem of rank minimization. A great many types of rank minimization problems have been successfully attacked in recent years with *semidefinite programming* techniques (see [7] for examples). A complete characterization of conditions for success of the nuclear norm approach, or "trace-heuristic", in this context is not know, though Recht provided in [8] a probabilistic characterization of success for particular classes of rank minimization problems.

Optimization in certain quantum physics problems is done over feasible regions of operators on Hilbert spaces, and so NC variables are useful there. Several examples and a general framework for such problems are presented in [6], where the semidefinite programming relaxations of Lasserre are extended to the NC setting. Motivation for the study of NC polynomials from control theory is discussed in [2].

2. Polynomials, associated matrices and sums of squares

To any symmetric polynomial $p \in \mathbb{R}\langle X, X^* \rangle$ we can associate a real, symmetric matrix M with the property

$$V^* M V = p$$

where $V^* = (X^{\alpha^*})_{|\alpha| \leq d}$, and V is the column vector $(X^\alpha)_{|\alpha| \leq d}$ (with the monomials in graded lexicographical order). The matrix M is not unique, in fact the set of all such matrices (for a fixed p) forms an affine space which we will denote $\mathcal{M}_p$.

By the rank of p, we mean the minimum of $\operatorname{rank}(M)$ over all $M \in \mathcal{M}_p$. For a positive polynomial, this minimum is to be taken over only the PSD matrices. The following lemma helps us obtain a lower bound on rank

Lemma 1. *If A is a symmetric matrix satisfying $V^* A V = 0$, then the $(2n)^d \times (2n)^d$ lower right submatrix of A is the zero matrix.*

Proof. Let B denote the block in question, and $\hat{V}$ the tautological vector of just the monomials of degree d. Then $V^* A V = 0$ implies that $\hat{V}^* B \hat{V} = 0$ as well since the product $\hat{V}^* B \hat{V}$ yields exactly the degree $2d$ terms of the polynomial $V^* A V$. But the entries of B are exactly the coefficients of the distinct monomials in $\hat{V}^* B \hat{V}$, hence B is the zero matrix.

The lemma above shows that there is no freedom in choosing the block corresponding to the degree $2d$ terms of the polynomial. Since the rank of this block gives a lower bound on the rank of the whole matrix, taking the block to be the $(2n)^d \times (2n)^d$ identity yields a polynomial with rank at least $(2n)^d$.

2.1. Positive polynomials and sums of squares

In the commutative case it is well known that the cone of positive polynomials properly contains the SOS cone. Motzkin's polynomial $M(x, y) = 1 + x^2 y^4 + y^2 x^4 - 3x^2 y^2$ is the first known example of a positive polynomial outside the SOS cone, and was discovered decades after Hilbert proved the existence of such polynomials.

In contrast, the NC setting offers the nice result, proved by Helton in [3], that any positive polynomial is a sum of squares. Here, a square takes the form $f^* f$, so that obviously a sum of squares is positive in the sense defined above. In order to understand the SOS representation of a positive polynomial, we use the matrix representation introduced above. The following lemma leads us to the semidefinite programming formulation of the rank minimization problem.

Lemma 2. *A polynomial p is matrix positive exactly when it can be expressed $p = V^* M V$, with M a PSD matrix. The rank of p is exactly the minimum number of squares over all SOS representations of p.*

The proof is straightforward. It follows that the minimum number of squares for a positive p is

$$\min \quad \text{rank } X$$

$$\text{s.t. } V^* X V = p,$$

$$X \succeq 0$$

which can be calculated efficiently (but not always accurately), by minimizing instead the trace of X.

As a simple example of this problem consider the polynomial $P = 1 + X^* X + XX^*$, clearly a SOS. The polynomial P is a sum of 3 squares, but can be expressed as a sum of 2 squares (and no fewer). To see why we parameterize the affine space $\mathcal{M}_P$ by the single parameter $t \in \mathbb{R}$. As usual $V = (1, X, X^*)^T$, and so $P = V^* V = X^* X + XX^* + 1$. Defining

$$M = \begin{pmatrix} 0 & 1 & -1 \\ 1 & 0 & 0 \\ -1 & 0 & 0 \end{pmatrix}$$

we get $\mathcal{M}_P = \{I + tM | t \in \mathbb{R}\}$, and find the minimal SOS representation

$$P = \left(X + \frac{\sqrt{2}}{2}\right)^* \left(X + \frac{\sqrt{2}}{2}\right) + \left(X^* - \frac{\sqrt{2}}{2}\right)^* \left(X^* - \frac{\sqrt{2}}{2}\right)$$

on the boundary of the region where $I + tM \succ 0$. Note that in this example trace is constant on $\mathcal{M}_P \cap PSD$, and that the given solution is obtained by maximizing t over $\{t | I + tM \succeq 0\}$.

3. The Examples

Pfister's Theorem (proved in [5]) gives a bound on the number of rational functions
in the SOS representation of a PSD polynomial. The bound is remarkable because
it does not depend on the degree of the polynomial in question. D'Angelo and Lebl
proved in [1] that this result fails for Hermitian polynomials. We'll show that it
fails for noncommutative polynomials. The first theorem below is needed for the
second. It is easy to check that the polynomial S below has rank $(2n)^d$, but more
is true.

Theorem 3. *Suppose that* $q \in \mathbb{R}\langle X, X^* \rangle$ *and define* $S = \sum_{|\alpha|=d} X^{\alpha^*} X^\alpha$. *Then*
$p = q^* S q$ *has rank at least* $(2n)^d$. *Here,* $(2n)^d$ *is the dimension of* $\text{span}\{X^\alpha\}_{|\alpha|=d}$

Proof. Since p is matrix positive, it is a sum of squares, and so we may write
$p = V^* M V$, appending V with the necessary monomials. Let q be such that $q^* S q =$
p, and write $q = \sum_\alpha q_\alpha X^\alpha$. Let $\hat{\alpha}$ be maximal, with respect to lexicographical
ordering, among all α such that $q_\alpha \neq 0$.

We have $V^* M V = p = q^* S q = q^* (\sum_{|\alpha|=d} X^{\alpha^*} X^\alpha) q = \sum_{|\alpha|=d} (X^\alpha q)^* (X^\alpha q)$.
For each α, write $X^\alpha q = Q_\alpha V$, where Q_α is the row vector of the coefficients of
$X^\alpha q$. Forming the matrix Q whose rows are the Q_α we get $V^* M V = p = V^* Q^* Q V$,
hence $V^* (M - Q^* Q) V = 0$.

The polynomials $X^\alpha q$ form a linearly independent set, and in fact have the
distinct leading terms $q_{\hat{\alpha}} X^{\alpha \circ \hat{\alpha}}$. It follows that the last $(2n)^{d+\deg(q)}$ columns of Q
form a block of rank at least $(2n)^d$. Writing Q in block form $Q = \begin{bmatrix} A & B \end{bmatrix}$ where
B is a $(2n)^d \times (2n)^{d+\deg(q)}$ matrix, we compute

$$p = V^* Q^T Q V = V^* \begin{bmatrix} A^T \\ B^T \end{bmatrix} \begin{bmatrix} A & B \end{bmatrix} V = V^* \begin{bmatrix} A^T A & A^T B \\ B^T A & B^T B \end{bmatrix} V.$$

The V above includes all monomials up to degree $(2n)^{d+\deg(q)}$. Since $V^*(M - Q^* Q)V = 0$, we know from the lemma that M cannot differ from $Q^* Q$ in its
$(2n)^{d+\deg(q)} \times (2n)^{d+\deg(q)}$ lower right block; this block equals $B^T B$. Therefore M,
an arbitrary matrix representation for p, has rank at least $(2n)^d$. □

Alternatively, one might ask whether a Pfister's Theorem holds for products
of the usual form. Consider what it would take for $q^* q S$ to be a SOS. Because $q^* q$ is
symmetric, we note that since SOS are symmetric we must have $q^* q S = (q^* q S)^* =
S q^* q$, so that $q^* q$ and S commute. Since we evaluate these polynomials on tuples
of matrices, it is tempting to treat them as symmetric matrices. In particular, one
might guess that if two of them commute, then they are both polynomials in a
third polynomial. This happens to be true, and it follows from the following more
general theorem from combinatorics:

Theorem 4 (Bergman's Centralizer Theorem). *Let K be a field, and $K\langle X \rangle$ the ring
of polynomials over K in noncommuting variables $X_1, \ldots, X_n$. Then the centralizer
of a nonscalar element in $K\langle X \rangle$ is isomorphic to $K[t]$ for a single variable t.*

The proof is a bit lengthy and can be found in [4]. It uses the fact that such a centralizer is integrally closed in its field of fractions together with an easier result in the formal series setting:

Theorem 5 (Cohn's Centralizer Theorem). *Let K be a field and $K\langle\langle X\rangle\rangle$ the ring for formal power series over K in noncommuting variables $X_1, \ldots, X_n$. Then the centralizer of a nonscalar element in $K\langle\langle X\rangle\rangle$ is isomorphic to $K[t]$ for a single variable t.*

These theorems apply despite the superficial difference that we are working with indeterminates $X_1, \ldots, X_n, X_1^*, \ldots, X_n^*$ for which $(X_i^*)^* = X_i$; there are no *polynomial* relations among them, and so we can regard them as $2n$ noncommuting variables $Y_1, \ldots, Y_{2n}$. Armed with Theorem 3.2, we are ready to give the counterexample:

Theorem 6. *If $p \in \mathbb{R}\langle X, X^*\rangle$, a matrix positive polynomial, is of the form $q^* q S$ with $S = \sum_{|\alpha|=d} X^{\alpha^*} X^\alpha$, then $\mathrm{rank}(p) \geq (2n)^d$.*

Proof. We will use the previous Theorem 3.1 together with Bergman's Centralizer Theorem. The main difficulty lies in showing that under the hypotheses, $q^* q$ is actually a polynomial in S.

Invoking the centralizer theorem we write $q^* q = f(h(X, X^*))$ and $S = g(h(X, X^*))$ for $h(X, X^*) \in \mathbb{R}\langle X, X^*\rangle$ and $f(t), g(t) \in \mathbb{R}[t]$. It follows from the equation $S = g(h(X, X^*))$ that g must have degree 1. To see why, write

$$h(X, X^*) = c_1 X^{\alpha_1} + \cdots + c_l X^{\alpha_l} + (\text{lower degree terms}), \qquad g(t) = a_k t^k + \cdots + a_0$$

with $c_j, a_i \in \mathbb{R}$. We note that each term $X^{\alpha_{j_1}} \cdots X^{\alpha_{j_k}}$ is symmetric since it must be one of the monomials $X^{\alpha^*} X^\alpha$ in S. Supposing $k > 1$, we have always that $\alpha_{j_1} = \alpha_{j_k}^*$. This implies that there is just one α_j, which is certainly not the case. Therefore $\deg(g) = 1$ and we write $g(t) = at + b$ so that $S = g(h(X, X^*)) = ah(X, X^*) + b$ or $h(X, X^*) = 1/a(S - b)$.

Now we have $q^* q = f(1/a(S - b)) = r(S)$ for some polynomial $r(t) \in \mathbb{R}[t]$. Since $r(S)$ has rank equal to 1 (it can be expressed as a single noncommutative square), it follows that $r(t)$ is of even degree. If not, write $r(t) = r_{2k+1} t^{2k+1} + \cdots + r_0$ with $r_{2k+1} \neq 0$. Then $r(S) = r_{2k+1} S^{2k+1} + (\text{lower degree terms})$ and we have by Theorem 3.1 that $S^{2k+1} = S^k S S^k$ and therefore $r(S)$ itself has rank at least $(2n)^d > 1$, a contradiction. Finally, $tr(t)$ has odd degree and therefore another application of Theorem 3.1 lets us conclude that $p = Sr(S)$ has rank at least $(2n)^d$.   □

References

[1] John P. D'Angelo and Jiri Lebl, *Pfister's theorem fails in the hermitian case*, to appear in Proc. Amer. Math. Soc.

[2] J.W. Helton, F. Dell Kronewitter, W.M. McEneaney, and Mark Stankus, *Singularly perturbed control systems using non-commutative computer algebra*, Internat. J. Robust Nonlinear Control **10** (2000), no. 11-12, 983–1003, George Zames commemorative issue. MR 1786378 (2001i:93067)

[3] J. William Helton, *"Positive" noncommutative polynomials are sums of squares*, Ann. of Math. (2) **156** (2002), no. 2, 675–694. MR 1933721 (2003k:12002)

[4] M. Lothaire, *Combinatorics on words*, Cambridge Mathematical Library, Cambridge University Press, Cambridge, 1997, with a foreword by Roger Lyndon and a preface by Dominique Perrin, Corrected reprint of the 1983 original, with a new preface by Perrin. MR 1475463 (98g:68134)

[5] Albrecht Pfister, *Zur Darstellung definiter Funktionen als Summe von Quadraten*, Invent. Math. **4** (1967), 229–237. MR 0222043 (36 #5095)

[6] S. Pironio, M. Navascués, and A. Acín, *Convergent relaxations of polynomial optimization problems with noncommuting variables*, SIAM J. Optim. **20** (2010), no. 5, 2157–2180. MR 2650843

[7] Benjamin Recht, Maryam Fazel, and Pablo A. Parrilo, *Guaranteed minimum-rank solutions of linear matrix equations via nuclear norm minimization*, SIAM Rev. **52** (2010), no. 3, 471–501. MR 2680543

[8] Benjamin Recht, Weiyu Xu, and Babak Hassibi, *Necessary and sufficient conditions for success of the nuclear norm heuristic for rank minimization*, CDC, IEEE, 2008, pp. 3065–3070.

Martin Harrison
Department of Mathematics
University of California Santa Barbara
Santa Barbara, CA 93106, USA
e-mail: martin@math.ucsb.edu

Operator Theory:
Advances and Applications, Vol. 222, 195–219
© 2012 Springer Basel

Free Analysis, Convexity and LMI Domains

J. William Helton[1], Igor Klep[2] and Scott McCullough[3]

Abstract. This paper concerns *free analytic maps* on *noncommutative domains*. These maps are free analogs of classical holomorphic functions in several complex variables, and are defined in terms of noncommuting variables amongst which there are no relations – they are *free* variables. Free analytic maps include vector-valued polynomials in free (noncommuting) variables and form a canonical class of mappings from one noncommutative domain $\mathcal{D}$ in say g variables to another noncommutative domain $\tilde{\mathcal{D}}$ in $\tilde{g}$ variables. Motivated by determining the possibilities for mapping a nonconvex noncommutative domain to a convex noncommutative domain, this article focuses on rigidity results for free analytic maps. Those obtained to date parallel and are often stronger than those in several complex variables. For instance, a proper free analytic map between noncommutative domains is one-one and, if $\tilde{g} = g$, free biholomorphic. Making its debut here is a free version of a theorem of Braun-Kaup-Upmeier: between two freely biholomorphic bounded circular noncommutative domains there exists a *linear* biholomorphism. An immediate consequence is the following *nonconvexification* result: if two bounded circular noncommutative domains are freely biholomorphic, then they are either both convex or both not convex. Because of their roles in systems engineering, *linear matrix inequalities* (LMIs) and noncommutative domains defined by an LMI (*LMI domains*) are of particular interest. As a refinement of above the nonconvexification result, if a bounded circular noncommutative domain $\mathcal{D}$ is freely biholomorphic to a bounded circular LMI domain, then $\mathcal{D}$ is itself an LMI domain.

Mathematics Subject Classification. Primary: 46L52, 47A56, 32A05, 46G20; secondary: 47A63, 32A10, 14P10.

Keywords. noncommutative set and function,analytic map, proper map, rigidity, linear matrix inequality, several complex variables, free analysis, free real algebraic geometry.

1) Research supported by NSF grants DMS-0700758, DMS-0757212, and the Ford Motor Co.
2) Supported by the Faculty Research Development Fund (FRDF) of The University of Auckland (project no. 3701119). Partially supported by the Slovenian Research Agency (project no. J1-3608 and program no. P1-0222). The article was written while the author was visiting the University of Konstanz.
3) Research supported by the NSF grant DMS-0758306.

1. Introduction

The notion of an analytic, free or noncommutative, map arises naturally in free probability, the study of noncommutative (free) rational functions [BGM06, Vo04, Vo10, SV06, MS11, KVV–], and systems theory [HBJP87]. In this paper rigidity results for such functions paralleling those for their classical commutative counterparts are presented. Often in the noncommutative (nc) setting such theorems have cleaner statements than their commutative counterparts. Among these we shall present the following:

(1) a *continuous* free map is *analytic* (§2.4) and hence admits a *power series* expansion (§2.5);

(2) if f is a *proper* analytic free map from a noncommutative domain in g variables to another in $\tilde{g}$ variables, then f is *injective* and $\tilde{g} \geq g$. If in addition $\tilde{g} = g$, then f is onto and has an inverse which is itself a (proper) analytic free map (§3.1). This injectivity conclusion contrasts markedly with the classical case where a (commutative) *proper* analytic function f from one domain in $\mathbb{C}^g$ to another in $\mathbb{C}^g$, need not be injective, although it must be onto.

(3) A free Braun-Kaup-Upmeier theorem (§5). A free analytic map f is called a free biholomorphism if f has an inverse f^{-1} which is also a free analytic map. As an extension of a theorem from [BKU78], two bounded, circular, noncommutative domains are freely *biholomorphic* if and only if they are freely *linearly* biholomorphic.

(4) Of special interest are free analytic mappings from or to or both from and to noncommutative domains defined by linear matrix inequalities, or LMI domains. Several additional recent results in this direction, as well as a concomitant free *convex Positivstellensatz* (§6.3), are also included.

Thus this article is largely a survey. The results of items (1), (2), and (4) appear elsewhere. However, the main result of (3) is new. Its proof relies on the existence of power series expansions for analytic free maps, a topic we discuss as part of (1) in §2.5 below. Our treatment is modestly different from that found in [Vo10, KVV–].

For the classical theory of commutative proper analytic maps see D'Angelo [DAn93] or Forstnerič [Fo93]. We assume the reader is familiar with basics of several complex variables as given, e.g., in Krantz [Kr01].

1.1. Motivation

One of the main advances in systems engineering in the 1990's was the conversion of a set of problems to *linear matrix inequalities (LMIs)*, since LMIs, up to modest size, can be solved numerically by semidefinite programs [SIG98]. A large class of linear systems problems are described in terms of a signal-flow diagram Σ plus L^2 constraints (such as energy dissipation). Routine methods convert such problems into noncommutative polynomial inequalities of the form $p(X) \succeq 0$ or $p(X) \succ 0$.

Instantiating specific systems of linear differential equations for the "boxes" in the system flow diagram amounts to substituting their coefficient matrices for variables in the polynomial p. Any property asserted to be true must hold when matrices of any size are substituted into p. Such problems are referred to as *dimension-free*. We emphasize, the polynomial p itself is determined by the signal-flow diagram Σ.

Engineers vigorously *seek convexity*, since optima are global and convexity lends itself to numerics. Indeed, there are over a thousand papers trying to convert linear systems problems to convex ones and the only known technique is the rather blunt trial and error instrument of trying to guess an LMI. Since having an LMI is seemingly more restrictive than convexity, there has been the hope, indeed expectation, that some practical class of convex situations has been missed.

Hence a main goal of this line of research has been to determine which *changes of variables* can produce convexity from nonconvex situations. As we shall see below, a free analytic map between noncommutative domains cannot produce convexity from a nonconvex set, at least under a circularity hypothesis. Thus we think the implications of our results here are negative for linear systems engineering; for dimension-free problems the evidence here is that there is no convexity beyond the obvious.

1.2. Reader's guide

The definitions as used in this paper are given in the following section §2, which contains the background on *noncommutative domains* and on *free maps* at the level of generality needed for this paper. As we shall see, free maps that are continuous are also analytic (§2.4). We explain, in §2.5, how to associate a power series expansion to an analytic free map using the noncommutative Fock space. One typically thinks of free maps as being analytic, but in a weak sense. In §3 we consider *proper* free maps and give several rigidity theorems. For instance, proper analytic free maps are injective (§3.1) and, under mild additional assumptions, tend to be linear (see §4 and §5 for precise statements). Results paralleling classical results on analytic maps in several complex variables, such as the Carathéodory-Cartan-Kaup-Wu (CCKW) Theorem, are given in §4. A new result – a free version of the Braun-Kaup-Upmeier (BKU) theorem – appears in §5. A brief overview of further topics, including links to references, is given in §6. Most of the material presented in this paper has been motivated by problems in systems engineering, and this was discussed briefly above in §1.1.

2. Free maps

This section contains the background on noncommutative sets and on *free maps* at the level of generality needed for this paper. Since power series are used in §5, included at the end of this section is a sketch of an argument showing that continuous free maps have formal power series expansions. The discussion borrows heavily from the recent basic work of Voiculescu [Vo04, Vo10] and of Kalyuzhnyi-Verbovetskiĭ and Vinnikov [KVV–], see also the references therein. These papers contain a more power series based approach to free maps and for more on this one can see Popescu [Po06, Po10], or also [HKMS09, HKM11a, HKM11b].

2.1. Noncommutative sets and domains

Fix a positive integer g. Given a positive integer n, let $M_n(\mathbb{C})^g$ denote g-tuples of $n \times n$ matrices. Of course, $M_n(\mathbb{C})^g$ is naturally identified with $M_n(\mathbb{C}) \otimes \mathbb{C}^g$.

A sequence $\mathcal{U} = (\mathcal{U}(n))_{n \in \mathbb{N}}$, where $\mathcal{U}(n) \subseteq M_n(\mathbb{C})^g$, is a **noncommutative set** if it is **closed with respect to simultaneous unitary similarity**; i.e., if $X \in \mathcal{U}(n)$ and U is an $n \times n$ unitary matrix, then

$$U^*XU = (U^*X_1U, \ldots, U^*X_gU) \in \mathcal{U}(n); \qquad (1)$$

and if it is **closed with respect to direct sums**; i.e., if $X \in \mathcal{U}(n)$ and $Y \in \mathcal{U}(m)$ implies

$$X \oplus Y = \begin{bmatrix} X & 0 \\ 0 & Y \end{bmatrix} \in \mathcal{U}(n+m). \qquad (2)$$

Noncommutative sets differ from the fully matricial $\mathbb{C}^g$-sets of Voiculescu [Vo04, Section 6] in that the latter are closed with respect to simultaneous similarity, not just simultaneous *unitary* similarity. Remark 2 below briefly discusses the significance of this distinction for the results on proper analytic free maps in this paper.

The noncommutative set $\mathcal{U}$ is a **noncommutative domain** if each $\mathcal{U}(n)$ is nonempty, open and connected. Of course the sequence $M(\mathbb{C})^g = (M_n(\mathbb{C})^g)$ is itself a noncommutative domain. Given $\varepsilon > 0$, the set $\mathcal{N}_\varepsilon = (\mathcal{N}_\varepsilon(n))$ given by

$$\mathcal{N}_\varepsilon(n) = \left\{ X \in M_n(\mathbb{C})^g : \sum X_j X_j^* \prec \varepsilon^2 \right\} \qquad (3)$$

is a noncommutative domain which we call the **noncommutative ε-neighborhood of 0 in $\mathbb{C}^g$**. The noncommutative set $\mathcal{U}$ is **bounded** if there is a $C \in \mathbb{R}$ such that

$$C^2 - \sum X_j X_j^* \succ 0 \qquad (4)$$

for every n and $X \in \mathcal{U}(n)$. Equivalently, for some $\lambda \in \mathbb{R}$, we have $\mathcal{U} \subseteq \mathcal{N}_\lambda$. Note that this condition is stronger than asking that each $\mathcal{U}(n)$ is bounded.

Let $\mathbb{C}\langle x \rangle = \mathbb{C}\langle x_1, \ldots, x_g \rangle$ denote the $\mathbb{C}$-algebra freely generated by g noncommuting letters $x = (x_1, \ldots, x_g)$. Its elements are linear combinations of words in x and are called (analytic) **polynomials**. Given an $r \times r$ matrix-valued polynomial $p \in M_r(\mathbb{C}) \otimes \mathbb{C}\langle x_1, \ldots, x_g \rangle$ with $p(0) = 0$, let $\mathcal{D}(n)$ denote the connected component of

$$\{ X \in M_n(\mathbb{C})^g : I + p(X) + p(X)^* \succ 0 \} \qquad (5)$$

containing the origin. The sequence $\mathcal{D} = (\mathcal{D}(n))$ is a noncommutative domain which is semi-algebraic in nature. Note that $\mathcal{D}$ contains an $\varepsilon > 0$ neighborhood of 0, and that the choice

$$p = \frac{1}{\varepsilon} \begin{bmatrix} 0_{g \times g} & \begin{matrix} x_1 \\ \vdots \\ x_g \end{matrix} \\ 0_{1 \times g} & 0_{1 \times 1} \end{bmatrix}$$

gives $\mathcal{D} = \mathcal{N}_\varepsilon$. Further examples of natural noncommutative domains can be generated by considering noncommutative polynomials in both the variables $x =$

$(x_1, \ldots, x_g)$ and their formal adjoints, $x^* = (x_1^*, \ldots, x_g^*)$. For us the motivating case of domains is determined by linear matrix inequalities (LMIs).

2.2. LMI domains

A special case of the noncommutative domains are those described by a linear matrix inequality. Given a positive integer d and $A_1, \ldots, A_g \in M_d(\mathbb{C})$, the linear matrix-valued polynomial

$$L(x) = \sum A_j x_j \in M_d(\mathbb{C}) \otimes \mathbb{C}\langle x_1, \ldots, x_g \rangle \qquad (6)$$

is a **(homogeneous) linear pencil**. Its adjoint is, by definition, $L(x)^* = \sum A_j^* x_j^*$. Let

$$\mathcal{L}(x) = I_d + L(x) + L(x)^*.$$

If $X \in M_n(\mathbb{C})^g$, then $\mathcal{L}(X)$ is defined by the canonical substitution,

$$\mathcal{L}(X) = I_d \otimes I_n + \sum A_j \otimes X_j + \sum A_j^* \otimes X_j^*,$$

and yields a symmetric $dn \times dn$ matrix. The inequality $\mathcal{L}(X) \succ 0$ for tuples $X \in M(\mathbb{C})^g$ is a **linear matrix inequality (LMI)**. The sequence of solution sets $\mathcal{D}_{\mathcal{L}}$ defined by

$$\mathcal{D}_{\mathcal{L}}(n) = \{X \in M_n(\mathbb{C})^g : \mathcal{L}(X) \succ 0\} \qquad (7)$$

is a noncommutative domain which contains a neighborhood of 0. It is called a **noncommutative (nc) LMI domain**. It is also a particular instance of a noncommutative semialgebraic set.

2.3. Free mappings

Let $\mathcal{U}$ denote a noncommutative subset of $M(\mathbb{C})^g$ and let $\tilde{g}$ be a positive integer. A **free map** f from $\mathcal{U}$ into $M(\mathbb{C})^{\tilde{g}}$ is a sequence of functions $f[n] : \mathcal{U}(n) \to M_n(\mathbb{C})^{\tilde{g}}$ which **respects direct sums**: for each n, m and $X \in \mathcal{U}(n)$ and $Y \in \mathcal{U}(m)$,

$$f(X \oplus Y) = f(X) \oplus f(Y); \qquad (8)$$

and **respects similarity**: for each n and $X, Y \in \mathcal{U}(n)$ and invertible $n \times n$ matrix Γ such that

$$X\Gamma = (X_1\Gamma, \ldots, X_g\Gamma) = (\Gamma Y_1, \ldots, \Gamma Y_g) = \Gamma Y \qquad (9)$$

we have

$$f(X)\Gamma = \Gamma f(Y). \qquad (10)$$

Note if $X \in \mathcal{U}(n)$ it is natural to write simply $f(X)$ instead of the more cumbersome $f[n](X)$ and likewise $f : \mathcal{U} \to M(\mathbb{C})^{\tilde{g}}$.

We say f **respects intertwining maps** if $X \in \mathcal{U}(n)$, $Y \in \mathcal{U}(m)$, $\Gamma : \mathbb{C}^m \to \mathbb{C}^n$, and $X\Gamma = \Gamma Y$ implies $f[n](X)\Gamma = \Gamma f[m](Y)$. The following proposition gives an alternate characterization of free maps. Its easy proof is left to the reader (alternately, see [HKM11b, Proposition 2.2]).

Proposition 1. *Suppose $\mathcal{U}$ is a noncommutative subset of $M(\mathbb{C})^g$. A sequence $f = (f[n])$ of functions $f[n] : \mathcal{U}(n) \to M_n(\mathbb{C})^{\tilde{g}}$ is a free map if and only if it respects intertwining maps.*

Remark 2. Let $\mathcal{U}$ be a noncommutative domain in $M(\mathbb{C})^g$ and suppose $f : \mathcal{U} \to M(\mathbb{C})^{\tilde{g}}$ is a free map. If $X \in \mathcal{U}$ is similar to Y with $Y = S^{-1}XS$, then we can define $f(Y) = S^{-1}f(X)S$. In this way f naturally extends to a free map on $\mathcal{H}(\mathcal{U}) \subseteq M(\mathbb{C})^g$ defined by

$$\mathcal{H}(\mathcal{U})(n) = \{Y \in M_n(\mathbb{C})^g : \text{there is an } X \in \mathcal{U}(n) \text{ such that } Y \text{ is similar to } X\}.$$

Thus if $\mathcal{U}$ is a domain of holomorphy, then $\mathcal{H}(\mathcal{U}) = \mathcal{U}$.

On the other hand, because our results on proper analytic free maps to come depend strongly upon the noncommutative set $\mathcal{U}$ itself, the distinction between noncommutative sets and fully matricial sets as in [Vo04] is important. See also [HM+, HKM+, HKM11b].

We close this subsection with a simple observation:

Proposition 3. *If $\mathcal{U}$ is a noncommutative subset of $M(\mathbb{C})^g$ and $f : \mathcal{U} \to M(\mathbb{C})^{\tilde{g}}$ is a free map, then the range of f, equal to the sequence $f(\mathcal{U}) = \big(f[n](\mathcal{U}(n))\big)$, is itself a noncommutative subset of $M(\mathbb{C})^{\tilde{g}}$.*

2.4. A continuous free map is analytic

Let $\mathcal{U} \subseteq M(\mathbb{C})^g$ be a noncommutative set. A free map $f : \mathcal{U} \to M(\mathbb{C})^{\tilde{g}}$ is **continuous** if each $f[n] : \mathcal{U}(n) \to M_n(\mathbb{C})^{\tilde{g}}$ is continuous. Likewise, if $\mathcal{U}$ is a noncommutative domain, then f is called **analytic** if each $f[n]$ is analytic. This implies the existence of directional derivatives for all directions at each point in the domain, and this is the property we use often. Somewhat surprising, though easy to prove, is the following:

Proposition 4. *Suppose $\mathcal{U}$ is a noncommutative domain in $M(\mathbb{C})^g$.*

(1) *A continuous free map $f : \mathcal{U} \to M(\mathbb{C})^{\tilde{g}}$ is analytic.*

(2) *If $X \in \mathcal{U}(n)$, and $H \in M_n(\mathbb{C})^g$ has sufficiently small norm, then*

$$f\begin{bmatrix} X & H \\ 0 & X \end{bmatrix} = \begin{bmatrix} f(X) & f'(X)[H] \\ 0 & f(X) \end{bmatrix}. \tag{11}$$

We shall not prove this here and refer the reader to [HKM11b, Proposition 2.5] for a proof. The equation (11) appearing in item (2) will be greatly expanded upon in §2.5 immediately below, where we explain how every free analytic map admits a convergent power series expansion.

2.5. Analytic free maps have a power series expansion

It is shown in [Vo10, Section 13] that a free analytic map f has a formal power series expansion in the noncommuting variables, which indeed is a powerful way to think of free analytic maps. Voiculescu also gives elegant formulas for the coefficients of the power series expansion of f in terms of clever evaluations of f. Convergence properties for bounded free analytic maps are studied in [Vo10, Sections 14–16]; see also [Vo10, Section 17] for a bad unbounded example. Also, Kalyuzhnyi-Verbovetskiĭ and Vinnikov [KVV–] are developing general results based on very weak hypotheses with the conclusion that f has a power series expansion and is thus a free analytic map. An early study of noncommutative mappings is given in [Ta73]; see also [Vo04].

Given a positive integer $\tilde{g}$, a **formal power series** F in the variables $x = \{x_1, \ldots, x_g\}$ with coefficients in $\mathbb{C}^{\tilde{g}}$ is an expression of the form

$$F = \sum_{w \in \langle x \rangle} F_w w$$

where the $F_w \in \mathbb{C}^{\tilde{g}}$, and $\langle x \rangle$ is the free monoid on x, i.e., the set of all words in the noncommuting variables x. (More generally, the F_w could be chosen to be operators between two Hilbert spaces. With the choice of $F_w \in \mathbb{C}^{\tilde{g}}$ and with some mild additional hypothesis, the power series F determines a free map from some noncommutative ε-neighborhood of 0 in $M(\mathbb{C})^g$ into $M(\mathbb{C})^{\tilde{g}}$.)

Letting $F^{(m)} = \sum_{|w|=m} F_w w$ denote the **homogeneous of degree** m **part** of F,

$$F = \sum_{m=0}^{\infty} \sum_{|w|=m} F_w w = \sum_m F^{(m)}. \tag{12}$$

Proposition 5. *Let $\mathcal{V}$ denote a noncommutative domain in $M(\mathbb{C})^g$ which contains some ε-neighborhood of the origin, $\mathcal{N}_\varepsilon$. Suppose $f = (f[n])$ is a sequence of analytic functions $f[n] : \mathcal{V}(n) \to M_n(\mathbb{C})^{\tilde{g}}$. If there is a formal power series F such that for $X \in \mathcal{N}_\varepsilon$ the series $F(X) = \sum_m F^{(m)}(X)$ converges in norm to $f(X)$, then f is a free analytic map $\mathcal{V} \to M(\mathbb{C})^{\tilde{g}}$.*

The following lemma will be used in the proof of Proposition 5.

Lemma 6. *Suppose W is an open connected subset of a locally connected metric space X and $o \in W$. Suppose $o \in W_1 \subset W_2 \subset \cdots$ is a nested increasing sequence of open subsets of W and let W_j^o denote the connected component of W_j containing o. If $\cup W_j = W$, then $\cup W_j^o = W$.*

Proof. Let $U = \cup W_j^o$. If U is a proper subset of W, then $V = W \setminus U$ is neither empty nor open. Hence, there is a $v \in V$ such that $N_\delta(v) \cap U \neq \emptyset$ for every $\delta > 0$. Here $N_\delta(v)$ is the δ neighborhood of v.

There is an N so that if $n \geq N$, then $v \in W_n$. There is a $\delta > 0$ such that $N_\delta(v)$ is connected, and $N_\delta(v) \subset W_n$ for all $n \geq N$. There is an M so that $N_\delta(v) \cap W_m^o \neq \emptyset$ for all $m \geq M$. In particular, since both $N_\delta(v)$ and W_m^o are connected, $N_\delta(v) \cup W_m^o$ is connected. Hence, for n large enough, $N_\delta(v) \cup W_m^o$ is both connected and a subset of W_m. This gives the contradiction $N_\delta(v) \subset W_m^o$. $\qquad\square$

Proof of Proposition 5. For notational convenience, let $\mathcal{N} = \mathcal{N}_\varepsilon$. For each n, the formal power series F determines an analytic function $\mathcal{N}(n) \to M_n(\mathbb{C})^{\tilde{g}}$ which agrees with $f[n]$ (on $\mathcal{N}(n)$). Moreover, if $X \in \mathcal{N}(n)$ and $Y \in \mathcal{N}(m)$, and $X\Gamma = \Gamma Y$, then $F(X)\Gamma = \Gamma F(Y)$. Hence $f[n](X)\Gamma = \Gamma f[m](Y)$.

Fix $X \in \mathcal{V}(n)$, $Y \in \mathcal{V}(m)$, and suppose there exists $\Gamma \neq 0$ such that $X\Gamma = \Gamma Y$. For each positive integer j let

$$\mathcal{W}_j = \left\{ (A, B) \in \mathcal{V}(n) \times \mathcal{V}(m) : \begin{bmatrix} I & -\frac{1}{j}\Gamma \\ 0 & I \end{bmatrix} \begin{bmatrix} A & 0 \\ 0 & B \end{bmatrix} \begin{bmatrix} I & \frac{1}{j}\Gamma \\ 0 & I \end{bmatrix} \in \mathcal{V}(n+m) \right\}$$

$$\subset \mathcal{V}(n) \oplus \mathcal{V}(m).$$

Note that $\mathcal{W}_j$ is open since $\mathcal{V}(n+m)$ is. Further, $\mathcal{W}_j \subset \mathcal{W}_{j+1}$ for each j; for j large enough, $(0,0) \in \mathcal{W}_j$; and $\cup \mathcal{W}_j = W := \mathcal{V}(n) \oplus \mathcal{V}(m)$. By Lemma 6, $\cup \mathcal{W}_j^o = W$, where $\mathcal{W}_j^o$ is the connected component of $\mathcal{W}_j$ containing $(0,0)$. Hence, $(X, Y) \in \mathcal{W}_j^o$ for large enough j which we now fix. Let $\mathcal{Y} \subset \mathcal{W}_j$ be a connected neighborhood of $(0,0)$ with $\mathcal{Y} \subset \mathcal{N}(n) \oplus \mathcal{N}(m)$.

We have analytic functions $G, H : \mathcal{W}_j^o \to M_{m+n}(\mathbb{C}^g)$ defined by

$$G(A, B) = \begin{bmatrix} I & -\frac{1}{j}\Gamma \\ 0 & I \end{bmatrix} \begin{bmatrix} f(n)(A) & 0 \\ 0 & f(m)(B) \end{bmatrix} \begin{bmatrix} I & \frac{1}{j}\Gamma \\ 0 & I \end{bmatrix}$$

$$H(A, B) = f(n+m)\left(\begin{bmatrix} I & -\frac{1}{j}\Gamma \\ 0 & I \end{bmatrix} \begin{bmatrix} A & 0 \\ 0 & B \end{bmatrix} \begin{bmatrix} I & \frac{1}{j}\Gamma \\ 0 & I \end{bmatrix} \right).$$

For $(A, B) \in \mathcal{Y}$ we have $G(A, B) = H(A, B)$ from above. By analyticity and the connectedness of $\mathcal{W}_j^o$, this shows $G(A, B) = H(A, B)$ on $\mathcal{W}_j^o$.

Since $(X, Y) \in \mathcal{W}_j^o$ we obtain the equality $G(X, Y) = H(X, Y)$, which gives, using $X\Gamma - \Gamma Y = 0$,

$$f \begin{bmatrix} X & 0 \\ 0 & Y \end{bmatrix} = \begin{bmatrix} f(X) & \frac{1}{j}(f(X)\Gamma - \Gamma f(Y)) \\ 0 & f(Y) \end{bmatrix}.$$

Thus $f(X)\Gamma - \Gamma f(Y) = 0$ and we conclude that f respects intertwinings and hence is a free map.   □

If $\mathcal{V}$ is a noncommutative set, a free map $f : \mathcal{V} \to M(\mathbb{C})^{\tilde{g}}$ is **uniformly bounded** provided there is a C such that $\|f(X)\| \leq C$ for every $n \in \mathbb{N}$ and $X \in \mathcal{V}(n)$.

Proposition 7. *If $f : \mathcal{N}_\varepsilon \to M(\mathbb{C})^{\tilde{g}}$ is a free analytic map then there is a formal power series*

$$F = \sum_{w \in \langle x \rangle} F_w w = \sum_{m=0}^{\infty} \sum_{|w|=m} F_w w \qquad (13)$$

which converges on $\mathcal{N}_\varepsilon$ and such that $F(X) = f(X)$ for $X \in \mathcal{N}_\varepsilon$.

Moreover, if f is uniformly bounded by C, then the power series converges uniformly in the sense that for each m, $0 \leq r < 1$, and tuple $T = (T_1, \dots, T_g)$ of operators on Hilbert space satisfying $\sum T_j T_j^* \prec r^2 \varepsilon^2 I$, we have

$$\left\| \sum_{|w|=m} F_w \otimes T^w \right\| \leq C r^m.$$

In particular, $\|F_w\| \leq \frac{C}{\varepsilon^n}$ for each word w of length n.

Remark 8. Taking advantage of polynomial identities for $M_n(\mathbb{C})$, the article [Vo10] gives an example of a formal power series G which converges for every tuple X of matrices, but has 0 radius of convergence in the sense that for every $r > 0$ there exists a tuple of operators $X = (X_1, \cdots, X_g)$ with $\sum X_j^* X_j < r^2$ for which $G(X)$ fails to converge.

2.6. The Fock space and creation operators

The **noncommutative Fock space**, denoted $\mathcal{F}_g$, is the Hilbert space with orthonormal basis $\langle x \rangle$. For $1 \leq j \leq g$, the operators $S_j : \mathcal{F}_g \to \mathcal{F}_g$ determined by $S_j w = x_j w$ for words $w \in \langle x \rangle$ are called the **creation operators**. It is readily checked that each S_j is an isometry and

$$I - P_0 = \sum S_j S_j^*,$$

where P_0 is the projection onto the one-dimensional subspace of $\mathcal{F}_g$ spanned by the empty word $\emptyset$. As is well known [Fr84, Po89], the creation operators serve as a universal model for row contractions. We state a precise version of this result suitable for our purposes as Proposition 9 below.

Fix a positive integer ℓ. A tuple $X \in M_n(\mathbb{C})^g$ is *nilpotent of order $\ell + 1$* if $X^w = 0$ for any word w of length $|w| > \ell$. Let $\mathcal{P}_\ell$ denote the subspace of $\mathcal{F}_g$ spanned by words of length at most ℓ; $\mathcal{P}_\ell$ has dimension

$$\sigma(\ell) = \sum_{j=0}^{\ell} g^j.$$

Let $V_\ell : \mathcal{P}_\ell \to \mathcal{F}_g$ denote the inclusion mapping and let

$$V_\ell^* S V_\ell = V_\ell^* (S_1, \ldots, S_g) V_\ell = (V_\ell^* S_1 V_\ell, \ldots, V_\ell^* S_g V_\ell).$$

As is easily verified, the subspace $\mathcal{P}_\ell$ is invariant for each S_j^* (and thus semi-invariant (i.e., the orthogonal difference of two invariant subspaces) for S_j). Hence, for a polynomial $p \in \mathbb{C}\langle x_1, \ldots, x_g \rangle$,

$$p(V_\ell^* S V_\ell) = V_\ell^* p(S) V_\ell.$$

In particular,

$$\sum_j (V_\ell^* S_j V_\ell)(V_\ell^* S_j^* V_\ell) \prec V_\ell^* \sum_j S_j S_j^* V_\ell = V_\ell^* P_0 V_\ell.$$

Hence, if $|z| < \varepsilon$, then $V_\ell^* z S V_\ell$ is in $\mathcal{N}_\varepsilon$, the ε-neighborhood of 0 in $M(\mathbb{C})^g$.

The following is a well-known algebraic version of a classical dilation theorem. The proof here follows along the lines of the de Branges-Rovnyak construction of the coisometric dilation of a contraction operator on Hilbert space [RR85].

Proposition 9. *Fix a positive integer ℓ and let $T = V_\ell^* S V_\ell$. If $X \in M_n(\mathbb{C})^g$ is nilpotent of order ℓ and if $\sum X_j X_j^* \prec r^2 I_n$ then there is an isometry $V : \mathbb{C}^n \to \mathbb{C}^n \otimes \mathcal{P}_\ell$ such that $V X_j^* = r(I \otimes T_j^*) V$, where I is the identity on $\mathbb{C}^n$.*

Proof. We give a de Branges-Rovnyak style proof. By scaling, assume that $r = 1$. Let

$$R = \sum X_j X_j^*.$$

Thus, by hypothesis $0 \preceq R \prec I$. Let

$$D = (I - \sum T_j T_j^*)^{\frac{1}{2}}.$$

The matrix D is known as the **defect** and, by hypothesis, is strictly positive definite. Moreover,

$$\sum_{|w| \le \ell} X^w DD(X^w)^* = I - \sum_{|w| = \ell+1} X^w (X^w)^* = I. \tag{14}$$

Define V by

$$V\gamma = \sum_w D(T^w)^* \gamma \otimes w.$$

The equality of equation (14) shows that V is an isometry. Finally

$$\begin{aligned} V X_j^* \gamma &= \sum_{|w| \le \ell-1} D(X^w)^* X_j^* \gamma \otimes w = \sum_{|w| \le \ell-1} D(X^{x_j w})^* \gamma \otimes w \\ &= T_j^* \sum_{|w| \le \ell-1} D(X^{x_j w})^* \gamma \otimes x_j w \\ &= S_j^* \Big(D\gamma + \sum_k \sum_{|w| \le \ell-1} D(T^{x_k w})^* \gamma \otimes x_k w \Big) \\ &= S_j^* V \gamma. \end{aligned}$$

$\square$

2.7. Creation operators meet free maps

In this section we determine formulas for the coefficients F_w of Proposition 7 of the power series expansion of f in terms of the creation operators S_j. Formulas for the F_w are also given in [Vo10, Section 13] and in [KVV–], where they are obtained by clever substitutions and have nice properties. Our formulas in terms of the familiar creation operators and related algebra provide a slightly different perspective and impose an organization which might prove interesting.

Lemma 10. *Fix a positive integer ℓ and let $T = V_\ell^* S V_\ell$ as before. If $f : M(\mathbb{C})^g \to M(\mathbb{C})^{\tilde{g}}$ is a free map, then there exists, for each word w of length at most ℓ, a vector $F_w \in \mathbb{C}^{\tilde{g}}$ such that*

$$f(T) = \sum_{|w| \le \ell} F_w \otimes T^w.$$

Given $u, w \in \langle x \rangle$, we say u **divides** w **(on the right)**, denoted $u|w$, if there is a $v \in \langle x \rangle$ such that $w = uv$.

Proof. Fix a word w of length at most ℓ. Define $F_w \in \mathbb{C}^{\tilde{g}}$ by

$$\langle F_w, \mathbf{y} \rangle = \langle \emptyset, f(T)^* \mathbf{y} \otimes w \rangle, \quad \mathbf{y} \in \mathbb{C}^{\tilde{g}}.$$

Given a word $u \in \mathcal{P}_\ell$ of length k, let R_u denote the operator of *right* multiplication by u on $\mathcal{P}_\ell$. Thus, R_u is determined by $R_u v = vu$ if $v \in \langle x \rangle$ has length at most $\ell - k$, and $R_u v = 0$ otherwise. Routine calculations show

$$T_j R_u = R_u T_j.$$

Hence, for the free map f, $f(T)R_u = R_u f(T)$. Thus, for words u, v of length at most ℓ and $\mathbf{y} \in \mathbb{C}^{\tilde{g}}$,

$$\langle u, f(T)^* \mathbf{y} \otimes v \rangle = \langle R_u \emptyset, f(T)^* \mathbf{y} \otimes v \rangle = \langle \emptyset, f(T)^* \mathbf{y} \otimes R_u^* v \rangle.$$

It follows that

$$\langle f(T)^* \mathbf{y} \otimes v, u \rangle = \begin{cases} \langle \mathbf{y}, F_\alpha \rangle & \text{if } v = \alpha u \\ 0 & \text{otherwise.} \end{cases} \tag{15}$$

On the other hand, if $v = wu$, then $(T^w)^* v = u$ and otherwise, $(T^w)^* v$ is orthogonal to u. Thus,

$$\left\langle \sum F_w^* \otimes (T^*)^w \mathbf{y} \otimes v, u \right\rangle = \begin{cases} F_w^* \mathbf{y} & \text{if } v = wu \\ 0 & \text{otherwise.} \end{cases} \tag{16}$$

Comparing equations (15) and (16) completes the proof. $\qquad\qquad\qquad\square$

Lemma 11. *Fix a positive integer ℓ and, as in Proposition 9, let $T = V_\ell^* S V_\ell$ act on $\mathcal{P}_\ell$. Suppose $V : \mathbb{C}^n \to \mathbb{C}^n \otimes \mathcal{P}_\ell$ is an isometry and $X \in M_n(\mathbb{C})^g$. If $f : M(\mathbb{C})^g \to M(\mathbb{C})^{\tilde{g}}$ is a free map and $VX^* = (I \otimes T^*)V$, then*

$$f(X) = V^* \big(I \otimes f(T) \big) V.$$

Proof. Taking adjoints gives $XV^* = V^*(I \otimes T)$. From the definition of a free map,

$$f(X)V^* = V^*(I \otimes f(T)).$$

Applying V on the right and using the fact that V is an isometry completes the proof. $\qquad\qquad\qquad\square$

Remark 12. Iterating the intertwining relation $VX^* = (I \otimes T^*)V$, it follows that, $V(X^w)^* = (I \otimes (T^w)^*)V$. In particular, if F is formal power series, then $F(X^*)V = VF(I \otimes T^*)$.

A free map $f : M(\mathbb{C})^g \to M(\mathbb{C})^{\tilde{g}}$ is **homogeneous of degree** ℓ if for all $X \in M(\mathbb{C})^g$ and $z \in \mathbb{C}$, $f(zX) = z^\ell f(X)$.

Lemma 13. *Suppose $f : M(\mathbb{C})^g \to M(\mathbb{C})^{\tilde{g}}$ is a free map. If f is continuous and homogeneous of degree ℓ, then there exists, for each word w of length ℓ, a vector $F_w \in \mathbb{C}^{\tilde{g}}$ such that*

$$f(X) = \sum_{|w| = \ell} F_w \otimes X^w \quad \text{for all } X \in M(\mathbb{C})^g.$$

Proof. Write $T = V_\ell^* S V_\ell$. Let n and $X \in M_n(\mathbb{C})^g$ be given and assume $\sum X_j X_j^* \prec I$. Let J denote the nilpotent Jordan block of size $(\ell+1) \times (\ell+1)$. Thus the entries of J are zero, except for the ℓ entries along the first super diagonal which are all 1. Let $Y = X \otimes J$. Then Y is nilpotent of order $\ell+1$ and $\sum Y_j Y_j^* \prec I$. By Proposition 9, there is an isometry $V : \mathbb{C}^n \otimes \mathbb{C}^{\ell+1} \to (\mathbb{C}^n \otimes \mathbb{C}^{\ell+1}) \otimes \mathcal{P}_\ell$ such that

$$VY^* = (I \otimes T^*)V.$$

By Theorem 11, $f(Y) = V^*(I \otimes f(T))V$. From Lemma 10 there exists, for words w of length at most ℓ, vectors $F_w \in \mathbb{C}^{\tilde{g}}$ such that $f(T) = \sum_{|w| \le \ell} F_w \otimes T^w$. Because f is a free map, $f(I \otimes T) = I \otimes f(T)$. Hence,

$$f(Y) = \sum_{|w| \le \ell} F_w \otimes V^*(I \otimes T^w)V = \sum_{|w| \le \ell} F_w \otimes Y^w = \sum_{m=0}^{\ell} \left(\sum_{|w|=m} F_w \otimes X^w \right) \otimes J^m.$$

Replacing X by zX and using the homogeneity of f gives,

$$z^\ell f(Y) = \sum_{m=0}^{\ell} \left(\sum_{|w|=m} F_w \otimes X^w \right) \otimes z^m J^m$$

It follows that

$$f(Y) = \left(\sum_{|w|=\ell} F_w \otimes X_w \right) \otimes J^\ell. \tag{17}$$

Next suppose that $E = D + J$, where D is diagonal with distinct entries on the diagonal. Thus there exists an invertible matrix Z such that $ZE = DZ$. Because f is a free map, $f(X \otimes D) = \oplus f(d_j X)$, where d_j is the jth diagonal entry of D. Because of the homogeneity of f,

$$f(X \otimes D) = \oplus d_j^\ell f(X) = f(X) \otimes D^\ell.$$

Hence,

$$f(X \otimes E) = (I \otimes Z^{-1}) f(X \otimes D)(I \otimes Z) = (I \otimes Z^{-1}) f(X) \otimes D^\ell (I \otimes Z) = f(X) \otimes E^\ell.$$

Choosing a sequence of D's which converge to 0, so that the corresponding E's converge to J, and using the continuity of f yields $f(Y) = f(X) \otimes J^\ell$. A comparison with (17) proves the lemma. $\qquad\square$

2.8. The proof of Proposition 7

Let $f : \mathcal{N}_\varepsilon \to M(\mathbb{C})^{\tilde{g}}$ be a free analytic map. Given $X \in M_n(\mathbb{C})^g$, there is a disc $D_X = \{z \in \mathbb{C} : |z| < r_X\}$ such that $zX \in \mathcal{N}_\varepsilon$ for $z \in D_X$. By analyticity of f, the function $D_X \ni z \mapsto f(zX)$ is analytic (with values in $M_n(\mathbb{C})^{\tilde{g}}$) and thus has a power series expansion,

$$f(zX) = \sum_m A_m z^m.$$

These $A_m = A_m(X)$ are uniquely determined by X and hence there exist functions $f^{(m)}[n](X) = A_m(X)$ mapping $M_n(\mathbb{C})^g$ to $M_n(\mathbb{C})^{\tilde{g}}$. In particular, if $X \in \mathcal{N}_\varepsilon(n)$, then

$$f(X) = \sum f^{(m)}[n](X). \tag{18}$$

Lemma 14. *For each m, the sequence $(f^{(m)}[n])_n$ is a continuous free map*

$$M(\mathbb{C})^g \to M(\mathbb{C})^{\tilde{g}}.$$

Moreover, $f^{(m)}$ is homogeneous of degree m.

Proof. Suppose $X, Y \in M(\mathbb{C})^g$ and $X\Gamma = \Gamma Y$. For $z \in D_X \cap D_Y$,

$$\sum f^{(m)}(X)\Gamma z^m = f(zX)\Gamma = \Gamma f(zY) = \sum \Gamma f^{(m)}(Y)z^m.$$

Thus $f^{(m)}(X)\Gamma = \Gamma f^{(m)}(Y)$ for each m and thus each $f^{(m)}$ is a free map. Since $f[n]$ is continuous, so is $f^{(m)}[n]$ for each n.

Finally, given X and $w \in \mathbb{C}$, for z of sufficiently small modulus,

$$\sum f^{(m)}(wX)z^m = f(z(wX)) = f(zwX) = \sum f^{(m)}(X)w^m z^m.$$

Thus $f^{(m)}(wX) = w^m f^{(m)}(X)$. $\square$

Returning to the proof of Proposition 7, for each m, let F_w for a word w with $|w| = m$, denote the coefficients produced by Lemma 13 so that

$$f^{(m)}(X) = \sum_{|w|=m} F_w \otimes X^w.$$

Substituting into equation (18) completes the proof of the first part of the Proposition 7.

Now suppose that f is uniformly bounded by C on $\mathcal{N}$. If $X \in \mathcal{N}$, then

$$C \geq \left\| \frac{1}{2\pi} \int f(\exp(it)X) \exp(-imt)\, dt \right\| = \| f^{(m)}(X) \|.$$

In particular, if $0 < r < 1$, then $\| f^{(m)}(rX) \| \leq r^m C$.

Let $T = V_m^* S V_m$ as in Subsection 2.6. In particular, if $\delta < \varepsilon$, then $\delta T \in \mathcal{N}$ and thus

$$C^2 \geq \| f^{(m)}(\delta T)\emptyset \|^2 = \delta^{2m} \sum_{|v|=m} \|F_v\|^2.$$

Thus, $\|F_v\| \leq \frac{C}{\delta^m}$ for all $0 < \delta < \varepsilon$ and words v of length m and the last statement of Proposition 7 follows. $\square$

3. Proper free maps

Given noncommutative domains $\mathcal{U}$ and $\mathcal{V}$ in $M(\mathbb{C})^g$ and $M(\mathbb{C})^{\tilde{g}}$ respectively, a free map $f : \mathcal{U} \to \mathcal{V}$ is **proper** if each $f[n] : \mathcal{U}(n) \to \mathcal{V}(n)$ is proper in the sense that if $K \subset \mathcal{V}(n)$ is compact, then $f^{-1}(K)$ is compact. In particular, for all n, if (z_j) is a sequence in $\mathcal{U}(n)$ and $z_j \to \partial\mathcal{U}(n)$, then $f(z_j) \to \partial\mathcal{V}(n)$. In the case $g = \tilde{g}$ and both f and f^{-1} are (proper) free analytic maps we say f is a **free biholomorphism**.

3.1. Proper implies injective

The following theorem was established in [HKM11b, Theorem 3.1]. We will not give the proof here but instead record a few corollaries below.

Theorem 15. *Let $\mathcal{U}$ and $\mathcal{V}$ be noncommutative domains containing 0 in $M(\mathbb{C})^g$ and $M(\mathbb{C})^{\tilde{g}}$, respectively and suppose $f : \mathcal{U} \to \mathcal{V}$ is a free map.*

(1) *If f is proper, then it is one-to-one, and $f^{-1} : f(\mathcal{U}) \to \mathcal{U}$ is a free map.*
(2) *If, for each n and $Z \in M_n(\mathbb{C})^{\tilde{g}}$, the set $f[n]^{-1}(\{Z\})$ has compact closure in $\mathcal{U}$, then f is one-to-one and moreover, $f^{-1} : f(\mathcal{U}) \to \mathcal{U}$ is a free map.*
(3) *If $g = \tilde{g}$ and $f : \mathcal{U} \to \mathcal{V}$ is proper and continuous, then f is biholomorphic.*

Corollary 16. *Suppose $\mathcal{U}$ and $\mathcal{V}$ are noncommutative domains in $M(\mathbb{C})^g$. If $f : \mathcal{U} \to \mathcal{V}$ is a free map and if each $f[n]$ is biholomorphic, then f is a free biholomorphism.*

Proof. Since each $f[n]$ is biholomorphic, each $f[n]$ is proper. Thus f is proper. Since also f is a free map, by Theorem 15(3) f is a free biholomorphism. $\qquad\square$

Corollary 17. *Let $\mathcal{U} \subset M(\mathbb{C})^g$ and $\mathcal{V} \subset M(\mathbb{C})^{\tilde{g}}$ be noncommutative domains. If $f : \mathcal{U} \to \mathcal{V}$ is a proper free analytic map and if $X \in \mathcal{U}(n)$, then $f'(X) : M_n(\mathbb{C})^g \to M_n(\mathbb{C})^{\tilde{g}}$ is one-to-one. In particular, if $g = \tilde{g}$, then $f'(X)$ is a vector space isomorphism.*

Proof. Suppose $f'(X)[H] = 0$. We scale H so that $\begin{bmatrix} X & H \\ 0 & X \end{bmatrix} \in \mathcal{U}$. From Proposition 4,

$$f \begin{bmatrix} X & H \\ 0 & X \end{bmatrix} = \begin{bmatrix} f(X) & f'(X)[H] \\ 0 & f(X) \end{bmatrix} = \begin{bmatrix} f(X) & 0 \\ 0 & f(X) \end{bmatrix} = f \begin{bmatrix} X & 0 \\ 0 & X \end{bmatrix}.$$

By the injectivity of f established in Theorem 15, $H = 0$. $\qquad\square$

Remark 18. Let us note that Theorem 15 is sharp as explained in [HKM11b, §3.1]: absent more conditions on the noncommutative domains $\mathcal{U}$ and $\mathcal{V}$, nothing beyond free biholomorphic can be concluded about f.

A natural condition on a noncommutative domain $\mathcal{U}$, which we shall consider in §5, is circularity. However, we first proceed to give some free analogs of well-known results from several complex variables.

4. Several analogs to classical theorems

The conclusion of Theorem 15 is sufficiently strong that most would say that it does not have a classical analog. Combining it with classical several complex variable theorems yields free analytic map analogs. Indeed, hypotheses for these analytic free map results are weaker than their classical analogs would suggest.

4.1. A free Carathéodory-Cartan-Kaup-Wu (CCKW) Theorem

The commutative Carathéodory-Cartan-Kaup-Wu (CCKW) Theorem [Kr01, Theorem 11.3.1] says that if f is an analytic self-map of a bounded domain in $\mathbb{C}^g$ which fixes a point P, then the eigenvalues of $f'(P)$ have modulus at most one. Conversely, if the eigenvalues all have modulus one, then f is in fact an automorphism; and further if $f'(P) = I$, then f is the identity. The CCKW Theorem together with Corollary 16 yields Corollary 19 below. We note that Theorem 15 can also be thought of as a noncommutative CCKW theorem in that it concludes, like the CCKW Theorem does, that a map f is biholomorphic, but under the (rather different) assumption that f is proper.

Most of the proofs in this section are skipped and can be found in [HKM11b, §4].

Corollary 19 ([HKM11b, Corollary 4.1]). *Let $\mathcal{D}$ be a given bounded noncommutative domain which contains 0. Suppose $f : \mathcal{D} \to \mathcal{D}$ is an free analytic map. Let ϕ denote the mapping $f[1] : \mathcal{D}(1) \to \mathcal{D}(1)$ and assume $\phi(0) = 0$.*

(1) *If all the eigenvalues of $\phi'(0)$ have modulus one, then f is a free biholomorphism; and*

(2) *if $\phi'(0) = I$, then f is the identity.*

Note a classical biholomorphic function f is completely determined by its value and differential at a point (cf. a remark after [Kr01, Theorem 11.3.1]). Much the same is true for free analytic maps and for the same reason.

Proposition 20. *Suppose $\mathcal{U}, \mathcal{V} \subset M(\mathbb{C})^g$ are noncommutative domains, $\mathcal{U}$ is bounded, both contain 0, and $f, g : \mathcal{U} \to \mathcal{V}$ are proper free analytic maps. If $f(0) = g(0)$ and $f'(0) = g'(0)$, then $f = g$.*

Proof. By Theorem 15 both f and g are free biholomorphisms. Thus $h = f \circ g^{-1} : \mathcal{U} \to \mathcal{U}$ is a free biholomorphism fixing 0 with $h[1]'(0) = I$. Thus, by Corollary 19, h is the identity. Consequently $f = g$. $\qquad\square$

4.2. Circular domains

A subset S of a complex vector space is **circular** if $\exp(it)s \in S$ whenever $s \in S$ and $t \in \mathbb{R}$. A noncommutative domain $\mathcal{U}$ is circular if each $\mathcal{U}(n)$ is circular.

Compare the following theorem to its commutative counterpart [Kr01, Theorem 11.1.2] where the domains $\mathcal{U}$ and $\mathcal{V}$ are the same.

Theorem 21. *Let $\mathcal{U}$ and $\mathcal{V}$ be bounded noncommutative domains in $M(\mathbb{C})^g$ and $M(\mathbb{C})^{\tilde{g}}$, respectively, both of which contain 0. Suppose $f : \mathcal{U} \to \mathcal{V}$ is a proper free analytic map with $f(0) = 0$. If $\mathcal{U}$ and the range $\mathcal{R} := f(\mathcal{U})$ of f are circular, then f is linear.*

The domain $\mathcal{U} = (\mathcal{U}(n))$ is **weakly convex** (a stronger notion of convex for a noncommutative domain appears later) if each $\mathcal{U}(n)$ is a convex set. Recall a set $C \subseteq \mathbb{C}^g$ is convex, if for every $X, Y \in C$, $\frac{X+Y}{2} \in C$.

Corollary 22. *Let $\mathcal{U}$ and $\mathcal{V}$ be bounded noncommutative domains in $M(\mathbb{C})^g$ both of which contain 0. Suppose $f : \mathcal{U} \to \mathcal{V}$ is a proper free analytic map with $f(0) = 0$. If both $\mathcal{U}$ and $\mathcal{V}$ are circular and if one is weakly convex, then so is the other.*

This corollary is an immediate consequence of Theorem 21 and the fact (see Theorem 15(3)) that f is onto $\mathcal{V}$.

We admit the hypothesis that the range $\mathcal{R} = f(\mathcal{U})$ of f in Theorem 21 is circular seems pretty contrived when the domains $\mathcal{U}$ and $\mathcal{V}$ have a different number of variables. On the other hand if they have the same number of variables it is the same as $\mathcal{V}$ being circular since by Theorem 15, f is onto.

Proof of Theorem 21. Because f is a proper free map it is injective and its inverse (defined on $\mathcal{R}$) is a free map by Theorem 15. Moreover, using the analyticity of f, its derivative is pointwise injective by Corollary 17. It follows that each $f[n] : \mathcal{U}(n) \to M_n(\mathbb{C})^{\tilde{g}}$ is an embedding [GP74, p. 17]. Thus, each $f[n]$ is a homeomorphism onto its range and its inverse $f[n]^{-1} = f^{-1}[n]$ is continuous.

Define $F : \mathcal{U} \to \mathcal{U}$ by

$$F(x) := f^{-1}\big(\exp(-i\theta)f(\exp(i\theta)x)\big) \tag{19}$$

This function respects direct sums and similarities, since it is the composition of maps which do. Moreover, it is continuous by the discussion above. Thus F is a free analytic map.

Using the relation $\exp(i\theta)f(F(x)) = f(\exp(i\theta))$ we find

$$\exp(i\theta)f'(F(0))F'(0) = f'(0).$$

Since $f'(0)$ is injective, $\exp(i\theta)F'(0) = I$. It follows from Corollary 19(2) that $F(x) = \exp(i\theta)x$ and thus, by (19), $f(\exp(i\theta)x) = \exp(i\theta)f(x)$. Since this holds for every θ, it follows that f is linear. $\qquad\square$

If f is not assumed to map 0 to 0 (but instead fixes some other point), then a proper self-map need not be linear. This follows from the example we discuss in §5.1.

5. A free Braun-Kaup-Upmeier (BKU) Theorem

Noncommutative domains $\mathcal{U}$ and $\mathcal{V}$ are **freely biholomorphic** if there exists a free biholomorphism $f : \mathcal{U} \to \mathcal{V}$. In this section we show how a theorem of Braun-Kaup-Upmeier [BKU78, KU76] can be used to show that bounded circular noncommutative domains that are freely biholomorphic are (freely) linearly biholomorphic.

Definition 23. Given a domain $D \subset \mathbb{C}^g$, let Aut(D) denote the group of all biholomorphic maps from D to D. Note that D is circular if and only if Aut(D) contains all rotations; i.e., all maps of the form $z \mapsto \exp(i\theta)z$ for $\theta \in \mathbb{R}$.

Let $\mathcal{D} = (\mathcal{D}(n))$ be a circular noncommutative domain. Thus each $\mathcal{D}(n)$ is open, connected, contains 0 and is invariant under rotations. The set $\mathcal{D}(1) \subset \mathbb{C}^g$ is in particular a circular domain in the classical sense and moreover $\mathrm{Aut}(\mathcal{D}(1))$ contains all rotations.

Theorem 24 (A free BKU Theorem). *Suppose $\mathcal{U}$ and $\mathcal{D}$ are bounded, circular noncommutative domains which contain noncommutative neighborhoods of 0. If $\mathcal{U}$ and $\mathcal{D}$ are freely biholomorphic, then there is a linear (free) biholomorphism $\lambda : \mathcal{D} \to \mathcal{U}$.*

A noncommutative domain $\mathcal{D}$ containing 0 is **convex** if it is closed with respect to conjugation by contractions; i.e., if $X \in \mathcal{D}(n)$ and C is a $m \times n$ contraction, then

$$CXC^* = (CX_1C^*, CX_2C^*, \ldots, CX_gC^*) \in \mathcal{D}(m).$$

It is not hard to see, using the fact that noncommutative domains are also closed with respect to direct sums, that each $\mathcal{D}(n)$ is itself convex. In the case that $\mathcal{D}$ is semialgebraic, then in fact an easy argument shows that the converse is true: if each $\mathcal{D}(n)$ is convex ($\mathcal{D}$ is weakly convex), then $\mathcal{D}$ is convex. (What is used here is that the domain is closed with respect to restrictions to reducing subspaces.) In fact, in the case that $\mathcal{D}$ is semialgebraic and convex, it is equivalent to being an LMI, cf. [HM+] for precise statements and proofs; the topic is also addressed briefly in §6.1 below. As an important corollary of Theorem 24, we have the following nonconvexification result.

Corollary 25. *Suppose $\mathcal{U}$ is a bounded circular noncommutative domain which contains a noncommutative neighborhood of 0.*

(1) *If $\mathcal{U}$ is freely biholomorphic to a bounded circular weakly convex noncommutative domain that contains a noncommutative neighborhood of 0, then $\mathcal{U}$ is itself convex.*

(2) *If $\mathcal{U}$ is freely biholomorphic to a bounded circular LMI domain, then $\mathcal{U}$ is itself an LMI domain.*

Proof. It is not hard to see that an LMI domain does in fact contain a noncommutative neighborhood of the origin. Thus, both statements of the corollary follow immediately from the theorem. □

Note that the corollary is in the free spirit of the main result of [KU76].

Remark 26. A main motivation for our line of research was investigating *changes of variables* with an emphasis on achieving convexity. Anticipating that the main result from [HM+] applies in the present context (see also §6.1), if $\mathcal{D}$ is a convex, bounded, noncommutative semialgebraic set then it is an LMI domain. In this way, the hypothesis in the last statement of the corollary could be rephrased as: if $\mathcal{U}$ is freely biholomorphic to a bounded circular convex noncommutative semialgebraic set, then $\mathcal{U}$ is itself an LMI domain. In the context of §1.1, the conclusion is that in this circumstance domains biholomorphic to bounded, convex, circular basic semialgebraic sets are already in fact determined by an LMI. Hence there no nontrivial changes of variables in this setting.

For the reader's convenience we include here the version of [BKU78, Theorem 1.7] needed in the proof of Theorem 24. Namely, the case in which the ambient domain is $\mathbb{C}^g$. Closed here means closed in the topology of uniform convergence on compact subsets. A bounded domain $D \subset \mathbb{C}^g$ is symmetric if for each $z \in D$ there is an involutive $\varphi \in \text{Aut}(D)$ such that z is an isolated fixed point of φ [Hg78].

Theorem 27 ([BKU78]). *Suppose* $S \subset \mathbb{C}^g$ *is a bounded circular domain and* $G \subset \text{Aut}(S)$ *is a closed subgroup of* $\text{Aut}(S)$ *which contains all rotations. Then*

(1) *there is a closed ($\mathbb{C}$-linear) subspace* M *of* $\mathbb{C}^g$ *such that* $A := S \cap M = G(0)$ *is the orbit of the origin.*

(2) *A is a bounded symmetric domain in M and coincides with*

$$\{z \in S : G(z) \text{ is a closed complex submanifold of } S\}.$$

In particular two bounded circular domains are biholomorphic if and only if they are linearly biholomorphic.

We record the following simple lemma before turning to the proof of Theorem 24.

Lemma 28. *Let* $D \subset \mathbb{C}^g$ *be a bounded domain and suppose* (φ_j) *is a sequence from* $\text{Aut}(D)$ *which converges uniformly on compact subsets of D to* $\varphi \in \text{Aut}(D)$.

(1) $\varphi_j^{-1}(0)$ *converges to* $\varphi^{-1}(0)$;

(2) *If the sequence (φ_j^{-1}) converges uniformly on compact subsets of D to ψ, then* $\psi = \varphi^{-1}$.

Proof. (1) Let $\varepsilon > 0$ be given. The sequence (φ_j^{-1}) is a uniformly bounded sequence and is thus locally equicontinuous. Thus, there is a $\delta > 0$ such that if $\|y - 0\| < \delta$, then $\|\varphi_j^{-1}(y) - \varphi_j^{-1}(0)\| < \varepsilon$. On the other hand, $(\varphi_j(\varphi^{-1}(0)))_j$ converges to 0, so for large enough j, $\|\varphi_j(\varphi^{-1}(0)) - 0\| < \delta$. With $y = \varphi_j(\varphi^{-1}(0))$, it follows that $\|\varphi_j(\varphi^{-1}(0)) - 0\| < \varepsilon$.

(2) Let $f = \varphi(\psi)$. From the first part of the lemma, $\psi(0) = \varphi^{-1}(0)$ and hence $f(0) = 0$. Moreover, $f'(0) = \varphi'(\psi(0))\psi'(0)$. Now φ_j' converges uniformly on compact sets to φ'. Since also $\varphi_j'(\psi(0))$ converges to $\varphi'(\psi(0))$, it follows that $\varphi_j'(\varphi_j^{-1}(0))$ converges to $\varphi'(\psi(0))$. On the other hand, $I = \varphi_j'(\varphi_j^{-1}(0))(\varphi_j^{-1})'(0)$. Thus, $f'(0) = I$ and we conclude, from a theorem of Carathéodory-Cartan-Kaup-Wu (see Corollary 19), that f is the identity. Since φ has an (nc) inverse, $\varphi^{-1} = \psi$. $\square$

Definition 29. Let $\text{Aut}_{nc}(\mathcal{D})$ denote the free automorphism group of the non-commutative domain $\mathcal{D}$. Thus $\text{Aut}_{nc}(\mathcal{D})$ is the set of all free biholomorphisms $f : \mathcal{D} \to \mathcal{D}$. It is evidently a group under composition. Note that $\mathcal{D}$ is circular implies $\text{Aut}_{nc}(\mathcal{D})$ contains all rotations. Given $g \in \text{Aut}_{nc}(\mathcal{D})$, let $\tilde{g} \in \text{Aut}(\mathcal{D}(1))$ denote its commutative collapse; i.e., $\tilde{g} = g[1]$.

Lemma 30. *Suppose $\mathcal{D}$ is a bounded noncommutative domain containing 0. Assume $f, h \in \text{Aut}_{\text{nc}}(\mathcal{D})$ satisfy $\tilde{f} = \tilde{h}$. Then $f = h$.*

Proof. Note that $F = h^{-1} \circ f \in \text{Aut}_{\text{nc}}(\mathcal{D})$. Further, since $\tilde{F} = x$ (the identity), F maps 0 to 0 and $\tilde{F}'(0) = I$. Thus, by Corollary 19, $F = x$ and therefore $h = f$. $\quad\square$

Lemma 31. *Suppose $\mathcal{D}$ is a noncommutative domain which contains a noncommutative neighborhood of 0, and $\mathcal{U}$ is a bounded noncommutative domain. If $f_m : \mathcal{D} \to \mathcal{U}$ is a sequence of free analytic maps, then there is a free analytic map $f : \mathcal{D} \to \mathcal{U}$ and a subsequence (f_{m_j}) of (f_m) which converges to f uniformly on compact sets.*

Proof. By hypothesis, there is an $\varepsilon > 0$ such that $\mathcal{N}_\varepsilon \subset \mathcal{D}$ and there is a $C > 0$ such that each $X \in \mathcal{U}$ satisfies $\|X\| \leq C$. Each f_m has power series expansion,

$$f_m = \sum \hat{f}_m(w) w$$

with $\|\hat{f}_m(w)\| \leq \frac{C}{\varepsilon^n}$, where n is the length of the word w, by Proposition 7. Moreover, by a diagonal argument, there is a subsequence f_{m_j} of f_m so that $\hat{f}_{m_j}(w)$ converges to some $\hat{f}(w)$ for each word w. Evidently, $\|\hat{f}(w)\| \leq \frac{C}{\varepsilon^n}$ and thus,

$$f = \sum \hat{f}(w) w$$

defines a free analytic map on the noncommutative $\frac{\varepsilon}{g}$-neighborhood of 0. (See 5.)

  We claim that f determines a free analytic map on all of $\mathcal{D}$ and moreover (f_{m_j}) converges to this f uniformly on compact sets; i.e., for each n and compact set $K \subset \mathcal{D}(n)$, the sequence $(f_{m_j}[n])$ converges uniformly to $f[n]$ on K.

  Conserving notation, let $f_j = f_{m_j}$. Fix n. The sequence $f_j[n] : \mathcal{D}(n) \to \mathcal{D}(n)$ is uniformly bounded and hence each subsequence (g_k) of $(f_j[n])$ has a further subsequence (h_ℓ) which converges uniformly on compact subsets to some analytic function $h : \mathcal{D}(n) \to \mathcal{U}(n)$. On the other hand, (h_ℓ) converges to $f[n]$ on the $\frac{\varepsilon}{g}$-neighborhood of 0 in $\mathcal{D}(n)$ and thus $h = f[n]$ on this neighborhood. It follows that $f[n]$ extends to be analytic on all of $\mathcal{D}(n)$. It follows that $(f_j[n])$ itself converges uniformly on compact subsets of $\mathcal{D}(n)$. In particular, $f[n]$ is analytic.

  To see that f is a *free* analytic function (and not just that each $f(n)$ is analytic), suppose $X\Gamma = \Gamma Y$. Then $f_j(X)\Gamma = \Gamma f_j(Y)$ for each j and hence the same is true in the limit. $\quad\square$

Lemma 32. *Suppose $\mathcal{D}$ is a bounded noncommutative domain which contains a noncommutative neighborhood of 0. Suppose (h_n) is a sequence from $\text{Aut}_{\text{nc}}(\mathcal{D})$. If $\tilde{h}_n$ converges to $g \in \text{Aut}(\mathcal{D}(1))$ uniformly on compact sets, then there is $h \in \text{Aut}_{\text{nc}}(\mathcal{D})$ such that $\tilde{h} = g$ and a subsequence (h_{n_j}) of (h_n) which converges uniformly on compact sets to h.*

Proof. By the previous lemma, there is a subsequence (h_{n_j}) of (h_n) which converges uniformly on compact subsets of $\mathcal{D}$ to a free map h. With $H_j = h_{n_j}^{-1}$, another application of the lemma produces a further subsequence, (H_{j_k}) which converges uniformly on compact subsets of $\mathcal{D}$ to some free map H. Hence, without loss of generality, it may be assumed that both (h_j) and (h_j^{-1}) converge (in each dimension) uniformly on compact sets to h and H respectively.

From Lemma 28, $\tilde{H}$ is the inverse of $\tilde{h} = g$. Thus, letting f denote the analytic free mapping $f = h \circ H$, it follows that $\tilde{f}$ is the identity and so by Corollary 19, f is itself the identity. Similarly, $H \circ h$ is the identity. Thus, h is a free biholomorphism and thus an element of $\mathrm{Aut}_{nc}(\mathcal{D})$. □

Proposition 33. *If $\mathcal{D}$ is a bounded noncommutative domain containing an ε-neighborhood of 0, then the set $\{\tilde{h} : h \in \mathrm{Aut}_{nc}(\mathcal{D})\}$ is a closed subgroup of $\mathrm{Aut}(\mathcal{D}(1))$.*

Proof. We must show if $h_n \in \mathrm{Aut}_{nc}(\mathcal{D})$ and $\tilde{h_n}$ converges to some $g \in \mathrm{Aut}(\mathcal{D}(1))$, then there is an $h \in \mathrm{Aut}_{nc}(\mathcal{D})$ such that $\tilde{h} = g$. Thus the proposition is an immediate consequence of the previous result, Lemma 32. □

Proof of Theorem 24. In the BKU Theorem 27, first choose $S = \mathcal{D}(1)$ and let

$$G = \{\tilde{f} : f \in \mathrm{Aut}_{nc}(\mathcal{D})\}.$$

Note that G is a subgroup of $\mathrm{Aut}(S)$ which contains all rotations. Moreover, by Proposition 33, G is closed. Thus Theorem 27 applies to G. Combining the two conclusions of the theorem, it follows that $G(0)$ is a closed complex submanifold of D.

Likewise, let $T = \mathcal{U}(1)$ and let

$$H = \{\tilde{h} : h \in \mathrm{Aut}_{nc}(\mathcal{U})\}$$

and note that H is a closed subgroup of $\mathrm{Aut}(T)$ containing all rotations. Consequently, Theorem 27 also applies to H.

Let $\psi : \mathcal{D} \to \mathcal{U}$ denote a given free biholomorphism. In particular, $\tilde{\psi} : S \to T$ is biholomorphic. Observe, $H = \{\tilde{\psi} \circ g \circ \tilde{\psi}^{-1} : g \in G\}$.

The set $\tilde{\psi}(G(0))$ is a closed complex submanifold of S, since $\tilde{\psi}$ is biholomorphic. On the other hand, $\tilde{\psi}(G(0)) = H(\tilde{\psi}(0))$. Thus, by (ii) of Theorem 27 applied to H and T, it follows that $\tilde{\psi}(0) \in H(0)$. Thus, there is an $h \in \mathrm{Aut}_{nc}(\mathcal{U})$ such that $\tilde{h}(\tilde{\psi}(0)) = 0$. Now $\varphi = h \circ \psi : \mathcal{D} \to \mathcal{U}$ is a free biholomorphism between bounded circular noncommutative domains and $\varphi(0) = 0$. Thus, φ is linear by Theorem 21. □

5.1. A concrete example of a nonlinear biholomorphic self-map on an nc LMI domain

It is surprisingly difficulty to find proper self-maps on LMI domains which are not linear. In this section we present the only (up to trivial modifications) univariate example, of which we are aware. Of course, by Theorem 21 the underlying domain cannot be circular. In two variables, it can happen that two LMI domains are linearly equivalent and yet there is a nonlinear biholomorphism between them taking 0 to 0. We conjecture this cannot happen in the univariate case.

Let $A = \begin{bmatrix} 1 & 1 \\ 0 & 0 \end{bmatrix}$ and let $\mathcal{L}$ denote the univariate 2×2 linear pencil,

$$\mathcal{L}(x) := I + Ax + A^* x^* = \begin{bmatrix} 1 + x + x^* & x \\ x^* & 1 \end{bmatrix}.$$

Let $\mathcal{D}_{\mathcal{L}} = \{X : \|X - 1\| < \sqrt{2}\}$. For $\theta \in \mathbb{R}$ consider

$$f_\theta(x) := \frac{\exp(i\theta)x}{1 + x - \exp(i\theta)x}.$$

Then $f_\theta : \mathcal{D}_{\mathcal{L}} \to \mathcal{D}_{\mathcal{L}}$ is a proper free analytic map, $f_\theta(0) = 0$, and $f_\theta'(0) = \exp(i\theta)$. Conversely, every proper free analytic map $f : \mathcal{D}_{\mathcal{L}} \to \mathcal{D}_{\mathcal{L}}$ fixing the origin equals one of the f_θ.

For proofs we refer to [HKM11b, §5.1].

6. Miscellaneous

In this section we briefly overview some of our other, more algebraic, results dealing with convexity and LMIs. While many of these results do have analogs in the present setting of complex scalars and analytic variables, they appear in the literature with real scalars and symmetric free noncommutative variables.

Let $\mathbb{R}\langle x \rangle$ denote the $\mathbb{R}$-algebra freely generated by g noncommuting letters $x = (x_1, \dots, x_g)$ with the involution $*$ which, on a word $w \in \langle x \rangle$, reverses the order; i.e., if

$$w = x_{i_1} x_{i_2} \cdots x_{i_k}, \tag{20}$$

then

$$w^* = x_{i_k} \cdots x_{i_2} x_{i_1}.$$

In the case $w = x_j$, note that $x_j^* = x_j$ and for this reason we sometimes refer to the variables as **symmetric**.

Let $\mathbb{S}_n^g$ denote the g-tuples $X = (X_1, \dots, X_g)$ of $n \times n$ symmetric real matrices. A word w as in equation (20) is evaluated at X in the obvious way,

$$w(X) = X_{i_1} X_{i_2} \cdots X_{i_k}.$$

The evaluation extends linearly to polynomials $p \in \mathbb{R}\langle x \rangle$. Note that the involution on $\mathbb{R}\langle x \rangle$ is compatible with evaluation and matrix transpose in that $p^*(X) = p(X)^*$.

Given r, let $M_r \otimes \mathbb{R}\langle x \rangle$ denote the $r \times r$ matrices with entries from $\mathbb{R}\langle x \rangle$. The evaluation on $\mathbb{R}\langle x \rangle$ extends to $M_r \otimes \mathbb{R}\langle x \rangle$ by simply evaluating entrywise; and the involution extends too by $(p_{j,\ell})^* = (p_{\ell,j}^*)$.

A polynomial $p \in M_r \otimes \mathbb{R}\langle x \rangle$ is **symmetric** if $p^* = p$ and in this case, $p(X)^* = p(X)$ for all $X \in \mathbb{S}_n^g$. In this setting, the analog of an LMI is the following. Given d and symmetric $d \times d$ matrices, the symmetric matrix-valued degree one polynomial,

$$L = I - \sum A_j x_j$$

is a **monic linear pencil.** The inequality $L(X) \succ 0$ is then an LMI. Less formally, the polynomial L itself will be referred to as an LMI.

6.1. nc convex semialgebraic is LMI

Suppose $p \in M_r \otimes \mathbb{R}\langle x \rangle$ and $p(0) = I_r$. For each positive integer n, let

$$\mathcal{P}_p(n) = \{X \in \mathbb{S}_n^g : p(X) \succ 0\},$$

and define $\mathcal{P}_p$ to be the sequence (graded set) $(\mathcal{P}_p(n))_{n=1}^{\infty}$. In analogy with classical real algebraic geometry we call sets of the form $\mathcal{P}_p$ **noncommutative basic open semialgebraic sets**. (Note that it is not necessary to explicitly consider intersections of noncommutative basic open semialgebraic sets since the intersection $\mathcal{P}_p \cap \mathcal{P}_q$ equals $\mathcal{P}_{p \oplus q}$.)

Theorem 34 ([HM+]). *Every convex bounded noncommutative basic open semialgebraic set $\mathcal{P}_p$ has an LMI representation; i.e., there is a monic linear pencil L such that $\mathcal{P}_p = \mathcal{P}_L$.*

Roughly speaking, Theorem 34 states that nc semialgebraic and convex equals LMI. Again, this result is much cleaner than the situation in the classical commutative case, where the gap between convex semialgebraic and LMI is large and not understood very well, cf. [HV07].

6.2. LMI inclusion

The topic of our paper [HKM+] is LMI inclusion and LMI equality. Given LMIs L_1 and L_2 in the same number of variables it is natural to ask:

(Q_1) does one dominate the other, that is, does $L_1(X) \succeq 0$ imply $L_2(X) \succeq 0$?

(Q_1) are they mutually dominant, that is, do they have the same solution set?

As we show in [HKM+], the domination questions (Q_1) and (Q_2) have elegant answers, indeed reduce to semidefinite programs (SDP) which we show how to construct. A positive answer to (Q_1) is equivalent to the existence of matrices V_j such that

$$L_2(x) = V_1^* L_1(x) V_1 + \cdots + V_\mu^* L_1(x) V_\mu. \tag{21}$$

As for (Q_2) we show that L_1 and L_2 are mutually dominant if and only if, up to certain redundancies described in the paper, L_1 and L_2 are unitarily equivalent.

A basic observation is that these LMI domination problems are equivalent to the complete positivity of certain linear maps τ from a subspace of matrices to a matrix algebra.

6.3. Convex Positivstellensatz

The equation (21) can be understood as a linear Positivstellensatz, i.e., it gives an algebraic certificate for $L_2|_{\mathcal{D}_{L_1}} \succeq 0$. Our paper [HKM+^{2}] greatly extends this to nonlinear L_2. To be more precise, suppose L is a monic linear pencil in g variables and let $\mathcal{D}_L$ be the corresponding nc LMI. Then a symmetric noncommutative polynomial $p \in \mathbb{R}\langle x \rangle$ is *positive semidefinite* on $\mathcal{D}_L$ if and only if it has a weighted sum of squares representation with optimal degree bounds. Namely,

$$p = s^* s + \sum_j^{\text{finite}} f_j^* L f_j, \tag{22}$$

where s, f_j are vectors of noncommutative polynomials of degree no greater than $\frac{\deg(p)}{2}$. (There is also a bound, coming from a theorem of Carathéodory on convex sets in finite-dimensional vector spaces and depending only on the degree of p, on the number of terms in the sum.) This result contrasts sharply with the commutative setting, where the degrees of s, f_j are vastly greater than $\deg(p)$ and assuming only p nonnegative yields a clean Positivstellensatz so seldom that the cases are noteworthy [Sc09].

The main ingredient of the proof is a solution to a noncommutative moment problem, i.e., an analysis of rank preserving extensions of truncated noncommutative Hankel matrices. For instance, any such *positive definite* matrix M_k of "degree k" has, for each $m \geq 0$, a positive semidefinite Hankel extension M_{k+m} of degree $k + m$ and the same rank as M_k. For details and proofs see [HKM+2].

6.4. Further topics

The reader who has made it to this point may be interested in some of the surveys, and the references therein, on various aspects of noncommutative (free) real algebraic geometry, and free positivity.

The article [HP07] treats positive noncommutative polynomials as a part of the larger tapestry of spectral theory and optimization. In [HKM12] this topic is expanded with further Positivstellensätze and computational aspects. The survey [dOHMP09] provides a serious overview of the connection between noncommutative convexity and systems engineering. The note [HMPV09] emphasizes the theme, as does the body of this article, that convexity in the noncommutative setting appears to be no more general than LMI. Finally, a tutorial with numerous exercises emphasizing the role of the middle matrix and border vector representation of the Hessian of a polynomial in analyzing convexity is [HKM+3].

Acknowledgment and dedication

The second and third author appreciate the opportunity provided by this volume to thank Bill for many years of his most generous friendship. Our association with Bill and his many collaborators and friends has had a profound, and decidedly positive, impact on our lives, both mathematical and personal. We are looking forward to many ??s to come.

References

[BGM06]   J.A. Ball, G. Groenewald, T. Malakorn: Bounded Real Lemma for Structured Non-Commutative Multidimensional Linear Systems and Robust Control, *Multidimens. Syst. Signal Process.* **17** (2006) 119–150

[BKU78]   R. Braun, W. Kaup, H. Upmeier: On the automorphisms of circular and Reinhardt domains in complex Banach spaces, *Manuscripta Math.* **25** (1978) 97–133

[DAn93]   J.P. D'Angelo: *Several complex variables and the geometry of real hypersurfaces*, CRC Press, 1993 [dOHMP09] M.C. de Oliveira, J.W. Helton, S. McCullough, M. Putinar: Engineering systems and free semi-algebraic geometry, In: *Emerging applications of algebraic geometry,* 17–61, IMA Vol. Math. Appl. **149**, Springer, 2009

[Fo93]   F. Forstnerič: Proper holomorphic mappings: a survey. In: *Several complex variables (Stockholm, 1987/1988)* 297-363, Math. Notes **38**, Princeton Univ. Press, 1993

[Fr84]   A.E. Frazho: Complements to models for noncommuting operators, *J. Funct. Anal.* **59** (1984) 445–461

[GP74]   V. Guillemin, A. Pollack: *Differential Topology,* Prentice-Hall, 1974

[Hg78]   S. Helgason: *Differential geometry, Lie groups, and symmetric spaces,* Pure and Applied Mathematics **80**, Academic Press, Inc., 1978

[HBJP87] J.W. Helton, J.A. Ball, C.R. Johnson, J.N. Palmer: *Operator theory, analytic functions, matrices, and electrical engineering,* CBMS Regional Conference Series in Mathematics **68**, AMS, 1987

[HKM11a] J.W. Helton, I. Klep, S. McCullough: Analytic mappings between noncommutative pencil balls, *J. Math. Anal. Appl.* **376** (2011) 407–428

[HKM11b] J.W. Helton, I. Klep, S. McCullough: Proper Analytic Free Maps, *J. Funct. Anal.* **260** (2011) 1476–1490

[HKM12]   J.W. Helton, I. Klep, S. McCullough: Convexity and Semidefinite Programming in dimension-free matrix unknowns, In: *Handbook of Semidefinite, Conic and Polynomial Optimization* edited by M. Anjos and J.B. Lasserre, 377–405, Springer, 2012

[HKM+]   J.W. Helton, I. Klep, S. McCullough: The matricial relaxation of a linear matrix inequality, to appear in *Math. Program.*
          http://arxiv.org/abs/1003.0908

[HKM+^{2}] J.W. Helton, I. Klep, S. McCullough: The convex Positivstellensatz in a free algebra, *preprint* http://arxiv.org/abs/1102.4859

[HKM+^{3}] J.W. Helton, I. Klep, S. McCullough: Free convex algebraic geometry, to appear in *Semidefinite Optimization and Convex Algebraic Geometry* edited by G. Blekherman, P. Parrilo, R. Thomas, SIAM, 2012

[HKMS09] J.W. Helton, I. Klep, S. McCullough, N. Slinglend: Non-commutative ball maps, *J. Funct. Anal.* **257** (2009) 47–87

[HM+]   J.W. Helton, S. McCullough: Every free basic convex semi-algebraic set has an LMI representation, *preprint* http://arxiv.org/abs/0908.4352

[HMPV09] J.W. Helton, S. McCullough, M. Putinar, V. Vinnikov: Convex matrix inequalities versus linear matrix inequalities, *IEEE Trans. Automat. Control* **54** (2009), no. 5, 952–964

[HP07]   J.W. Helton, M. Putinar: Positive polynomials in scalar and matrix variables, the spectral theorem, and optimization, *Operator theory, structured matrices, and dilations,* 229–306, Theta Ser. Adv. Math. **7**, Theta, 2007

[HV07]   J.W. Helton, V. Vinnikov: Linear matrix inequality representation of sets, *Comm. Pure Appl. Math.* **60** (2007) 654–674

[KVV–]   D. Kalyuzhnyi-Verbovetskiĭ, V. Vinnikov: Foundations of noncommutative function theory, *in preparation*

[KU76]   W. Kaup, H. Upmeier: Banach spaces with biholomorphically equivalent unit balls are isomorphic, *Proc. Amer. Math. Soc.* **58** (1976) 129–133

[Kr01]   S.G. Krantz: *Function Theory of Several Complex Variables*, AMS, 2001

[MS11]   P.S. Muhly, B. Solel: Progress in noncommutative function theory, *Sci. China Ser. A* **54** (2011) 2275–2294

[Po89]   G. Popescu: Isometric Dilations for Infinite Sequences of Noncommuting Operators, *Trans. Amer. Math. Soc.* **316** (1989) 523–536

[Po06]   G. Popescu: Free holomorphic functions on the unit ball of $\mathcal{B}(\mathcal{H})^n$, *J. Funct. Anal.* **241** (2006) 268–333

[Po10]   G. Popescu: Free holomorphic automorphisms of the unit ball of $B(H)^n$, *J. reine angew. Math.* **638** (2010) 119–168

[RR85]   M. Rosenblum, J. Rovnyak: *Hardy Classes and Operator Theory,* Oxford University Press, 1985

[Sc09]   C. Scheiderer: Positivity and sums of squares: a guide to recent results. In: *Emerging applications of algebraic geometry* 271–324, IMA Vol. Math. Appl. **149**, Springer, 2009

[SIG98] R.E. Skelton, T. Iwasaki, K.M. Grigoriadis: *A unified algebraic approach to linear control design*, Taylor & Francis Ltd., 1998

[SV06]   D. Shlyakhtenko, D.-V. Voiculescu: Free analysis workshop summary: American Institute of Mathematics,
http://www.aimath.org/pastworkshops/freeanalysis.html

[Ta73]   J.L. Taylor: Functions of several noncommuting variables, *Bull. Amer. Math. Soc.* **79** (1973) 1–34

[Vo04]   D.-V. Voiculescu: Free analysis questions I: Duality transform for the coalgebra of $\partial_{X:B}$, *International Math. Res. Notices* **16** (2004) 793–822

[Vo10]   D.-V. Voiculescu: Free Analysis Questions II: The Grassmannian Completion and the Series Expansions at the Origin, *J. reine angew. Math.* **645** (2010) 155–236

J. William Helton
Department of Mathematics
University of California San Diego, USA
e-mail: helton@math.ucsd.edu

Igor Klep
Department of Mathematics
The University of Auckland, New Zealand
e-mail: igor.klep@auckland.ac.nz

Scott McCullough
Department of Mathematics
University of Florida
Gainesville, FL 32611-8105, USA
e-mail: sam@math.ufl.edu

Operator Theory:
Advances and Applications, Vol. 222, 221–231
© 2012 Springer Basel

Traces of Commutators of Integral Operators – the Aftermath

Roger Howe

For Bill Helton, with cordial regards

Abstract. We review the papers of J. Helton with R. Howe on traces of commutators of Hilbert space operators, and survey their influence in the 35 years since publication.

Mathematics Subject Classification. Primary: 47-02;
secondary: 01-02, 30-02, 47B47.

Keywords. Trace of commutators, almost commuting operators, Toeplitz operator.

1. Introduction

In 1972–73, I had the pleasure of collaborating with Bill Helton on a pair of papers [HH1, HH2] studying systems of self-adjoint Hilbert space operators that "almost commute", in the sense that their commutators were of the Schatten-von Neumann trace class, with some attention also to more general systems. This work was part of a very lively research activity at SUNY Stony Brook in operator theory, including the work of Brown, Douglas and Fillmore on extensions of commutative C^*-algebras by the compact operators, which led to KK-theory [BDF, Kas, KS, R], and the work of Pincus on his principal function and mosaic [Pin], to which our results were strongly related. All of this work has had a significant afterlife, with citations continuing to the present. It is another pleasure to review here some of the work our papers have influenced. In doing this, I emphasize the non-systematic nature of this survey, and apologize in advance to authors whose relevant papers I have failed to identify.

2. Trace form

I will begin with a brief summary of the results of [HH1] and [HH2]. In [HH1], the main object of study is a pair A, B, of self-adjoint (bounded, on a Hilbert space) operators whose commutator

$$[A, B] = AB - BA$$

belongs to the trace class. These could be the real and imaginary parts of a non-self adjoint operator $T = A + iB$, whose "self-commutator" $[T^*, T] = -2i[A, B]$ is trace class; in which case T would be a normal operator modulo the ideal of trace class (or, a fortiori, compact) operators.

A prime example of an operator with trace class self-commutator is the unilateral shift S: for a Hilbert space H with orthonormal basis $\{\mathbf{e}_i : i = 0, 1, 2, \ldots, \infty\}$, define $S(\mathbf{e}_i) = \mathbf{e}_{i+1}$. Then $[S^*, S]$ is orthogonal projection to the line through $\mathbf{e}_0$.

The study of S is tightly bound up with the theory of Toeplitz operators. Recall that, if $H^2(\mathbf{T})$ denotes the Hardy space of the circle – the space of all square integrable boundary values of functions holomorphic on the unit disk, and if f is continuous function on $\mathbf{T}$, then the *Toeplitz operator* T_f associated to f is the operator defined by

$$T_f(h) = P^+(fh), \qquad h \in H^2(\mathbf{T}),$$

where P^+ denotes orthogonal projection of $L^2(\mathbf{T})$ to H^2, and fh is just the usual pointwise product of f and h. The precise relationship between Toeplitz operators and the shift S is:

a) the $S = T_z$; and
b) the Toeplitz operators define a cross section to the compact operators in the C^* algebra generated by S.

In thinking about Toeplitz operators, I had noticed the formula

$$\text{trace}[T_f, T_g] = \frac{1}{2\pi i} \int_D J(\tilde{f}, \tilde{g}) dx dy, \tag{1}$$

where $\tilde{f}$ and $\tilde{g}$ are continuous extensions of f and g to the unit disk D, and

$$J(\tilde{f}, \tilde{g}) = \frac{\partial \tilde{f}}{\partial x} \frac{\partial \tilde{g}}{\partial y} - \frac{\partial \tilde{g}}{\partial x} \frac{\partial \tilde{f}}{\partial y}$$

is the Jacobian of $\tilde{f}$ and $\tilde{g}$. Thus, if f and g (and $\tilde{f}$ and $\tilde{g}$) are real valued, the integral in formula (1) represents the area of the image of D under the mapping defined by

$$\begin{bmatrix} x \\ y \end{bmatrix} \rightarrow \begin{bmatrix} \tilde{f}\left(\begin{bmatrix} x \\ y \end{bmatrix}\right) \\ \tilde{g}\left(\begin{bmatrix} x \\ y \end{bmatrix}\right) \end{bmatrix}.$$

When I showed this formula to Bill, he recognized immediately that, for any almost commuting pair A, B, of operators, the bilinear form

$$B(p,q) = \text{trace}[p(A,B), q(A,B)] \qquad (2)$$

(where p and q are non-commutative polynomials in A and B) should be an important invariant of the algebra generated by A and B. So we joined forces to study this bilinear form, and were quite delighted when, after several weeks of daily discussions, we realized that a direct generalization of the formula for Toeplitz operators would hold for any almost commuting pair. We could represent the form B as

$$B(\tilde{p}, \tilde{q}) = \int_{\mathbf{R}^2} J(\tilde{p}, \tilde{q}) d\mu, \qquad (3a)$$

where $\tilde{p}$ and $\tilde{q}$ are now the commutative polynomials associated to p and q, and $d\mu$ is a suitable compactly supported Borel measure on the plane. Moreover, $d\mu$ is supported on the spectrum of $T = A + iB$, and at a point z of the spectrum not in the essential spectrum, we could show that

$$d\mu = -\frac{1}{2\pi i} \text{Ind}(T - z). \qquad (3b)$$

Here $\text{Ind}(A)$ means the Fredholm index of the operator A. Precisely, if A is Fredholm, that is, invertible modulo the compact operators, then

$$\text{Ind}(A) = \dim \ker A - \dim \text{coker} A. \qquad (3c)$$

Here dim coker refers to the fact that the image of A will be a closed subspace of finite codimension, and this is what is signified by dim coker.

In the proof of formula (3), a key role is played by a feature of B that we called the *collapsing property*. This says that, if p and q are both functions of a third polynomial h, then $B(\tilde{p}, \tilde{q})$ vanishes (for the obvious reason that two polynomials in the same operator commute with each other). The collapsing property is the condition that enforces the representation given in (3). We had a very awkward argument to pass from the collapsing property to (3). We were very much indebted to Nolan Wallach, now Bill's colleague at UCSD, for supplying a simple proof, which we reproduced in [HH1] as *Wallach's lemma*.

We also gave in [HH1] a more abstract version of (3), in which we considered a collection of self-adjoint operators A_i, the commutator of any two of which was assumed to be trace class. The resulting bilinear form then gave a relative (de Rham) cohomology class on the joint spectrum of the A_i.

Finally, somewhat as an afterthought, we gave a related formula for the multiplicative commutators of the one-parameter groups gotten by exponentiating of the operators. Suppose that $f(A, B) = f$ is a function of A and B that is a Fredholm operator of index 0. Then for some finite rank operator F, the perturbation $f' = f + F$ is invertible. Consider two such operators f' and g'. Then the multiplicative commutator $\{f', g'\} = f'g'f'^{-1}g'^{-1}$ will differ from the identity operator by a trace class operator. This means that the determinant $\det\{f', g'\}$ exists. Also, it is easy to see that it does not depend on the choices of F and G, but only on the functions f and g. It turns out that there is a formula analogous to (3):

$$\delta(f,g) = \det\{f', g'\} = \exp\left(\int_{\mathbf{R}^2} \frac{J(\tilde{p}, \tilde{q})}{fg} d\mu\right) \tag{4a}$$

In [HH1], we reproduced a proof due to Pincus for functions of the form $f = e^h$ for some other smooth function h. The proof was based on the general formula

$$\det\left(e^A e^B e^{-A} e^{-B}\right) = e^{\text{trace}[A,B]} \tag{4b}$$

for a pair of almost commuting operators, with the determinant being defined for operators of the form $I + C$, with C trace class. This formula was established using the first few terms of the Baker-Campbell-Hausdorff formula. I do not know if Pincus ever published his proof separately. The formula was proven in general by Larry Brown in a paper [Br], appearing in the same volume as [HH1].

The paper [HH2] presents an attempt, only partly successful, to extend the results of [HH1] to more general situations. We defined a notion of *cryptointegral algebra of dimension n*. Instead of assuming trace class commutators of any pair of operators, we assumed only that commutators were in a certain Schatten-von Neumann p-class, and that a certain $2n$-multilinear functional (the complete anti-symmetrization of $2n$ elements) was of trace class. Unfortunately, we could give a representation analogous to (3) for the resulting multilinear form only for dimension 1 and dimension 2. However, we did show that the algebra of pseudodifferential operators of order 0 on a manifold satisfied our conditions to be a cryptointegral algebra, and for this algebra, we gave a nice formula for the form, in terms of integrating over the cotangent sphere bundle of the manifold. We also knew that an algebra of Toeplitz operators on the $(2n - 1)$-sphere, the boundary of the unit ball in $\mathbf{C}^n$, also qualified as a cryptointegral algebra, and that the multilinear form had a similarly nice representation, in terms of integration over the sphere. This uniform analysis of Toeplitz and pseudodifferential operators also showed that the index theorem of U. Venugopalkrishna [V], for Toeplitz operators on S^{2n-1} could be folded into the Atiyah-Singer Index Theorem. The details of these facts about higher-dimensional Toeplitz operators were published later, in [H].

3. Afterwards

Here is the beginning of the review of our paper [HH1] in *Mathematical Reviews*, by James Deddens.

> The authors develop a theory of traces of commutators of self-adjoint operators, which gives explicit but elegant formulas for computation. The work is related to index theory of Atiyah and Singer, the *g*-function of Pincus, the Brown-Douglas-Fillmore theory of extensions, and the trace norm estimates of Berger and Shaw, and Putnam.

As this statement indicates, our results tied in very closely with contemporary work in operator theory, and especially, with our colleagues at Stony Brook.

Indeed, Joel Pincus took an immediate interest in our results, and with his student Richard Carey showed [CP1] that the measure $d\mu$ of formula (3) had the form

$$d\mu = \frac{1}{\pi} g(x, y) dx dy, \tag{5}$$

where g was a function that Pincus had constructed some years earlier in studying a pair of non-commuting self-adjoint operators, and named the *principal function* [Pin]. This link gave a natural, intrinsic interpretation for the principal function, and also established that the measure $d\mu$ was absolutely continuous with respect to Lebesgue measure. Bill and I did not have strong control of $d\mu$ on the essential spectrum of T except in the case when $[A, B]$ has finite rank, in which case we gave a simple argument showing that the Radon-Nkodym derivative $\frac{d\mu}{dx dy}$ with respect to Lebesgue measure was bounded by half the rank of $[A, B]$. A few years later, Pincus and Carey generalized the combined theory to the case of type II factors [CP2].

This immediate folding in of our work into the Pincus machine somewhat complicates the job of tracing the influence of [HH1], since that may be direct, or indirect through the use of the trace-invariant properties of the Pincus principal function. In this note, we restrict attention primarily to papers that directly cite [HH1, HH2] and do not try to sort out the indirect influence. Of course, some papers cite both our results and the Carey-Pincus or Pincus-Xia papers that incorporated the trace functional.

Our results also fit nicely into the theory of hyponormal operators, on which there is a very substantial literature [MP1, X1, X2]. Recall that an operator T is *hyponormal* provided that the self-commutator $[T^*, T]$ is a non-negative operator. In particular, the unilateral shift operator S is hyponormal. In particular, our result meshed very well with the result of C. Berger and B. Shaw [BS], proved almost at the same time, that a hyponormal operator satisfying a natural finiteness condition automatically had trace class self-commutator. Hence, both from the concrete and abstract points of view, their were many examples to which our formula applied.

Finally, our theory complemented the more general and abstract work of L. Brown, R.G. Douglas and P. Fillmore, who considered extensions of a commutative C^* algebra by the compact operators. A pair of self-adjoint operators will generate such an extension when their commutator $[A, B]$ is compact. Our results showed that the stronger assumption that $[A, B]$ be trace class allowed one to obtain more refined information. Although our term *cryptointegral* has been quietly dropped, the theme that Schatten p-class conditions on commutators are worth paying attention to has been refined and amplified over the years [Do, DoPY, Go, Wa, Co1, Co2]. The philosophy has become that the requirement of trace class of Schatten p-class commutators amounted to a kind of "smoothness" condition, so that this kind of analysis requirement amounts to doing differential topology rather than continuous topology. The p for which commutators are in T^p is related to the dimension of the algebra. More specifically, the trace class condition is related to the fact that the essential spectrum of S is the unit circle, which is one-dimensional. (Or more generally, for a smooth function f, the essential spectrum of the Toeplitz operator T_f is just the image of the circle under f.) Later papers formulated the notion of dimension of an extension. Saying that the dimension is p is more or less equivalent to requiring commutators to belong to the Schatten-von Neumann class T^p [Si1]. Eventually these ideas were incorporated into non-commutative geometry as formulated by A. Connes [Co1, Co2], and formed part of the motivation for his definition of cyclic cohomology [Co1]. The tool of cyclic cohomology allowed Connes to define invariants corresponding to any cocycle on the spectrum of the algebra, not just the Euler class (i.e., integration over the whole space), which was what was realized by our antisymmetrization form on the pseudodifferential operators. An alternative approach to non-commutative geometry discussed by M. Kapranov [Kap], based on expansions modulo commutator ideals, likewise cites [HH2].

These themes are still playing out, in operator theory, in non-commutative geometry, and also in some more surprising places. We briefly survey a few examples.

p-hyponormal operators. A series of recent papers by M. Chō and several co-authors establishes trace formulas for commutators of operators that generalize hyponormal operators, for example, the class of *p-hyponormal* operators [ChHL, ChHK, ChHKL, ChGHY, ChH]. For $0 < p < 1$, an operator T on a Hilbert space is called p-hyponormal if the inequality

$$(T^*T)^p \geq (TT^*)^p$$

holds. The property of p-hyponormality clearly extends the long-studied property of hyponormality. Cho and his collaborators have been extending results about hyponormal operators to this larger class, including the theory of the trace form, as in the papers cited.

Toeplitz operators on pseudoconvex domains. Trace invariants of operators on pseudoconvex domains have also attracted continuing study. In [EGZ], Englis, Guo and Zhang study commutators of Toeplitz operators on the unit ball in $\mathbf{C}^d$. These commutators are in the Schatten d-class T^d, but the behavior of their eigenvalues is more regular than that of arbitrary T^n operators. They belong to the Macaev class $L^{d,\infty}$ [Co1]. This means that the product of d such commutators belongs to $L^{1,\infty}$, and therefore has a trace in the sense of Dixmier [Di]. Englis, Guo and Zhang compute the Dixmier trace of such a product of commutators: it is just the integral over the unit ball of the product of the Jacobians of the symbols of the Toeplitz operators in question (times $\frac{1}{d}$!). This result is formally consistent with our formula for the complete antisymmetrization, but two different notions of trace are being used in the two formulas. In [EZ], Englis and Zhang generalize their formula to arbitrary pseudo-convex domains. The formula for the general case is considerably more complex than for the ball, and involves geometric invariants of the domain.

Szegö's formula, and orthogonal polynomials. The formulas (4) have found continuing employment in the theory of Toeplitz operators, and related questions about orthogonal polynomials on the circle. The connection comes mainly through Szegö's limit formula for Toeplitz determinants and related phenomena [Wi]. There has been continuing interest in Szegö's formula, and the formula (4b) has been found a useful tool for proving it. The possibility of a connection of (4) with Szegö was already suggested in [HH1], and this was substantiated precisely not long after by Widom [Wi]. Related arguments have echoed through the literature [BE, BH, BW, Ehr]. This has also found interesting connections to the theory of orthogonal polynomials on the unit circle, especially through the Geronimo-Case-Borodin-Okounkov formula [GC, BO], which has also been a popular result for simplification [Bö, BöW]. A derivation of the GCBO formula using (4b) is given in [Si2], along with an account of the history. A simplified proof of formula (4b) itself has been given by Böttcher [Bö], who used differentiation to avoid the Baker-Campbell-Hausdorff formula.

Quadrature domains. Finally, I would like to mention a surprising connection found by M. Putinar [Put2, Put3, Put4, Put5], [GP1, GP2] of the formulas (3) and (4) to the theory of *quadrature domains* which have been the subject of a large literature since the 1970s [AS, Sh]. A quadrature domain is a region Ω in the complex plane for which there is a formula

$$\int_D f(z)dxdy = u(f), \tag{6}$$

for any function holomorphic on Ω, where u is a distribution of finite support. That is, $u(f)$ is a weighted sum of the values of f and a finite number of its derivatives at a finite number of points. Cauchy's Formula says that disks (the interiors of a circle around a point) are quadrature domains, but there are many others. There is a notion of *degree* of a quadrature domain, defined in terms of the distribution u. Quadrature domains of degree 1 are those for which u is a multiple of the point mass at a given point. It is known that the only quadrature domains of degree one are the disks. However, there are many quadrature domains of higher degree, and the behavior of quadrature domains becomes more intricate as the degree increases.

In general, for an almost normal operator T, the trace functional, or equivalently, the principal function, does not determine the unitary equivalence class of T. However, when the self-commutator $[T^*, T]$ has rank one, say

$$[T^*, T](\vec{v}) = \gamma(\vec{v}, \xi)\xi \tag{7}$$

for some constant γ and some unit vector ξ, and all vectors $\vec{v}$ in the Hilbert space, then in fact, T is determined by the trace functional. Indeed, you can write down explicit models for T in terms of the principal function. Moreover, any function with compact support and taking values between 0 and 1 can be the principal function for some T. Thus, one has a bijection between operators with rank one self-commutator and compactly-supported, $[0, 1]$-valued measurable functions in the plane. This has been understood since the beginnings of the theory, see [Pin].

In particular, given a compact domain $\Omega \subset \mathbf{C}$, there is a unique (up to unitary equivalence) operator T_Ω satisfying (7), and whose principal function is χ_Ω, the characteristic function of Ω. Putinar [Put3] has shown that Ω is a quadrature domain if and only if the T^* invariant subspace generated by ξ is finite dimensional. The dimension of this space is in fact the degree of the quadrature domain. Thus, in the notation of §1, for the shift operator S, the vector ξ is just e_0, and e_0 spans the kernel of S^*. Operator theory has had a symbiotic relationship with complex analysis since its origins. This consonance between the refined analysis of the structure of planar domains and the simplest non-normal operators gives a further example of this symbiosis, and provides a satisfying conclusion to this brief survey.

References

[AS]   D. Aharonov and H. Shapiro, Domains on which analytic functions satisfy quadrature identities. J. Analyse Math. 30 (1976), 39–73.

[AKKM] S. Albeverio, K. Makarov, Konstantin and A. Motovilov, Graph subspaces and the spectral shift function, Canad. J. Math. **55** (2003), 449–503.

[BE]   E. Basor and T. Ehrhardt, Asymptotics of determinants of Bessel operators. Comm. Math. Phys. **234** (2003), 491–516.

[BH]   E. Basor and J.W. Helton, A new proof of the Szegö limit theorem and new results for Toeplitz operators with discontinuous symbol, J. Operator Theory **3** (1980), 23–39.

[BW]   E. Basor and H. Widom, On a Toeplitz determinant identity of Borodin and Okounkov, Integral Equations Operator Theory **37** (2000), no. 4, 397–401.

[BD]   P. Baum and R.G. Douglas, Relative K homology and C^* algebras. K-Theory **5** (1991), 1–46.

[BS]   C. Berger and B. Shaw, Selfcommutators of multicyclic hyponormal operators are always trace class, Bull. Amer. Math. Soc. 79 (1973), 1193–1199.

[Bö]   A. Böttcher, On the determinant formulas by Borodin, Okounkov, Baik, Deift and Rains, in Toeplitz matrices and singular integral equations (Pobershau, 2001), Oper. Theory Adv. Appl. **135**, Birkhäuser, Basel, 200, 91–99.

[BöW]   A. Böttcher and H. Widom, Szegö via Jacobi, Linear Algebra Appl.**419** (2006), 656–667.

[BO]   A. Borodin and A. Okounkov, A Fredholm determinant formula for Toeplitz determinants. Integral Equations Operator Theory **37** (2000), 386–396.

[Br]   L. Brown, The determinant invariant for operators with trace class self commutator, Proceedings of a Conference on Operator Theory (Dalhousie Univ., Halifax, N.S., 1973), Lecture Notes in Math., **345**, Springer, Berlin, 1973, 210–228.

[BDF]   L. Brown, R.G. Douglas and P. Fillmore, Unitary equivalence modulo the compact operators and extensions of C^*-algebras. Proceedings of a Conference on Operator Theory (Dalhousie Univ., Halifax, N.S., 1973), Lecture Notes in Math. **345**, Springer, Berlin, 1973, 53–128.

[CP1]   R. Carey and J. Pincus, An exponential formula for determining functions, Indiana Univ. Math. J. **23**, (1974), 1031–1042.

[CP2]   R. Carey and J. Pincus, Mosaics, principal functions, and mean motion in von Neumann algebras. Acta Math. **138** (1977), 153–218.

[ChHL] M. Chō, T. Huruya and C. Li. Trace formulae associated with the polar decomposition of operators, Math. Proc. R. Ir. Acad **105A** (2005), 57–69.

[ChHK] M. Chō and T. Huruya, Trace formulae of p-hyponormal operators II, Hokkaido Math. J. **35** (2006), 247–259.

[ChHKL] M. Chō, T. Huruya, A. Kim and C. Li, Principal functions for high powers of operators, Tokyo. J. Math. **29** (2006), 111–116.

[ChGHY] M. Chō, M. Giga, T. Huruya and T. Yamasaki, A remark on support of the principal function for class A operators, Integral Equations and Operator Theory **57** (2007), 303–308.

[ChH]   M. Chō and T. Huruya, Trace formula for partial isometry case, Tokyo J. Math **32**(2009), 27–32.

[Co1]   A. Connes, Noncommutative differential geometry. Inst. Hautes Études Sci. Publ. Math. No. **62** (1985), 257–360.

[Co2]   A. Connes, Noncommutative Geometry, Academic Press, Inc., San Diego, CA, 1994.

[Di]   J. Dixmier, Existence de traces non normales. (French) C. R. Acad. Sci. Paris Sr. A-B **262**(1966), A1107–A1108.

[Do]   R.G. Douglas, On the smoothness of elements of Ext, in *Topics in modern operator theory* (Timisoara/Herculane, 1980), Operator Theory: Adv. Appl. **2**, Birkhäuser, Basel-Boston, Mass., 1981, 63–69.

[DoPY] R.G. Douglas, V. Paulsen and K. Yan, Operator theory and algebraic geometry. Bull. Amer. Math. Soc. (N.S.) **20** (1989), 67–71.

[DoV]   R.G. Douglas and D. Voiculescu, On the smoothness of sphere extensions. J. Operator Theory **6** (1981), 103–111.

[Ehr]   T. Ehrhardt, A generalization of Pincus' formula and Toeplitz operator determinants, Arch. Math. (Basel) **80** (2003), 302–309.

[EGZ]   M. Englis, K. Guo and G. Zhang, Toeplitz and Hankel operators and Dismier traces on the unit ball of $\mathbf{C}^n$, Proc. Amer. Math. Soc. **137** (2009), 3669–3678.

[EZ]   M. Englis and G. Zhang, Hankel operators and the Dismier trace on strictly pseudoconvex domains, Doc. Math **15** (2010), 601–622.

[GC]   J. Geronimo and K. Case, Scattering theory and polynomials orthogonal on the unit circle. J. Math. Phys. **20** (1979), 299–310.

[Go]   G. Gong, Smooth extensions for finite CW complexes. Trans. Amer. Math. Soc. **342** (1994), 343–358.

[GP1]   B. Gustafsson and M. Putinar, An exponential transform and regularity of free boundaries in two dimensions, Ann. Sc. Norm. Super. Pisa **26** (1998), 507–543.

[GP2]   B. Gustafsson and M. Putinar, Linear analysis of quadrature domains. II. (English summary) Israel J. Math. **119** (2000), 187–216.

[GP3]   B. Gustafsson and M. Putinar, Selected topics on quadrature domains. (English summary) Phys. D **235** (2007), 90–100.

[HH1]   J.W. Helton and R. Howe, Integral operators: commutators, traces, index and homology, Proceedings of a Conference on Operator Theory (Dalhousie University, Halifax, N.S. 1973), Lecture Notes in Math. **345**, Springer, Berlin, 1973, 141–209.

[HH2]   J.W. Helton and R. Howe, Traces of commutators of integral operators, Acta Math. **135** (1975), 271–305.

[H]   R. Howe, Quantum mechanics and partial differential equations. J. Funct. Anal. **38** (1980), 188–254.

[JLO]   A. Jaffe, A. Lesniewski and K. Osterwalder, Quantum K-theory. I. The Chern character, Comm. Math. Phys. **118** (1988), 1–14.

[KS]   J. Kaminker and C. Schochet, K-theory and Steenrod homology: applications to the Brown-Douglas-Fillmore theory of operator algebras. Trans. Amer. Math. Soc. **227** (1977), 63–107.

[Kap]   M. Kapranov, Noncommutative geometry based on commutator expansions. J. Reine Angew. Math. **505** (1998), 73–118.

[Kas]   G.G. Kasparov, Topological invariants of elliptic operators. I. K-homology. (Russian) Math. USSR-Izv. **9** (1975), 751–792 (1976).

[Ki1]   F. Kittaneh, Inequalities for the Schatten p-norm. Glasgow Math. J. **26** (1985), 141–143.

[Ki2]   F. Kittaneh, Some trace class commutators of trace zero. Proc. Amer. Math. Soc. **113** (1991), 855–661.

[MP1]   M. Martin and M. Putinar, A unitary invariant for hyponormal operators. J. Funct. Anal. **73** (1987), 297–323.

[MP2]   M. Martin and M. Putinar, Lectures on hyponormal operators. Operator Theory: Advances and Applications **39**. Birkhäuser Verlag, Basel, 1989.

[Pin]   J.D. Pincus, Commutators and systems of singular integral equations. I, Acta Math. **121** (1968), 219–249.

[Put1] M. Putinar, Hyponormal operators and eigendistributions. Advances in invariant subspaces and other results of operator theory (Timisoara and Herculane, 1984), Oper. Theory Adv. Appl., **17**, Birkhäuser, Basel, 1986, 249–273.

[Put2] M. Putinar, Linear analysis of quadrature domains. Ark. Mat. **33** (1995), 357–376.

[Put3] M. Putinar, Extremal solutions of the two-dimensional L-problem of moments, J. Funct. Anal. **136** (1996), 331–364.

[Put4] M. Putinar, Extremal solutions of the two-dimensional L-problem of moments II, J. Approx. Theory **92** (1998), 38–58.

[Put5] M. Putinar, Linear analysis of quadrature domains. III. (English summary) J. Math. Anal. Appl. **239** (1999), 101–117.

[R]   J. Rosenberg, K and KK: topology and operator algebras. Operator theory: operator algebras and applications, Part 1 (Durham, NH, 1988), Proc. Sympos. Pure Math. **51**, Part 1, Amer. Math. Soc., Providence, RI, 1990, 445–489.

[Sh]   H. Shapiro, The Schwarz function and its generalization to higher dimensions. University of Arkansas Lecture Notes in the Mathematical Sciences, 9. John Wiley & Sons, Inc., New York, 1992.

[Si1] B. Simon, Trace ideals and their applications, London Mathematical Society Lecture Note Series **35**. Cambridge University Press, Cambridge-New York, 1979.

[Si2] B. Simon, Orthogonal polynomials on the unit circle. Part 2. Spectral theory. American Mathematical Society Colloquium Publications, 54, Part 2. American Mathematical Society, Providence, RI, 2005.

[V]   U. Venugopalkrishna, Fredholm operators associated with strongly pseudoconvex domains in $\mathbf{C}^n$, J. Functional Analysis **9** (1972), 349–373.

[Wa]   X. Wang, Voiculescu Theorem, Sobolev lemma and extensions of smooth algebras, Bull. Amer. Math. Soc. (New Series) **27** (1992), 292–297.

[Wi]   H. Widom, Asymptotic behavior of block Toeplitz matrices and determinants. II. Advances in Math. **21** (1976), 1–29

[X1]   D. Xia, Trace formula for almost Lie algebra of operators and cyclic one-cocycles, Nonlinear and convex analysis (Santa Barbara, Calif., 1985), Lecture Notes in Pure and Appl. Math. **107**, Dekker, New York, 1987, 299–308.

[X2]   D. Xia, Trace formulas for almost commuting operators, cyclic cohomology and subnormal operators, Integral Equations Operator Theory **14** (1991), 276–298.

[Y1]   R. Yang, Operator theory in the Hardy space over the bidisk. III, J. Funct. Anal. **186** (2001), 521–545.

[Y2]   R. Yang, A trace formula for isometric pairs, Proc. Amer. Math. Soc. **131** (2003), 533–541.

Roger Howe
Yale University
Mathematics Department
PO Box 208283
New Haven, CT 06520-8283, USA
e-mail: howe@math.yale.edu

Operator Theory:
Advances and Applications, Vol. 222, 233–246
© 2012 Springer Basel

Information States in Control Theory: From Classical to Quantum

M.R. James

Dedicated to Bill Helton

Abstract. This paper is concerned with the concept of *information state* and its use in optimal feedback control of classical and quantum systems. The use of information states for *measurement* feedback problems is summarized. Generalization to fully quantum coherent feedback control problems is considered.

Mathematics Subject Classification. 93E03, 81Q93.

Keywords. Quantum control, information state.

1. Introduction

This paper is dedicated to Bill Helton, with whom I had the honor and pleasure of collaborating in the topic area of nonlinear H^∞ control theory, [13]. We developed in some detail the application of information state methods to the nonlinear H^∞ control problem, [19, 18, 1]. In this paper I review the information state concept for classical output feedback optimal control problems, and then discuss extensions of this concept to quantum feedback control problems, [16, 17, 21].

Feedback is one of the most fundamental ideas in control engineering. Feedback control is a critical enabler for technological development, Figure 1. From its origins in steam engine governors, through applications in electronics, aerospace, robotics, telecommunications and elsewhere, the use of feedback control has been essential in shaping our modern world. In the 20th century, quantum technology, through semiconductor physics and microchips, made possible the information age. New developments in quantum technology, which include quantum information and computing, precise metrology, atom lasers, and quantum electromechanical

Research supported by the Australian Research Council.

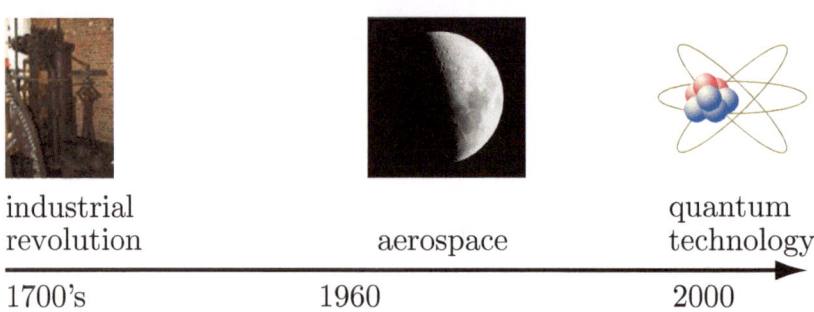

FIGURE 1. Feedback control timeline.

systems, further exploit quantum phenomena and hold significant promise for the future.

Optimization is basic to many fields and is widely used to design control systems. Optimization based control system design requires specification of (i) the *objective* of the control system, and (ii) the *information* available to the control system. In a feedback system, Figure 2, control actions are determined on the basis of information gained as the system operates. A key issue is how to represent information in a feedback loop. The concept of *information state* was introduced for this purpose, [22]. An information state is a statistic[1] that takes into account the performance objective in a feedback loop.

In quantum science and technology, the extraction of information about a system, and the use of this information for estimation and control, is a topic of fundamental importance. The postulates of quantum mechanics specify the random nature of quantum measurements, and over a period of decades quantum measurement theory has led to a well-developed theory of quantum conditional expectation and quantum filtering, [3, 4, 7, 6, 29]. Quantum filtering theory may be used as a framework for *measurement feedback* optimal control of quantum systems, and we summarize how this is done in Section 3. In particular, we highlight the role of information states in this context. However, quantum measurement necessarily involves the loss of quantum information, which may not be desirable. Fortunately, feedback in quantum systems need not involve measurement. In fully quantum coherent feedback, the physical system being controlled, as well as the device used for the controller, are quantum systems. For instance, optical beams may be used to interconnect quantum devices and enable the transmission of quantum information from one system to another, thereby serving as "quantum wires". To my knowledge, to date there has been no extension of information states to fully quantum coherent feedback optimal control, although it has been a topic of discussion. Instead, direct methods have been employed for special situations, [21, 25]. One of the key obstacles that makes optimal fully quantum coherent feedback

---

[1]In statistics, a *statistic* is a measure of some attribute of a data sample.

control difficult is the general failure of conditioning onto non-commuting physical observables, a difficulty of fundamentally quantum mechanical origin (conditioning works successfully when measurements are used as then commuting observables are involved). Section 4 discusses a possible means for abstracting the notion of information state may provide a suitable means for approaching the solution of optimal fully quantum feedback control problems in the context of a concrete example.

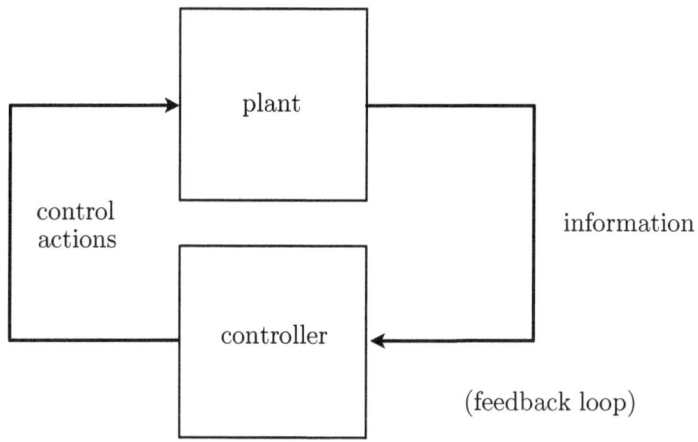

FIGURE 2. Information flow in a feedback loop.

2. Classical output feedback optimal control

In many situations, information available to the controller is often partial, and subject to noise. In this section we look at a standard scenario using stochastic models, and show how information states can be found for two types of performance criteria.

Consider the following Ito stochastic differential equation model

$$dx = f(x, u)dt + g(x)dw \tag{1}$$
$$dy = h(x)dt + dv \tag{2}$$

where (i) u is the control input signal, (ii) y is the observed output signal, (iii) x is a vector of internal state variables, and (iv) w and v are independent standard Wiener processes. Note that $x(t)$ is a Markov process (given u) with generator

$$\mathcal{L}^u(\phi) = f(\cdot, u)\phi' + \frac{1}{2}g^2\phi''$$

The system is shown schematically in Figure 3.

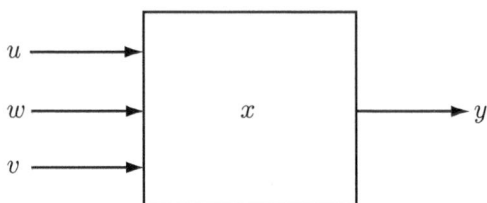

FIGURE 3. A partially observed stochastic system with control input u and observed output y. The internal state x is not directly accessible.

The control signal u is determined by the controller K using information contained in the observation signal y. The controller is to operate in real-time, so the controller is *causal*:

$u(t)$ depends on $y(s)$, $0 \leq s \leq t$

In other words, $u(t)$ is adapted to $\mathscr{Y}_t = \sigma\{y(s), 0 \leq s \leq t\}$, and we may write $u(t) = K_t(y(s), 0 \leq s \leq t)$, as in Figure 4.

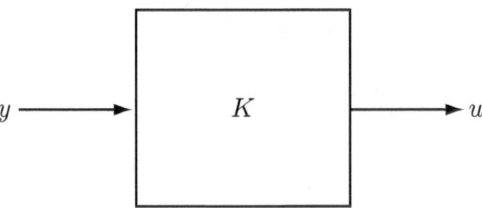

FIGURE 4. A controller maps measurement records to control actions in a causal manner.

For a controller K define the performance objective

$$J(K) = \mathbf{E}[\int_0^T L(x(s), u(s))ds + \Phi(x(T))] \tag{3}$$

where (i) $L(x, u)$ and $\Phi(x)$ are suitably chosen cost functions reflecting the desired objective (e.g., regulation to a nominal state, say 0), and (ii) $\mathbf{E}$ denotes expectation with respect to the underlying probability distributions.

The optimal control problem is to minimize $J(K)$ over all admissible controllers K. This is a *partially observed* stochastic optimal control problem: $J(K)$ is expressed in terms of the state x which is not directly accessible. In order to solve this problem, we now re-express $J(K)$ in terms of a new 'state' that is accessible.

Using basic properties of conditional expectation, we have

$$J(K) = \mathbf{E}[\int_0^T L(x(s), u(s))ds + \Phi(x(T))] \tag{4}$$

$$= \mathbf{E}[\mathbf{E}[\int_0^T L(x(s), u(s))ds + \Phi(x(T))|\mathscr{Y}_T]] \tag{5}$$

$$= \mathbf{E}[\int_0^T \tilde{L}(\pi_s, u(s))ds + \tilde{\Phi}(\pi_T)] \tag{6}$$

where π_t is the *conditional state*

$$\pi_t(\phi) = \mathbf{E}[\phi(x(t))|\mathscr{Y}_t] \tag{7}$$

and

$$\tilde{L}(\pi, u) = \pi(L(\cdot, u)), \quad \tilde{\Phi}(\pi) = \pi(\Phi). \tag{8}$$

The conditional state π_t has the following relevant properties: (i) π_t is adapted to $\mathscr{Y}_t$, (ii) the objective is expressed in terms of π_t, (iii) π_t is a Markov process (given u), with dynamics

$$d\pi_t(\phi) = \pi_t(\mathcal{L}^{u(t)}(\phi))dt + (\pi_t(\phi h) - \pi_t(\phi)\pi_t(h))(dy(t) - \pi_t(h)dt), \tag{9}$$

the equation for nonlinear filtering [9, Chapter 18]. The conditional state π_t is an example of an *information state*, [22].

An information state enables *dynamic programming* methods to be used to solve the optimization problem. Indeed, the *value function* is defined by

$$V(\pi, t) = \inf_K \mathbf{E}_{\pi,t}\left[\int_t^T \tilde{L}(\pi_s, u(s))ds + \tilde{\Phi}(\pi_T)\right], \tag{10}$$

for which the corresponding dynamic programming equation is

$$\frac{\partial}{\partial t}V(\pi, t) + \inf_u\{\tilde{\mathcal{L}}^u V(\pi, t) + \tilde{L}(\pi, u)\} = 0, \tag{11}$$

$$V(\pi, T) = \tilde{\Phi}(\pi).$$

Here, $\tilde{\mathcal{L}}^u$ is the generator for the process π_t.

If the dynamic programming equation has a suitably smooth solution, then the optimal *feedback* control function

$$\mathbf{u}^\star(\pi, t) = \operatorname{argmin}_u\{\tilde{\mathcal{L}}^u V(\pi, t) + \tilde{L}(\pi, u)\}$$

determines the optimal controller $K^\star$:

$$d\pi_t(\phi) = \pi_t(\mathcal{L}^{u(t)}(\phi))dt + (\pi_t(\phi h) - \pi_t(\phi)\pi_t(h))(dy(t) - \pi_t(h)dt) \tag{12}$$

$$u(t) = \mathbf{u}^\star(\pi_t, t) \tag{13}$$

The optimal controller $K^\star$ has the well-known *separation structure*, where the dynamical part (the filtering equation (12) for the information state π_t) is concerned with estimation, and an optimal control part $\mathbf{u}^\star$ (13), which determines control actions from the information state. In the special case of Linear-Quadratic-Gaussian control, the conditional state is Gaussian, with conditional mean and covariance given by the Kalman filter, while the optimal feedback $\mathbf{u}^\star$ is linear with the gain determined from the control LQR Riccati equation.

An alternative performance objective is the *risk-sensitive* performance objective [15, 27, 5, 19], defined for a controller K by

$$J(K) = \mathbf{E}\left[\exp\left(\mu\left\{\int_0^T L(x(s), u(s))ds + \Phi(x(T))\right\}\right)\right],\tag{14}$$

where $\mu > 0$ is a risk parameter. Due to the exponential we cannot use the conditional state as we did above. Instead, we define an unnormalized *risk-sensitive conditional state*

$$\sigma_t^\mu(\phi) = \mathbf{E}^0\left[\exp\left(\mu\left\{\int_0^t L(x(s), u(s))ds\right\}\right)\Lambda_t\phi(x(t))|\mathscr{Y}_t\right]\tag{15}$$

which includes the cost function $L(x, u)$. Here, the reference expectation is defined by

$$\mathbf{E}^0[\cdot] = \mathbf{E}[\cdot\Lambda_T^{-1}],$$

where

$$d\Lambda_t = \Lambda_t h(x(t))dy(t), \quad \Lambda_0 = 1.$$

The risk-sensitive state σ_t^μ evolves according to

$$d\sigma_t^\mu(\phi) = \sigma_t^\mu((\mathcal{L}^{u(t)} + \mu L(\cdot, u(t)))\phi)dt + \sigma_t^\mu(h)dy(t).\tag{16}$$

The performance objective may then be expressed as

$$J(K) = \mathbf{E}^0[\sigma_T^\mu(e^{\mu\Phi})].\tag{17}$$

Thus σ_t^μ is an *information state* for the risk-sensitive optimal control problem, and we may use this quantity in dynamic programming.

The *value function* for the risk-sensitive problem is defined by

$$V^\mu(\sigma, t) = \inf_K \mathbf{E}_{\sigma,t}[\sigma_T^\mu(e^{\mu\Phi})].\tag{18}$$

The corresponding dynamic programming equation is

$$\frac{\partial}{\partial t}V^\mu(\sigma, t) + \inf_u\{\tilde{\mathcal{L}}^{\mu,u}V^\mu(\sigma, t)\} = 0,$$
$$V^\mu(\sigma, T) = \sigma(\exp(\mu\Phi)),\tag{19}$$

where $\tilde{\mathcal{L}}^{\mu,u}$ is the generator for the process σ_t^μ. The optimal risk-sensitive feedback control function is

$$\mathbf{u}^{\mu,\star}(\sigma, t) = \mathrm{argmin}_u\{\tilde{\mathcal{L}}^{\mu,u}V(\sigma, t)\}\tag{20}$$

and so the *optimal risk-sensitive controller* $K^\star$ is given by

$$d\sigma_t^\mu(\phi) = \sigma_t^\mu((\mathcal{L}^{u(t)} + \mu L(\cdot, u(t)))\phi)dt + \sigma_t^\mu(h)dy(t)\tag{21}$$
$$u(t) = \mathbf{u}^{\mu,\star}(\sigma_t^\mu, t).\tag{22}$$

Again, the optimal controller consists of a dynamical equation (21) and a control function (22), but estimation is not separated from control due to the cost term appearing in the filter (21).

3. Quantum measurement feedback optimal control

In this section we consider an extension of the optimal control results of the previous section to quantum systems. A schematic representation of the *measurement feedback* system is shown in Figure 5, where the classical system K is the unknown controller to be determined.

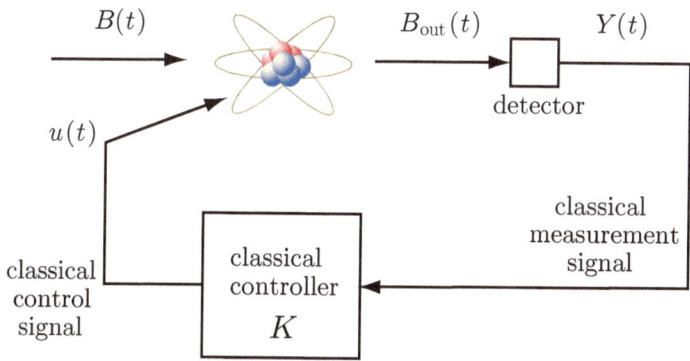

FIGURE 5. An open quantum system controlled by a classical signal $u(t)$ and interacting with a quantum field. The output component of the field is continuously monitored producing an observation process $Y(t)$.

In what follows we make use of *quantum stochastic differential equation* (QSDE) models for open quantum systems [14, 10, 26, 11], and the theory of *quantum filtering* [3, 4, 7, 6, 29]. The *state* of an open quantum system is specified by a state ρ_0 for the system (say atom) and a state for the environment, say the vacuum state Φ for the field. *Quantum expectation* $\mathbb{E}$ is given by $\mathbb{E}[X \otimes F] = \text{Tr}[(\rho_0 \otimes \Phi)(X \otimes F)] = \text{Tr}[\rho_0 X]\text{Tr}[\Phi F]$ for system operators X and field operators F. Here, ρ_0 and Φ are density operators defined on the appropriate subspaces (system and environment).

In the QSDE framework for open quantum systems, dynamical evolution is determined by the *Schrödinger* equation

$$dU(t) = \{LdB^*(t) - L^*dB(t) - (\frac{1}{2}L^*L + iH(u))dt\}U(t) \qquad (23)$$

for a unitary operator $U(t)$, where $B(t)$ is a *quantum Wiener process*. System operators X and output field $B_{\text{out}}(t)$ evolve according to the Heisenberg equations

$$X(t) = j_t(X) = U^*(t)(X \otimes I)U(t) \qquad (24)$$
$$B_{\text{out}}(t) = U^*(t)(I \otimes B(t))U(t) \qquad (25)$$

A standard measurement device (e.g., homodyne detector) is used to measure the following quadrature observable of the output field (see Figure 5):

$$Y(t) = B_{\text{out}}(t) + B_{\text{out}}^*(t). \qquad (26)$$

For each t, the operator $Y(t)$ is self-adjoint, and for different times t_1, t_2, the operators $Y(t_1)$ and $Y(t_2)$ commute, and so by the spectral theorem [6] $Y(t)$ is equivalent to a classical stochastic process (physically, a photocurrent measurement signal).

Using the quantum Ito rule, the system process $X(t) = j_t(X)$ – a *quantum Markov process* (given u) – and output process $Y(t)$ are given by

$$dj_t(X) = j_t(\mathcal{L}^{u(t)}(X))dt + dB^*(t)j_t([X, L]) + j_t([L^*, X])dB(t) \qquad (27)$$
$$dY(t) = j_t(L + L^*)dt + dB(t) + dB^*(t) \qquad (28)$$

where

$$\mathcal{L}^u(X) = -i[X, H(u)] + \frac{1}{2}L^*[X, L] + \frac{1}{2}[L^*, X]L. \qquad (29)$$

We denote by $\mathscr{Y}_t$ the *commutative* $*$-algebra of operators generated by the observation process $Y(s), 0 \leq s \leq t$. Since $j_t(X)$ commutes with all operators in $\mathscr{Y}_t$, the *quantum conditional expectation*

$$\pi_t(X) = \mathbb{E}[j_t(X)|\mathscr{Y}_t] \qquad (30)$$

is well defined. The differential equation for $\pi_t(X)$ is called the *quantum filter* [3, 4, 7, 6]:

$$d\pi_t(X) = \pi_t(\mathcal{L}^{u(t)}(X))dt \qquad (31)$$
$$+ (\pi_t(XL + L^*X) - \pi_t(X)\pi_t(L + L^*))(dY(t) - \pi_t(L + L^*)dt)$$

We now consider a quantum measurement feedback optimal control problem defined as follows. For a measurement feedback controller K define the performance objective [17][2]

$$J(K) = \mathbb{E}\left[\int_0^T C_1(s)ds + C_2(T)\right], \qquad (32)$$

where (i) $C_1(t) = j_t(C_1(u(t)))$ and $C_2(t) = j_t(C_2)$ are non-negative observables, and (ii) $\mathbb{E}$ denotes quantum expectation with respect to the underlying states for the system and field (vacuum). The *measurement feedback quantum optimal control problem* is to minimize $J(K)$ over all measurement feedback controllers K, Figure 5. Note that information about the system observables is not directly accessible, and so this is a partially observed optimal control problem.

Using standard properties of quantum conditional expectation, the performance objective can be expressed in terms of the quantum conditional state π_t as follows:

$$J(K) = \mathbb{E}\left[\int_0^T \pi_s(C_1(u(s)))ds + \pi_T(C_2)\right]. \qquad (33)$$

---

[2]Earlier formulations of quantum measurement feedback optimal control problems were specified directly in terms of conditional states [2, 8].

Then dynamic program may be used to solve this problem, as in the classical case. The *optimal measurement feedback controller* has the separation form

$$d\pi_t(X) = \pi_t(\mathcal{L}^{u(t)}(X))dt \tag{34}$$
$$+ (\pi_t(XL + L^*X) - \pi_t(X)\pi_t(L + L^*))(dY(t) - \pi_t(L + L^*)dt),$$
$$u(t) = \mathbf{u}^*(\pi_t, t), \tag{35}$$

where the feedback function $\mathbf{u}^*$ is determined from the solution to a dynamic programming equation, see [17]. Again the conditional state π_t serves as an *information state*, this time for a quantum *measurement feedback* problem.

The risk-sensitive performance criterion (14) may be extended to the present quantum context as follows, [17, 28]. Let $R(t)$ be defined by

$$\frac{dR(t)}{dt} = \frac{\mu}{2}C_1(t)R(t), \quad R(0) = I. \tag{36}$$

Then define the risk-sensitive cost to be

$$J^\mu(K) = \mathbb{E}[R^*(T)e^{\mu C_2(T)}R(T)]. \tag{37}$$

This definition accommodates in a natural way the observables in the running cost, which need not commute in general.

To solve this quantum risk-sensitive problem, we proceed as follows. Define $V(t)$ by

$$dV(t) = \left\{ LdZ(t) + \left(-\frac{1}{2}L^*L - iH(u(t)) + \frac{\mu}{2}C_1(u(t)) \right) dt \right\} V(t), \quad V(0) = I,$$

where $Z(t) = B(t) + B^*(t)$ (equivalent to a standard Wiener process with respect to the vacuum field state). The process $V(t)$ commutes with all operators in the commutative $*$-algebra $\mathscr{Z}_t$ generated by $Z(s), 0 \le s \le t$. We then have

$$J^\mu(K) = \mathbb{E}[V^*(T)e^{\mu C_2}V(T)]. \tag{38}$$

Next, define an unnormalized risk-sensitive conditional state

$$\sigma_t^\mu(X) = U^*(t)\mathbb{E}[V^*(t)XV(t)|\mathscr{Z}_t]U(t) \tag{39}$$

which evolves according to

$$d\sigma_t^\mu(X) = \sigma_t^\mu((\mathcal{L}^{u(t)} + \mu C_1(u(t)))X))dt + \sigma^\mu(XL + L^*X)dY(t) \tag{40}$$

Then we have

$$J^\mu(K) = \mathbb{E}^0[\sigma_T^\mu(e^{\mu C_2})], \tag{41}$$

and so σ_t^μ serves as an *information state*, and the optimal risk-sensitive control problem may be solved using dynamic programming.

The *optimal risk-sensitive measurement feedback controller* has the form

$$d\sigma_t^\mu(X) = \sigma_t^\mu((\mathcal{L}^{u(t)} + \mu C_1(u(t)))X))dt + \sigma_t^\mu(L + L^*)dY(t) \tag{42}$$
$$u(t) = \mathbf{u}^{\mu*}(\sigma_t^\mu, t), \tag{43}$$

where the feedback function $\mathbf{u}^{\mu*}$ is determined from the solution to a dynamic programming equation, see [16, 17].

The inclusion of a cost term in a quantum conditional state σ_t^μ appears to be new to physics, [16, 17, 28]. This state depends on (i) information gained as the system evolves (knowledge), and (ii) the objective of the closed loop feedback system (purpose).

4. Coherent quantum feedback control

An important challenge for control theory is to develop ways of *designing* signal-based coherent feedback systems in order to meet performance specifications, [30], [31], [21], [23], [25], [12], [24], [20]. While a detailed discussion of signal-based coherent feedback control design is beyond the scope of this article, we briefly describe an example from [21], [23]. In this example, the plant is a cavity with three mirrors defining three field channels. The problem was to design a coherent feedback system to minimize the influence of one input channel w on an output channel z, Figure 6. That is, if light is shone onto the mirror corresponding to the

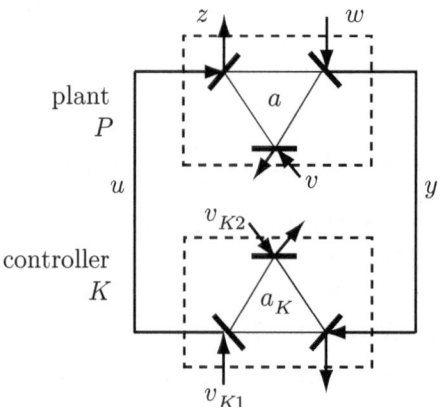

FIGURE 6. Coherent feedback control example, showing plant a and controller a_K cavity modes, together with performance quantity z and the "disturbance" input w. The coherent signals u and y are used to transfer quantum information between the plant and the controller. The feedback system was designed to minimize the intensity of the light at the output z when an optical signal is applied at the input w.

input channel w, we would like the output channel z to be dark. This is a simple example of robust control, where z may be regarded as a performance quantity (to be minimized in magnitude), while w plays the role of an external disturbance. In [21], it was shown how such problems could be solved systematically by extending methods from classical robust control theory, and importantly, taking into account the physical realization of the coherent controller as a quantum system. Indeed, the controller designed turned out to be another cavity, with mirror transmissivity parameters determined using mathematical methods. This approach was validated by experiment [23].

Classical output feedback H^∞ control problems can be solved through the use of a suitable information state, [18, 13]. However, there is no known information state for the quantum coherent H^∞ problem discussed above, and we now consider this matter more closely to see what concepts might be suitable for coherent feedback quantum control.

Referring to Figure 6, the plant P and controller K are connected by directional quantum signals u and y (beams of light). Such quantum signals may carry quantum information, and measurement need not be involved. The H^∞ objective for the feedback network $P \wedge K$ is of the form

$$\mathbb{E}_{P \wedge K} \left[V(t) - V - \int_0^t S(r) dr \right] \leq 0 \qquad (44)$$

where V is a storage function and S is an observable representing the supply rate for the input signal w and a performance variable z (see [21, 20] for general definitions of storage functions and supply rates). The storage function V is a non-negative self-adjoint operator (observable). For example, for an optical cavity we may take $V = a^* a$, where a and a^* are respectively the annihilation and creation operators of the cavity mode (note that V has spectrum $0, 1, 2, \ldots$, each value corresponds to a possible number of quanta (photons) in the cavity). A crucial difference between the fully quantum coherent feedback and the measurement feedback situation discussed in Section 3 is that the algebra of operators $\mathscr{Y}_t$ generated by the plant output process $y(s), 0 \leq s \leq t$, is not commutative in general, and so a conditioning approach cannot be expected to work.

The controller K shown in Figure 6 is an open quantum system that involves additional quantum noise inputs v_K. These additional quantum noise terms are needed to ensure that K is realizable as an open quantum system, and may be thought of as a "quantum randomization"(cf. classical randomized strategies). The controller maps quantum signals as follows:

$$K : B_{K,in} = \begin{bmatrix} y \\ v_{K1} \\ v_{K2} \end{bmatrix} \mapsto B_{K,out} = \begin{bmatrix} z_{K1} \\ u \\ z_{K2} \end{bmatrix} \qquad (45)$$

As an open system not connected to the plant P, the controller K has unitary dynamics given by a unitary operator $U_K(t)$ satisfying

$$dU_K(t) = \left\{ L_K dB_{K,in}^*(t) - L_K^\dagger dB_{K,in}(t) - \left(\frac{1}{2} L_K^\dagger L_K + i H_K \right) dt \right\} U_K(t), \qquad (46)$$
$$U_K(0) = I,$$

where $L_K = (L_{K0}, L_{K1}, L_{K2})^T$ and H_K are the physical parameters determining the controller K (an optical cavity, Figure 6). This means that the input and output fields of the controller are related by

$$B_{K,\text{out}}(t) = U_K^*(t)B_{K,\text{in}}(t)U_K(t), \tag{47}$$

while the internal controller operators X_K evolves according to

$$X_K(t) = j_{K,t}(X_K) = U_K^*(t)X_K U_K(t).$$

In particular, the control field $u(t)$ is given by

$$u(t) = U_K^*(t)v_{K1}(t)U_K(t), \tag{48}$$

or in differential form,

$$du(t) = j_{K,t}(L_{K1})dt + dv_{K1}(t) \tag{49}$$

Thus the controller K is an open quantum system specified as follows:

$$K : \begin{cases} \text{dynamics eq. (46)} \\ u(t) \text{ determined by (48) or (49)} \end{cases} \tag{50}$$

The controller K has the property that it satisfies a performance objective of the form

$$\mathbb{E}_K\left[V_K(t) - V_K - \int_0^t S_K(r)dr \right] \leq 0, \tag{51}$$

and indeed a key step in classical approaches is such a reformulation of the original objective (44). The expression (51) does not (directly) involve the plant P, and S_K is a suitable supply rate defined for the controller and the signals u and y. The expectation is with respect to a state of the controller and not the plant. Furthermore, this property ensures that, when the controller K is connected to the plant P, the feedback system $P \wedge K$ satisfies the objective (44). In this way, the open system defining the controller K serves as an *information system*, generalizing the concept of information state discussed in previous sections.

5. Conclusion

In this paper I have described how information states may be used to solve classical and quantum *measurement* feedback optimal control problems. Conditional expectation is a key mathematical tool that enables suitable information states to be defined. However, for fully quantum coherent feedback optimal control problems, the signals in the feedback loop are in general non-commutative quantum signals, and standard methods involving conditioning are not applicable. Accordingly, I suggest that a concept of *information system* abstracting the notion of information state may provide a suitable means for approaching the solution of optimal fully quantum feedback control problems. Future work will be required to develop this idea further.

References

[1] T. Basar and P. Bernhard. H^∞-Optimal Control and Related Minimax Design Problems: A Dynamic Game Approach. Birkhäuser, Boston, second edition, 1995.

[2] V.P. Belavkin. On the theory of controlling observable quantum systems. Automation and Remote Control, 44(2):178–188, 1983.

[3] V.P. Belavkin. Quantum continual measurements and a posteriori collapse on CCR. Commun. Math. Phys., 146:611–635, 1992.

[4] V.P. Belavkin. Quantum stochastic calculus and quantum nonlinear filtering. J. Multivariate Analysis, 42:171–201, 1992.

[5] A. Bensoussan and J.H. van Schuppen. Optimal control of partially observable stochastic systems with an exponential-of-integral performance index. SIAM Journal on Control and Optimization, 23:599–613, 1985.

[6] L. Bouten, R. van Handel, and M.R. James. An introduction to quantum filtering. SIAM J. Control and Optimization, 46(6):2199–2241, 2007.

[7] H. Carmichael. An Open Systems Approach to Quantum Optics. Springer, Berlin, 1993.

[8] A.C. Doherty and K. Jacobs. Feedback-control of quantum systems using continuous state-estimation. Phys. Rev. A, 60:2700, 1999.

[9] R.J. Elliott. Stochastic Calculus and Applications. Springer Verlag, New York, 1982.

[10] C.W. Gardiner and M.J. Collett. Input and output in damped quantum systems: Quantum stochastic differential equations and the master equation. Phys. Rev. A, 31(6):3761–3774, 1985.

[11] C.W. Gardiner and P. Zoller. Quantum Noise. Springer, Berlin, 2000.

[12] J. Gough and M.R. James. The series product and its application to quantum feedforward and feedback networks. IEEE Trans. Automatic Control, 54(11):2530–2544, 2009.

[13] J.W. Helton and M.R. James. Extending H^∞ Control to Nonlinear Systems: Control of Nonlinear Systems to Achieve Performance Objectives, volume 1 of Advances in Design and Control. SIAM, Philadelphia, 1999.

[14] R.L. Hudson and K.R. Parthasarathy. Quantum Ito's formula and stochastic evolutions. Commun. Math. Phys., 93:301–323, 1984.

[15] D.H. Jacobson. Optimal stochastic linear systems with exponential performance criteria and their relation to deterministic differential games. IEEE Transactions on Automatic Control, 18(2):124–131, 1973.

[16] M.R. James. Risk-sensitive optimal control of quantum systems. Phys. Rev. A, 69:032108, 2004.

[17] M.R. James. A quantum Langevin formulation of risk-sensitive optimal control. J. Optics B: Semiclassical and Quantum, Special Issue on Quantum Control, 7(10):S198–S207, 2005.

[18] M.R. James and J.S. Baras. Robust H_∞ output feedback control for nonlinear systems. IEEE Transactions on Automatic Control, 40:1007–1017, 1995.

[19] M.R. James, J.S. Baras, and R.J. Elliott. Risk-sensitive control and dynamic games for partially observed discrete-time nonlinear systems. IEEE Transactions on Automatic Control, 39:780–792, 1994.

[20] M.R. James and J. Gough. Quantum dissipative systems and feedback control design by interconnection. *IEEE Trans Auto. Control*, 55(8):1806–1821, August 2010.

[21] M.R. James, H. Nurdin, and I.R. Petersen. H^∞ control of linear quantum systems. *IEEE Trans Auto. Control*, 53(8):1787–1803, 2008.

[22] P.R. Kumar and P. Varaiya. *Stochastic Systems: Estimation, Identification and Adaptive Control*. Prentice-Hall, Englewood Cliffs, NJ, 1986.

[23] H. Mabuchi. Coherent-feedback quantum control with a dynamic compensator. *Phys. Rev. A*, 78(3):032323, 2008.

[24] H. Nurdin, M.R. James, and A.C. Doherty. Network synthesis of linear dynamical quantum stochastic systems. *SIAM J. Control and Optim.*, 48(4):2686–2718, 2009.

[25] H. Nurdin, M.R. James, and I.R. Petersen. Coherent quantum LQG control. *Automatica*, 45:1837–1846, 2009.

[26] K.R. Parthasarathy. *An Introduction to Quantum Stochastic Calculus*. Birkhäuser, Berlin, 1992.

[27] P. Whittle. Risk-sensitive linear/ quadratic/ Gaussian control. *Advances in Applied Probability*, 13:764–777, 1981.

[28] S.D. Wilson, C. D'Helon, A.C. Doherty, and M.R. James. Quantum risk-sensitive control. In *Proc. 45th IEEE Conference on Decision and Control*, pages 3132–3137, December 2006.

[29] H.M. Wiseman and G.J. Milburn. *Quantum Measurement and Control*. Cambridge University Press, Cambridge, UK, 2010.

[30] M. Yanagisawa and H. Kimura. Transfer function approach to quantum control-part I: Dynamics of quantum feedback systems. *IEEE Trans. Automatic Control*, (48):2107–2120, 2003.

[31] M. Yanagisawa and H. Kimura. Transfer function approach to quantum control-part II: Control concepts and applications. *IEEE Trans. Automatic Control*, (48):2121–2132, 2003.

M.R. James
ARC Centre for Quantum Computation
and Communication Technology
Research School of Engineering
Australian National University
Canberra, ACT 0200, Australia
e-mail: Matthew.James@anu.edu.au

Operator Theory:
Advances and Applications, Vol. 222, 247–258
© 2012 Springer Basel

Convex Hulls of Quadratically Parameterized Sets With Quadratic Constraints

Jiawang Nie

Dedicated to Bill Helton on the occasion of his 65th birthday.

Abstract. Let V be a semialgebraic set parameterized as
$$\{(f_1(x), \ldots, f_m(x)) : x \in T\}$$
for quadratic polynomials $f_0, \ldots, f_m$ and a subset T of $\mathbb{R}^n$. This paper studies semidefinite representation of the convex hull $\operatorname{conv}(V)$ or its closure, i.e., describing $\operatorname{conv}(V)$ by projections of spectrahedra (defined by linear matrix inequalities). When T is defined by a single quadratic constraint, we prove that $\operatorname{conv}(V)$ is equal to the first-order moment type semidefinite relaxation of V, up to taking closures. Similar results hold when every f_i is a quadratic form and T is defined by two homogeneous (modulo constants) quadratic constraints, or when all f_i are quadratic rational functions with a common denominator and T is defined by a single quadratic constraint, under some proper conditions.

Mathematics Subject Classification. 14P10, 90C22, 90C25.

Keywords. Convex sets, convex hulls, homogenization, linear matrix inequality, parametrization, semidefinite representation.

1. Introduction

A basic question in convex algebraic geometry is to find convex hulls of semialgebraic sets. A typical class of semialgebraic sets is parameterized by multivariate polynomial functions defined on some sets. Let $V \subset \mathbb{R}^m$ be a set parameterized as

$$V = \{(f_1(x), \ldots, f_m(x)) : x \in T\} \tag{1}$$

with every $f_i(x)$ being a polynomial and T a semialgebraic set in $\mathbb{R}^n$. We are interested in finding a representation for the convex hull $\operatorname{conv}(V)$ of V or its closure,

The research was partially supported by NSF grants DMS-0757212 and DMS-0844775.

based on $f_1, \ldots, f_m$ and T. Since V is semialgebraic, $\mathrm{conv}(V)$ is a convex semialgebraic set. Thus, one wonders whether $\mathrm{conv}(V)$ is representable by a spectrahedron or its projection, i.e., as a feasible set of *semidefinite programming (SDP)*. A *spectrahedron* of $\mathbb{R}^k$ is a set defined by a linear matrix inequality (LMI) like

$$L_0 + w_1 L_1 + \cdots + w_k L_k \succeq 0$$

for some constant symmetric matrices $L_0, \ldots, L_k$. Here the notation $X \succeq 0$ (resp. $X \succ 0$) means the matrix X is positive semidefinite (resp. definite). Equivalently, a spectrahedron is the intersection of a positive semidefinite cone and an affine linear subspace. Not every convex semialgebraic set is a spectrahedron, as found by Helton and Vinnikov [7]. Actually, they [7] proved a necessary condition called *rigid convexity* for a set to be a spectrahedron. They also proved that rigid convexity is sufficient in the two-dimensional case. Typically, projections of spectrahedra are required in representing convex sets (if so, they are also called *semidefinite representations*). It has been found that a very general class of convex sets are representable as projections of spectrahedra, as shown in [4, 5]. The proofs used sum of squares (SOS) type representations of polynomials that are positive on compact semialgebraic sets, as given by Putinar [15] or Schmüdgen [16]. More recent work about semidefinite representations of convex semialgebraic sets can be found in [6, 9, 10, 11, 12].

A natural semidefinite relaxation for the convex hull $\mathrm{conv}(V)$ can be obtained by using the moment approach [9, 13]. To describe it briefly, we consider the simple case that $n = 1$, $T = \mathbb{R}$ and $(f_1(x), f_2(x), f_3(x)) = (x^2, x^3, x^4)$ with $m = 3$. The most basic moment type semidefinite relaxation of $\mathrm{conv}(V)$ in this case is

$$R = \left\{ (y_2, y_3, y_4) : \begin{bmatrix} 1 & y_1 & y_2 \\ y_1 & y_2 & y_3 \\ y_2 & y_3 & y_4 \end{bmatrix} \succeq 0 \text{ for some } y_1 \in \mathbb{R} \right\}.$$

The underlying idea is to replace each monomial x^i by a lifting variable y_i and to pose the LMI in the definition of R, which is due to the fact that

$$\begin{bmatrix} 1 \\ x \\ x^2 \end{bmatrix} \begin{bmatrix} 1 \\ x \\ x^2 \end{bmatrix}^T = \begin{bmatrix} 1 & x & x^2 \\ x & x^2 & x^3 \\ x^2 & x^3 & x^4 \end{bmatrix} \succeq 0 \qquad \forall\, x \in \mathbb{R}.$$

If $n = 1$, the sets R and $\mathrm{conv}(V)$ (or their closures) are equal (cf. [13]). When $T = \mathbb{R}^n$ with $n > 1$, we have similar results if every f_i is quadratic or every f_i is quartic but $n = 2$ (cf. [8]). However, in more general cases, similar results typically do not exist anymore.

In this paper, we consider the special case that every f_i is quadratic and T is a quadratic set of $\mathbb{R}^n$. When T is defined by a single quadratic constraint, we will show that the first-order moment type semidefinite relaxation represents $\mathrm{conv}(V)$ or its closure as the projection of a spectrahedron (Section 2). This is also true when every f_i is a quadratic form and T is defined by two homogeneous (modulo constants) quadratic constraints (Section 3), or when all f_i are quadratic rational functions with a common denominator and T is defined by a single quadratic constraint (Section 4), under some proper conditions.

Notations. The symbol $\mathbb{R}$ (resp. $\mathbb{R}_+$) denotes the set of (resp. nonnegative) real numbers. For a symmetric matrix, $X \prec 0$ means X is negative definite $(-X \succ 0)$; $\bullet$ denotes the standard Frobenius inner product in matrix spaces; $\|\cdot\|_2$ denotes the standard 2-norm. The superscript ^{T} denotes the transpose of a matrix; $\overline{K}$ denotes the closure of a set K in a Euclidean space, and $\mathrm{conv}(K)$ denotes the convex hull of K. Given a function $q(x)$ defined on $\mathbb{R}^n$, denote

$$S(q) = \{x \in \mathbb{R}^n : q(x) \geq 0\}, \quad E(q) = \{x \in \mathbb{R}^n : q(x) = 0\}.$$

2. A single quadratic constraint

Suppose $V \subset \mathbb{R}^m$ is a semialgebraic set parameterized as

$$V = \{(f_1(x), \ldots, f_m(x)) : x \in T\} \tag{1}$$

where every $f_i(x) = a_i + b_i^T x + x^T F_i x$ is quadratic and $T \subseteq \mathbb{R}^n$ is defined by a single quadratic inequality $q(x) \geq 0$ or equality $q(x) = 0$. The a_i, b_i, F_i are vectors or symmetric matrices of proper dimensions. Similarly, write

$$q(x) = c + d^T x + x^T Q x.$$

For every $x \in T$, it always holds that for $X = xx^T$

$$f_i(x) = a_i + b_i^T x + F_i \bullet X, \quad q(x) = c + d^T x + Q \bullet X \geq 0, \quad \begin{bmatrix} 1 & x^T \\ x & X \end{bmatrix} \succeq 0.$$

Clearly, when $T = S(q)$, the convex hull $\mathrm{conv}(V)$ of V is contained in the convex set

$$\mathcal{W}_{\mathrm{in}} = \left\{ (a_1 + b_1^T x + F_1 \bullet X, \ldots, a_m + b_m^T x + F_m \bullet X) \;\middle|\; \begin{matrix} \begin{bmatrix} 1 & x^T \\ x & X \end{bmatrix} \succeq 0, \\ c + d^T x + Q \bullet X \geq 0 \end{matrix} \right\}.$$

When $T = E(q)$, the convex hull $\mathrm{conv}(V)$ is then contained in the convex set

$$\mathcal{W}_{\mathrm{eq}} = \left\{ (a_1 + b_1^T x + F_1 \bullet X, \ldots, a_m + b_m^T x + F_m \bullet X) \;\middle|\; \begin{matrix} \begin{bmatrix} 1 & x^T \\ x & X \end{bmatrix} \succeq 0, \\ c + d^T x + Q \bullet X = 0 \end{matrix} \right\}.$$

Both $\mathcal{W}_{\mathrm{in}}$ and $\mathcal{W}_{\mathrm{eq}}$ are projections of spectrahedra. One wonders whether $\mathcal{W}_{\mathrm{in}}$ or $\mathcal{W}_{\mathrm{eq}}$ is equal to $\mathrm{conv}(V)$. Interestingly, this is almost always true, as given below.

Theorem 1. *Let $V, T, \mathcal{W}_{\mathrm{in}}, \mathcal{W}_{\mathrm{eq}}, q$ be defined as above, and $T \neq \emptyset$.*

(i) *Let $T = S(q)$. If T is compact, then $\mathrm{conv}(V) = \mathcal{W}_{\mathrm{in}}$; otherwise, $\overline{\mathrm{conv}(V)} = \overline{\mathcal{W}_{\mathrm{in}}}$.*

(ii) *Let $T = E(q)$. If T is compact, then $\mathrm{conv}(V) = \mathcal{W}_{\mathrm{eq}}$; otherwise, $\overline{\mathrm{conv}(V)} = \overline{\mathcal{W}_{\mathrm{eq}}}$.*

To prove the above theorem, we need a result on quadratic moment problems. A *quadratic moment sequence* is a triple $(t, z, Z) \in \mathbb{R} \times \mathbb{R}^n \times \mathbb{R}^{n \times n}$ with Z symmetric. We say (t, z, Z) admits a representing measure supported on T if there exists a positive Borel measure μ with its support $\text{supp}(\mu) \subseteq T$ and

$$t = \int 1 \, d\mu, \quad z = \int x \, d\mu, \quad Z = \int xx^T \, d\mu.$$

Denote by $\mathscr{R}(T)$ the set of all such quadratic moment sequences (t, z, Z) satisfying the above.

Theorem 2 ([2, Theorems 4.7, 4.8]). *Let* $q(x) = c + d^T x + x^T Q x$, $T = S(q)$ *or* $E(q)$ *be nonempty, and* (t, z, Z) *be a quadratic moment sequence satisfying*

$$\begin{bmatrix} 1 & z^T \\ z & Z \end{bmatrix} \succeq 0, \quad \begin{cases} c + d^T z + Q \bullet Z \geq 0, & \text{if } T = S(q); \\ c + d^T z + Q \bullet Z = 0, & \text{if } T = E(q). \end{cases}$$

(i) *If* $S(q)$ *is compact, then* $(t, z, Z) \in \mathscr{R}(S(q))$; *otherwise,* $(t, z, Z) \in \overline{\mathscr{R}(S(q))}$.
(ii) *If* $E(q)$ *is compact, then* $(t, z, Z) \in \mathscr{R}(E(q))$; *otherwise,* $(t, z, Z) \in \overline{\mathscr{R}(E(q))}$.

Proof of Theorem 1. (i) We have already seen that $\text{conv}(V) \subseteq \mathcal{W}_{\text{in}}$, which clearly implies $\overline{\text{conv}(V)} \subseteq \overline{\mathcal{W}_{\text{in}}}$. Suppose (x, X) is a pair satisfying the conditions in $\mathcal{W}_{\text{in}}$.

If $T = S(q)$ is compact, by Theorem 2, the quadratic moment sequence $(1, x, X)$ admits a representing measure supported in T. By the Bayer-Teichmann Theorem [1], the triple $(1, x, X)$ also admits a measure having a finite support contained in T. So, there exist $u_1, \ldots, u_r \in T$ and scalars $\lambda_1 > 0, \ldots, \lambda_r > 0$ such that

$$\begin{bmatrix} 1 & x^T \\ x & X \end{bmatrix} = \lambda_1 \begin{bmatrix} 1 & u_1^T \\ u_1 & u_1 u_1^T \end{bmatrix} + \cdots + \lambda_r \begin{bmatrix} 1 & u_r^T \\ u_r & u_r u_r^T \end{bmatrix}.$$

The above implies that

$$(a_1 + b_1^T x + F_1 \bullet X, \ldots, a_m + b_m^T x + F_m \bullet X) = \sum_{i=1}^{r} \lambda_i (f_1(u_i), \ldots, f_m(u_i)).$$

Clearly, $\lambda_1 + \cdots + \lambda_r = 1$. So, $\mathcal{W}_{\text{in}} \subseteq \text{conv}(V)$ and hence $\mathcal{W}_{\text{in}} = \text{conv}(V)$.

If $T = S(q)$ is noncompact, the quadratic moment sequence $(1, x, X) \in \overline{\mathscr{R}(T)}$, and

$$(1, x, X) = \lim_{k \to \infty} (1, x^{(k)}, X^{(k)}), \quad \text{with every } (1, x^{(k)}, X^{(k)}) \in \mathscr{R}(T).$$

As we have seen in the above, every

$$(a_1 + b_1^T x^{(k)} + F_1 \bullet X^{(k)}, \ldots, a_m + b_m^T x^{(k)} + F_m \bullet X^{(k)}) \in \text{conv}(V).$$

This implies

$$(a_1 + b_1^T x + F_1 \bullet X, \ldots, a_m + b_m^T x + F_m \bullet X) \in \overline{\text{conv}(V)}.$$

So, $\mathcal{W}_{\text{in}} \subseteq \overline{\text{conv}(V)}$ and consequently $\overline{\mathcal{W}_{\text{in}}} = \overline{\text{conv}(V)}$.
(ii) can be proved in the same way as for (i). □

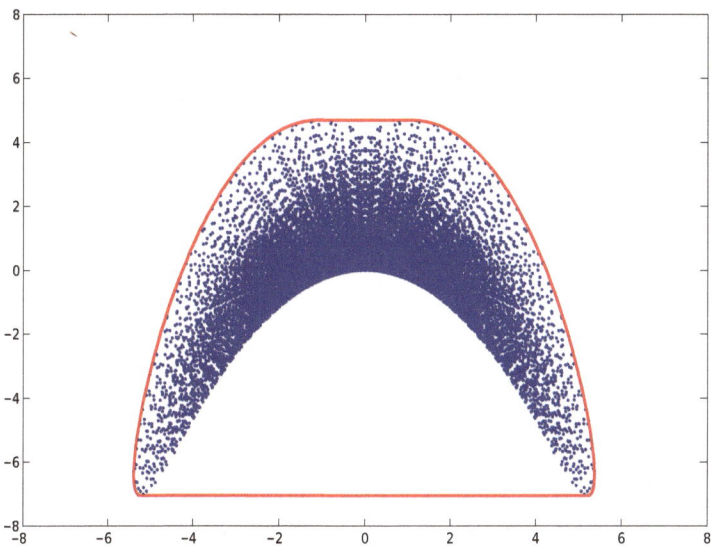

FIGURE 1. The dotted area is the set V in Example 3, and the outer curve is the boundary of the convex hull $\operatorname{conv}(V)$.

Example 3. *Consider the parametrization*

$$V = \left\{(3x_1 - 2x_2 - 4x_3, 5x_1x_2 + 7x_1x_3 - 9x_2x_3) : \|x\|_2 \leq 1\right\}.$$

The set V is drawn in the dotted area of Figure 1. By Theorem 1, the convex hull $\operatorname{conv}(V)$ is given by the semidefinite representation

$$\left\{ \begin{pmatrix} 3x_1 - 2x_2 - 4x_3 \\ 5X_{12} + 7X_{13} - 9X_{23} \end{pmatrix} \middle| \begin{bmatrix} 1 & x_1 & x_2 & x_3 \\ x_1 & X_{11} & X_{12} & X_{13} \\ x_2 & X_{12} & X_{22} & X_{23} \\ x_3 & X_{13} & X_{23} & X_{33} \end{bmatrix} \succeq 0, \\ 1 - X_{11} - X_{22} - X_{33} \geq 0 \right\}.$$

The boundary of the above set is the outer curve in Figure 1. One can easily see that $\operatorname{conv}(V)$ is correctly given by the above semidefinite representation. □

3. Two homogeneous constraints

Suppose $V \subset \mathbb{R}^m$ is a semialgebraic set parameterized as

$$V = \{(x^T A_1 x, \ldots, x^T A_m x) : x \in T\}. \tag{1}$$

Here, every A_i is a symmetric matrix and T is defined by two homogeneous (modulo constants) inequalities/equalities $h_j(x) \geq 0$ or $h_j(x) = 0$, $j = 1, 2$. Write

$$h_1(x) = x^T B_1 x - c_1, \quad h_2(x) = x^T B_2 x - c_2,$$

for symmetric matrices B_1, B_2. The set T is one of the four cases:

$$E(h_1) \cap E(h_2), \quad S(h_1) \cap E(h_2), \quad E(h_1) \cap S(h_2), \quad S(h_1) \cap S(h_2).$$

Note the relations:

$$x^T A_i x = A_i \bullet (xx^T) \quad (1 \leq i \leq m), \quad xx^T \succeq 0,$$
$$x^T B_1 x = B_1 \bullet (xx^T), \quad x^T B_2 x = B_2 \bullet (xx^T).$$

If we replace xx^T by a symmetric matrix $X \succeq 0$, then V, as well as $\mathrm{conv}(V)$, is contained respectively in the following projections of spectrahedra:

$$\begin{aligned}
\mathcal{H}_{e,e} &= \{(A_1 \bullet X, \ldots, A_m \bullet X) : X \succeq 0, B_1 \bullet X = c_1, B_2 \bullet X = c_2\}, \\
\mathcal{H}_{i,e} &= \{(A_1 \bullet X, \ldots, A_m \bullet X) : X \succeq 0, B_1 \bullet X \geq c_1, B_2 \bullet X = c_2\}, \\
\mathcal{H}_{e,i} &= \{(A_1 \bullet X, \ldots, A_m \bullet X) : X \succeq 0, B_1 \bullet X = c_1, B_2 \bullet X \geq c_2\}, \\
\mathcal{H}_{i,i} &= \{(A_1 \bullet X, \ldots, A_m \bullet X) : X \succeq 0, B_1 \bullet X \geq c_1, B_2 \bullet X \geq c_2\}.
\end{aligned} \tag{2}$$

To analyze whether they represent $\mathrm{conv}(V)$ respectively, we need the following conditions for the four cases:

$$\begin{cases}
C_{e,e} : \exists (\mu_1, \mu_2) \in \mathbb{R} \times \mathbb{R}, \ s.t. \quad \mu_1 B_1 + \mu_2 B_2 \prec 0, \\
C_{i,e} : \exists (\mu_1, \mu_2) \in \mathbb{R}_+ \times \mathbb{R}, \ s.t. \quad \mu_1 B_1 + \mu_2 B_2 \prec 0, \\
C_{e,i} : \exists (\mu_1, \mu_2) \in \mathbb{R} \times \mathbb{R}_+, \ s.t. \quad \mu_1 B_1 + \mu_2 B_2 \prec 0, \\
C_{i,i} : \exists (\mu_1, \mu_2) \in \mathbb{R}_+ \times \mathbb{R}_+, \ s.t. \quad \mu_1 B_1 + \mu_2 B_2 \prec 0.
\end{cases} \tag{3}$$

Theorem 4. *Let $V \neq \emptyset, \mathcal{H}_{e,e}, \mathcal{H}_{i,e}, \mathcal{H}_{e,i}, \mathcal{H}_{i,i}$ be defined as above. Then we have*

$$\mathrm{conv}(V) = \begin{cases}
\mathcal{H}_{e,e}, & \text{if } T = E(h_1) \cap E(h_2) \text{ and } C_{e,e} \text{ holds;} \\
\mathcal{H}_{i,e}, & \text{if } T = S(h_1) \cap E(h_2) \text{ and } C_{i,e} \text{ holds;} \\
\mathcal{H}_{e,i}, & \text{if } T = E(h_1) \cap S(h_2) \text{ and } C_{e,i} \text{ holds;} \\
\mathcal{H}_{i,i}, & \text{if } T = S(h_1) \cap S(h_2) \text{ and } C_{i,i} \text{ holds.}
\end{cases} \tag{4}$$

Proof. We just prove for the case that $T = S(h_1) \cap S(h_2)$ and condition $C_{i,i}$ holds. The proof is similar for the other three cases. The condition $C_{i,i}$ implies that there exist $\mu_1 \geq 0, \mu_2 \geq 0, \epsilon > 0$ such that for all $x \in T$

$$-\mu_1 c_1 - \mu_2 c_2 \geq x^T(-\mu_1 B_1 - \mu_2 B_2)x \geq \epsilon \|x\|_2^2.$$

So, T and $\mathrm{conv}(V)$ are compact. Clearly, $\mathrm{conv}(V) \subseteq \mathcal{H}_{i,i}$. We need to show $\mathcal{H}_{i,i} \subseteq \mathrm{conv}(V)$. Suppose otherwise it is false, then there exists a symmetric matrix Z satisfying

$$(A_1 \bullet Z, \ldots, A_m \bullet Z) \notin \mathrm{conv}(V), \quad B_1 \bullet Z \geq c_1, \quad B_2 \bullet Z \geq c_2, \quad Z \succeq 0.$$

Because $\mathrm{conv}(V)$ is a closed convex set, by the Hahn-Banach theorem, there exists a vector $(\ell_0, \ell_1, \ldots, \ell_m) \neq 0$ satisfying

$$\ell_1 x^T A_1 x + \cdots + \ell_m x^T A_m x \geq \ell_0 \quad \forall x \in T,$$
$$\ell_1 A_1 \bullet Z + \cdots + \ell_m A_m \bullet Z < \ell_0.$$

Consider the SDP problem

$$p^* := \min \quad \ell_1 A_1 \bullet X + \cdots + \ell_m A_m \bullet X$$
$$\text{s.t.} \quad X \succeq 0,\ B_1 \bullet X \geq c_1,\ B_2 \bullet X \geq c_2. \tag{5}$$

Its dual optimization problem is

$$\max \quad c_1 \lambda_1 + c_2 \lambda_2$$
$$\text{s.t.} \quad \sum_i \ell_i A_i - \lambda_1 B_1 - \lambda_2 B_2 \succeq 0,\ \lambda_1 \geq 0, \lambda_2 \geq 0. \tag{6}$$

The condition $C_{i,i}$ implies that the dual problem (6) has nonempty interior. So, the primal problem (5) has an optimizer. Define $\tilde{A}_0, \tilde{B}_1, \tilde{B}_2$ and a new variable Y as:

$$\tilde{A}_0 = \begin{bmatrix} \sum_{i=1}^{m} \ell_i A_i & 0 & 0 \\ 0 & 0 & 0 \\ 0 & 0 & 0 \end{bmatrix}, \quad \tilde{B}_1 = \begin{bmatrix} B_1 & 0 & 0 \\ 0 & -1 & 0 \\ 0 & 0 & 0 \end{bmatrix}, \tag{7}$$

$$\tilde{B}_2 = \begin{bmatrix} B_2 & 0 & 0 \\ 0 & 0 & 0 \\ 0 & 0 & -1 \end{bmatrix}, \quad Y = \begin{bmatrix} X & Y_{12} \\ Y_{12}^T & Y_{22} \end{bmatrix}. \tag{8}$$

They are all $(n+2) \times (n+2)$ symmetric matrices. Clearly, the primal problem (5) is equivalent to

$$p^* := \min \quad \tilde{A}_0 \bullet Y$$
$$\text{s.t.} \quad Y \succeq 0,\ \tilde{B}_1 \bullet Y = c_1,\ \tilde{B}_1 \bullet Y = c_2. \tag{9}$$

It must also have an optimizer. By Theorem 2.1 of Pataki [14], (9) has an extremal solution U of rank r satisfying

$$\tfrac{1}{2} r(r+1) \leq 2.$$

So, we must have $r = 1$ and can write $Y = vv^T$. Let $u = v(1:n)$. Then $u \in T$ and

$$p^* = \ell_1 u^T A_1 u + \cdots + \ell_m u^T A_m u \geq \ell_0.$$

However, Z is also a feasible solution of (5), and we get the contradiction

$$p^* \leq \ell_1 A_1 \bullet Z + \cdots + \ell_m A_m \bullet Z < p^*.$$

Therefore, $\mathcal{H}_{i,i} \subseteq \text{conv}(V)$ and they must be equal. □

Example 5. *Consider the parameterization*

$$V = \left\{ \begin{pmatrix} 2x_1^2 - 3x_2^2 - 4x_3^2 \\ 5x_1 x_2 - 7x_1 x_3 - 9x_2 x_3 \end{pmatrix} \ \middle|\ \begin{array}{l} x_1^2 - x_2^2 - x_3^2 = 0, \\ 1 - x^T x \geq 0 \end{array} \right\}.$$

The set V is drawn in the dotted area of Figure 2. By Theorem 4, the convex hull conv(V) is given by the following semidefinite representation

$$\left\{ \begin{pmatrix} 2X_{11} - 3X_{22} - 4X_{33} \\ 5X_{12} - 7X_{13} - 9X_{23} \end{pmatrix} \ \middle|\ \begin{bmatrix} X_{11} & X_{12} & X_{13} \\ X_{12} & X_{22} & X_{23} \\ X_{13} & X_{23} & X_{33} \end{bmatrix} \succeq 0,\ \begin{array}{l} X_{11} - X_{22} - X_{33} = 0, \\ 1 - X_{11} - X_{22} - X_{33} \geq 0 \end{array} \right\}.$$

The convex region described above is surrounded by the outer curve in Figure 2, which is clearly the convex hull of the dotted area. □

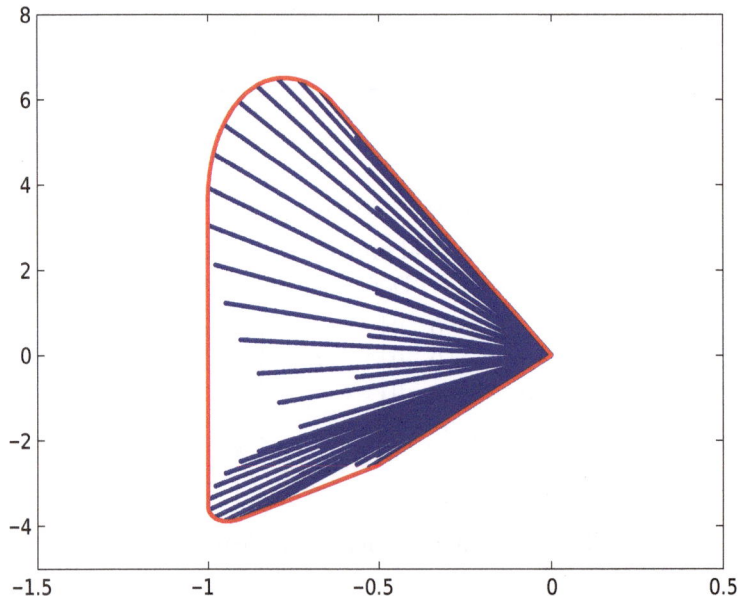

FIGURE 2. The dotted area is the set V in Example 5, and the outer curve surrounds its convex hull.

The conditions like $C_{i,i}$ can not be removed in Theorem 4. We show this by a counterexample.

Example 6. *Consider the quadratically parameterized set*

$$V = \{(x_1 x_2, x_1^2) : 1 - x_1 x_2 \geq 0, 1 + x_2^2 - x_1^2 \geq 0\},$$

which is motivated by Example 4.4 of [3]. *The condition $C_{i,i}$ is clearly not satisfied. The semidefinite relaxation $\mathcal{H}_{i,i}$ for* conv(V) *is*

$$\{(X_{12}, X_{11}) : X \succeq 0, 1 - X_{12} \geq 0, 1 + X_{22} - X_{11} \geq 0\}.$$

They are not equal, and neither are their closures. This is because V is bounded above in the direction $(1,1)$, while $\mathcal{H}_{i,i}$ is unbounded (cf. [3, *Example 4.4]). So,* $\overline{\text{conv}(V)} \neq \overline{\mathcal{H}_{i,i}}$ *for this example, which is due to the failure of the condition $C_{i,i}$.* □

4. Rational parametrization

Consider the rationally parameterized set

$$U \quad = \quad \left\{\left(\frac{f_1(x)}{f_0(x)}, \ldots, \frac{f_m(x)}{f_0(x)}\right) : x \in T\right\} \tag{1}$$

with all $f_0, \ldots, f_m$ being polynomials and T a semialgebraic set in $\mathbb{R}^n$. Assume $f_0(x)$ is nonnegative on T and every f_i/f_0 is well defined on T, i.e., the limit

$\lim_{x \to z} f_i(x)/f_0(x)$ exists whenever f_0 vanishes at $z \in T$. The convex hull $\operatorname{conv}(U)$ would be investigated through the homogenization

$$P \quad = \quad \left\{ \left(f_1^h(x^h), \ldots, f_m^h(x^h) \right) : f_0^h(x^h) = 1, \; x^h \in T^h \right\}. \tag{2}$$

Here $x^h = (x_0, x_1, \ldots, x_n)$ is an augmentation of x and

$$f_i^h(x^h) \quad = \quad x_0^d f_i(x/x_0) \qquad (d = \max_i \deg(f_i))$$

is a homogenization of $f_i(x)$, and T^h is the homogenization of T defined as

$$T^h = \overline{\{ x^h : x_0 > 0, x/x_0 \in T \}}. \tag{3}$$

The relation between $\operatorname{conv}(V)$ and $\operatorname{conv}(P)$ is given as below.

Proposition 7. *Suppose $f_0(x)$ is nonnegative on T and does not vanish on a dense subset of T, and every f_i/f_0 is well defined on T. Then*

$$\overline{\operatorname{conv}(U)} \quad = \quad \overline{\operatorname{conv}(P)}. \tag{4}$$

Moreover, if $T^h \cap \{ f_0^h(x^h) = 1 \}$ and T are compact and $f_0(x)$ is positive on T, then

$$\operatorname{conv}(U) \quad = \quad \operatorname{conv}(P). \tag{5}$$

Proof. Let T_1 be a dense subset of T such that $f_0(x) > 0$ for all $x \in T_1$. Clearly,

$$\overline{\operatorname{conv}(U)} = \overline{\operatorname{conv} \left\{ \left(\frac{f_1^h(x^h)}{f_0^h(x^h)}, \ldots, \frac{f_m^h(x^h)}{f_0^h(x^h)} \right) : x^h \in T_1^h \right\}}.$$

Since every f_i^h is homogeneous, we can scale such that $f_0^h(x^h) = 1$. Then,

$$\overline{\operatorname{conv}(U)} = \overline{\operatorname{conv} \left\{ \left(f_1^h(x^h), \ldots, f_m^h(x^h) \right) : f_0^h(x^h) = 1, \; x^h \in T_1^h \right\}}.$$

The density of T_1 in T and the above imply (4).

When T is compact and $f_0(x)$ is positive on T, $\operatorname{conv}(U)$ is compact. The $\operatorname{conv}(P)$ is also compact when $T^h \cap \{ f_0^h(x^h) = 1 \}$ is compact. Thus, (5) follows from (4). □

Remark: Suppose $d = \max_i \deg(f_i)$ is even and T is defined by polynomials of even degrees. If

$$T^h \cap \{ f_0^h(x^h) = 1 \} = \overline{\{ x^h : x_0 > 0, \; f_0^h(x^h) = 1, \; x/x_0 \in T \}},$$

then we can remove the condition $x_0 > 0$ in the definition of T^h in (3) and Proposition 7 still holds.

If every f_i in (1) is quadratic, T is defined by a single quadratic inequality, and f_0 is nonnegative on T, then a semidefinite representation for the convex hull $\operatorname{conv}(U)$ or its closure can be obtained by applying Proposition 7 and Theorem 4.

Suppose $T = \{x : g(x) \geq 0\}$, with $g(x)$ being quadratic. Write every $f_i^h(x^h) = (x^h)^T F_i\, x^h$ and $g^h(x^h) = (x^h)^T G\, x^h$. Then

$$\overline{\mathrm{conv}(P)} = \mathrm{conv}\left\{\left((x^h)^T F_1\, x^h, \ldots, (x^h)^T F_m\, x^h\right) : \begin{array}{c} (x^h)^T F_0\, x^h = 1, \\ x_0 > 0, (x^h)^T G\, x^h \geq 0 \end{array}\right\}.$$

(6)

The forms f_i^h and g^h are all quadratic. If

$$\{x^h : (x^h)^T F_0 x^h = 1, \ (x^h)^T G x^h \geq 0\}$$
$$= \overline{\{x^h : x_0 > 0, \ (x^h)^T F_0 x^h = 1, \ (x^h)^T G x^h \geq 0\}},$$

then the condition $x_0 > 0$ can be removed from the right-hand side of (6), and we get

$$\overline{\mathrm{conv}(P)} = \mathrm{conv}\left\{\left((x^h)^T F_1\, x^h, \ldots, (x^h)^T F_m\, x^h\right) : \begin{array}{c} (x^h)^T F_0\, x^h = 1, \\ (x^h)^T G\, x^h \geq 0 \end{array}\right\}.$$

(7)

If there exist $\mu_1 \in \mathbb{R}$ and $\mu_2 \in \mathbb{R}_+$ satisfying $\mu_1 F_0 + \mu_2 G \prec 0$, then a semidefinite representation for $\overline{\mathrm{conv}(P)}$ can be obtained by applying Theorem 4. The case $T = \{x : g(x) = 0\}$ is defined by a single quadratic equality is similar.

Example 8. *Consider the quadratically rational parametrization:*

$$U = \left\{\left(\frac{x_1^2 + x_2^2 + x_3^2 + x_1 + x_2 + x_3}{1 + x^T x}, \frac{x_1 x_2 + x_1 x_3 + x_2 x_3}{1 + x^T x}\right) : x_1^2 + x_2^2 + x_3^2 \leq 1\right\}.$$

The dotted area in Figure 2 is the set U above. The set P in (2) is

$$P = \left\{\left(\begin{array}{c} x_1^2 + x_2^2 + x_3^2 + x_0(x_1 + x_2 + x_3) \\ x_1 x_2 + x_1 x_3 + x_2 x_3 \end{array}\right) \,\middle|\, \begin{array}{c} x_0^2 + x_1^2 + x_2^2 + x_3^2 = 1, \\ x_0^2 - x_1^2 - x_2^2 - x_3^2 \geq 0 \end{array}\right\}.$$

By Theorem 4, the convex hull $\mathrm{conv}(P)$ is given by the semidefinite representation

$$\left\{\left(\begin{array}{c} X_{11} + X_{22} + X_{33} + X_{01} + X_{02} + X_{03} \\ X_{12} + X_{13} + X_{23} \end{array}\right) \,\middle|\, \begin{array}{l} \begin{bmatrix} X_{00} & X_{01} & X_{02} & X_{03} \\ X_{01} & X_{11} & X_{12} & X_{13} \\ X_{02} & X_{12} & X_{22} & X_{23} \\ X_{03} & X_{13} & X_{23} & X_{33} \end{bmatrix} \succeq 0, \\ X_{00} + X_{11} + X_{22} + X_{33} = 1, \\ X_{00} - X_{11} - X_{22} - X_{33} \geq 0 \end{array}\right\}.$$

The convex region described above is surrounded by the outer curve in Figure 3, which also surrounds the convex hull of the dotted area. Since T is compact and the denominator $1+x^T x$ is strictly positive, $\mathrm{conv}(U) = \mathrm{conv}(P)$ by Proposition 7. □

References

[1] C. Bayer and J. Teichmann. The proof of Tchakaloff's Theorem. *Proc. Amer. Math. Soc.*, 134(2006), 3035–3040.

[2] L. Fialkow and J. Nie. Positivity of Riesz functionals and solutions of quadratic and quartic moment problems. *J. Functional Analysis*, Vol. 258, No. 1, pp. 328–356, 2010.

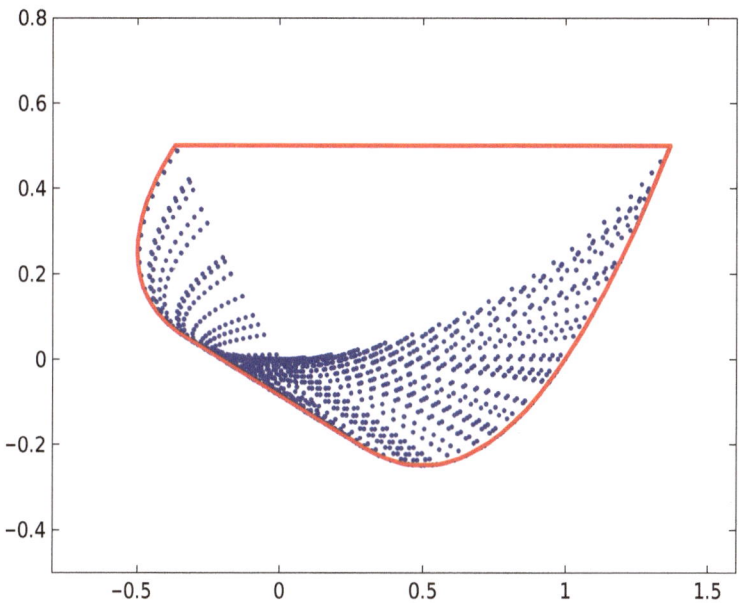

FIGURE 3. The dotted area is the set U in Example 8, and the outer curve is the boundary of its convex hull.

[3] S. He, Z. Luo, J. Nie and S. Zhang. Semidefinite Relaxation Bounds for Indefinite Homogeneous Quadratic Optimization. *SIAM Journal on Optimization*, Vol. 19, No. 2, pp. 503–523, 2008.

[4] J.W. Helton and J. Nie. Semidefinite representation of convex sets. *Mathematical Programming*, Ser. A, Vol. 122, No. 1, pp. 21–64, 2010.

[5] J.W. Helton and J. Nie. Sufficient and necessary conditions for semidefinite representability of convex hulls and sets. *SIAM Journal on Optimization*, Vol. 20, No. 2, pp. 759–791, 2009.

[6] J.W. Helton and J. Nie. Structured semidefinite representation of some convex sets. *Proceedings of 47th IEEE Conference on Decision and Control*, pp. 4797–4800, Cancun, Mexico, Dec. 9–11, 2008.

[7] W. Helton and V. Vinnikov. Linear matrix inequality representation of sets. *Comm. Pure Appl. Math.* 60 (2007), No. 5, pp. 654–674.

[8] D. Henrion. Semidefinite representation of convex hulls of rational varieties. *Acta Applicandae Mathematicae*, Vol. 115, No. 3, pp. 319–327, 2011.

[9] J. Lasserre. Convex sets with semidefinite representation. *Mathematical Programming*, Vol. 120, No. 2, pp. 457–477, 2009.

[10] J. Lasserre. Convexity in semi-algebraic geometry and polynomial optimization. *SIAM Journal on Optimization*, Vol. 19, No. 4, pp. 1995–2014, 2009.

[11] J. Nie. First order conditions for semidefinite representations of convex sets defined by rational or singular polynomials. *Mathematical Programming*, Ser. A, Vol. 131, No. 1, pp. 1–36, 2012.

[12] J. Nie. Polynomial matrix inequality and semidefinite representation. *Mathematics of Operations Research*, Vol. 36, No. 3, pp. 398–415, 2011.

[13] P. Parrilo. Exact semidefinite representation for genus zero curves. Talk at the Banff workshop "Positive Polynomials and Optimization", Banff, Canada, October 8–12, 2006.

[14] G. Pataki. On the rank of extreme matrices in semidefinite programs and the multiplicity of optimal eigenvalues. *Mathematics of Operations Research*, 23 (2), 339–358, 1998.

[15] M. Putinar. Positive polynomials on compact semi-algebraic sets, *Ind. Univ. Math. J.* 42 (1993) 203–206.

[16] K. Schmüdgen. The K-moment problem for compact semialgebraic sets. *Math. Ann.* **289** (1991), 203–206.

Jiawang Nie
Department of Mathematics
University of California
9500 Gilman Drive
La Jolla, CA 92093, USA
e-mail: njw@math.ucsd.edu

Operator Theory:
Advances and Applications, Vol. 222, 259–277
© 2012 Springer Basel

Computing Linear Matrix Representations of Helton-Vinnikov Curves

Daniel Plaumann, Bernd Sturmfels and Cynthia Vinzant

Dedicated to Bill Helton on the occasion of his 65th birthday

Abstract. Helton and Vinnikov showed that every rigidly convex curve in the real plane bounds a spectrahedron. This leads to the computational problem of explicitly producing a symmetric (positive definite) linear determinantal representation for a given curve. We study three approaches to this problem: an algebraic approach via solving polynomial equations, a geometric approach via contact curves, and an analytic approach via theta functions. These are explained, compared, and tested experimentally for low degree instances.

Mathematics Subject Classification. Primary: 14Q05; secondary: 14K25.

Keywords. Plane curves, symmetric determinantal representations, spectrahedra, linear matrix inequalities, hyperbolic polynomials, theta functions.

1. Introduction

The Helton-Vinnikov Theorem [16] gives a geometric characterization of two-dimensional spectrahedra. They are precisely the subsets of $\mathbb{R}^2$ that are bounded by rigidly convex algebraic curves, here called *Helton-Vinnikov curves*. These curves are cut out by *hyperbolic polynomials* in three variables, as discussed in [18]. This theorem is a refinement of a result from classical algebraic geometry which states that every homogeneous polynomial in three variables can be written as

$$f(x,y,z) = \det(Ax + By + Cz) \qquad (1)$$

where A, B and C are symmetric matrices. Here the coefficients of f and the matrix entries are complex numbers. When the coefficients of f are real then it is desirable to find A, B and C with real entries. The representations relevant for spectrahedra

Daniel Plaumann was supported by the Alexander-von-Humboldt Foundation through a Feodor Lynen postdoctoral fellowship, hosted at UC Berkeley. Bernd Sturmfels and Cynthia Vinzant acknowledge support by the U.S. National Science Foundation (DMS-0757207 and DMS-0968882).

are the *real definite representations*, which means that the linear span of the real matrices A, B and C contain a positive definite matrix. Such a representation is possible if and only if the corresponding curve $\{(x : y : z) \in \mathbb{P}^2_{\mathbb{R}} : f(x, y, z) = 0\}$ is *rigidly convex*. This condition means that the curve has the maximal number of nested ovals, namely, there are $d/2$ resp. $(d-1)/2$ nested ovals when the degree d of f is even resp. odd. The innermost oval bounds a spectrahedron.

Two linear matrix representations $Ax + By + Cz$ and $A'x + B'y + C'z$ of the same plane curve are said to be *equivalent* if they lie in the same orbit under conjugation, i.e., if there exists an invertible complex matrix U that satisfies

$$U \cdot (Ax + By + Cz) \cdot U^T \quad = \quad A'x + B'y + C'z.$$

We call an equivalence class of complex representations *real* (resp. *real definite*) if it contains a real (resp. real definite) representative. Deciding whether a given complex representation is equivalent to a real or real definite one is rather difficult.

We shall see that the number of equivalence classes of complex representations (1) is finite, and, for smooth curves, the precise number is known (Thm. 1). Using more general results of Vinnikov [24], we also derive the number of real and real definite equivalence classes. If a Helton-Vinnikov curve is smooth then the number of real definite equivalence classes equals 2^g, where $g = \binom{d-1}{2}$ is the genus.

This paper concerns the computational problem of constructing one representative from each equivalence class for a given polynomial $f(x, y, z)$. As a warm-up example, consider the following elliptic curve in Weierstrass normal form:

$$f(x, y, z) \quad = \quad (x + ay)(x + by)(x + cy) - xz^2.$$

Here a, b, c are distinct non-zero reals. This cubic has precisely three inequivalent linear symmetric determinantal representations over $\mathbb{C}$, given by the matrices

$$\begin{bmatrix} x + ay & z\sqrt{\frac{b}{b-c}} & z\sqrt{\frac{c}{c-b}} \\ z\sqrt{\frac{b}{b-c}} & x + by & 0 \\ z\sqrt{\frac{c}{c-b}} & 0 & x + cy \end{bmatrix}, \begin{bmatrix} x + ay & z\sqrt{\frac{a}{a-c}} & 0 \\ z\sqrt{\frac{a}{a-c}} & x + by & z\sqrt{\frac{c}{c-a}} \\ 0 & z\sqrt{\frac{c}{c-a}} & x + cy \end{bmatrix}, \begin{bmatrix} x + ay & 0 & z\sqrt{\frac{a}{a-b}} \\ 0 & x + by & z\sqrt{\frac{b}{b-a}} \\ z\sqrt{\frac{a}{a-b}} & z\sqrt{\frac{b}{b-a}} & x + cy \end{bmatrix}$$

All three matrices are non-real if a, b and c have the same sign, and otherwise two of the matrices are real. For instance, if $a < 0$ and $0 < b < c$ then the first two matrices are real. In that case, the cubic is a Helton-Vinnikov curve, and its bounded region, when drawn in the affine plane $\{x = 1\}$, is the spectrahedron

$$\left\{ (y, z) \in \mathbb{R}^2 \ : \ \begin{bmatrix} 1 + ay & z\sqrt{\frac{a}{a-c}} & 0 \\ z\sqrt{\frac{a}{a-c}} & 1 + by & z\sqrt{\frac{c}{c-a}} \\ 0 & z\sqrt{\frac{c}{c-a}} & 1 + cy \end{bmatrix} \succeq 0 \right\}.$$

The symbol "$\succeq$" means that the matrix is positive semidefinite. This spectrahedron is depicted in Figure 1 for the parameter values $a = -1$, $b = 1$ and $c = 2$.

This article is organized as follows. In Section 2 we translate (1) into a system of polynomial equations in the matrix entries of A, B, C, we determine the

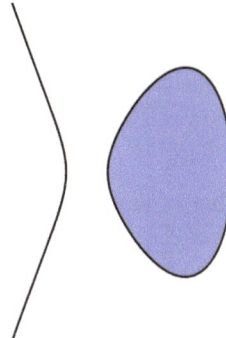

FIGURE 1. A cubic Helton-Vinnikov curve and its spectrahedron

number of solutions (in Theorem 1), and we discuss practical aspects of computing these solutions using both symbolic and numeric software. Section 3 is devoted to geometric constructions for obtaining the representation (1). Following Dixon [10], these require finding contact curves of degree $d-1$ for the given curve of degree d.

An explicit formula for (1) appears in the article of Helton and Vinnikov [16, Eq. 4.2]. That formula requires the numerical evaluation of Abelian integrals and theta functions. In Section 4, we explain the Helton-Vinnikov formula, and we report on our computational experience with the implementations of [7, 8, 9] in the Maple package algcurves. In Section 5 we focus on the case of quartic polynomials and relate our results in [21] to the combinatorics of theta functions. Smooth quartics have 36 inequivalent representations (1). In the Helton-Vinnikov case, twelve of these are real, but only eight are real definite. One of our findings is an explicit quartic in $\mathbb{Q}[x, y, z]$ that has all of its 12 real representations over $\mathbb{Q}$.

2. Solving polynomial equations

Our given input is a homogeneous polynomial $f(x, y, z)$ of degree d, usually over $\mathbb{Q}$. We assume for simplicity that the corresponding curve in the complex projective plane is smooth, we normalize so that $f(x, 0, 0) = x^d$, and we further assume that the factors of the binary form $f(x, y, 0) = \prod_{i=1}^{d}(x + \beta_i y)$ are distinct. Under these hypotheses, every equivalence class of representations (1) contains a representative where A is the identity matrix and B is the diagonal matrix with entries $\beta_1 < \beta_2 < \cdots < \beta_d$. This follows from the linear algebra fact that any two quadratic forms with distinct eigenvalues can be diagonalized simultaneously over $\mathbb{C}$. See Section IX.3 in Greub's text book [13] or the proof of Theorem 4.3 in [21].

After fixing the choices $A = \operatorname{diag}(1, 1, \ldots, 1)$ and $B = \operatorname{diag}(\beta_1, \beta_2, \ldots, \beta_d)$ for the first two matrices, we are left with the problem of finding the $\binom{d+1}{2}$ entries of the symmetric matrix $C = (c_{ij})$. By equating the coefficients of all terms $x^\alpha y^\beta z^\gamma$ with $\gamma \geq 1$ on both sides of (1), we obtain a system of $\binom{d+1}{2}$ polynomial equations in the $\binom{d+1}{2}$ unknowns c_{ij}. More precisely, the coefficient of $x^\alpha y^\beta z^\gamma$ in (1) leads to an equation of degree γ in the c_{ij}. We are thus faced with the problem of solving a square system of polynomial equations. The expected number of complex solutions of that system is, according to Bézout's Theorem,

$$1^d \cdot 2^{d-1} \cdot 3^{d-2} \cdot 4^{d-3} \cdot \ldots \cdot (d-1)^2 \cdot d. \tag{2}$$

This estimate overcounts the number of equivalence classes of representations (1) because we can conjugate the matrix $Ax + By + Cz$ by a diagonal matrix whose entries are $+1$ or -1. This conjugation does not change A or B but it leads to 2^{d-1} distinct matrices C all of which are equivalent. Hence, we can expect the number of inequivalent linear matrix representations (1) to be bounded above by

$$3^{d-2} \cdot 4^{d-3} \cdot \ldots \cdot (d-1)^2 \cdot d. \tag{3}$$

We shall refer to this number as the *Bézout bound* for our problem.

It is a result in classical algebraic geometry that the number of complex solutions to our equations is finite, and the precise number of solutions is in fact known as well. The following theorem summarizes both what is known for arbitrary smooth curves over $\mathbb{C}$ and what can be shown for Helton-Vinnikov curves over $\mathbb{R}$:

Theorem 1. *The number of equivalence classes of linear symmetric determinantal representations* (1) *of a generic smooth curve of degree d in the projective plane is*

$$2^{\binom{d-1}{2}-1} \cdot \left(2^{\binom{d-1}{2}} + 1\right), \tag{4}$$

unless $d \geq 11$ and d is congruent to ± 3 modulo 8, when the number drops by one. In the case of a Helton-Vinnikov curve, the number of real equivalence classes of symmetric linear determinantal representations (1) *is either $2^{\binom{d-1}{2}-1}(2^{\lceil \frac{d}{2}\rceil-1}+1)$ or one less. The number of real definite equivalence classes is precisely $2^{\binom{d-1}{2}}$.*

Sketch of Proof. The equivalence classes of representations (1) correspond to *ineffective even theta characteristics* [3] on a smooth curve of genus $g = \binom{d-1}{2}$. The number of even theta characteristics is $2^{g-1}(2^g + 1)$, and all even theta characteristics are ineffective for $d \leq 5$ and $d \equiv 0, 1, 2, 4, 6, 7 \bmod 8$. In all other cases there is precisely one effective even theta characteristic, provided the curve is generic. This was shown by Meyer-Brandis in his 1998 diploma thesis [19], and it refines results known classically in algebraic geometry [11, Chapters 4–5]. The count of real and real definite representations will be proved at the end of Section 4. □

The following table lists the numbers in (3) and (4) for small values of d:

degree d	2	3	4	5	6	7
genus g	0	1	3	6	10	15
Bézout bound	1	3	36	2160	777600	1959552000
True number	1	3	36	2080	524800	536887296

This table shows that computing all solutions to our equations is a challenge when $d \geq 6$. Below we shall discuss some computer experiments we conducted for $d \leq 5$.

As before, we fix A to be the $d \times d$ identity matrix, denoted Id_d, and we fix B to be the diagonal matrix with entries $\beta_1 < \cdots < \beta_d$. We also fix the diagonal entries of C since these are determined by solving the d linear equations that arise by comparing the coefficient of any of the d monomials $x^i y^{d-i-1} z$ in (1). They are expressed in terms of f and the β_i by the following explicit formula:

$$c_{ii} \;=\; \beta_i \cdot \frac{\frac{\partial f}{\partial z}(-\beta_i, 1, 0)}{\frac{\partial f}{\partial y}(-\beta_i, 1, 0)} \qquad \text{for } i = 1, 2, \ldots, d. \tag{5}$$

We are thus left with a system of $\binom{d}{2}$ equations in the $\binom{d}{2}$ off-diagonal unknowns c_{ij}. In order to remove the extraneous factor of 2^{d-1} in the Bézout bound (2) coming from sign changes on the rows and columns of C, we can perform a multiplicative change of coordinates as follows: $x_{1j} = c_{1j}^2$ for $j = 1, 2, \ldots, d$ and $x_{ij} = c_{1i} c_{1j} c_{ij}$ for $2 \le i < j \le d$. This translates our system of polynomial equations in the c_{ij} into a system of Laurent polynomial equations in the x_{ij}, and each solution to the latter encodes an equivalence class of 2^{d-1} solutions to the former.

Example 2. Let $d = 4$. We shall illustrate the two distinct formulations of the system of equations to be solved. We fix a quartic Helton-Vinnikov polynomial

$$f(x,y,z) \;=\; \det \begin{bmatrix} x + \beta_1 y + \gamma_{11} z & \gamma_{12} z & \gamma_{13} z & \gamma_{14} z \\ \gamma_{12} z & x + \beta_2 y + \gamma_{22} z & \gamma_{23} z & \gamma_{24} z \\ \gamma_{13} z & \gamma_{23} z & x + \beta_3 y + \gamma_{33} z & \gamma_{34} z \\ \gamma_{14} z & \gamma_{24} z & \gamma_{34} z & x + \beta_4 y + \gamma_{44} z \end{bmatrix}$$

where β_i and γ_{jk} are rational numbers. From the quartic $f(x,y,z)$ alone we can recover the β_i and the diagonal entries γ_{jj} as described above. Our aim is now to compute **all** points $(c_{12}, c_{13}, c_{14}, c_{23}, c_{24}, c_{34}) \in \mathbb{C}^6$ that satisfy the identity

$$\det \begin{bmatrix} x + \beta_1 y + \gamma_{11} z & c_{12} z & c_{13} z & c_{14} z \\ c_{12} z & x + \beta_2 y + \gamma_{22} z & c_{23} z & c_{24} z \\ c_{13} z & c_{23} z & x + \beta_3 y + \gamma_{33} z & c_{34} z \\ c_{14} z & c_{24} z & c_{34} z & x + \beta_4 y + \gamma_{44} z \end{bmatrix} = f(x,y,z).$$

The coefficient of z^4 gives one equation of degree 4 in the six unknowns c_{ij}, the coefficients of xz^3 and yz^3 give two cubic equations, and the coefficient of $x^2 z^2, xyz^2$ and $y^2 z^2$ give three quadratic equations in the c_{ij}. The number of solutions in $\mathbb{C}^6$ to this system of equations is equal to $2^3 3^2 4 = 288$. These solutions can be found using symbolic software, such as Singular [6]. However, the above formulation has the disadvantage that each equivalence class of solutions appears eight times.

We note that, for generic choices of β_i, γ_{jk}, all solutions lie in the torus $(\mathbb{C}^*)^6$ where $\mathbb{C}^* = \mathbb{C} \backslash \{0\}$, and we shall now assume that this is the case. Then the 8-fold redundancy can be removed by working with the following invariant coordinates:

$$x_{12} = c_{12}^2, \;\; x_{13} = c_{13}^2, \;\; x_{14} = c_{14}^2,$$

$$x_{23} = c_{12} c_{13} c_{23}, \;\; x_{24} = c_{12} c_{14} c_{24}, \;\; x_{34} = c_{13} c_{14} c_{34}.$$

We rewrite our six equations in these coordinates by performing the substitution:

$$c_{12} = x_{12}^{1/2} , \ c_{13} = x_{13}^{1/2} , \ c_{14} = x_{14}^{1/2} ,$$

$$c_{23} = \frac{x_{23}}{x_{12}^{1/2} x_{13}^{1/2}} , \ c_{24} = \frac{x_{24}}{x_{12}^{1/2} x_{14}^{1/2}} , \ c_{34} = \frac{x_{34}}{x_{13}^{1/2} x_{14}^{1/2}} .$$

This gives six Laurent polynomial equations in six unknowns x_{12}, x_{13}, x_{14}, x_{23}, x_{24}, x_{34}. They have precisely 36 solutions in $(\mathbb{C}^*)^6$, one for each equivalence class.   □

   While the solution of the above equations using symbolic Gröbner-based software is easy for $d = 4$, we found that this is no longer the case for $d \geq 5$. For $d = 5$, it was necessary to employ tools from numerical algebraic geometry, and we found that Bertini [4] works well for our purpose. The computation reported below is due to Charles Chen, an undergraduate student at UC Berkeley. This was part of Chen's term project in convex algebraic geometry during Fall 2010.

   For a concrete example, let us consider the following polynomial which defines a smooth Helton-Vinnikov curve of degree $d = 5$:

$$f(x,y,z) = x^5 + 3x^4y - 2x^4z - 5x^3y^2 - 12x^3z^2 - 15x^2y^3 + 10x^2y^2z - 28x^2yz^2 + 14x^2z^3$$

$$+ 4xy^4 - 6xy^2z^2 - 12xyz^3 + 26xz^4 + 12y^5 - 8y^4z - 32y^3z^2 + 16y^2z^3 + 48yz^4 - 24z^5.$$

The symmetric linear determinantal representation we seek has the form

$$\begin{bmatrix} x+y & 0 & 0 & 0 & 0 \\ 0 & x+2y & 0 & 0 & 0 \\ 0 & 0 & x-y & 0 & 0 \\ 0 & 0 & 0 & x-2y & 0 \\ 0 & 0 & 0 & 0 & x+3y-2z \end{bmatrix} + C \cdot z,$$

where $C = (c_{ij})$ is an unknown symmetric 5×5-matrix with zeros on the diagonal. This leads to a system of 10 polynomial equations in the 10 unknowns c_{ij}, namely, 4 quadrics, 3 cubics, 2 quartics and one quintic. The number of complex solutions equals $16 \cdot 2080 = 33280$, which is less than the Bézout bound of $2^4 \cdot 3^3 \cdot 4^2 \cdot 5 = 16 \cdot 2160 = 34560$. One of the 33280 solutions is the following integer matrix, which we had used to construct $f(x, y, z)$ in the first place:

$$C = \begin{bmatrix} 0 & 2 & 1 & 0 & 0 \\ 2 & 0 & 0 & 0 & 1 \\ 1 & 0 & 0 & 2 & 1 \\ 0 & 0 & 2 & 0 & -1 \\ 0 & 1 & 1 & -1 & 0 \end{bmatrix}$$

Of course, the other 15 matrices in the same equivalence class have the same friendly integer entries. The other $16 \cdot 2079 = 33264$ complex solutions were found numerically using the software Bertini [4]. Of these, $16 \cdot 63$ are real. Chen's code, based on Bertini, outputs one representative per class. One of the real solutions is

$$C \approx \begin{bmatrix} 0 & 1.8771213868 & 0.1333876113 & 0.3369345269 & 0.2151885297 \\ 1.8771213868 & 0 & 1.3262201851 & 0.1725327846 & 1.0570303927 \\ 0.1333876113 & 1.3262201851 & 0 & -2.0093203944 & -0.8767796987 \\ 0.3369345269 & 0.1725327846 & -2.0093203944 & 0 & -0.7659896773 \\ 0.2151885297 & 1.0570303927 & -0.8767796987 & -0.7659896773 & 0 \end{bmatrix}.$$

3. Constructing contact curves

A result in classical algebraic geometry states that the equivalence classes of symmetric linear determinantal representations of a plane curve of degree d are in one-to-one correspondence with certain systems of contact curves of degree $d - 1$. Following [11, Prop. 4.1.6], we now state this in precise terms. Suppose that our given polynomial is $f = \det(M)$ where $M = (\ell_{ij})$ is a symmetric $d \times d$-matrix of linear forms in x, y, z. We can then form the $d \times d$ adjoint matrix, $\mathrm{adj}(M)$, whose entry m_{ij} is the (i,j)th $(d-1)$-minor of M multiplied by $(-1)^{i+j}$. For any vector of parameters $u = (u_1, u_2, \ldots, u_d)^T$, we consider the degree $d-1$ polynomial

$$
g_u(x,y,z) \;=\; u^T \mathrm{adj}(M) u \;=\;
\begin{bmatrix} u_1 \\ u_2 \\ \vdots \\ u_d \end{bmatrix}^T
\begin{bmatrix} m_{11} & m_{12} & \cdots & m_{1d} \\ m_{12} & m_{22} & \cdots & m_{2d} \\ \vdots & \vdots & \ddots & \vdots \\ m_{1d} & m_{2d} & \cdots & m_{dd} \end{bmatrix}
\begin{bmatrix} u_1 \\ u_2 \\ \vdots \\ u_d \end{bmatrix}. \tag{6}
$$

The curve $\mathcal{V}(g_u)$ has degree $d-1$, and it is a *contact curve*, which means that all intersection points of $\mathcal{V}(f)$ and $\mathcal{V}(g_u)$ have even multiplicity, generically multiplicity 2. To see this, we use [11, Lemma 4.1.7], which states that, for any $u, v \in \mathbb{C}^d$,

$$
g_u(x,y,z) \cdot g_v(x,y,z) \;-\; (u^T \mathrm{adj}(M) v)^2 \;\in\; \langle f \rangle \quad \text{in } \mathbb{C}[x,y,z].
$$

In particular, for $u = e_i$, $v = e_j$, this shows that both $\mathcal{V}(m_{ii})$ and $\mathcal{V}(m_{jj})$ are contact curves, and $\mathcal{V}(m_{ij})$ meets $\mathcal{V}(f)$ in their $d(d-1)$ contact points.

We say that two contact curves $\mathcal{V}(g_1)$ and $\mathcal{V}(g_2)$ of degree r lie in the same *system* if there exists another curve $\mathcal{V}(h)$ of degree r that meets $\mathcal{V}(f)$ precisely in the $r \cdot d$ points $\mathcal{V}(f, g_1) \cup \mathcal{V}(f, g_2)$. A system of contact curves is called *syzygetic* if it contains a polynomial of the form $\ell^2 g$, where ℓ is linear and g is a contact curve of degree $r-2$, and *azygetic* otherwise. A contact curve of $\mathcal{V}(f)$ is called syzygetic, resp. azygetic, if it lies in a system that is syzygetic, resp. azygetic.

Dixon [10] proved that the contact curves $\mathcal{V}(g_u)$ are azygetic and all azygetic contact curves of degree $d-1$ appear as g_u for some determinantal representation $f = \det(M)$. In particular, he gives a method of constructing a determinantal representation M for f starting from one azygetic contact curve of degree $d-1$.

The input to Dixon's algorithm is an azygetic contact curve g of degree $d-1$ of the given curve f of degree d. Given the two polynomials f and g, the algorithm constructs the matrix $\widehat{M} = \mathrm{adj}(M)$ in (6). It proceeds as follows. Since $\mathcal{V}(g)$ meets $\mathcal{V}(f)$ in $d(d-1)/2$ points, the vector space of polynomials of degree $d-1$ vanishing at these points (without multiplicity) has dimension d. Let $m_{11} = g$ and extend m_{11} to a basis $\{m_{11}, m_{12}, \ldots, m_{1d}\}$ of this vector space. For $i, j \in \{2, 3, \ldots, d\}$, the polynomial $m_{1i} m_{1j}$ vanishes to order two on $\mathcal{V}(f, m_{11})$, so it lies in the ideal $\langle m_{11}, f \rangle$. Using the Extended Buchberger Algorithm, one finds a degree $d-1$ polynomial m_{ij} such that $m_{1i} m_{1j} - m_{11} m_{ij} \in \langle f \rangle$. The $d \times d$-matrix $\widehat{M} = (m_{ij})$ has rank 1 modulo $\langle f \rangle$, therefore its 2×2-minors are multiples of f. This implies that the adjoint matrix $\mathrm{adj}(\widehat{M}) = \det(\widehat{M}) \cdot \widehat{M}^{-1}$ has the form $\lambda f^{d-2} \cdot M$ where

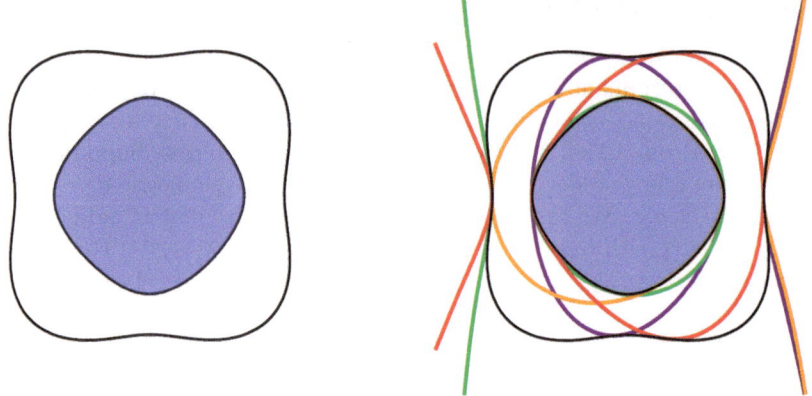

FIGURE 2. A quartic Helton-Vinnikov curve and four contact cubics

$\lambda \in \mathbb{C}\backslash\{0\}$ and M is a symmetric matrix of linear forms with $\det(M) = f$. One could run through this construction starting from a syzygetic contact curve, but the resulting matrix $\widehat{M}$ would have determinant zero.

The main challenge with Dixon's algorithm is to construct its input polynomial g. Suitable contact curves are not easy to find. A symbolic implementation of the algorithm may involve large field extensions, and we found it equally difficult to implement numerically. For further discussions see [19, §2.2] and [21, §2].

Remark 3. Starting from a *real* azygetic contact curve g, one can use Dixon's method to produce a *real* determinantal representation of f. A determinantal representation M is equivalent to its conjugate $\overline{M}$ if and only if the system of contact curves $\{u^T \text{adj}(M)u : u \in \mathbb{C}^4\} \subset \mathbb{C}[x, y, z]_3$ is real, i.e., invariant under conjugation. The representation M is equivalent to a real matrix if and only if this system contains a real contact curve. By [14, Prop 2.2], if the curve $\mathcal{V}(f)$ has real points, then these two notions of reality agree. However, this approach does not easily reveal whether an equivalence class contains a real definite representative.

Example 4. The following Helton-Vinnikov quartic was studied in [21, Ex. 4.1]:

$$f(x, y, z) \quad = \quad 2x^4 + y^4 + z^4 - 3x^2y^2 - 3x^2z^2 + y^2z^2.$$

It is shown on the left in Figure 2. It has a symmetric determinantal representation

$$f(x, y, z) \quad = \quad \det \begin{bmatrix} ux + y & 0 & az & bz \\ 0 & ux - y & cz & dz \\ az & cz & x + y & 0 \\ bz & dz & 0 & x - y \end{bmatrix}, \tag{7}$$

where $a = -0.5746\ldots$, $b = 1.0349\ldots$, $c = 0.6997\ldots$, $d = 0.4800\ldots$ and $u = \sqrt{2}$ are the coordinates of a real zero of the following maximal ideal in $\mathbb{Q}[a, b, c, d, u]$:

$\langle u^2 - 2, 256d^8 - 384d^6u + 256d^6 - 384d^4u + 672d^4 - 336d^2u + 448d^2 - 84u + 121,$
$\quad 23c + 7584d^7u + 10688d^7 - 5872d^5u - 8384d^5 + 1806d^3u + 2452d^3 - 181du - 307d,$
$\quad 23b + 5760d^7u + 8192d^7 - 4688d^5u - 6512d^5 + 1452d^3u + 2200d^3 - 212du - 232d,$
$\quad 23a - 1440d^7u - 2048d^7 + 1632d^5u + 2272d^5 - 570d^3u - 872d^3 + 99du + 81d\,\rangle.$

The principal 3×3-minors of the 4×4-matrix in (7) are Helton-Vinnikov polynomials of degree 3. They are the four contact cubics shown on the right in Figure 2. $\square$

In summary, Dixon's method furnishes an explicit bijection between equivalence classes of symmetric determinantal representations (1) of a fixed curve $\mathcal{V}(f)$ of degree d and azygetic systems of contact curves of $\mathcal{V}(f)$ of degree $d - 1$.

For $d = 3$, there is another geometric approach to finding representations (1). We learned this from Didier Henrion who attributes it to Frédéric Han. Suppose we are given a general homogeneous cubic $f(x, y, z)$. We first compute the Hessian

$$\mathrm{Hes}(f) \quad = \quad \det \begin{bmatrix} \partial^2 f/\partial x^2 & \partial^2 f/\partial x \partial y & \partial^2 f/\partial x \partial z \\ \partial^2 f/\partial x \partial y & \partial^2 f/\partial y^2 & \partial^2 f/\partial y \partial z \\ \partial^2 f/\partial x \partial z & \partial^2 f/\partial y \partial z & \partial^2 f/\partial z^2 \end{bmatrix}.$$

This is also a cubic polynomial, and hence so is the linear combination $t \cdot f + \mathrm{Hes}(f)$, where t is a parameter. We now take the Hessian of that new cubic, with the aim of recovering f. It turns out that we can do this by solving a cubic equation in t.

Proposition 5. *There exist precisely three points $(s, t) \in \mathbb{C}^2$ such that*

$$s \cdot f \quad = \quad \mathrm{Hes}\big(t \cdot f + \mathrm{Hes}(f)\big). \tag{8}$$

The resulting three symmetric determinantal representations of f are inequivalent.

Proof. The statement is invariant under linear changes of coordinates in $\mathbb{P}^2$, so, by [1, Lemma 1], we may assume that the given cubic is in *Hesse normal form*:

$$f(x, y, z) \quad = \quad x^3 + y^3 + z^3 - mxyz.$$

In that case, the result follows from the discussion in [17, page 139]. Alternatively, we can solve the equations obtained by comparing coefficients in (8). This leads to

$t^3 - (12m^4 + 2592m)t - 16m^6 + 8640m^3 + 93312 = 0 \qquad$ and
$s = (12m^4 + 2592m)t^2 + (48m^6 - 25920m^3 - 279936)t + 48m^8 + 20736m^5 + 2239488m^2.$

This has three solutions $(s, t) \in \mathbb{C}^3$. The resulting representations (1) are inequivalent because the Hessian normal form of a PGL$(3, \mathbb{C})$-orbit of cubics is unique. $\square$

4. Evaluating theta functions

The proof of the Helton-Vinnikov Theorem relies on a formula, stated in [16, Eq. 4.2], that gives a positive definite determinantal representation of a Helton-Vinnikov curve in terms of theta functions and the period matrix of the curve. Our aim in this section is to explain this formula and to report on computational experiments with it. Numerical algorithms for computing theta functions, period matrices and Abelian integrals have become available in recent years through work of Bobenko, Deconinck, Heil, van Hoeij, Patterson, Schmies, and others [7, 8, 9]. There exists an implementation in `Maple`, and we used that for our computations. Our `Maple` worksheet that evaluates the Helton-Vinnikov formula can be found at

$$\text{www.math.uni-konstanz.de/}\sim\text{plaumann/theta.html} \qquad (9)$$

Before stating the Helton-Vinnikov formula, we review the basics on theta functions. Our emphasis will be on clearly defining the ingredients of the formula rather than explaining the underlying theory. For general background see [20]. Fix $g \in \mathbb{N}$ and let $\mathcal{H}^g$ be the *Siegel upper half-space*, which consists of all complex, symmetric $g \times g$-matrices whose imaginary part is positive definite. The *Riemann theta function* is the holomorphic function on $\mathbb{C}^g \times \mathcal{H}^g$ defined by the exponential series

$$\theta(\mathbf{u}, \Omega) \;=\; \sum_{m \in \mathbb{Z}^g} \exp\big(\pi i (m^T \Omega m + 2m^T \mathbf{u})\big),$$

where $i = \sqrt{-1}$, $\mathbf{u} = (u_1, \ldots, u_g) \in \mathbb{C}^g$ and $\Omega \in \mathcal{H}^g$. We will only need to consider $\theta(\mathbf{u}, \Omega)$ as a function in $\mathbf{u}$, for a fixed matrix Ω, so we may drop Ω from the notation. In other words, we define $\theta \colon \mathbb{C}^g \to \mathbb{C}$ by $\theta(\mathbf{u}) = \theta(\mathbf{u}, \Omega)$. The theta function is quasi-periodic with respect to the lattice $\mathbb{Z}^g + \Omega \mathbb{Z}^g \subset \mathbb{C}^g$, which means

$$\theta(\mathbf{u} + \mathbf{m} + \Omega \mathbf{n}) \;=\; \exp\big(\pi i (-2\mathbf{n}^T \mathbf{u} - \mathbf{n}^T \Omega \mathbf{n})\big) \cdot \theta(\mathbf{u}), \quad \text{for all } \mathbf{m}, \mathbf{n} \in \mathbb{Z}^g.$$

A *theta characteristic* is a vector $\epsilon = \mathbf{a} + \Omega \mathbf{b} \in \mathbb{C}^g$ with $\mathbf{a}, \mathbf{b} \in \{0, \frac{1}{2}\}^g$. The function

$$\theta[\epsilon](\mathbf{u}) \;=\; \exp\big(\pi i (\mathbf{a}^T \Omega \mathbf{a} + 2\mathbf{a}^T (\mathbf{u} + \mathbf{b}))\big) \cdot \theta(\mathbf{u} + \Omega \mathbf{a} + \mathbf{b})$$

is the *theta function with characteristic* ϵ. There are 2^{2g} different theta characteristics, indexed by ordered pairs $(2\mathbf{a}, 2\mathbf{b})$ of binary vectors in $\{0, 1\}^g$. We also use the notation $\theta \begin{bmatrix} 2\mathbf{a} \\ 2\mathbf{b} \end{bmatrix}(\mathbf{u})$ for the function $\theta[\epsilon](\mathbf{u})$. For $\epsilon = 0$ we simply recover $\theta(\mathbf{u})$.

Let $f \in \mathbb{C}[x, y, z]_d$ be a homogeneous polynomial of degree d. Assume that the projective curve $X = \mathcal{V}_{\mathbb{C}}(f)$ is smooth and thus a compact Riemann surface of genus $g = \frac{1}{2}(d-1)(d-2)$. Let $(\omega_1', \ldots, \omega_g')$ be a basis of the g-dimensional complex vector space of holomorphic 1-forms on X, and let $\alpha_1, \ldots, \alpha_g, \beta_1, \ldots, \beta_g$ be closed 1-cycles on X that form a *symplectic basis* of $H_1(X, \mathbb{Z}) \cong \mathbb{Z}^g \times \mathbb{Z}^g$. This means that the intersection numbers of these cycles on X satisfy $(\alpha_j \cdot \alpha_k) = (\beta_j \cdot \beta_k) = 0$, and $(\alpha_j \cdot \beta_k) = \delta_{jk}$ for all $j, k = 1, \ldots, g$. The *period matrix* of the curve X with respect to these bases is the complex $g \times 2g$-matrix $(\Omega_1 \,|\, \Omega_2)$ whose entries are

$$(\Omega_1)_{jk} = \int_{\alpha_k} \omega_j' \quad \text{and} \quad (\Omega_2)_{jk} = \int_{\beta_k} \omega_j', \quad \text{for } j, k = 1, \ldots, g.$$

The $g \times g$-matrices Ω_1 and Ω_2 are invertible. Performing the coordinate change

$$(\omega_1, \ldots, \omega_g) \;=\; (\omega_1', \ldots, \omega_g') \cdot (\Omega_1^{-1})^T$$

leads to a basis in which the period matrix is of the form $(\mathrm{Id}_g \,|\, \Omega_1^{-1}\Omega_2)$. The basis $\omega = (\omega_1, \ldots, \omega_g)$ is called a *normalized basis of differentials* and depends uniquely on the symplectic homology basis. The $g \times g$-matrix $\Omega = \Omega_1^{-1}\Omega_2$ is symmetric and lies in the Siegel upper half-space $\mathcal{H}^g$. It is called the *Riemann period matrix* of the polynomial $f(x, y, z)$ with respect to the homology basis $(\alpha_1, \ldots, \alpha_g, \beta_1, \ldots, \beta_g)$.

With the given polynomial f we have now associated a system $\{\theta[\epsilon](\,\cdot\,, \Omega)\}$ of 2^{2g} theta functions with characteristics. A theta characteristic $\epsilon = \mathbf{a} + \Omega\mathbf{b}$ is called *even* (resp. *odd*) if the scalar product $(2\mathbf{a})^T(2\mathbf{b})$ of its binary vector labels is an even (resp. odd) integer. This is equivalent to $\theta[\epsilon]$ being an even (resp. odd) function in $\mathbf{u}$. In symbols, we have $\theta[\epsilon](-\mathbf{u}) = (-1)^{4\mathbf{a}^T\mathbf{b}}\theta[\epsilon](\mathbf{u})$. Changing symplectic bases of $H_1(X, \mathbb{Z})$ corresponds to the right-action of the symplectic group $\mathrm{Sp}_{2g}(\mathbb{Z})$ on the period matrix $(\Omega_1 \,|\, \Omega_2)$. This action will permute the theta characteristics. In particular, there is no distinguished even theta characteristic 0.

Finally, we define the *Abel-Jacobi map* by $\phi(P) = (\int_{P_0}^{P} \omega_1, \ldots, \int_{P_0}^{P} \omega_g)^T$ for $P \in X$, where $P_0 \in X$ is any fixed base point. This is a holomorphic map, but it is well defined only up to the period lattice $\Lambda = \mathbb{Z}^g + \Omega\mathbb{Z}^g \subset \mathbb{C}^g$. In other words, the Abel-Jacobi map is a holomorphic map $\phi\colon X \to \mathrm{Jac}(X) = \mathbb{C}^g/\Lambda$.

We are now ready to state the formula for (1) in terms of theta functions.

Theorem 6 (Helton-Vinnikov [16]). *Let $f \in \mathbb{R}[x, y, z]_d$ with $f(1, 0, 0) = 1$ and let X denote $\mathcal{V}_{\mathbb{C}}(f) \subset \mathbb{P}^2$. We make the following two assumptions:*

1. *The curve X is a non-rational Helton-Vinnikov curve with the point $(1 : 0 : 0)$ inside its innermost oval. The latter means that, for all $v \in \mathbb{R}^3 \backslash \{0\}$, the univariate polynomial $f(v + t \cdot (1, 0, 0)) \in \mathbb{R}[t]$ has only real zeros.*

2. *The d real intersection points of X with the line $\{z = 0\}$ are distinct non-singular points $Q_1, \ldots, Q_d$, with coordinates $Q_i = (-\beta_j : 1 : 0)$ where $\beta_j \neq 0$.*

Then $f(x, y, z) = \det(\mathrm{Id}_d x + By + Cz)$ where $B = \mathrm{diag}(\beta_1, \ldots, \beta_d)$ and C is real symmetric with diagonal entries c_{jj} as in (5). The off-diagonal entries of C are

$$c_{jk} = \frac{\beta_k - \beta_j}{\theta[\delta](0)} \cdot \frac{\theta[\delta]\big(\phi(Q_k) - \phi(Q_j)\big)}{\theta[\epsilon]\big(\phi(Q_k) - \phi(Q_j)\big)} \sqrt{\frac{\omega \cdot \nabla\theta[\epsilon](0)}{-d(z/y)}(Q_j)} \sqrt{\frac{\omega \cdot \nabla\theta[\epsilon](0)}{-d(z/y)}(Q_k)}. \quad (10)$$

Here ϵ is an arbitrary odd theta characteristic and δ is a suitable even theta characteristic with $\theta[\delta](0) \neq 0$. The theta functions are taken with respect to a normalized basis of differentials $\omega = (\omega_1, \ldots, \omega_g)$, and $\phi\colon X \to \mathrm{Jac}(X)$ is the Abel-Jacobi map.

The remarkable expression for the constants c_{jk} in (10) does not depend on the choice of the odd characteristic ϵ. If the curve X is smooth, then all equivalence classes of symmetric determinantal representations are obtained when δ runs through all non-vanishing even theta characteristics. The proof of Theorem 6 given in [16] is only an outline. It relies heavily on earlier results on Riemann surfaces due to Ball and Vinnikov in [2, 24]. As we found these not easy to read, we were particularly pleased to be able to verify Theorem 6 with our experiments.

The Helton-Vinnikov formula (10) remains valid when X is a singular curve. In that case the period matrix, the differentials, and the Abel-Jacobi map are meant to be defined on the desingularization of X, a compact Riemann surface of genus g with $g < \frac{1}{2}(d-1)(d-2)$. The formula holds as stated, but one no longer obtains all equivalence classes of symmetric determinantal representations.

The Riemann period matrix, the theta functions, their directional derivatives, and the Abel-Jacobi-map can all be evaluated numerically in recent versions of Maple. When computing the expressions under the square roots, note that both the numerator and denominator are 1-forms on X. Every holomorphic 1-form on the curve X can be written as $r \cdot du$, where $u = z/x$, $v = y/x$, and r is a rational function in u and v. The algcurves package in Maple will compute ω in this form, so we obtain $\omega_j = r_j(u,v) \cdot du$. To evaluate the 1-form $d(z/y)$, we set $h(u,v) = f(1,v,u)$ and use the identity $dh(u,v) = \frac{\partial h}{\partial u}du + \frac{\partial h}{\partial v}dv = 0$. This implies

$$d(z/y) \;=\; d(u/v) \;=\; \frac{1}{v}du - \frac{u}{v^2}dv \;=\; \left(\frac{1}{v} + \frac{u\frac{\partial h}{\partial u}}{v^2\frac{\partial h}{\partial v}}\right)du, \tag{11}$$

so that $d(z/y)(Q_j) = -\beta_j du$. Under the square root signs in (10), the factor du appears in the numerator ω and also in the denominator (11), and we cancel it. Hence the expressions under the square roots are rational functions, namely $r(u,v) \cdot \nabla\theta[\epsilon](0)$ divided by the expression in parentheses on the right in (11), where $r(u,v)$ is the vector of rational functions $r_j(u,v)$.

While the evaluation of theta functions is numerically stable, we found the computation of the period matrix and the Abel-Jacobi map to be more fragile. Computing the d vectors $\phi(Q_j)$ is also by far the most time-consuming step. Nonetheless, Maple succeeded in correctly evaluating the Helton-Vinnikov formula for a wide range of curves with $d \leq 4$, and for some of degree $d = 5$. However, the off-diagonal entries in (10) we found in our computations were sometimes wrong by a constant factor (independent of j, k), for reasons we do not currently understand.

For a concrete example take the quartic in Example 4. Using the formula (10) we obtained all eight definite determinantal representations $\det(\mathrm{Id}_d x + By + Cz)$. Our Maple code runs for a few minutes and finds all solutions accurately with a precision of 20 digits. We verified this using the prime ideal in Example 4.

The case of smooth quintics (genus 6) is already a challenge. With the help of Bernard Deconinck, we were able to compute a determinantal representation (10) with an error of less than 10^{-3} for the quintic polynomial at the end of Section 2. However, the representation we obtained was not real (see Remark 9).

We conclude this section with the proof of the second part of Theorem 1 and discuss what are the suitable choices for the even theta characteristic δ in Theorem 6 that will lead to real and real definite equivalence classes of representations. Note that a representation obtained from the theorem is real definite if and only if the matrix C is real. The real non-definite equivalence classes of representations correspond to the case when C is a non-real matrix for which there exists a matrix $U \in \mathrm{GL}_d(\mathbb{C})$ such that UU^T, UBU^T, and UCU^T have all real entries. Whether such U exists for given complex symmetric matrices B and C is not at all obvious. An explicit example is given in Example 11.

For the proof of Theorem 1, we repeat the relevant part of the statement:

Theorem 7. *Let $f \in \mathbb{R}[x, y, z]_d$ be homogeneous and assume that the projective curve $X = \mathcal{V}_\mathbb{C}(f)$ is a smooth Helton-Vinnikov curve. The number of real equivalence classes of symmetric determinantal representations is generically either*

$$2^{g-1}(2^{k-1} + 1)$$

or one less, where $k = \lceil \frac{d}{2} \rceil$ is the number of connected components of the set of real points $X(\mathbb{R})$. Of these real equivalence classes, exactly 2^g are definite.

Proof. The result follows from work of Vinnikov [24] on self-adjoint determinantal representations, which we apply here to our situation. By [24, Prop. 2.2], a symplectic basis of $H_1(X, \mathbb{Z})$ can be chosen in such a way that the Riemann period matrix Ω satisfies $\Omega + \overline{\Omega} = H$, where H is a $g \times g$ block diagonal matrix of rank $r = g - k + 1$ with $r/2$ blocks $\begin{bmatrix} 0 & 1 \\ 1 & 0 \end{bmatrix}$ in the top left corner and all other entries zero.

The linear symmetric determinantal representation obtained by Theorem 6 from an even theta characteristic δ is equivalent to a real one if and only if δ is real, i.e., invariant under the action of complex conjugation on the g-dimensional torus $\mathrm{Jac}(X)$. Since $X(\mathbb{R}) \neq \emptyset$, any conjugation-invariant divisor class on X contains a real divisor (see [14, Prop. 2.2]). From such a divisor, one can construct a symmetric determinantal representation (see [3] or [23]). When the symplectic basis of $H_1(X, \mathbb{Z})$ is chosen as above, the action of complex conjugation on $\mathrm{Jac}(X)$ is given by $\zeta = \mathbf{u} + \Omega \mathbf{v} \mapsto \overline{\zeta} = \mathbf{u} + \Omega(H\mathbf{u} - \mathbf{v})$ (see [24, Prop. 2.2]). For the even theta characteristic $\delta = \mathbf{a} + \Omega \mathbf{b}$, $\mathbf{a}, \mathbf{b} \in \{0, \frac{1}{2}\}^g$, the condition $\delta = \overline{\delta}$ in $\mathrm{Jac}(X) = \mathbb{C}^g / (\mathbb{Z}^g + \Omega \mathbb{Z}^g)$ thus becomes

$$H\mathbf{a} \equiv 2\mathbf{b} \mod \mathbb{Z}^g.$$

This happens if and only if $a_1 = \cdots = a_r = 0$. Counting the possible choices of $a_{r+1}, \ldots, a_g$ and $\mathbf{b}$, we conclude that there are exactly $2^{g-1}(2^{k-1} + 1)$ even real theta characteristics. From the first part of Theorem 1, we know that when $d \equiv \pm 3 \mod 8$, exactly one even theta characteristic vanishes, i.e., $\theta[\delta](0) = 0$. All other even theta characteristics are non-vanishing and therefore correspond to determinantal representations. Furthermore, by [24, Thm. 6.1], an even theta characteristic $\delta = \mathbf{a} + \Omega \mathbf{b}$ will correspond to a real definite equivalence class if and only if $\mathbf{a} = 0$, and all such δ are always non-vanishing [24, Cor. 4.3]. Thus there are exactly 2^g definite representations, since $\mathbf{b}$ can be any element of $\{0, \frac{1}{2}\}^g$.   □

Example 8. When $g = 3$ and a homology basis has been picked as above, the even real theta characteristics are given by the 12 binary labels

$$\begin{bmatrix} 000 \\ 000 \end{bmatrix} \begin{bmatrix} 000 \\ 001 \end{bmatrix} \begin{bmatrix} 000 \\ 010 \end{bmatrix} \begin{bmatrix} 000 \\ 011 \end{bmatrix} \begin{bmatrix} 000 \\ 100 \end{bmatrix} \begin{bmatrix} 000 \\ 101 \end{bmatrix} \begin{bmatrix} 000 \\ 110 \end{bmatrix} \begin{bmatrix} 000 \\ 111 \end{bmatrix} \begin{bmatrix} 001 \\ 000 \end{bmatrix} \begin{bmatrix} 001 \\ 010 \end{bmatrix} \begin{bmatrix} 001 \\ 100 \end{bmatrix} \begin{bmatrix} 001 \\ 110 \end{bmatrix}. \quad (12)$$

The first eight labels correspond to the definite representations and the last four correspond to the non-definite real equivalence classes of representations.

Remark 9. The characterization of real and real definite even theta characteristics provided by the proof of Theorem 7 depends on the choice of a particular symplectic homology basis. Unfortunately, the current `Maple` code for computing the period matrix does not give the user any control over the homology basis. This makes it hard to find real representations using Theorem 6 in any systematic way.

5. Quartic curves revisited

In this section we focus on the case of smooth quartic curves, studied in detail in [21], so we now fix $d = 4$ and $g = 3$. Quartic curves are special because they have contact lines, i.e., bitangents, and we can explicitly write down higher degree contact curves as products of bitangents. This was exploited in [21, §2], where we used azygetic triples of bitangents as our input to Dixon's algorithm (Section 3).

Plane quartics are canonical embeddings of genus 3 curves [11], and there is a close connection between contact curves and theta functions. The 28 bitangents of the curve are in bijection with the 28 odd theta characteristics $\epsilon = \mathbf{a} + \Omega\mathbf{b}$, and this will be made explicit in (15) below. The 36 azygetic systems of contact cubics correspond to the 36 even theta characteristics. As seen in Example 8, of the resulting 36 determinantal representations, 12 are real, but only 8 are definite. We can also derive the number 12 from the combinatorics of the bitangents as in [21].

Proposition 10. *A smooth Helton-Vinnikov quartic $\mathcal{V}(f)$ has exactly 12 inequivalent representations $f = \det(Ax + By + Cz)$ with A, B, C symmetric and real.*

Proof. This is a special case of Theorem 1, however, we here give an alternative proof using the setup of [21]. Let M be a symmetric linear determinantal representation of f and $\mathcal{M}$ the system of contact cubics $\{u^T \mathrm{adj}(M)u\} \subset \mathbb{C}[x,y,z]_3$. The representation M is equivalent to its conjugate $\overline{M}$ if and only if the system $\mathcal{M}$ is real, i.e., invariant under conjugation. The representation M is equivalent to a real matrix if and only if $\mathcal{M}$ contains a real cubic. The matrix M induces a labeling of the $28 = \binom{8}{2}$ bitangents, b_{ij}, with $1 \le i < j \le 8$. The system $\mathcal{M}$ is real if and only if conjugation acts on the bitangents via this labeling, that is, there exists $\pi \in S_8$ such that $\overline{b_{ij}} = b_{\pi(i)\pi(j)}$. Since f is a Helton-Vinnikov polynomial, this permutation will be the product of four disjoint transpositions (see [21, Table 1]).

Suppose $\mathcal{M}$ is real, with permutation $\pi \in S_8$. The other 35 representations (1) correspond to the $\binom{8}{4}/2$ partitions of $\{1, \ldots, 8\}$ into two sets of size 4. If $I|I^c$ is such a partition then the corresponding system of contact cubics contains 56 products of three bitangents, namely $b_{ij}b_{ik}b_{i\ell}$ and $b_{im}b_{jm}b_{k\ell}$ where i, j, k, l, m are distinct and $\{i, j, k, l\} = I$ or I^c. This system is real if and only if π fixes the partition $I|I^c$. There are exactly 11 such partitions: if $\pi = (12)(34)(56)(78)$, they are

$$\begin{array}{cccccc} 1234|5678, & 1256|3478, & 1278|3456, & 1357|2468, & 1358|2467, & 1368|2457 \\ 1367|2458, & 1457|2368, & 1458|2367, & 1467|2358, & \text{and} & 1468|2357. \end{array} \quad (13)$$

Together with the system $\mathcal{M}$, there are 12 real systems of azygetic contact cubics.

Next, we will show that each of these systems actually contains a real cubic. To do this, we use contact conics, as the product of a bitangent with a contact conic is a contact cubic. By [21, Lemma 6.7], there exists a real bitangent $b \in \mathbb{R}[x, y, z]_1$ and a real system of contact conics $\mathcal{Q} \subset \mathbb{C}[x, y, z]_2$ such that their product $\{b \cdot q : q \in \mathcal{Q}\}$ lies in the system $\mathcal{M} \subset \mathbb{C}[x, y, z]_3$. Furthermore, by [21, Prop. 6.6], since $\mathcal{V}_{\mathbb{R}}(f)$ is nonempty, every real system of contact conics $\mathcal{Q}$ to f contains a real conic q. The desired real contact cubic is the product $b \cdot q$. □

The technique in the last paragraph of the above proof led us to the following result: *There exists a smooth Helton-Vinnikov quartic $f \in \mathbb{Q}[x, y, z]_4$ that has 12 inequivalent determinantal representations (1) over the field $\mathbb{Q}$ of rational numbers.*

Example 11. The special rational Helton-Vinnikov quartic we found is

$$f(x, y, z) = 93081x^4 + 53516x^3y - 73684x^2y^2 + -31504xy^3 + 9216y^4$$
$$- 369150x^2z^2 - 159700xyz^2 + 57600y^2z^2 + 90000z^4.$$

This polynomial satisfies $f(x, y, z) = \det(M)$ where

$$M = \begin{bmatrix} 50x & -25x & -26x - 34y - 25z & 9x + 6y + 15z \\ -25x & 25x & 27x + 18y - 20z & -9x - 6y \\ -26x - 34y - 25z & 27x + 18y - 20z & 108x + 72y & -18x - 12y \\ 9x + 6y + 15z & -9x - 6y & -18x - 12y & 6x + 4y \end{bmatrix}.$$

This representation is definite because the matrix M is positive definite at the point $(1 : 0 : 0)$. Hence $\mathcal{V}(f)$ is a Helton-Vinnikov curve with this point in its inner convex oval. Rational representatives for the other seven definite classes are found at our website (9), along with representatives for the four non-definite real classes. One of them is the matrix

$$M_{1468} = \begin{bmatrix} 25x & 0 & -32x + 12y & -60z \\ 0 & 25x & 10z & 24x + 16y \\ -32x + 12y & 10z & 6x + 4y & 0 \\ -60z & 24x + 16y & 0 & 6x + 4y \end{bmatrix}. \tag{14}$$

We have $\det(M_{1468}) = 4 \cdot f(x, y, z)$, and this matrix is neither positive definite nor negative definite for any real values of x, y, z. Any equivalent representation of a multiple of f in the form $\det(\mathrm{Id}_4x + By + Cz)$ considered in Sections 2 and 4 cannot have all entries of C real. One such representation, for a suitable $U \in \mathrm{GL}_4(\mathbb{C})$, is

$$U^T M_{1468} U = \begin{bmatrix} x + \frac{64}{71}y & 0 & -\frac{23}{1349}\sqrt{26980}\,i\,z & \frac{-51}{1633}\sqrt{16330}\,z \\ 0 & x + \frac{2}{3}y & -\frac{2}{19}\sqrt{570}\,z & \frac{4}{23}\sqrt{345}\,i\,z \\ -\frac{23}{1349}\sqrt{26980}\,i\,z & -\frac{2}{19}\sqrt{570}\,z & x - \frac{4}{19}y & 0 \\ \frac{-51}{1633}\sqrt{16330}\,z & \frac{4}{23}\sqrt{345}\,i\,z & 0 & x - \frac{18}{23}y \end{bmatrix}.$$

The correspondence between bitangents and odd theta characteristics can be understood abstractly via the isomorphism between the Jacobian of X and the divisor class group $\mathrm{Cl}^0(X)$, and can be turned into an explicit formula for the bitangents. Let $u = z/x$, $v = y/x$ and write $h(u, v) = f(1, v, u)$. Then a basis for the three-dimensional complex vector space of holomorphic 1-forms on X is given by

$$(\omega_1, \omega_2, \omega_3) = \left(\frac{du}{\partial h / \partial v}, \frac{vdu}{\partial h / \partial v}, \frac{udu}{\partial h / \partial v} \right).$$

Let $(\Omega_1 | \Omega_2)$ be the period matrix of f with respect to $(\omega_1, \omega_2, \omega_3)$ and any symplectic basis of $H_1(X, \mathbb{Z})$. Then given an odd theta characteristic $\epsilon = \mathbf{a} + \Omega \mathbf{b}$, the corresponding bitangent is defined by the linear form

$$b_\epsilon(x, y, z) = \left(\nabla \theta[\epsilon](0) \right)^T \cdot \Omega_1^{-1} \cdot (x, y, z)^T, \tag{15}$$

where $\nabla \theta[\epsilon]$ is the gradient of $\theta[\epsilon]$ in the three complex variables z_1, z_2, z_3. For the proof, see Dolgachev [11, Section 5.5.4]. This holds independently of the symplectic basis of $H_1(X, \mathbb{Z})$, but a change of that basis will permute the bitangents.

The formula (15) can be evaluated using the `Maple` code described in Section 4. This allows us to compute the 8×8-bitangent matrix (b_{ij}) of [21, Eq. 3.4] directly from the Riemann period matrix Ω of the curve X, using a technique due to Riemann described in [15, §2]. In this manner, one computes the symmetric determinantal representations (1) of the curve X directly from the period matrix Ω. This computation seems to be a key ingredient in constructing explicit three-phase solutions of the Kadomtsev-Petviashvili equation [12], and we hope that the combinatorial tools developed here and in [21] will be useful for integrable systems.

One of the earliest papers on algorithms for theta functions in genus three was written by Arthur Cayley in 1897. In [5] he gives a concrete bijection between the bitangents b_{ij} of a plane quartic and the odd theta characteristics, and also between the classes $I | I^c$ of determinantal representations and the even theta characteristics. We here reproduce a relabeled version of the table in Cayley's article:

2b\2a	000	100	010	110	001	101	011	111
000	$\emptyset$	1238	1267	1245	**1468**	1578	1356	1347
100	**1234**	48	1235	35	**1457**	16	1378	27
010	**1256**	1247	57	46	**1367**	1345	23	18
110	**1278**	37	68	1236	**1358**	25	14	1567
001	**1357**	1346	1478	1568	12	38	67	45
101	**1368**	26	1456	17	34	1248	58	1235
011	**1458**	1678	13	28	56	47	1257	1246
111	**1467**	15	24	1348	78	1237	1268	36

Here a partition $I | I^c$ of $\{1, \ldots, 8\}$ is represented by the 4-tuple I which contains the index 1. For instance, the 4-tuple 1238 corresponds to the even theta characteristic $\begin{bmatrix} 100 \\ 000 \end{bmatrix}$. Each partition $I | I^c$ represents a Cremona transformation leading to a new representation (1) as described in [21, §3]. The twelve 4-tuples marked in bold face are the real equivalence classes, and this gives a bijection between the lists in (12) and in (13). Likewise, the pairs ij in Cayley's table represent bitangents b_{ij} and the corresponding odd theta characteristics. For instance, the odd characteristic

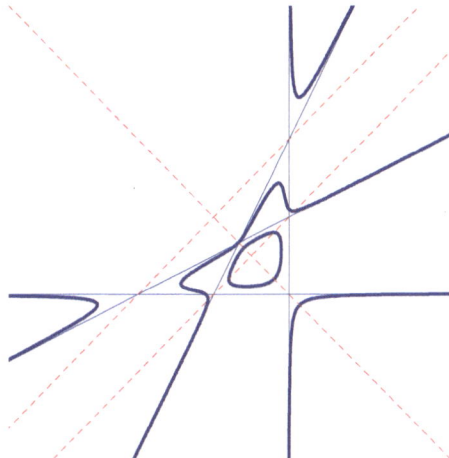

FIGURE 3. Degeneration of a Helton-Vinnikov quartic into four lines

$\begin{bmatrix} 100 \\ 111 \end{bmatrix}$ represents the bitangent b_{15}. In this manner, we can parametrize the 28 bitangents of all plane quartics explicitly with odd theta functions.

Experts in moduli of curves will be quick to point out that this parametrization should extend from smooth curves to all stable curves. This is indeed the case. For instance, four distinct lines form a stable Helton-Vinnikov quartic such as

$$f(x, y, z) \;\; = \;\; xyz(x + y + z).$$

The bitangent matrix (b_{ij}) of this reducible curve has 7 distinct non-zero entries:

$$\begin{bmatrix}
0 & z & y & y+z & x & x+z & x+y & x{+}y{+}z \\
z & 0 & y+z & y & x+z & x & x{+}y{+}z & x+y \\
y & y+z & 0 & z & x+y & x{+}y{+}z & x & x+z \\
y+z & y & z & 0 & x{+}y{+}z & x+y & x+z & x \\
x & x+z & x+y & x{+}y{+}z & 0 & z & y & y+z \\
x+z & x & x{+}y{+}z & x+y & z & 0 & y+z & y \\
x+y & x{+}y{+}z & x & x+z & y & y+z & 0 & z \\
x{+}y{+}z & x+y & x+z & x & y+z & y & z & 0
\end{bmatrix}.$$

All principal 4×4-minors of this 8×8-matrix are multiples of $f(x, y, z)$, most of them non-zero. They are all in the same equivalence class, which is real but not definite. The entries in the bitangent matrix indicate a partition of the 28 odd theta characteristics into seven groups of four. For instance, the antidiagonal entry $x + y + z$ corresponds to the four entries $18, 27, 36$ and 45 in Cayley's table, and hence to the four odd theta characteristics $\begin{bmatrix} 010 \\ 111 \end{bmatrix}$, $\begin{bmatrix} 100 \\ 111 \end{bmatrix}$, $\begin{bmatrix} 111 \\ 111 \end{bmatrix}$ and $\begin{bmatrix} 001 \\ 111 \end{bmatrix}$.

If we consider a family of smooth quartics that degenerates to the reducible quartic $f(x, y, z)$, then its bitangent matrix will degenerate to the above 8×8-matrix, and hence the 28 distinct bitangents of the smooth curve bunch up in seven clusters of four. This degeneration is visualized in Figure 3. Among the seven limit bitangents are the three lines spanned by pairs of intersection points.

Algebraically, such a degenerating family can be described as a curve over a field with a valuation, such as the field of *real Puiseux series* $\mathbb{R}\{\{\epsilon\}\}$. The notions of spectrahedra and Helton-Vinnikov curves makes perfect sense over the real closed field $\mathbb{R}\{\{\epsilon\}\}$. This has been investigated from the perspective of tropical geometry by David Speyer, who proved in [22] that tropicalized Helton-Vinnikov curves are precisely *honeycomb curves*. We believe that the tropicalization in [22] offers yet another approach to constructing linear determinantal representations (1), in addition to the three methods presented here, and we hope to return to this topic.

Acknowledgment

We wish to thank Charles Chen, Bernard Deconinck, Didier Henrion, Chris Swierczewski and Victor Vinnikov for discussions and computational contributions that were very helpful to us in the preparation of this article.

References

[1] M. Artebani and I. Dolgachev: The Hesse pencil of a plane cubic curve. *Enseign. Math.* **55**, 235–273, 2009.

[2] J.A. Ball and V. Vinnikov: Zero-pole interpolation for matrix meromorphic functions on a compact Riemann surface and a matrix Fay trisecant identity. *Amer. J. Math.*, **121** (4), 841–888, 1999.

[3] A. Beauville: Determinantal hypersurfaces. *Michigan Math. Journal*, **48**, 39–64, 2000.

[4] D. Bates, J. Hauenstein, A. Sommese, and C. Wampler: BERTINI: Software for Numerical Algebraic Geometry, http://www.nd.edu/~sommese/bertini/, (2010).

[5] A. Cayley: Algorithm for the characteristics of the triple ϑ-functions. *Journal für die reine und angewandte Mathematik*, **87**, 165–169, 1879.

[6] W. Decker, G.-M. Greuel, G. Pfister, and H. Schönemann: SINGULAR: A computer algebra system for polynomial computations, www.singular.uni-kl.de (2010).

[7] B. Deconinck, M. Heil, A. Bobenko, M. van Hoeij, and M. Schmies: Computing Riemann theta functions. *Mathematics of Computation*, **73** (247), 1417–1442, 2004.

[8] B. Deconinck and M.S. Patterson: Computing the Abel map. *Physica D*, **237** (24), 3214–3232, 2008.

[9] B. Deconinck and M. van Hoeij: Computing Riemann matrices of algebraic curves. *Physica D*, **152/153**, 28–46, 2001.

[10] A.C. Dixon: Note on the reduction of a ternary quantic to a symmetrical determinant. *Cambr. Proc.* **11**, 350–351, 1902.

[11] I. Dolgachev: *Classical Algebraic Geometry: A Modern View*, Cambridge University Press, 2012.

[12] B. Dubrovin, R. Flickinger and H. Segur: Three-phase solutions of the Kadomtsev-Petviashvili equation. *Stud. Appl. Math.* **99** (1997) 137–203.

[13] W. Greub: *Linear Algebra*. Springer-Verlag, New York, 4th edn., 1975. Graduate Texts in Math., No 23.

[14] B.H. Gross and J. Harris: Real algebraic curves. *Ann. Sci. École Norm. Sup.* (4), **14 (2)**, 157–182, 1981.

[15] J. Guardia: On the Torelli problem and Jacobian Nullwerte in genus three, *Michigan Mathematical Journal* **60**, 51–65, 2011.

[16] J.W. Helton and V. Vinnikov: Linear matrix inequality representation of sets. *Comm. Pure Appl. Math.*, **60 (5)**, 654–674, 2007.

[17] K. Hulek: *Elementary Algebraic Geometry*, Student Mathematical Library, Vol. 20, American Mathematical Society, Providence, RI, 2003.

[18] A. Lewis, P. Parrilo and M. Ramana: The Lax conjecture is true. *Proceedings Amer. Math. Soc.*, **133**, 2495–2499, 2005.

[19] T. Meyer-Brandis. Berührungssysteme und symmetrische Darstellungen ebener Kurven, 1998. Diplomarbeit, Universität Mainz, written under the supervision of D. van Straten, posted at http://enriques.mathematik.uni-mainz.de/straten/diploms

[20] D. Mumford: *Tata Lectures on Theta. I.* Modern Birkhäuser Classics. Birkhäuser, Boston, MA, 2007. Reprint of the 1983 edition.

[21] D. Plaumann, B. Sturmfels and C. Vinzant: Quartic curves and their bitangents, *Journal of Symbolic Computation* **46**, 712–733, 2011.

[22] D. Speyer: Horn's problem, Vinnikov curves, and the hive cone. *Duke Math. J.* **127** (2005), no. 3, 395–427.

[23] V. Vinnikov: Complete description of determinantal representations of smooth irreducible curves. *Linear Algebra Appl.*, **125**, 103–140, 1989.

[24] V. Vinnikov: Selfadjoint determinantal representations of real plane curves. *Mathematische Annalen*, **296 (3)**, 453–479, 1993.

Daniel Plaumann
Fachbereich Mathematik und Statistik
Universität Konstanz
D-78457 Konstanz, Germany
e-mail: Daniel.Plaumann@uni-konstanz.de

Bernd Sturmfels
Department of Mathematics
University of California
Berkeley, CA 94720, USA
e-mail: bernd@math.berkeley.edu

Cynthia Vinzant
Department of Mathematics
University of Michigan
Ann Arbor, MI 48109, USA
e-mail: vinzant@umich.edu

Operator Theory:
Advances and Applications, Vol. 222, 279–293

Positive Completion Problems Over C^*-algebras

L. Rodman and H.J. Woerdeman

Dedicated to J.W. Helton on the occasion of his 65th birthday

Abstract. We study how positive completion problems over matrices with complex entries generalize to the setting of matrices with C^*-algebra entries. In particular, it is observed that some C^*-algebras have the Toeplitz banded completion property but fail to have the completion property for the complete graph with one undirected edge missing. Positive completions in the framework of multi-level Toeplitz matrices are also studied. Many open problems are formulated.

Mathematics Subject Classification. 46L05 (47A57).

Keywords. Positive matrix completion, positive Toeplitz completion, C^*-algebra.

1. Introduction

Let $\mathcal{A}$ be a (not necessarily unital) complex C^*-algebra. *It will be assumed every-where that $\mathcal{A} \neq \{0\}$.* An element $a \in \mathcal{A}$ is called *positive* if $a = b^*b$ for some $b \in \mathcal{A}$, or equivalently if $a = x^2$ for some $x = x^* \in \mathcal{A}$. If $\mathcal{A}$ is unital, then $a \in \mathcal{A}$ is called *strictly positive* if $a = b^*b$ for some invertible $b \in \mathcal{A}$.

All graphs Γ in this paper are assumed to be undirected, with finite vertex set $V(\Gamma)$, without multiple edges, and with all loops (x,x), $x \in V(\Gamma)$ contained in the set of edges $E(\Gamma)$ of Γ. As Γ is undirected we have that $(x,y) \in E(\Gamma)$ if and only if $(y,x) \in E(\Gamma)$; here $x, y \in V(\Gamma)$. The condition that the graphs contain all loops is imposed mainly for transparency of statements of theorems and open problems; many statements can be easily generalized to graphs with not all loops (x,x) contained in $E(\Gamma)$ by considering the subgraph defined by the set of vertices $V_0(\Gamma) := \{x \in V(\Gamma) : (x,x) \in E(\Gamma)\}$.

Research of the second author is partially supported by NSF grant DMS 0901628.

For a given graph Γ, we label the vertices $V(\Gamma) = \{1, 2, \ldots, n\}$. The graph Γ_0 is called a *subgraph* of Γ if $V(\Gamma_0) \subseteq V(\Gamma)$ and for $i, j \in V(\Gamma_0)$ we have $(i, j) \in E(\Gamma_0)$ if and only if $(i, j) \in E(\Gamma)$. A complete subgraph of Γ is called a *clique*; that is, $K = (V(K), E(K))$ is a clique of Γ if $V(K) \subseteq V(\Gamma)$ and $E(K) = V(K) \times V(K) \subseteq E(\Gamma)$. If $\mathcal{A}$ is a C^*-algebra, then we let $M^{n \times n}(\mathcal{A})$ be the C^*-algebra of $n \times n$ matrices with entries in $\mathcal{A}$. Define the convex cone $M_+(\Gamma) \subseteq M^{n \times n}(\mathcal{A})$ by

$$M_+(\Gamma) := \left\{ [a_{i,j}]_{i,j=1}^n : \text{for every clique } K \text{ of } \Gamma \text{ the} \atop \text{matrix } [a_{i,j}]_{i,j \in K} \text{ is positive} \right\}.$$

In particular, all diagonal elements in $X \in M_+(\Gamma)$ are positive.

For a fixed C^*-algebra $\mathcal{A}$, consider the collection of graphs $G(\mathcal{A})$ with the following property: $\Gamma \in G(\mathcal{A})$ if and only if for every $X = [x_{i,j}]_{i,j=1}^n \in M_+(\Gamma)$ there exists $Y = [y_{i,j}]_{i,j=1}^n \in M^{n \times n}(\mathcal{A})$ such that

$$(i, j) \in E(\Gamma) \quad \Longrightarrow \quad y_{i,j} = x_{i,j}$$

and Y is positive. We say that the element Y with these properties is a *positive completion* of X with respect to Γ. As the direct sum of positive matrices is positive, $G(\mathcal{A})$ always contains disconnected unions of complete graphs; in some cases $G(\mathcal{A})$ contains nothing else as we will see in Theorem 10.

The following is a basic problem:

Open Problem 1. *For a given C^*-algebra $\mathcal{A}$, describe the set $G(\mathcal{A})$.*

Analogously, one defines *strictly positive completions* Y. Namely, assuming $\mathcal{A}$ is unital, let

$$M_{++}(\Gamma) := \left\{ [a_{i,j}]_{i,j=1}^n : \text{for every clique } K \text{ of } \Gamma \text{ the} \atop \text{matrix } [a_{i,j}]_{i,j \in K} \text{ is strictly positive} \right\},$$

and consider the collection of graphs $G_+(\mathcal{A})$ having the property as above for $G(\mathcal{A})$, but with $M_+(\Gamma)$ replaced by $M_{++}(\Gamma)$ and "positive" replaced by "strictly positive". In contrast with the positive completions case, the class $G_+(\mathcal{A})$ is known. The following result was proved in [13]:

Proposition 1. *A graph Γ is in $G_+(\mathcal{A})$ if and only if Γ is chordal.*

Recall that a graph Γ is said to be *chordal* if for every loop

$$(i_1, i_2), (i_2, i_3), \ldots, (i_{p-1}, i_p), (i_p, i_1) \in E(\Gamma)$$

of distinct elements $i_1, \ldots, i_p \in V(\Gamma)$, $p \geq 4$, there is a chord $(i_j, i_k) \in E(\Gamma)$ for some indices $j, k \in \{1, 2, \ldots, p\}$, $j < k - 1$. Equivalently, Γ is chordal if it does not contain minimal cycles of length ≥ 4. We denote by $\mathcal{C}h$ the collection of chordal graphs. The book [8] is a good source on chordal graphs.

The paper is organized as follows. In Section 2 we describe elementary properties of the set $G(\mathcal{A})$, as well as characterize the graphs in $G(\mathcal{A})$ for several classes of C^*-algebras $\mathcal{A}$. In Section 3 we study how $G(\mathcal{A})$ and $G(M^{m \times m}(\mathcal{A}))$ compare. In Section 4 we consider the special case of unital commutative C^*-algebras. In Sections 5 and 6 we study the completion problem in the setting of Toeplitz and multi-level Toeplitz matrices, respectively. Throughout the paper we suggest several open problems that require further research.

Notation: $\mathbb{C}$, $\mathbb{Z}$, and $\mathbb{N}$ stand for the field of complex numbers, the set of integers, and the set of positive integers, respectively.

2. The set $G(\mathcal{A})$

We start with elementary properties of positive elements.

Lemma 2. *Let $A_{11} \in M^{n \times n}(\mathcal{A})$, $A_{12} \in M^{n \times m}(\mathcal{A})$, $A_{21} \in M^{m \times n}(\mathcal{A})$, $A_{22} \in M^{m \times m}(\mathcal{A})$.*

(a) *If*
$$\begin{bmatrix} A_{11} & A_{12} \\ A_{21} & A_{22} \end{bmatrix} \in M^{(n+m) \times (n+m)}(\mathcal{A})$$
is positive, then A_{11} and A_{22} are positive.

(b) *The matrix*
$$\begin{bmatrix} A_{11} & 0 \\ 0 & A_{22} \end{bmatrix} \in M^{(n+m) \times (n+m)}(\mathcal{A})$$
is positive if and only if A_{11} and A_{22} are positive.

(c) *The matrix*
$$\begin{bmatrix} A_{11} & A_{12} \\ A_{21} & 0 \end{bmatrix} \in M^{(n+m) \times (n+m)}(\mathcal{A})$$
is positive if and only if $A_{12} = 0$, $A_{21} = 0$, and A_{11} is positive.

The proof is elementary using the fact that the set of positive elements is a convex cone; see, for instance, [12].

We introduce notation for the following special graphs: C_n is the loop on n vertices ($n \geq 1$); K_n the complete graph on n vertices ($n \geq 1$); $K_n^{(0)}$ is the complete graph in which exactly one (undirected) edge $(i, j)(= (j, i))$, $i \neq j$, is removed; here $n \geq 3$.

Proposition 3.
(a) *If $\Gamma \in G(\mathcal{A})$, then every subgraph of Γ belongs to $G(\mathcal{A})$.*
(b) *If $\Gamma_1, \Gamma_2 \in G(\mathcal{A})$, then the disconnected union of Γ_1 and Γ_2 also belongs to $G(\mathcal{A})$.*
(c) *$G(\mathcal{A})$ consists only of disconnected unions of complete graphs if and only if $K_3^{(0)} \notin G(\mathcal{A})$.*

Proof. (a) Let Γ_0 be a subgraph of Γ; we may assume $V(\Gamma) = \{1, 2, \ldots, n\}$, $V(\Gamma_0) = \{1, 2, \ldots, m\}$, $m < n$. Let $A \in M_+(\Gamma_0)$. Then
$$B := \begin{bmatrix} A & 0 \\ 0 & 0 \end{bmatrix} \in M_+(\Gamma).$$

If $Y = \begin{bmatrix} Y_{11} & Y_{12} \\ Y_{21} & Y_{22} \end{bmatrix} \in M^{n \times n}(\mathcal{A})$ is a positive completion of B with respect to Γ, then by Lemma 2 $Y_{11} \in M^{m \times m}(\mathcal{A})$ is a positive completion of A with respect to Γ_0.

(b) is clear.

(c) By part (a), we need only to show that if Γ is not a disconnected union of complete graphs, then $K_3^{(0)}$ is a subgraph of Γ. Indeed, there is a connected component Γ' of Γ which is not a complete graph. It is easy to see that Γ' contains $K_3^{(0)}$ as a subgraph. Indeed, let C be a maximal clique of Γ'. Because Γ' is connected, there exist $i \in V(\Gamma_0) \setminus V(C)$ and $k \in V(C)$ such that $(i,k) \in E(\Gamma_0)$. Because the clique C is maximal, there is $j \in V(C)$ such that $(i,j) \notin E(\Gamma_0)$. The subgraph defined by the vertices i,j,k is $K_3^{(0)}$. $\square$

Theorem 4. *For any C^*-algebra $\mathcal{A}$, we have $G(\mathcal{A}) \subseteq Ch$.*

Proof. Let Γ be a graph which is not chordal. Let $V(\Gamma) = \{1,2,\ldots,n\}$. Then by [9] there exists an $n \times n$ complex matrix $Q = [q_{i,j}]_{i,j=1}^n$ such that $[q_{i,j}]_{i,j \in K}$ is positive definite for every clique K of Γ, but

$$\max\{\text{minimal eigenvalue of } R\} < 0, \qquad (1)$$

where the maximum is taken over all matrices $R = [r_{i,j}]_{i,j=1}^n$, $r_{i,j} \in \mathbb{C}$, such that $(i,j) \in E(\Gamma) \implies r_{i,j} = q_{i,j}$. Let a be a nonzero positive element of $\mathcal{A}$, and consider

$$A = [q_{i,j}a]_{i,j=1}^n \in M^{n \times n}(\mathcal{A}).$$

Clearly, $A \in M_+(\Gamma)$. Let $Y = [y_{i,j}]_{i,j=1}^n \in M^{n \times n}(\mathcal{A})$ such that

$$(i,j) \in E(\Gamma) \implies y_{i,j} = q_{i,j}a.$$

We will verify that Y is not positive, thereby proving the theorem. Regard $\mathcal{A}$ as a norm closed *-subalgebra of $B(\mathcal{H})$, for a suitable Hilbert space $\mathcal{H}$. Then $M^{n \times n}(\mathcal{A})$ is a norm closed *-subalgebra of $M^{n \times n}(B(\mathcal{H}))$. As a is positive and nonzero there is a vector $x \in \mathcal{H}$ such that $\langle a(x), x \rangle > 0$ (indeed, take any x in the range of a). By scaling, we may choose x so that $\langle a(x), x \rangle = 1$. Let $\mathcal{M} \subseteq \mathcal{H}^n$ be the subspace spanned by the vectors

$$(x,0,\ldots,0), (0,x,0,\ldots,0), (0,0,0,\ldots,x).$$

Then the compression of A to $\mathcal{M}$ is unitarily similar to Q. Letting $P_{\mathcal{M}}$ denote the orthogonal projection on $\mathcal{M}$, we see by (1) that $P_{\mathcal{M}}(Y)|_{\mathcal{M}}$ has negative eigenvalues, hence Y cannot be positive. $\square$

Corollary 5. *Assume that $K_n^{(0)} \in G(\mathcal{A})$ for $n \geq 3$. Then $G(\mathcal{A}) = Ch$.*

Proof. It follows from the well-known perfect elimination scheme property of chordal graphs (see, e.g., [8]) that $G(\mathcal{A}) \supseteq Ch$. The other inclusion follows from Theorem 4. $\square$

For many C^*-algebras we have $G(\mathcal{A}) = Ch$; this was proved for complex numbers in [9]. It turns out that all von Neumann algebras have this property. Note that if $G(\mathcal{A}) = Ch$, then all closed two-sided *-ideals and corresponding factor algebras of $\mathcal{A}$ also have this property; see [13].

The following question, posed in [13], is still open:

Open Problem 2. *Characterize C^*-algebras $\mathcal{A}$ for which $G(\mathcal{A}) = \mathcal{C}h$.*

In [13] additional information regarding this open problem may be found. As subproblems, we propose the following:

Open Problem 3. *Characterize C^*-algebras $\mathcal{A}$ for which $G(\mathcal{A})$ consists only of disconnected unions of complete graphs.*

Open Problem 4. *Give examples (if they exist) of C^*-algebras $\mathcal{A}$ with the property that $G(\mathcal{A}) \neq \mathcal{C}h$ and $G(\mathcal{A})$ contains a graph which is not a disconnected union of complete graphs.*

3. $G(\mathcal{A})$ versus $G(M^{m \times m}(\mathcal{A}))$

The following result addresses the relation between $G(\mathcal{A})$ and $G(M^{m \times m}(\mathcal{A}))$.

Theorem 6. *For every C^*-algebra $\mathcal{A}$ we have that*

$$G(M^{m \times m}(\mathcal{A})) \subseteq G(\mathcal{A}), \qquad m \in \mathbb{N}. \tag{2}$$

Conversely, if $G(\mathcal{A})$ consists only of disconnected unions of complete graphs then so does $G(M^{m \times m}(\mathcal{A}))$. Also, if $G(\mathcal{A}) = \mathcal{C}h$, then $G(M^{m \times m}(\mathcal{A})) = \mathcal{C}h$ as well.

Proof. Let $\Gamma \in G(M^{m \times m}(\mathcal{A}))$, and let $[a_{i,j}]_{i,j=1}^n \in M_+(\Gamma)$. Put

$$A_{i,j} = \begin{bmatrix} a_{i,j} & 0 & \cdots & 0 \\ 0 & 0 & \cdots & 0 \\ \vdots & \vdots & & \vdots \\ 0 & 0 & \cdots & 0 \end{bmatrix} \in M^{m \times m}(\mathcal{A}).$$

Then $[A_{i,j}]_{i,j=1}^n \in M_+(\Gamma)$, and thus there exists a positive $Y = [Y_{i,j}]_{i,j=1}^n$ with $Y_{i,j} = A_{i,j}$ whenever $(i,j) \in E(\Gamma)$. By Lemma 2(c) we obtain that $Y_{i,j}$ must be of the form

$$Y_{i,j} = \begin{bmatrix} y_{i,j} & 0 & \cdots & 0 \\ 0 & 0 & \cdots & 0 \\ \vdots & \vdots & & \vdots \\ 0 & 0 & \cdots & 0 \end{bmatrix}, \quad 1 \leq i, j \leq n,$$

and that $[y_{i,j}]_{i,j=1}^n$ is positive with $y_{i,j} = a_{i,j}$ whenever $(i,j) \in E(\Gamma)$. This shows that $\Gamma \in G(\mathcal{A})$, proving the first part of the theorem.

When $G(\mathcal{A})$ consists only of disconnected unions of complete graphs, it follows from (2) and the fact that every $G(\mathcal{B})$ contains disconnected unions of complete graphs, that $G(M^{m \times m}(\mathcal{A}))$ also consists only of disconnected unions of complete graphs.

Next, suppose that $G(\mathcal{A}) = \mathcal{C}h$. Let Γ be a chordal graph with n vertices, and $[A_{i,j}]_{i,j=1}^n \in M_+(\Gamma)$ with $A_{i,j} = [a_{k,l}^{(i,j)}]_{k,l=1}^m \in M^{m \times m}(\mathcal{A})$. Introduce the graph $\widehat{\Gamma}$ obtained from Γ by replacing every vertex v_i of Γ by a complete graph $K_m^{(i)}$ with m

vertices. In addition, for all $(i,j) \in E(\Gamma)$ put an edge between every vertex in $K_m^{(i)}$ and $K_m^{(j)}$. This results in a chordal graph with nm vertices. Viewing $[A_{i,j}]_{i,j=1}^n = [[a_{k,l}^{(i,j)}]_{k,l=1}^m]_{i,j=1}^n$ as a member of $M^{nm \times nm}(\mathcal{A})$, and since $G(\mathcal{A}) = \mathcal{C}h$, we have that there is a positive $[[y_{k,l}^{(i,j)}]_{k,l=1}^m]_{i,j=1}^n \in M^{nm \times nm}(\mathcal{A})$ with $y_{k,l}^{(i,j)} = a_{k,l}^{(i,j)}$ whenever there is a corresponding edge in $\widehat{\Gamma}$, which exactly happens when $(i,j) \in \Gamma$. But then $\Gamma \in G(M^{m \times m}(\mathcal{A}))$ follows. Thus, $\mathcal{C}h \subseteq G(M^{m \times m}(\mathcal{A}))$, hence $\mathcal{C}h = G(M^{m \times m}(\mathcal{A}))$ by Theorem 4. $\qquad\square$

The above result provides some evidence that the answer to the following problem may be affirmative.

Open Problem 5. *Is it always true that $G(\mathcal{A}) = G(M^{m \times m}(\mathcal{A}))$, $m \in \mathbb{N}$?*

Note that if no examples exist that address Open Problem 4, then we have a positive answer to Open Problem 5. In fact, the problems are equivalent in the following sense.

Theorem 7. *If $\mathcal{A}$ is such that $G(\mathcal{A}) = G(M^{m \times m}(\mathcal{A}))$ for all $m \in \mathbb{N}$, then either $G(\mathcal{A})$ consists only of disconnected unions of complete graphs or $G(\mathcal{A}) = \mathcal{C}h$.*

Proof. Suppose that $G(\mathcal{A}) = G(M^{m \times m}(\mathcal{A}))$ for all $m \in \mathbb{N}$ and that $G(\mathcal{A})$ does not consist only of disconnected unions of complete graphs. Then, by Proposition 3(c), we have that $K_3^{(0)} \in G(\mathcal{A})$. We claim that $K_n^{(0)} \in G(\mathcal{A})$ for all $n \geq 3$, where $K_n^{(0)}$ is missing the edge $(1,n)$ (and $(n,1)$). Indeed, let

$$[a_{i,j}]_{i,j=1}^n \in M_+(K_n^{(0)}) \cap M^{n \times n}(\mathcal{A}).$$

Let now

$$B_{11} = \begin{bmatrix} a_{1,1} & 0 & \cdots & 0 \\ 0 & 0 & \cdots & 0 \\ \vdots & \vdots & & \vdots \\ 0 & 0 & \cdots & 0 \end{bmatrix}, B_{12} = B_{21}^* = \begin{bmatrix} a_{1,2} & \cdots & a_{1,n-1} \\ 0 & \cdots & 0 \\ \vdots & & \vdots \\ 0 & \cdots & 0 \end{bmatrix},$$

$$B_{22} = [a_{i,j}]_{i,j=2}^{n-1}, B_{23} = B_{32}^* = \begin{bmatrix} 0 & \cdots & 0 & a_{2,n} \\ \vdots & & \vdots & \vdots \\ 0 & \cdots & 0 & a_{n-1,n} \end{bmatrix}, B_{33} = \begin{bmatrix} 0 & \cdots & 0 & 0 \\ \vdots & & \vdots & \vdots \\ 0 & \cdots & 0 & 0 \\ 0 & \cdots & 0 & a_{n,n} \end{bmatrix}.$$

Then

$$\begin{bmatrix} B_{11} & B_{12} & 0 \\ B_{21} & B_{22} & B_{23} \\ 0 & B_{32} & B_{33} \end{bmatrix} \in M_+(K_3^{(0)}) \cap M^{3 \times 3}(M^{(n-2) \times (n-2)}(\mathcal{A})).$$

As $K_3^{(0)} \in G(\mathcal{A}) = G(M^{(n-2) \times (n-2)}(\mathcal{A}))$, there exists a positive $[B_{ij}]_{i,j=1}^3 \in M^{(n-2) \times (n-2)}(\mathcal{A})$. It follows from Lemma 2(c) that $B_{13} = B_{31}^*$ must be of the form

$$B_{11} = \begin{bmatrix} 0 & \cdots & 0 & a_{1,n} \\ 0 & \cdots & 0 & 0 \\ \vdots & & \vdots & \vdots \\ 0 & \cdots & 0 & 0 \end{bmatrix},$$

for some $a_{1,n} \in \mathcal{A}$ (that typically is different from $a_{1,n}$ given before). But then it follows that $(a_{ij})_{i,j=1}^n$ is positive in $M^{n \times n}(\mathcal{A})$. This shows that $K_n^{(0)} \in G(\mathcal{A})$. Applying Corollary 5 we may now conclude that $G(\mathcal{A}) = \mathcal{C}h$. $\qquad \square$

4. Unital commutative C^*-algebras

As is well known every unital commutative C^*-algebra is *-isomorphic to the C^*-algebra $C(X)$ consisting of continuous functions on a compact Hausdorff topological space X. As it turns out, the properties of $C(X)$ with respect to positive completions depend on topological properties of X.

The following example was given in [13]: Let $X = \overline{\mathbb{D}}$ be the closed unit disc in the complex plane, and consider the matrix

$$A := \begin{bmatrix} 1 & z & 0 \\ \overline{z} & |z|^2 & |z| \\ 0 & |z| & 1 \end{bmatrix} \in M^{3 \times 3}(C(\overline{\mathbb{D}})), \qquad z \in \overline{\mathbb{D}}. \tag{3}$$

Clearly, $A \in M_+(K_3^{(0)})$, where $K_3^{(0)}$ is realized with

$$E(K_3^{(0)}) = \{(1,1),(1,2),(2,1),(2,2),(2,3),(3,2),(3,3)\}.$$

However, there is no $y(z) \in C(\overline{\mathbb{D}})$ such that

$$\begin{bmatrix} 1 & z & y(z) \\ \overline{z} & |z|^2 & |z| \\ \overline{y(z)} & |z| & 1 \end{bmatrix}$$

is positive (indeed, for each $0 \neq z \in \overline{\mathbb{D}}$ the only option is $y(z) = z/|z|$, but this does not result in an element y of $C(\overline{\mathbb{D}})$). Hence $K_3^{(0)} \notin G(C(\overline{\mathbb{D}}))$, and therefore by Proposition 3(c), $G(C(\overline{\mathbb{D}}))$ consists only of disconnected unions of complete graphs. See Theorem 10 below for a generalization of this example.

Theorem 8. *Let $X = \beta(\mathcal{D})$ be the Čech-Stone compactification of a discrete topological space $\mathcal{D}$. Then $G(C(X)) = \mathcal{C}h$.*

For the proof we need the following fact:

Proposition 9. *Let X be a compact Hausdorff topological space.*
Then $A \in M^{n \times n}(C(X))$ is positive (in the sense of C^*-algebras) if and only if $A(x)$ is a positive semidefinite matrix for every $x \in X$.

Proof. The part "only if" is obvious. Assume now that $A(x)$ is positive semidefinite for every $x \in X$. Fix $\lambda \in \mathbb{C} \setminus [0, \infty)$. Then $A(x) - \lambda I$ is invertible for every $x \in X$, thus $\det(A(x) - \lambda I) \neq 0$. The formula

$$(A(x) - \lambda I)^{-1} = \frac{\text{algebraic adjoint } (A(x) - \lambda I)}{\det(A(x) - \lambda I)}$$

shows that $(A(x) - \lambda I)^{-1} \in M^{n \times n}(C(X))$. Therefore $\lambda \notin \sigma(A)$. Now by the well-known property that a selfadjoint element of a unital C^*-algebra is positive if and only if its spectrum is contained in $[0, \infty)$ (use, for example, [12, Theorem 2.2.1]) we obtain that A is positive. $\qquad\square$

Proof of Theorem 8. In view of Corollary 5 we need only to show that $K_n^{(0)} \in G(C(X))$ for $n \geq 3$ (with $(1, n)$ (and $(n, 1)$) as the missing edge). Let $A = [a_{i,j}]_{i,j=1}^n \in M^{n \times n}(C(X))$ $(n \geq 3)$ be such that $[a_{i,j}]_{i,j=1}^{n-1}$ and $[a_{i,j}]_{i,j=2}^n$ are both positive. Since $G(\mathbb{C})$ is the class of chordal graphs, for every $x \in \mathcal{D}$ there exists $b(x) \in \mathbb{C}$ such that the matrix

$$\begin{bmatrix} a_{1,1}(x) & a_{1,2}(x) & \cdots & a_{1,n-1}(x) & b(x) \\ a_{2,1}(x) & a_{2,2}(x) & \cdots & a_{2,n-1}(x) & a_{2,n}(x) \\ \vdots & \vdots & & \vdots & \vdots \\ a_{n-1,1}(x) & a_{n-1,2}(x) & \cdots & a_{n-1,n-1}(x) & a_{n-1,n}(x) \\ \overline{b(x)} & a_{n,2}(x) & \cdots & a_{n,n-1}(x) & a_{n,n}(x) \end{bmatrix}$$

is positive semidefinite. Clearly $\{b(x) : x \in \mathcal{D}\}$ is bounded; in fact,

$$|b(x)| \leq \max\{\|a_{1,1}\|, \|a_{n,n}\|\}.$$

Since every bounded complex-valued function on $\mathcal{D}$ can be extended to a continuous function on $\beta(\mathcal{D})$, there is $q \in C(\beta(\mathcal{D}))$ such that $q(x) = b(x)$ for every $x \in \mathcal{D}$. Let $Y \in M^{n \times n}(C(X))$ have its upper right corner q, lower left corner $\overline{q}$, and all other entries $y_{i,j}$ equal to $a_{i,j}$. Then $Y(x)$ is a positive semidefinite matrix for every $x \in \mathcal{D}$. Let now $x_0 \in \beta(\mathcal{D})$. Since $\mathcal{D}$ is dense in $\beta(\mathcal{D})$, there is a net $\{y_\alpha\}_{\alpha \in S}$ in $\mathcal{D}$ indexed by some directed set S such that $\{y_\alpha\}_{\alpha \in S}$ converges to x_0. Letting $\lambda_{\min}(M)$ denote the minimal eigenvalue of a hermitian matrix M, we have that $\lambda_{\min}(Y(y_\alpha)) \geq 0$ for all $n \in \mathbb{N}$ and all $\alpha \in S$. But then, by the continuity of $\lambda_{\min}(Y(x))$ as a function of $x \in \beta(\mathcal{D})$, we obtain $\lambda_{\min}(Y(x_0)) \geq 0$. Now, by Proposition 9, Y is positive. $\qquad\square$

Theorem 10. *Let X be a compact Hausdorff space with a converging sequence of distinct elements. Then $G(C(X))$ consists of disconnected unions of complete graphs.*

Proof. By Proposition 3(c) we need only to check that $K_3^{(0)} \notin G(C(X))$. But this was done in [4]. $\qquad\square$

According to [11], a description of compact Hausdorff spaces without converging sequences of distinct elements is an extremely difficult problem, nearly intractable.

5. Positive Toeplitz completions

We consider the banded Toeplitz positive completion problem, as follows. We say that $\mathcal{A}$ is in the *positive Toeplitz completion class* (in short, $\mathcal{A} \in \mathcal{PTC}$) if for all $0 < k < n$ and $[a_{i-j}]_{i,j=0}^{k-1}$ positive in $M^{n \times n}(\mathcal{A})$, there exists a positive Toeplitz $Y = [y_{i-j}]_{i,j=0}^{n-1} \in M^{n \times n}(\mathcal{A})$ with $y_l = a_l$, $|l| \leq k - 1$. Note that the positivity of $[a_{i-j}]_{i,j=0}^{k-1}$ is necessary for existence of a positive Y as above. This is an important particular case of the general positive completion problem, and has been much studied for $\mathcal{A} = B(\mathcal{H})$, the algebra of linear bounded operators on a Hilbert space $\mathcal{H}$; see [2] and references therein.

The basic problem here is:

Open Problem 6. *Describe the positive Toeplitz completion class* $\mathcal{PTC}$.

It is straightforward to see that every $\mathcal{A}$ with $G(\mathcal{A}) = \mathcal{C}h$ is in the class $\mathcal{PTC}$. The following result, in conjunction with Theorem 10, shows that there exist unital commutative C^*-algebras $\mathcal{A}$ in the class $\mathcal{PTC}$ for which $G(\mathcal{A})$ is the collection of disconnected unions of complete graphs.

Theorem 11. *Let X be a compact Hausdorff space. Then $C(X)$ is in the positive Toeplitz completion class.*

Proof. It is sufficient to consider the case $k = n - 1$ in the definition of $\mathcal{PTC}$. Let $[a_{i-j}(x)]_{i,j=0}^{n-2} \in M^{(n-1) \times (n-1)}(C(X))$ be positive. In particular,

$$a_\ell(x) = \overline{a_{-\ell}(x)}, \quad x \in X, \quad \ell = 0, 1, \ldots, n - 2.$$

For every $x \in X$, consider the unique $y = y(x) \in \mathbb{C}$, so that

$$B(x) := \begin{bmatrix} a_0(x) & a_{-1}(x) & \cdots & a_{-n+2}(x) & y \\ a_1(x) & a_0(x) & a_{-1}(x) & \cdots & a_{-n+2}(x) \\ \vdots & \ddots & \ddots & \ddots & \vdots \\ a_{n-2}(x) & \cdots & a_1(x) & a_0(x) & a_{-1}(x) \\ \overline{y} & a_{n-2}(x) & \cdots & a_1(x) & a_0(x) \end{bmatrix},$$

satisfies:

(1) $\left(\det [a_{i-j}(x)]_{i,j=0}^{n-3}\right) \cdot \det B(x) = \left(\det [a_{i-j}(x)]_{i,j=0}^{n-2}\right)^2$,

(2) the matrix $B(x)$ is positive semidefinite.

This particular completion is called the *central completion*; see for instance, [3], [2, Section 2.6]. Note also that y is bounded as a function of x:

$$|y|^2 \leq a_0(x)^2 \leq M := \max_{x \in X} a_0(x)^2.$$

We claim that y is a continuous function of $x \in X$. In view of Proposition 9 this will prove that B is positive and thereby the proof of Theorem 11 will be complete. Arguing by contradiction, assume $y(x)$ is not continuous at some $x_0 \in X$. (All notions and basic results on net convergence and subnets that we use here can be found in [10], for example.) Let $\{U_m\}_{m \in S}$ be a base of open neighborhoods of x_0, indexed by a directed set S with the direction $\succeq$ compatible with the containment of neighborhoods: if $m_1, m_2 \in S$ then $m_1 \succeq m_2$ if and only if $U_{m_1} \subseteq U_{m_2}$. It follows that there exist $\epsilon_0 > 0$ and a net $\{x_m\}_{m \in S}$ such that $x_m \in U_m$ for every $m \in S$ and

$$|y(x_m) - y(x_0)| \geq \epsilon_0 \quad \text{for all } m \in S. \tag{4}$$

Clearly the net $\{x_m\}_{m \in S}$ converges to x_0. Since $\{y(x_m)\}_{m \in S}$ is a net of uniformly bounded complex numbers, hence contained in a compact set of $\mathbb{C}$, there is a converging subnet of $\{y(x_m)\}_{m \in S}$. In other words, there exist a net $\{\tilde{y}_p\}_{p \in \tilde{S}}, \tilde{y}_p \in \mathbb{C}$ for all $p \in \tilde{S}$, where $\tilde{S}$ is a directed set with the direction given by $\geq$, and a function $\phi : \tilde{S} \to S$ with the following properties:

(3) $\tilde{y}_p = y(x_{\phi(p)})$ for every $p \in \tilde{S}$;

(4) for every $m \in S$ there is $q \in \tilde{S}$ such that

$$p \geq q, \ p \in \tilde{S} \quad \Longrightarrow \quad \phi(p) \succeq m.$$

(5) the net $\{\tilde{y}_p\}_{p \in \tilde{S}}$ converges to some $y_0 \in \mathbb{C}$.

In view of (3), we have for every $p \in \tilde{S}$:

$$\left(\det [a_{i-j}(x)]_{i,j=0}^{n-3}\right) \cdot \det \begin{bmatrix} a_0(x) & a_{-1}(x) & \cdots & a_{-n+2}(x) & y \\ a_1(x) & a_0(x) & a_{-1}(x) & \cdots & a_{-n+2}(x) \\ \vdots & \ddots & \ddots & \ddots & \vdots \\ a_{n-2}(x) & \cdots & a_1(x) & a_0(x) & a_{-1}(x) \\ \overline{y} & a_{n-2}(x) & \cdots & a_1(x) & a_0(x) \end{bmatrix}$$

$$= \left(\det [a_{i-j}(x)]_{i,j=0}^{n-2}\right)^2, \tag{5}$$

where $x = x_{\phi(p)}$ and $y = \tilde{y}_p$. It follows (using (4), (5), the convergence of the net $\{x_m\}_{m \in S}$ to x_0, and the continuity of the functions $a_j(x)$, $j = -n+2, \ldots, n-2$) that

$$\left(\det [a_{i-j}(x_0)]_{i,j=0}^{n-3}\right)$$

$$\cdot \det \begin{bmatrix} a_0(x_0) & a_{-1}(x_0) & \cdots & a_{n-2}(x_0) & y_0 \\ a_1(x_0) & a_0(x_0) & a_{-1}(x_0) & \cdots & a_{n-2}(x_0) \\ \vdots & \ddots & \ddots & \ddots & \vdots \\ a_{-n+2}(x_0) & \cdots & a_1(x_0) & a_0(x_0) & a_{-1}(x_0) \\ \overline{y_0} & a_{-n+2}(x_0) & \cdots & a_1(x_0) & a_0(x_0) \end{bmatrix}$$

$$= \left(\det [a_{i-j}(x_0)]_{i,j=0}^{n-2}\right)^2. \tag{6}$$

Indeed, fix $\epsilon > 0$. Then there exists $q \in \tilde{S}$ such that $|\tilde{y}_p - y_0| < \epsilon$ for all $p \geq q$, $p \in \tilde{S}$ (property (5)), there exists $m_0 \in S$ with the property that

$$|a_j(x_m) - a_j(x_0)| < \epsilon, \quad j = -n+2, \ldots, n-2, \quad \text{for all } m \in S \text{ such that } m \succeq m_0$$

(continuity of the a_j's at x_0), and there exists $q' \in \widetilde{S}$ such that

$$p \geq q', \; p \in \widetilde{S} \quad \Longrightarrow \quad \phi(p) \succeq m_0$$

(property 4)). Choose $q'' \in \widetilde{S}$ so that $q'' \succeq q, q'$. Then

$$p \geq q'', \; p \in \widetilde{S} \Longrightarrow |\widetilde{y}_p - y_0| < \epsilon, \; |a_j(x_{\phi(p)}) - a_j(x_0)| < \epsilon, \; j = -n+2, \ldots, n-2.$$

Since $\epsilon > 0$ is arbitrary, (6) follows from (5). A similar argument shows that

$$
\begin{bmatrix}
a_0(x_0) & a_{-1}(x_0) & \cdots & a_{-n+2}(x_0) & y_0 \\
a_1(x_0) & a_0(x_0) & a_{-1}(x_0) & \cdots & a_{-n+2}(x_0) \\
\vdots & \ddots & \ddots & \ddots & \vdots \\
a_{n-2}(x_0) & \cdots & a_1(x_0) & a_0(x_0) & a_{-1}(x_0) \\
\overline{y_0} & a_{n-2}(x_0) & \cdots & a_1(x_0) & a_0(x_0)
\end{bmatrix}
$$

is positive semidefinite (we use here also the continuity of the minimal eigenvalue of a hermitian matrix as a function of the matrix entries, analogously to the proof of Theorem 8). By the uniqueness of the central completion we must have $y_0 = y(x_0)$. But now (4) gives (using (3)): $|\widetilde{y}_p - y_0| \geq \epsilon_0$ for all $p \in \widetilde{S}$, a contradiction with (5). $\qquad\square$

It should be observed that the Toeplitz property is essential when we note that (1) and (2) above determine uniquely the central completion. Indeed, for

$$[a_{i,j}]_{i,j=1}^3 = \begin{bmatrix} 1 & 0 & ? \\ 0 & 0 & 0 \\ ? & 0 & 1 \end{bmatrix}$$

the unique central completion (found by choosing $? = 0$) is not uniquely determined by requiring that $[a_{i,j}]_{i,j=1}^3$ is positive and

$$a_{2,2} \cdot \det[a_{i,j}]_{i,j=1}^3 = \det[a_{i,j}]_{i,j=1}^2 \det[a_{i,j}]_{i,j=2}^3.$$

If $\mathcal{A}$ is unital, then strictly positive Toeplitz completions can be considered. Replacing "positive" by "strictly positive" in the definition of the $\mathcal{PTC}$ class, we obtain the definition of the *strictly positive Toeplitz completion class*, in short $\mathcal{PTC}_+$.

Proposition 12. $\mathcal{A} \in \mathcal{PTC}_+$ *for every unital C^*-algebra $\mathcal{A}$.*

Proof. As in the proof of Theorem 11, we consider only the case $k = n-1$ in the definition of $\mathcal{PTC}_+$; thus we are given that $[a_{i-j}]_{i,j=0}^{n-2}$ strictly positive. But then since $K_n^{(0)}$ is chordal we have by Proposition 1 that there exists a strictly positive $Y = [y_{i,j}]_{i,j=0}^{n-1}$ with $y_{i,j} = a_{i-j}$ for $|i - j| \leq n-2$, $0 \leq i, j \leq n-1$. As Y is automatically Toeplitz, we are done. $\qquad\square$

We conclude this section with the following open problem.

Open Problem 7. *Is it true that if $\mathcal{A} \in \mathcal{PTC}$, then also $M^{n \times n}(\mathcal{A}) \in \mathcal{PTC}$, for all $n \in \mathbb{N}$?*

6. Multi-level positive Toeplitz completions

A d-level Toeplitz matrix is a matrix of the form $[a_{k-l}]_{k,l\in\Lambda}$, where Λ is a finite set of $\mathbb{Z}^d$, and the entries are in a C^*-algebra $\mathcal{A}$. For instance, if

$$\Lambda = \{0,1\}^2 = \{(0,0),(0,1),(1,0),(1,1)\},$$

then we get the two level Toeplitz matrix

$$\begin{bmatrix} a_{00} & a_{0,-1} & a_{-1,0} & a_{-1,-1} \\ a_{01} & a_{00} & a_{-1,1} & a_{-1,0} \\ a_{10} & a_{1,-1} & a_{0,0} & a_{0,-1} \\ a_{11} & a_{10} & a_{01} & a_{00} \end{bmatrix}.$$

Since Λ is of the form $K \times L$ with $K, L \subseteq \mathbb{Z}$, this matrix is a (in this case, 2×2) block Toeplitz matrix where each of the blocks are themselves (2×2) Toeplitz matrices.

The following problem was studied before in the context of matrices over $\mathbb{C}$. Fix a C^*-algebra $\mathcal{A}$. We say that a finite subset Λ of $\mathbb{Z}^d$ *has the $\mathcal{A}$-extension property*, if whenever $[a_{i-j}]_{i,j\in\Lambda} \in M^{|\Lambda|\times|\Lambda|}(\mathcal{A})$ is positive and $K \supseteq \Lambda$ is a finite subset of $\mathbb{Z}^d$, then there exist $a_l \in \mathcal{A}$, $l \in (K-K)\setminus(\Lambda-\Lambda)$ so that $[a_{i-j}]_{i,j\in K} \in M^{|K|\times|K|}(\mathcal{A})$ is positive. (Here, we denote by $|Y|$ the cardinality of a finite set Y.) Clearly, for any $p \in \mathbb{Z}^d$ we have that Λ has the $\mathcal{A}$-extension property if and only if $p + \Lambda$ has the $\mathcal{A}$-extension property.

Theorem 11 shows that $\{0,\ldots,k-1\}$ has the $C(X)$-extension property. It is a classical result that $\{0,\ldots,k-1\}$ has the $\mathbb{C}$-extension property. The following known results all address the case when $\mathcal{A} = \mathbb{C}$.

Theorem 13. [5] *Let $d = 1$. Then $\Lambda \subseteq \mathbb{Z}$ has the $\mathbb{C}$-extension property if and only if $\Lambda = \{p + kq : k \in \{0,\ldots,n\}\}$ for some $p, q, n \in \mathbb{Z}$.*

Theorem 14. [1] *Let $d = 2$. Let $\Lambda \subseteq \mathbb{Z}^2$ be so that $\mathbb{Z}^2$ is the smallest group containing Λ. Then Λ has the $\mathbb{C}$-extension property if and only if for some $a \in \mathbb{Z}^2$ the set $\Lambda - a = \Phi(R)$ for some group-homomorphism $\Phi : \mathbb{Z}^2 \to \mathbb{Z}^2$ and some set R of the form $\{0\} \times \{0,\ldots,n\}$, $\{0,1\} \times \{0,\ldots,n\}$, or $(\{0,1\} \times \{0,\ldots,n\})\setminus\{(1,n)\}$, where $n \in \{0,1,2,\ldots\}$.*

Theorem 15. [14, 15] *For $n_1, n_2, n_3 \geq 1$ the set*

$$\{0,\ldots,n_1\} \times \{0,\ldots,n_2\} \times \{0,\ldots,n_3\} \subseteq \mathbb{Z}^3$$

does not have the $\mathbb{C}$-extension property.

It is easy to see that if Λ does not have the $\mathbb{C}$-extension property, then it also does not have the $\mathcal{A}$-extension property for any C^*-algebra $\mathcal{A}$ (indeed, if the positive matrix $[c_{i-j}]_{i,j\in\Lambda}$ with $c_k \in \mathbb{C}$ shows that Λ does not have the $\mathbb{C}$-extension property, then $[c_{i-j}a^*a]_{i,j\in\Lambda}$, where $a \in \mathcal{A} \setminus \{0\}$, will show that Λ does not have the $\mathcal{A}$-extension property).

We have the following observation.

Proposition 16. *Let $d = 1$ and suppose that $\mathcal{A}$ is in the class $\mathcal{PTC}$. Then $\Lambda \subseteq \mathbb{Z}$ has the $\mathcal{A}$-extension property if and only if $\Lambda = \{a + kb : k \in \{0, \ldots, n\}\}$ for some $a, b, n \in \mathbb{Z}$.*

Proof. Assume that $\mathcal{A}$ is in the class $\mathcal{PTC}$. By definition of this class it follows immediately that $\{0, \ldots, k\}$ has the $\mathcal{A}$-extension property. Next, if $\Lambda = \{0, n, 2n, \ldots, kn\}$ and $a_{ln} \in \mathcal{A}$, $-k, \ldots, k$, are given, one may then obtain a_{np}, $|p| > k$, so that $[a_{n(p-q)}]_{p,q \in K}$ is positive for every finite K. Setting $a_l = 0$ for $l \notin n\mathbb{Z}$ yields that $[a_{i-j}]_{i,j \in K}$ is positive for every finite K. Thus $\Lambda = \{0, n, 2n, \ldots, kn\}$ has the $\mathcal{A}$-extension property. But then it follows that $\Lambda = \{p + kq : k \in \{0, \ldots, n\}\}$ also has the $\mathcal{A}$-extension property. Finally, no other finite subsets can have the $C(X)$-extension property due to Theorem 13 (and the observation before the proposition). $\qquad\square$

Open Problem 8. *What C^*-algebras $\mathcal{A}$ have the property that Λ has the $\mathbb{C}$-extension property if and only if Λ has the $\mathcal{A}$-extension property?*

If we restrict Open Problem 8 to $\Lambda \subseteq \mathbb{Z}$ the question is the same as Open Problem 6.

Open Problem 9. *What if we restrict Open Problem 8 to $\Lambda \subseteq \mathbb{Z}^2$? Is $C(X)$, X a compact Hausdorff space, one of these?*

Let $\mathcal{A}$ be a unital C^*-algebra. We say that a finite subset Λ of $\mathbb{Z}^d$ has the $\mathcal{A}$-extension property in the strict case, if whenever $[a_{i-j}]_{i,j \in \Lambda}$ is strictly positive (in $M^{|\Lambda| \times |\Lambda|}(\mathcal{A})$) and $K \supseteq \Lambda$ is a finite subset of $\mathbb{Z}^d$, then there exist $a_l \in \mathcal{A}$, $l \in (K - K) \setminus (\Lambda - \Lambda)$ so that $[a_{i-j}]_{i,j \in K}$ is strictly positive.

Open Problem 10. *For which unital $\mathcal{A}$ is it true that Λ has the $\mathcal{A}$-extension property if and only if Λ has the $\mathcal{A}$-extension property in the strict case?*

As is clear from Theorem 14, there are relatively few subsets of $\mathbb{Z}^2$ that have the $\mathbb{C}$-extension property. For instance

$$\Lambda = (\{0, \ldots, n\} \times \{0, \ldots, m\}) \setminus \{(n, m)\}, \tag{7}$$

fails to have the $\mathbb{C}$-extension property when $n, m \geq 2$. Thus, the existence of a positive completion is not guaranteed simply by the positivity of $[a_{i-j}]_{i,j \in \Lambda}$. The following result gives sufficient conditions for the existence of a positive completion in the strictly positive case.

Theorem 17. *Let $\mathcal{A}$ be a unital C^*-algebra. Fix $m, n \in \mathbb{N}$, let*

$$\Lambda = (\{0, \ldots, n\} \times \{0, \ldots, m\}) \setminus \{(n, m)\}, \tag{8}$$

and let a_k, $k \in \Lambda - \Lambda$ be given. Suppose that:

(i) *$[a_{k-l}]_{k,l \in \Lambda}$ and $[a_{k-l}]_{k,l \in -\Lambda}$ are strictly positive in*

$$M^{(nm+n+m) \times (nm+n+m)}(\mathcal{A}),$$

(ii) $\Phi_1 \Phi^{-1} \Phi_2^* = \Phi_2^* \Phi^{-1} \Phi_1$;

(iii) $a_{-n,m} = [a_{k-l}]_{\substack{k=(0,m-1) \\ l\in\{1,\ldots,n\}\times\{0,\ldots,m-1\}}} \quad \Phi^{-1}[a_{k-l}]_{\substack{k\in\{0,\ldots,n-1\}\times\{1,\ldots,m\} \\ l=(n-1,0)}}.$

Here

$$\Phi = [a_{k-l}]_{k,l\in\{0,\ldots,n-1\}\times\{0,\ldots,m-1\}},$$

$$\Phi_1 = [a_{k-l}]_{k\in\{0,\ldots,n-1\}\times\{0,\ldots,m-1\},\ l\in\{1,\ldots,n\}\times\{0,\ldots,m-1\}},$$

$$\Phi_2 = [a_{k-l}]_{k\in\{0,\ldots,n-1\}\times\{0,\ldots,m-1\},\ l\in\{0,\ldots,n-1\}\times\{1,\ldots,m\}}.$$

Then for every finite $K \subseteq \mathbb{Z}^2$ that properly contains Λ there exist $a_l \in \mathcal{A}$, $l \in (K - K) \setminus (\Lambda - \Lambda)$ such that $[a_{k-l}]_{k,l\in K}$ is strictly positive in $M^{|K|\times|K|}(\mathcal{A})$.

Proof. For $\mathcal{A} = B(\mathcal{H})$, the C^*-algebra of bounded linear operators on a Hilbert space $\mathcal{H}$, the result is a direct consequence of [7, Theorem 1.1] (see also [2, Theorem 3.3.1]) after defining

$a_{n,m} = [a_{k-l}]_{\substack{k=(n,m) \\ l\in\{0,\ldots,n\}\times\{0,\ldots,m\}\setminus\{(0,0),(n,m)\}}}$

$\times [[a_{k-l}]_{k,l\in\{0,\ldots,n\}\times\{0,\ldots,m\}\setminus\{(0,0),(n,m)\}}]^{-1} [a_{k-l}]_{\substack{k\in\{0,\ldots,n\}\times\{0,\ldots,m\}\setminus\{(0,0),(n,m)\} \\ l=(0,0)}}.$

For the general case, use the fact that $\mathcal{A}$ is *-isomorphic to a norm-closed selfadjoint subalgebra of $B(\mathcal{H})$ for a suitable Hilbert space $\mathcal{H}$, and observe that in the proof of [7, Theorem 1.1] all the missing a_l's are constructed using formulas of the form $a_l = PQ^{-1}R$, where P, Q, and R are finite matrices with entries in $\mathcal{A}$. But then it follows immediately that $a_l \in \mathcal{A}$ for all l, and the proof is finished. □

Condition (i) in Theorem 17 is also a necessary condition for the existence of a strictly positive completion. Conditions (ii) and (iii) are necessary conditions in case the missing a_l's are required to appear as moments of an $\mathcal{A}$-valued Bernstein-Szegő measure; for details, see [6], [7] or [2, Section 3.3]. For multilevel Toeplitz matrices $[a_{i-j}]_{i,j\in\Lambda}$, where now $\Lambda \subseteq \mathbb{Z}^d$, a generalization of the above result is not known.

Open Problem 11. *Prove a result analogous to Theorem 17, where now K is an appropriate set in $\mathbb{Z}^d$ with $d \geq 3$. Better yet, prove a d-variable generalization of [7, Theorem 2.1] (or [2, Theorem 3.3.1]), which is the Bernstein-Szegő measure result.*

Acknowledgement

We thank David J. Lutzer for consultations concerning topology of compact Hausdorff spaces.

References

[1] Mihály Bakonyi and Geir Nævdal. The finite subsets of $\mathbb{Z}^2$ having the extension property. *J. London Math. Soc.* (2), 62(3):904–916, 2000.

[2] Mihály Bakonyi and Hugo J. Woerdeman. *Matrix completions, moments, and sums of Hermitian squares.* Princeton University Press, Princeton, NJ, 2011.

[3] Mihály Bakonyi and Dan Timotin. The central completion of a positive block operator matrix. *Operator theory, structured matrices, and dilations, Theta Ser. Adv. Math.*, 7:69–83, 2007; Theta, Bucharest.

[4] Kenneth R. Davidson. Lifting positive elements in C^*-algebras. *Integral Equations Operator Theory*, 14(2):183–191, 1991.

[5] Jean-Pierre Gabardo. Trigonometric moment problems for arbitrary finite subsets of $\mathbf{Z}^n$. *Trans. Amer. Math. Soc.*, 350(11):4473–4498, 1998.

[6] Jeffrey S. Geronimo and Hugo J. Woerdeman. Positive extensions, Fejér-Riesz factorization and autoregressive filters in two variables. *Ann. of Math.* (2), 160(3):839–906, 2004.

[7] Jeffrey S. Geronimo and Hugo J. Woerdeman. The operator-valued autoregressive filter problem and the suboptimal Nehari problem in two variables. *Integral Equations Operator Theory*, 53(3):343–361, 2005.

[8] Martin Charles Golumbic. *Algorithmic graph theory and perfect graphs*, volume 57 of *Annals of Discrete Mathematics*. Elsevier Science B.V., Amsterdam, second edition, 2004.

[9] Robert Grone, Charles R. Johnson, Eduardo M. de Sá, and Henry Wolkowicz. Positive definite completions of partial Hermitian matrices. *Linear Algebra Appl.*, 58:109–124, 1984.

[10] John L. Kelly. *General Topology*. D. Van Nostrand Company, Princeton, New Jersey, 1957.

[11] Jan van Mill, private communication.

[12] Gerard J. Murphy. C^*-*algebras and operator theory*. Academic Press Inc., Boston, MA, 1990.

[13] Vern I. Paulsen and Leiba Rodman. Positive completions of matrices over C^*-algebras. *J. Operator Theory*, 25(2):237–253, 1991.

[14] L.A. Sahnovič. Effective construction of noncontinuable Hermite-positive functions of several variables. *Funktsional. Anal. i Prilozhen.*, 14(4):55–60, 96, 1980.

[15] L.A. Sakhnovich. *Interpolation theory and its applications*, volume 428 of *Mathematics and its Applications*. Kluwer Academic, Dordrecht, 1997.

L. Rodman
Department of Mathematics
College of William and Mary
Williamsburg, VA, USA
e-mail: `lxrodm@math.wm.edu`

H.J. Woerdeman
Department of Mathematics
Drexel University
Philadelphia, PA, USA
e-mail: `hugo@math.drexel.edu`

Operator Theory:
Advances and Applications, Vol. 222, 295–309
© 2012 Springer Basel

Fractional-order Systems and the Internal Model Principle

Mir Shahrouz Takyar and Tryphon T. Georgiou

Dedicated to Bill Helton on his 65th birthday

Abstract. We consider the *fractional-integrator* as a feedback design element. It is shown, in a simple setting, that the fractional integrator ensures zero steady-state tracking. This observation should be contrasted with the typical formulation of the internal model principle which requires a *full integrator* in the loop for such a purpose. The use of a fractional integrator allows increased stability margin, trading-off phase margin against the rate of convergence to steady-state. A similar rationale can be applied to tracking sinusoidal signals. Likewise, in this case, fractional poles on the imaginary axis suffice to achieve zero steady-state following and disturbance rejection. We establish the above observations for cases with simple dynamics and conjecture that they hold in general. We also explain and discuss basic implementations of a fractional integrating element.

Mathematics Subject Classification. 93C80.

Keywords. Fractional calculus, internal model principle.

1. Introduction

Fractional derivatives have been used to model a wide range of physical systems. In particular, they are encountered in distributed parameter models of diffusive systems, delay lines, electromagnetic multipoles, turbulent flow, and many others (see, e.g., [1, 2, 3]). They are also used to account for characteristic of spectra with power decay that differs from the typical integer multiple of 20 [dB/dec]. Examples include $1/f$ noise, speech, music, electrocardiogram (ECG) signals, and

This work was supported in part by the National Science Foundation and by the Air Force Office for Scientific Research.

certain other natural processes (see, e.g., [4]). While fractional models are infinite-dimensional, their algebraic representation is rather compact. Moreover, the algebra and function theory of fractional integrators and fractional derivatives are well developed (e.g., [5, 6]). Fractional elements have been considered in control applications both, for modeling processes [7, 8, 9] as well as for controller design [10]. We refer to [11] for a review of the current literature as well as for a thorough list of references on the subject.

In this paper, we focus on a special class of dynamical systems which, in the frequency domain, are governed by a fractional power of the Laplace variable s, focusing on the fractional integrator. Our interest in this stems from the fact that, for control purposes, such elements provide enough gain to ensure steady-state tracking with a more efficient compromise between the convergence rate and phase lag of the loop gain. Indeed, in contrast to the typical statement of the internal model principle in textbooks and publications, a full integrator in the loop is not necessary for asymptotic tracking – a fractional integrator is sufficient. Such fractional elements provide infinite DC-gain which is what ensures zero steady-state error.

The transfer function of the fractional integrator in the Laplace domain is $1/s^{\alpha}$ for $0 < \alpha < 1$. For the most part we consider the case where the exponent α is equal to $1/2$. In general, $1/s^{\alpha}$ for $0 < \alpha < 1$ is positive-real very much like $1/s$. As a consequence, in principle, it is realizable with (infinitely many) passive elements as in a transmission line. Obviously, a truncated transmission line as we discuss later on, will provide an approximation. Yet, nothing precludes the physical realization of such elements by a more compact device, like a capacitor, based on a suitable geometry and material. Accordingly, envisioning the eventual physical embodiment of such a device, we often refer to $1/\sqrt{s}$ as a *half-capacitor*. We present and compare different finite-dimensional approximations. The range of frequencies with the required "−10 [dB/dec] attenuation along with a −45° phase lag" depends on the number and values of components used and the chosen geometry. The DC-gain, while large, is necessarily finite and hence, the ideal tracking response is not achieved. These factors and the corresponding tradeoffs in the fabrication process may play a role in a potential choice for such designs.

The contribution of this study consists in pointing out the significance and potential use of fractional integrators and "sinusoidal" fractional components as feedback design elements for steady-state tracking performance. Micro-electro-mechanical (MEMS) and integrated circuit realizations of such components may also present interesting possibilities for future research.

2. Fractional elements in a feedback loop

2.1. The half-integrator

We consider the simple control loop in Figure 1 with the unity feedback gain, and study the tracking behavior of this loop for a step command input r. The transfer

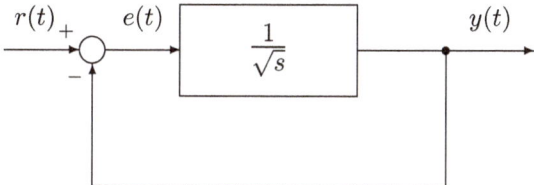

FIGURE 1. Half-integrator with negative unity feedback.

functions from r to the "tracking error" signal e and "output" y are

$$S(s) = \frac{\sqrt{s}}{\sqrt{s}+1}, \qquad \text{and} \qquad T(s) = \frac{1}{\sqrt{s}+1},$$

respectively. Figure 2 shows the step responses of the closed-loop system, to e and y, obtained via taking the inverse Laplace transform. As shown later on, $e(t)$ converges to zero while the output y follows the step input r. However, as expected, compared to a full-integrator in the place of $1/\sqrt{s}$, the convergence rate is considerably slower.

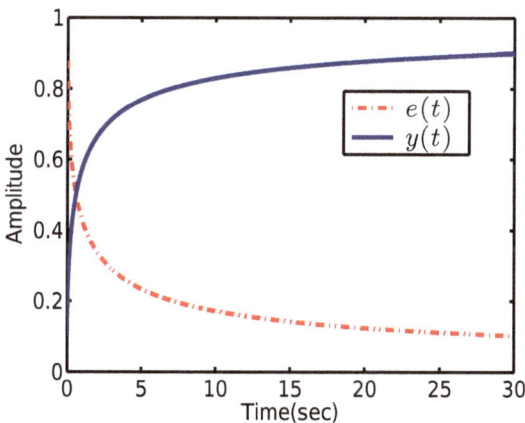

FIGURE 2. Error and output signals for a step input in the feedback system of Figure 1.

Assuming that r is a unity step input,

$$\mathcal{L}\{e(t)\} =: E(s)$$

$$= \frac{1}{s}S(s) = \frac{1}{\sqrt{s}} - \frac{1}{\sqrt{s}+1}.$$

Since

$$\mathcal{L}^{-1}\left\{\frac{1}{\sqrt{s}}\right\} = \frac{1}{\sqrt{\pi t}}$$

(see, e.g., [12, 6, 13]), and

$$\mathcal{L}^{-1}\left\{\frac{1}{\sqrt{s}+1}\right\} = \frac{1}{\sqrt{\pi t}} - \exp(t)\left(1 - \frac{2}{\sqrt{\pi}}\int_0^{\sqrt{t}}\exp(-\tau^2)d\tau\right),$$

it follows that

$$e(t) = \exp(t)\left(1 - \frac{2}{\sqrt{\pi}}\int_0^{\sqrt{t}}\exp(-\tau^2)d\tau\right).$$

Now, applying L'Hospital's rule, the steady-state value of the error is

$$\lim_{t\to\infty} e(t) = \lim_{t\to\infty}\frac{1 - \frac{2}{\sqrt{\pi}}\int_0^{\sqrt{t}}\exp(-\tau^2)d\tau}{\exp(-t)} = \lim_{t\to\infty}\frac{\frac{1}{\sqrt{\pi t}}\exp(-t)}{\exp(-t)} = 0.$$

Remark 1. In order to apply the "final value theorem" and evaluate $\lim_{t\to\infty} e(t) = \lim_{s\to 0} sE(s)$, the time function $e(t)$ must have a limit. The above steps establish precisely this for the special case at hand. Then, of course,

$$\lim_{s\to 0} sE(s) = \lim_{s\to 0}\frac{\sqrt{s}}{\sqrt{s}+1} = 0$$

gives the same value.

2.2. A sinusoidal tracking element

We carry out a similar analysis for yet another example of a fractional-order system – one with transfer function $1/\sqrt{s^2+\omega_o^2}$. This again is passive (positive real) and infinite-dimensional. Likewise the effect of such an element within a stable feedback loop is to ensure perfect tracking and disturbance rejection of sinusoidal signals at frequency ω_o. The loop gain has a singularity at ω_o, which creates a notch in the closed-loop response, which in turn suffices to ensure steady-state performance.

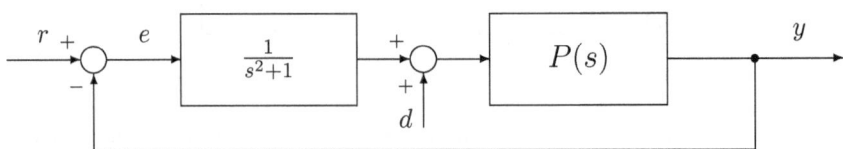

FIGURE 3. Traditional sinusoidal tracking.

The traditional "internal model" for sinusoidal disturbance rejection/tracking in a feedback system is shown in Figure 3. Here, $P(s)$ denotes the plant transfer function. Assuming a stable feedback loop, the infinite loop gain at $\omega_o = 1$ (rad/sec) due to $\frac{1}{s^2+1}$, ensures perfect tracking/rejection of sinusoidal inputs at that frequency. A possible down side is of course the phase that the poles at $s = \pm i$ introduce.

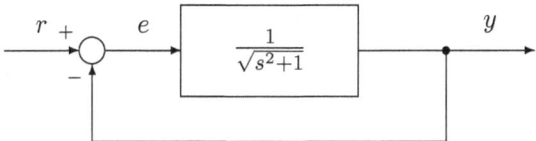

FIGURE 4. Fractional dynamics for sinusoidal tracking.

Consider now the feedback system shown in Figure 4 with trivial plant dynamics. The transfer function from r to the error signal e is

$$S(s) = \frac{\sqrt{s^2+1}}{\sqrt{s^2+1}+1}.$$

Then, for a sinusoidal input at frequency $\omega_o = 1$ (rad/sec),

$$E(s) = \frac{1}{s^2+1}S(s) = \frac{1}{\sqrt{s^2+1}} - \frac{1}{\sqrt{s^2+1}+1}.$$

The first term in the equation above is the Laplace transform of $J_0(t)$, the zero-order causal Bessel function of the first kind. The asymptotic expansion of the Bessel function $J_k(t)$ is well known (see [14]) and for large values of t, i.e., $t \gg |k^2 - \frac{1}{4}|$,

$$J_k(t) \approx \sqrt{\frac{2}{\pi t}} \cos(t - \frac{k\pi}{2} - \frac{\pi}{4}).$$

The second term can be shown to be analytic in the right half-plane and bounded on the imaginary axis. These together imply that, in the time domain, both terms of the error-signal approach zero at the steady-state and hence, $\lim_{t\to\infty} e(t) = 0$. As before, the phase introduced by the fractional element is half of that of $1/(s^2+1)$.

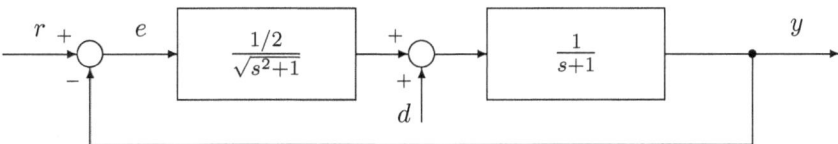

FIGURE 5. Sinusoidal tracking and disturbance rejection in a closed-loop system.

We offer yet another example of sinusoidal tracking using a fractional element as in Figure 5. This is interesting for one additional reason: the feedback system with two $\frac{1}{2}$-fractional poles at $\pm i$, a single pole at -1, and no zero, is stable (as we explain below). Yet, a similar feedback system with simple poles at $\pm i$ instead (e.g., with $P(s) = 1/(s+1)$ in Figure 3) is easily shown to be unstable by a "root-locus" argument. Thus, the two $\frac{1}{2}$-fractional poles are *considerably easier to* "stabilize" than simple poles on the imaginary axis!

Returning to the system shown in Figure 5,

$$S(s) = \frac{\sqrt{s^2+1}(s+1)}{\sqrt{s^2+1}(s+1) + 1/2},$$

while

$$E(s) = \frac{1}{s^2+1}S(s) = \frac{s+1}{\sqrt{s^2+1}\left(\sqrt{s^2+1}(s+1) + 1/2\right)}$$

$$= \underbrace{2\left(\frac{1}{\sqrt{s^2+1}} + \frac{s}{\sqrt{s^2+1}} - 1\right)}_{\text{first term}} - \underbrace{2\left(\frac{(s+1)^2}{\sqrt{s^2+1}(s+1) + 1/2} - 1\right)}_{\text{second term}}. \qquad (1)$$

By taking the inverse Laplace transform of the first term of $E(s)$ we arrive at

$$2(J_0(t) + J_0(t)'), \text{ for } t > 0,$$

where J_0 is a Bessel functions of the first kind of order zero and J_0' its derivative. The function $J_0(t)$ is in fact holomorphic on the plane minus the negative real axis, and both itself and its derivative approach zero as $t \to \infty$. In fact, $J_0(t)' = -J_1(t)$. Thus, the time-domain function corresponding to this term decays as $t^{-1/2}$ at infinity.

Regarding the second term, we note that the phase of $\sqrt{s^2+1}(s+1)$ is less than π in the right half of the complex plane. Hence, this second term of $E(s)$ in (1) is analytic in the right half-plane. In fact,

$$\frac{(s+1)^2}{\sqrt{s^2+1}(s+1) + 1/2} - 1 = \frac{\sqrt{s^2+1}(s+1)^3 - s^4 - 2s^3 - (5/2)s^2 - 3s - 5/4}{s^4 + 2s^3 + 2s^2 + 2s + 3/4}$$

$$=: F(s).$$

Hence, it is also bounded and continuous on $i\mathbb{R}$, and decays as $|s|^{-1}$ when $|s| \to \infty$. Further, it can be easily decomposed in the form

$$F(s) = \left(\sqrt{s^2+1} - s\right)G(s) + H(s),$$

where $G(s)$ and $H(s)$ are strictly proper rational transfer functions with all their poles in the left half-plane. The inverse Laplace transform of the fractional element $\sqrt{s^2+1} - s$ is $J_1(t)/t$ where J_1 is the Bessel function of the first kind of order one, as before. It can be shown that $J_1(t)/t$ is continuous, bounded and decays to zero as $t \to \infty$, and so does the inverse Laplace transform of $F(s)$. Thus, $\lim_{t\to\infty} e(t)$ exists and it is zero as well. Therefore, the control system in Figure 5 rejects sinusoidal disturbances at frequency $\omega_o = 1$ (rad/sec) as claimed.

To recap, reference tracking and disturbance rejection for sinusoidal signals in a closed-loop system is possible using fractional-order dynamics of the form $1/\sqrt{s^2+\omega_o^2}$. Thus, the dictum of the classical "internal model principle" (see [15], also [16] and the references therein) which requires that the loop gain must contain a model of the disturbance dynamics is not necessary. The "strength" of the pole at $\pm i\omega_o$ trades off phase against convergence rate to steady-state. It is conjectured that this observation holds in general, in that, infinite loop gain at 0 or at $\pm i w_o$ provided by a fractional integrator or a fractional element is sufficient to ensure steady-state tracking of corresponding reference signal for any plant dynamics. However, other issues including stability and finite time behavior of such systems need further investigation.

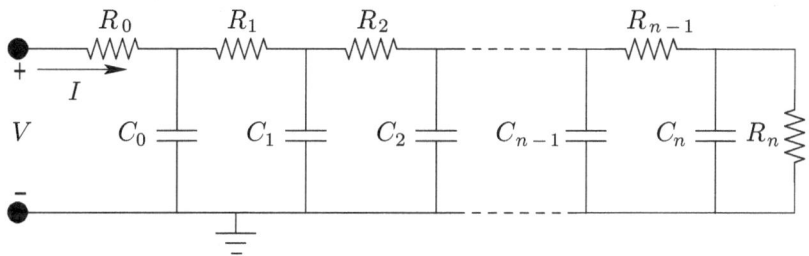

FIGURE 6. Truncated transmission line.

3. Implementation of a half-integrator

Typically, ladder networks have irrational transfer functions which are most naturally expressed in the form of continued fractions [17, 18, 5]. Approximation of such a network can be in the form of a "truncated transmission-line model," as shown in Figure 6. We present such an option for approximating a fractional integrator. Here, I represents the input (current) and V output (voltage). The corresponding impedance is

$$G(s) = \frac{V(s)}{I(s)} = R_0 + \cfrac{1}{C_1 s + \cfrac{1}{R_1 + \cfrac{1}{C_2 s + \cfrac{1}{R_2 + \cfrac{1}{\cdots + \cfrac{1}{C_n s + \frac{1}{R_n}}}}}}}. \tag{2}$$

We are interested in suitable choices for the resistors (R's) and capacitors (C's), so that the truncated transmission line can approximate the half-integrator over a given frequency band. We present three different sets of R_k's and C_k's for $k = 1, \ldots, n$ which can be used to this end. The first two sets of values are

1-st realization: $\begin{cases} C_1 = C_2 = C_3 = \cdots = C_n = C \\ R_1 = R_2 = R_3 = \cdots = R_n = R \;, \\ R_0 = \frac{1}{2}R \end{cases}$ (3a)

2-nd realization: $\begin{cases} C_k = k \; [\mu F], & \text{for } k = 1, 2, \ldots, n \\ R_k = k \; [\Omega], & \text{for } k = 1, 2, \ldots, n \;, \\ R_0 = 0 \end{cases}$ (3b)

while a third suggested realization is based on the Padé approximation and will be given later on.

3.1. Realization based on (3a)

The truncated transmission-line in Figure 6 with parameters as in (3a) and $n = 90, R = 100[\Omega]$, and $C = 100[\mu F]$ gives rise to impedance characteristics shown in Figure 7. It is seen that it approximates relatively accurately $1/\sqrt{s}$ over the mid-

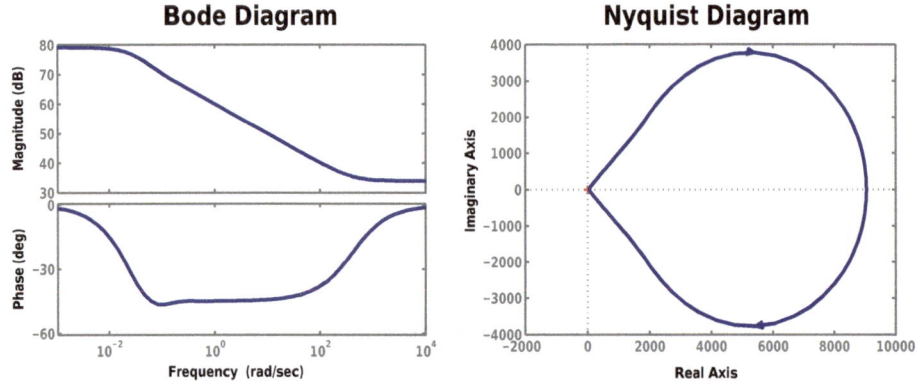

FIGURE 7. Bode and Nyquist plots of the transmission line in Figure 6 with components (3a), $n = 90, R = 100[\Omega]$, and $C = 100[\mu F]$.

range of frequencies, between 10^{-1} and 10^2 [rad/sec]. The attenuation is $\simeq -10$ [dB/dec] and the phase is $\simeq -45°$. Poles and zeros of such a circuit are real, lie in the left half of the complex plane, and interlace since this is an RC network; they are shown in Figure 8 for a smaller value of $n = 10$ for ease of inspection. The expression in (2) may be viewed as the nth level truncation of a "real J-fraction" (see [17]); it can be easily shown that all poles of this fraction are real, simple, and have positive residues. Furthermore, once again, they all lie on the negative half of the real axis, because the denominators of all kth approximants of (2), for $k = 1, \ldots, n$, are polynomials in s with positive coefficients.

We now explain why the truncated transmission line with the choice in (3a) provides indeed an approximation of a half-integrator. In fact, this choice of values results in

$$G(i\omega) \approx \sqrt{\frac{R}{C}} \frac{1}{\sqrt{i\omega}}, \quad \text{for} \quad \frac{6}{n^2 RC} \leq \omega \leq \frac{1}{6RC} \qquad (4)$$

to within 2% for $n \geq 10$. We begin by dividing out the leading terms and rewriting (2) as

$$\frac{G(s)}{R_0} = 1 + \cfrac{\frac{1}{R_0 C_1 s}}{1 + \cfrac{\frac{1}{R_1 C_1 s}}{1 + \cfrac{\frac{1}{R_1 C_2 s}}{1 + \cfrac{\frac{1}{R_2 C_2 s}}{\ddots + \cfrac{1}{1 + \frac{1}{R_n C_n s}}}}}} = 1 + \cfrac{2\gamma}{1 + \cfrac{\gamma}{1 + \cfrac{\gamma}{\ddots + \frac{1}{1+\gamma}}}}, \qquad (5)$$

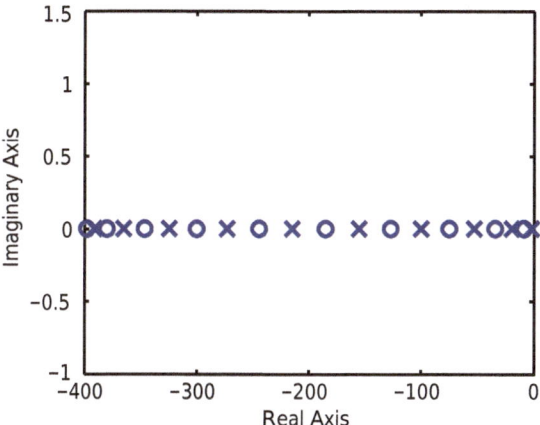

FIGURE 8. Pole-Zero configuration of the truncated transmission line in Figure 6 for $n = 10$.

where the R_k's and C_k's have been set as in (3a), and where $\gamma := 1/(RCs)$. This is a $2n$-term fraction. It can be simplified further using the following two lemmas.

Lemma 1. *Consider the continued fraction*

$$\mathcal{F}_1 = \cfrac{1}{1 - \cfrac{\rho_1}{1 + \rho_1 - \cfrac{\rho_2}{1 + \rho_2 - \cfrac{\rho_3}{1 + \rho_3 - \cfrac{\rho_4}{\ddots}}}}}.$$

The numerator of the nth level approximant of $\mathcal{F}_1$ is equal to the sum of the first n terms of the infinite series $1 + \sum_{p=1}^{\infty} \rho_1 \rho_2 \dots \rho_p$, while the denominator of the nth approximant is equal to one.

Proof. The proof follows by induction (see also [17, page 18]). $\qquad\qquad \square$

Lemma 2. *The nth level approximant of the continued fraction*

$$\mathcal{F}_2 = b + \cfrac{a}{b + \cfrac{a}{b + \cfrac{a}{\ddots}}} \tag{6}$$

with $a, b \neq 0$, can be written in the form of

$$u + v - \cfrac{v}{\cfrac{v}{u} + \cfrac{1}{\sum_{p=0}^{n-1} \left(\frac{v}{u}\right)^p}},$$

where u and v are solutions to the quadratic equation $x^2 - bx - a = 0$ and $|u| \geq |v|$.

Proof. Since $u + v = b$ and $uv = -a$, $\mathcal{F}_2$ in (6) can be written as

$$\mathcal{F}_2 = u + v - \cfrac{uv}{u + v - \cfrac{uv}{u+v-\cfrac{uv}{\ddots}}}.$$

Successive divisions by u lead to

$$\mathcal{F}_2 = u + v - v\cfrac{1}{\frac{v}{u}+1-\cfrac{v}{u+v-\cfrac{uv}{u+v-\cfrac{uv}{u+v-\frac{uv}{\ddots}}}}} = u + v - v\cfrac{1}{\frac{v}{u}+1-\cfrac{\frac{v}{u}}{1+\frac{v}{u}-\cfrac{\frac{v}{u}}{1+\frac{v}{u}-\frac{\frac{v}{u}}{\ddots}}}}. \quad (7)$$

The denominator of the last term in (7) consists of two parts: $\frac{v}{u}$ and

$$1 - \cfrac{\frac{v}{u}}{1+\frac{v}{u}-\cfrac{\frac{v}{u}}{1+\frac{v}{u}-\frac{\frac{v}{u}}{\ddots}}}.$$

The latter is a particular case of the inverse of $\mathcal{F}_1$ in Lemma 1 with parameters

$$\rho_1 = \rho_2 = \rho_3 = \ldots = \frac{v}{u}.$$

Application of Lemma 1 now gives the nth approximant of $\mathcal{F}_2$ as

$$u + v - \cfrac{v}{\frac{v}{u}+\cfrac{1}{1+\frac{v}{u}+(\frac{v}{u})^2+\cdots+(\frac{v}{u})^{n-1}}} = u + v - \cfrac{v}{\frac{v}{u}+\cfrac{1}{\sum_{p=0}^{n-1}(\frac{v}{u})^p}}. \qquad \square$$

Using the notation of Lemma 2, we take u and v to be the roots of $x^2 - x - \gamma$ with γ as in (5), i.e.,

$$u = \frac{1+\sqrt{1+4\gamma}}{2}, \quad v = \frac{1-\sqrt{1+4\gamma}}{2}. \qquad (8)$$

Since $u + v = 1$, we rewrite (5) as

$$\frac{G(s)}{R_0} = 1 + 2 \times \cfrac{-v}{\frac{v}{u}+\cfrac{1}{\sum_{p=0}^{2n-1}(\frac{v}{u})^p}} = 1 + 2 \times \cfrac{-v}{\frac{v}{u}+\frac{1-\frac{v}{u}}{1-(\frac{v}{u})^{2n}}}$$

$$= 1 + 2\left(\frac{u-v}{1-(\frac{v}{u})^{2n+1}} - u\right). \qquad (9)$$

Substituting R_0, u and v from (3a) and (8) into (9) we have that

$$\frac{G(s)}{R/2} = 1 + 2 \left(\frac{\sqrt{1+4\gamma}}{1 + \left(\frac{-1+\sqrt{1+4\gamma}}{1+\sqrt{1+4\gamma}}\right)^{2n+1}} - \frac{1+\sqrt{1+4\gamma}}{2} \right)$$

$$= 2\sqrt{1+4\gamma} \left(\frac{1}{1 + \left(\frac{-1+\sqrt{1+4\gamma}}{1+\sqrt{1+4\gamma}}\right)^{2n+1}} - \frac{1}{2} \right).$$

Further simplification of this last expression leads to

$$G(s)\sqrt{\frac{C}{R}}\sqrt{s} = \sqrt{1 + \frac{1}{4\gamma}} \left(\frac{1 - \left(\frac{-1+\sqrt{1+4\gamma}}{1+\sqrt{1+4\gamma}}\right)^{2n+1}}{1 + \left(\frac{-1+\sqrt{1+4\gamma}}{1+\sqrt{1+4\gamma}}\right)^{2n+1}} \right)$$

$$= \sqrt{1 + \frac{1}{4\gamma}} \tanh\left(\frac{2n+1}{2} \log(1 + \frac{2}{-1+\sqrt{1+4\gamma}}) \right). \qquad (10)$$

For $n > 10$ in (10), the term $\sqrt{1+1/4\gamma}$ is greater than 1, decreasing for $\gamma \in [6, n^2/6]$, and ranges to within 2% of unity ($\simeq 2\%$ at the left limit $\gamma = 6$). On the other hand, the term with the hyperbolic tangent is less than 1, also decreasing, and ranges to within 2% of unity ($\simeq 2\%$ at the right limit $\gamma = n^2/6$). Hence, the right-hand side of (10) is within 2% of unity as claimed. Finally, substituting $1/RCs$ for γ leads to (4), i.e., the truncated transmission line with quantities proposed in (3a) follows the characteristic of a half-integrator to within 2%, over the indicated frequency band.

3.2. Second realization

The truncated transmission-line model worked out in Section 3.1 conveys the basic idea of building a half-integrating element via passive circuitry. The particular choice of values for R's and C's in (3a) makes it a practical design. However, the convergence appears to be slow and, hence, a significant number of elements is needed to attain a reasonable approximation. Herein we discuss the alternative choice of parameters in (3b). Numerical verification suggests that for this choice of parameters the frequency response of the truncated transmission line converges numerically *considerably faster*. However, because the parameters increase with the index, the analysis is difficult and proof of convergence has not been established.

The frequency response of this circuit with the new R, C values matches that of a fractional integrator with exponent 1/2 quite accurately over a frequency band. Figure 9 shows the Bode and Nyquist plots of such a transmission line with 90 pairs of (C_k, R_k) for comparison. The response similarly matches that of $1/\sqrt{s}$ from 10^{-1} to 10^6 [rad/sec] to a similar accuracy as before. But now, interestingly, the approximation is over a frequency band which is 3 orders of magnitude wider.

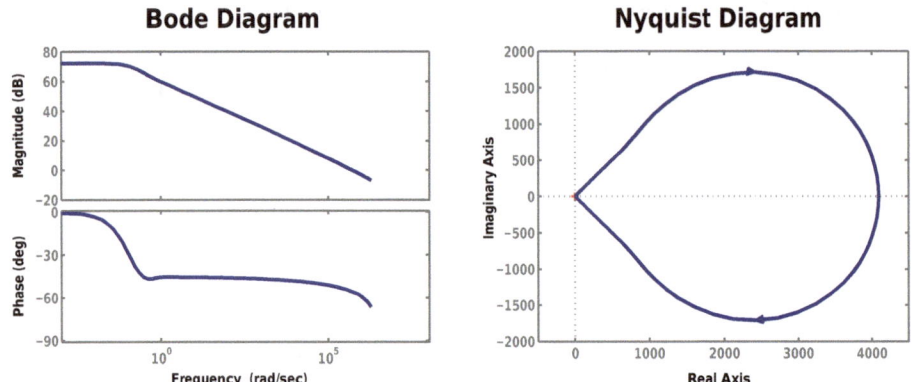

FIGURE 9. Bode and Nyquist plots of the truncated transmission line in Figure 6 with values given in (3b) for $n = 90$.

3.3. Third realization based on Padé approximation

We consider the Padé approximation of $1/\sqrt{s}$ about $s_0 = 1$. To this end, starting from the Taylor series expansion of $1/\sqrt{s}$ we match $l + m + 1$ terms by a rational function $p_{l,m}(s) := a_l(s)/b_m(s)$, where $a_l(s)$, $b_m(s)$ are polynomials of order l and m, respectively. To obtain their coefficients, one needs to solve linear equations (see, e.g., [19]). For the particular case of a fractional integrator $1/\sqrt{s}$ and $l = m = 5$, the Padé fraction becomes

$$p_{5,5}(s) = \frac{s^5 + 55s^4 + 330s^3 + 462s^2 + 165s + 11}{11s^5 + 165s^4 + 462s^3 + 330s^2 + 55s + 1}.$$

Euclidean division allows writing this as a truncated RC-ladder network

$$G_{\text{Padé}}(s) = p_{5,5}(s) = R_0 + \cfrac{1}{C_1 s + \cfrac{1}{R_1 + \cfrac{1}{C_2 s + \cfrac{1}{R_2 + \cfrac{1}{\ddots + \cfrac{1}{C_5 s + \frac{1}{R_5}}}}}}},$$

with $R_0 = \frac{1}{11}$, and

$$R_1 = \tfrac{200}{429}, \ R_2 = \tfrac{128}{143}, \ R_3 = \tfrac{4096}{2805}, \ R_4 = \tfrac{32768}{13585}, \ R_5 = \tfrac{262144}{46189},$$
$$C_1 = \tfrac{11}{40}, \ C_2 = \tfrac{429}{640}, \ C_3 = \tfrac{4719}{4096}, \ C_4 = \tfrac{60775}{32768}, \ C_5 = \tfrac{877591}{262144}. \tag{11}$$

The Bode and Nyquist plots are shown in Figure 10. The response approximates that of the ideal half-capacitor in the frequency range from 10^{-1} to 10^1 [rad/sec] to within .08 dB. We have drawn in Figure 10 the response for the $(5,5)$-Padé approximant for comparison.

The following comments are in order: while Padé approximation gives a very good approximation for a given number of components, it may not be suitable for integrated fabrication due to the required precision on the varying values of its

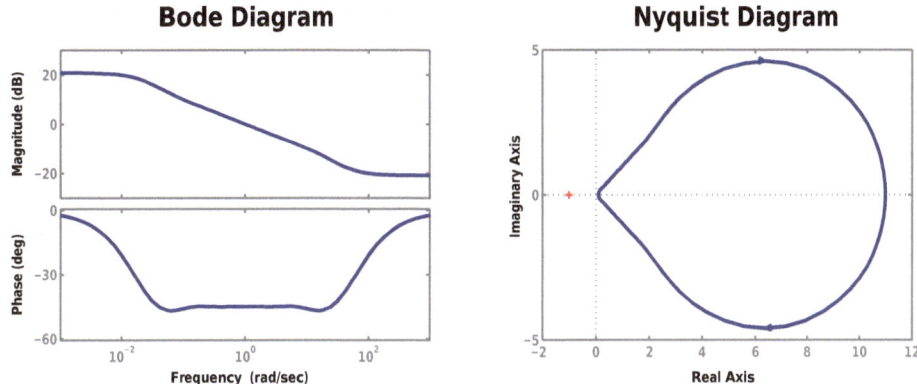

FIGURE 10. Bode and Nyquist plots of the truncated transmission line in Figure 6 with values given in (11).

components. For background and references on efficient design of RC-circuits see [20, 21, 22]. Further, the circuit appears to be quite sensitive to those values.

It should also be noted that all of the above approximations are good over a frequency range that does not include the origin. Consequently, their use in a feedback system will not achieve perfect tracking in response to a step reference. In particular, for the above example the steady-state error to a step input is $1/(1 + G(0))$ with

$$G(0) = R_0 + R_1 + \cdots + R_n.$$

Finally, likewise, approximation of elements of the form $1/\sqrt{s^2 + \omega_o^2}$ may again be based on truncated continued fractions and Padé fractions. For similar reasons, the quality of approximation will degrade close to the essential singularity on $i\mathbb{R}$, and consequently, steady-state performance will be met only approximately.

4. Recap

We underscore the relevance of a fractional integrator as a design element for set-point following in a control loop. The classical internal model principle requires an integrator in the feedback loop. The magnitude of an integrator in the frequency domain drops at a rate of 20 [dB/dec] and its phase is $-90°$ throughout. This may adversely affect stability and robustness of the closed-loop system as it directly affects the phase margin. Yet, a fractional integrator suffices for achieving steady-state tracking in set-point following and at the same time introduces a smaller phase lag (compared to that of an integrator). Similar conclusions carry through for steady-state tracking of sinusoidal inputs – this case requires fractional poles on the imaginary axis. Besides their relevance in feedback theory and steady-state tracking, such elements may find applications in communications. Finally, the realization of such elements with traditional circuitry and ladder circuits appears challenging. Yet, it is anticipated that alternative implementations of fractional elements will be possible using available technologies (off-chip and on-chip) for various frequency ranges.

Acknowledgement

We kindly thank Professor YangQuan Chen for providing us with certain key references.

References

[1] D. Duffy, *Transform methods for solving partial differential equations.* Boca Raton, FL: Crc Press, 2004.

[2] N. Engheta, "On fractional calculus and fractional multipoles in electromagnetism," *Antennas and Propagation, IEEE Transactions on,* vol. 44, no. 4, pp. 554–566, 1996.

[3] W. Chen, "Fractional and fractal derivatives modeling of turbulence," *eprint arXiv:nlin/0511066,* 2005.

[4] M. Ortigueira and J. Machado, "Special section: fractional calculus applications in signals and systems," *Signal Processing,* vol. 86, no. 10, 2006.

[5] K. Oldham and J. Spanier, *The fractional calculus,* ser. Mathematics in Science and Engineering, R. Bellman, Ed. New York: Academic Press, 1974, vol. 111.

[6] I. Podlubny, *Fractional differential equations,* ser. Mathematics in Science and Engineering. San Diego: Academic Press, 1999, vol. 198.

[7] J. Machado, "Discrete-time fractional-order controllers," *Fractional Calculus and Applied Analysis,* vol. 4, no. 1, pp. 47–66, 2001.

[8] Y. Chen, "Ubiquitous fractional order controls," in *Proc. of the second IFAC symposium on fractional derivatives and applications (IFAC FDA06, Plenary Paper).* Citeseer, 2006, pp. 19–21.

[9] A. Oustaloup, P. Melchior, P. Lanusse, O. Cois, and F. Dancla, "The CRONE toolbox for Matlab," in *Computer-Aided Control System Design, 2000. CACSD 2000. IEEE International Symposium on.* IEEE, 2000, pp. 190–195.

[10] K.J. Åström, "Model uncertainty and feedback," in *Iterative identification and control: advances in theory and applications,* P. Albertos and A.S. Piqueras, Eds. New York: Springer Verlag, Jan. 2002, pp. 63–97.

[11] B.M. Vinagre and Y. Chen, "Fractional calculus applications in automatic control and robotics," in *IEEE Proc. on Conference on Decision and Control,* Las Vegas, Dec. 2002, Tutorial Workshop – 316 pages.

[12] P.A. McCollum and B.F. Brown, *Laplace Transform Tables and Theorems.* New York: Holt, Rinehart and Winston, 1965.

[13] A.D. Poularikas, *The handbook of formulas and table for signal processing,* ser. The Electrical Engineering Handbook Series. New York: CRC Press LLC and IEEE Press, 1999, vol. 198.

[14] G.B. Arfken and H.J. Weber, *Mathematical Methods for Physicists,* 6th ed. San Diego: Harcourt, 2005.

[15] B.A. Francis and W.M. Wonham, "The internal model principle of control theory," *Automatica*, vol. 12, no. 5, pp. 457–465, Sept. 1976.

[16] E. Sontag, "Adaptation and regulation with signal detection implies internal model," *Systems & control letters*, vol. 50, no. 2, pp. 119–126, 2003.

[17] H.S. Wall, *Analytic theory of continued fractions*. New York: D. Van Nostrand Company, Inc., 1948.

[18] C. Brezinski, *History of continued fractions and Padé approximations*, ser. Computational Mathematics. New York: Springer-Verlag, 1991, vol. 12.

[19] G.A.J. Baker, *Essentials of Padé approximants*. New York: Academic Press, 1975.

[20] M. Ichise, Y. Nagayanagi, and T. Kojima, "An analog simulation of non-integer order transfer functions for analysis of electrode processes," *Journal of Electroanalytical Chemistry and Interfacial Electrochemistry*, vol. 33, no. 2, pp. 253–265, 1971.

[21] K. Oldham, "Semiintegral electroanalysis. Analog implementation," *Analytical Chemistry*, vol. 45, no. 1, pp. 39–47, 1973.

[22] G. Bohannan, "Analog realization of a fractional control element – revisited," 2002.

Mir Shahrouz Takyar and Tryphon T. Georgiou
Department of Electrical and Computer Engineering
University of Minnesota
Minneapolis, MN 55455, USA
e-mail: `shahrouz@umn.edu`
        `tryphon@umn.edu`

Operator Theory:
Advances and Applications, Vol. 222, 311–324
© 2012 Springer Basel

Optimal Mass Transport for Problems in Control, Statistical Estimation, and Image Analysis

Emmanuel Tannenbaum, Tryphon Georgiou and Allen Tannenbaum

This paper is dedicated to our dear friend and colleague,
Professor Bill Helton on the occasion of his 65th birthday.

Abstract. In this paper, we describe some properties of the Wasserstein-2 metric on the space of probability distributions of particular relevance to problems in control and signal processing. The resulting geodesics lead to interesting connections with Boltzmann entropy, heat equations (both linear and nonlinear), and suggest possible Riemannian structures on density functions. In particular, we observe similarities and connections with metrics originating in information geometry and prediction theory.

Mathematics Subject Classification. 34H05, 49J20.

Keywords. Optimal mass transport, information geometries, Riemannian metrics, spectral analysis.

1. Introduction

Optimal mass transport is an important problem with applications in numerous disciplines including econometrics, fluid dynamics, automatic control, transportation, statistical physics, shape optimization, expert systems, and meteorology [16, 17]. The problem was first formulated by the civil engineer Gaspar Monge in 1781, and concerned finding the optimal way, in the sense of minimal transportation cost, of moving a pile of soil from one site to another. Much later the problem was extensively analyzed by Kantorovich [13] with a focus on economic resource allocation, and so is now known as the Monge–Kantorovich (MK) or optimal mass transport (OMT) problem.

In the present work, via gradient flows defined by the Wasserstein 2-metric from optimal transport theory, we will relate some key concepts in information theory with mass transport that impacts problems in control, statistical estimation, and image analysis. A goal of our program is to understand the apparent relationship between mass transport, conservation laws, entropy functionals, on one hand and probability and power distributions and related metrics on the other.

Previously, we explored a differential-geometric structure for spectral density functions of discrete-time random processes and the present work aims along similar lines. Indeed, in [7] we introduced a Riemannian geometry for power spectral densities that is quite analogous to the Riemannian structure of information geometry which is used to study perturbations of probability density functions and relates to the Fisher information metric [1] – in contrast, the metric in [7] was motivated by a problem in prediction theory. In the present we are motivated by ideas in statistical mechanics and show how optimal mass transport leads in a straightforward way to a Riemannian structure on the space of probability measures. This structure will relate in a natural manner the Wasserstein metric, Fisher information, and Boltzmann entropy. Following Benamou and Brenier [4], optimal mass transport allows us to define geodesics on the space of probability measures and analogies will be drawn with the metric in [7] which is connected to problems in statistical prediction.

This paper is dedicated to our dear friend and colleague Dr. J. William Helton. Bill has been a inspiration to us over the years: he is a great scientist, scholar, mentor, and truly a pioneer with numerous seminal contributions in systems and control. **Happy Birthday, Bill (with many more to come)!**

2. Background on Boltzmann entropy and statistical mechanics

We begin by describing some standard material about Boltzmann entropy from statistical mechanics that will motivate our development in the sequel. For a complete discussion on such issues in statistical mechanics, see [15]. For simplicity, in this treatment we will work with probability distributions defined over a one-dimensional continuous index, though our results are readily generalizable to higher dimensions. Accordingly, consider a probability distribution, $\mu(x,t)$, defined over the real line. We have that $\mu(x,t) \geq 0$, that the integral of μ over the real line is 1, and we may also assume that $\lim_{x \to \pm\infty} \mu(x,t) = 0$.

To determine the time evolution of μ, we assume that this time evolution is governed by a transition rate function $w(x,y)$, defined over $\mathbb{R}^2$. We assume that $w(x,y) \geq 0$, and that $w(x,y) = w(y,x)$. Informally, $w(x,y)$ measures the probability per unit time that a particle at x will "hop" or be transported to y. This means that, during an infinitesimal time dt, the probability that a particle at x will hop to y is $w(x,y)dt$, and accordingly, integrates to 1. Symmetry means that a particle at x has the same chance of being transported to y as does a particle at y have to be transported to x, and leads to the principle of detailed balance,

$$\frac{\partial \mu(x,t)}{\partial t} = \int_{-\infty}^{\infty} w(x,y)[\mu(y,t) - \mu(x,t)]\, dy. \tag{1}$$

We now define the *entropy*, $S = S[\mu]$, of the distribution via,

$$S = -\int_{-\infty}^{\infty} \mu(x,t) \log \mu(x,t) \, dx.$$

For a given distribution that evolves in time, we may consider the time-variation of the corresponding values of S under the action of Equation (1). We have that,

$$\begin{aligned}
\frac{dS}{dt} &= -\int_{-\infty}^{\infty} (1 + \log \mu) \frac{\partial \mu}{\partial t} \, dx \\
&= -\int_{-\infty}^{\infty} \frac{\partial \mu}{\partial t} \log \mu \, dx \\
&= -\int_{-\infty}^{\infty} \int_{-\infty}^{\infty} w(x,y) \, (\mu(y,t) - \mu(x,t)) \log \mu(x) \, dx \, dy \\
&= \frac{1}{2} \int_{-\infty}^{\infty} \int_{-\infty}^{\infty} w(x,y) \lambda(x,y,t) \, dx \, dy
\end{aligned}$$

where

$$\lambda(x,y,t) = (\log \mu(y,t) - \log \mu(x,t))(\mu(y,t) - \mu(x,t)).$$

Since $\log x$ is an increasing function of x, both $\log \mu(y,t) - \log \mu(x,t)$ and $\mu(y,t) - \mu(x,t)$ have the same sign, and so their product is nonnegative, which implies that the overall expression is nonnegative. So, we have that $dS/dt \geq 0$, which is to say that the entropy is always either constant or increasing.

Now, over an infinitesimal time interval dt, we expect that transitions are "local", in the sense that transitions only occur between nearby regions. In a sense, we are assuming that a particle cannot "jump" from one point to another, but must be transported continuously. We would therefore like to work with a $w(x,y)$ that is only non-zero within an infinitesimal neighborhood of x. A natural function that suggests itself is the delta-function $\delta(x - y)$. Plugging this value for $w(x,y)$ into equation (1) gives

$$\frac{\partial \mu(x,t)}{\partial t} = \mu(x,t) - \mu(x,t) = 0,$$

which is not very interesting.

The next function to try would be $\partial \delta(x - y)/\partial x$. However, because $\partial \delta(x - y)/\partial x = -\partial \delta(y - x)/\partial y$, this is not a valid transition function. So now let us take another derivative and look at $\partial^2 \delta(x - y)/\partial x^2$. It turns out that this function does satisfy the desired symmetry property. However, there is a problem with this function, namely, it is not everywhere positive, which can be seen by considering the δ-function as the limit of unit-normalized Gaussians. We can make this explicit. A unit-normalized Gaussian centered at a point x has the functional form,

$$g(y;x) = \frac{1}{\sigma\sqrt{2\pi}} \exp[-\frac{(y-x)^2}{2\sigma^2}] \tag{2}$$

where σ is the standard deviation. We then have that $\delta(y - x) = \lim_{\sigma \to 0} g(y; x)$. Differentiating once, we obtain,

$$\frac{dg}{dy} = -\frac{y - x}{\sigma^2} \times \frac{1}{\sigma\sqrt{2\pi}} \exp\left[-\frac{(y - x)^2}{2\sigma^2}\right] \tag{3}$$

and differentiating twice we obtain,

$$\frac{d^2g}{dy^2} = \frac{1}{\sigma^4}((y - x)^2 - \sigma^2) \times \frac{1}{\sigma\sqrt{2\pi}} \exp\left[-\frac{(y - x)^2}{2\sigma^2}\right]. \tag{4}$$

If we define $\partial^2\delta(y - x)/\partial y^2 = \lim_{\sigma \to 0}(d^2g/dy^2)$, then since $d^2g/dy^2 < 0$ for $|y - x| < \sigma$, we have that the doubly-differentiated δ-function is negative within a standard deviation of the δ-function itself. In short, the point is that the doubly-differentiated δ-function has an infinitesimally small region where it is negative, and in fact it is $-\infty$ in this region, making its use problematic.

To work our way around this problem, let us consider $w(y, x)$ of the form $w(y - x)$. Substituting into Equation (1), we have,

$$\frac{\partial\mu(x,t)}{\partial t} = \int_{-\infty}^{\infty} w(u)[\mu(x + u, t) - \mu(x, t)]\, du.$$

Expanding $\mu(x + u, t)$ in a Taylor-series about x, we obtain,

$$\frac{\partial\mu(x,t)}{\partial t} = \int_{-\infty}^{\infty} w(u) \sum_{n=1}^{\infty} \frac{1}{n!} \frac{\partial^n \mu(x,t)}{\partial x^n} u^n\, du.$$

Because $w(u) = w(y - x) = w(y, x) = w(x, y) = w(x - y) = w(-u)$, the odd powers drop out of the integral, giving,

$$\frac{\partial\mu(x,t)}{\partial t} = \int_{-\infty}^{\infty} u^2 w(u)\left[\frac{1}{2}\frac{\partial^2\mu(x,t)}{\partial x^2} + \sum_{n=1}^{\infty} \frac{1}{(2(n+1))!}\frac{\partial^{2(n+1)}\mu(x,t)}{\partial x^{2(n+1)}}u^{2n}\right] du. \tag{5}$$

Now, if $w(u)$ is very narrowly distributed, then $w(u) \approx 0$ outside a very small neighborhood of $u = 0$. Within this neighborhood, the higher-order terms of the Taylor expansion are negligible. Outside this neighborhood, these terms are negligible as well, since $w(u) \approx 0$ and so do not contribute to the overall integral. Indeed, the only significant contribution to the integral is from the lowest-order term containing the second derivative of μ, and only in a very small neighborhood of $u = 0$. However, because outside this neighborhood the contribution to the integral is negligible, the integral of the lowest-order term over the small neighborhood is essentially equal to the integral over the whole real line. So, for narrowly distributed functions $w(u)$ that are positive everywhere, we have,

$$\frac{\partial\mu(x,t)}{\partial t} = \int_{-\infty}^{\infty} u^2 w(u)\frac{1}{2}\frac{\partial^2\mu(x,t)}{\partial x^2}\, du$$

$$= D\frac{\partial^2\mu(x,t)}{\partial x^2} = D\Delta\mu,$$

where $D \equiv (1/2) \int_{-\infty}^{\infty} u^2 w(u) \, du$ is the diffusivity coefficient. This is exactly the **linear heat equation**.

To use a specific value of $w(u)$, let us consider the function $d^2 g(y; x)/dy^2$ considered previously, but now let us simply remove the $-\sigma^2$ term in the parentheses. We then have that,

$$w(u; \sigma) = \frac{u^2}{\sigma^4} \times \frac{1}{\sigma \sqrt{2\pi}} \exp\left[-\frac{u^2}{2\sigma^2}\right] \tag{6}$$

where we consider this function for very small values of σ. It appears that such a function obeys the locality property discussed above, and it is certainly positive everywhere. Note that the justification for only considering the lowest-order term in Equation (5) for the specific function $w(u; \sigma)$ that we chose, note that each of the higher-order terms give a contribution of,

$$\frac{1}{(2n)!} \frac{\partial^{2n} \mu(x, t)}{\partial x^{2n}} \int_{-\infty}^{\infty} u^{2n} \frac{u^2}{\sigma^4} \times \frac{1}{\sigma \sqrt{2\pi}} \exp\left[-\frac{u^2}{2\sigma^2}\right] du$$

$$= \frac{\sigma^{2(n-1)}}{(2n)! \sqrt{2\pi}} \frac{\partial^{2n} \mu(x, t)}{\partial x^{2n}} \int_{-\infty}^{\infty} v^{2(n+1)} \exp\left[-\frac{v^2}{2}\right] dv$$

where $n \geq 2$. Note then that $2(n - 1) \geq 2$, and so the higher-order terms give a contribution that is of order σ^2 or higher, so that they vanish as $\sigma \to 0$, leaving only the contribution containing the second partial of μ with respect to x.

3. Monge–Kantorovich

A modern formulation of the Monge–Kantorovich Optimal Mass Transport (OMT) problem is as follows. Let Ω_0 and Ω_1 be two subdomains of $\mathbb{R}^d$, with smooth boundaries, each with a positive density function, μ_0 and μ_1, respectively. We assume

$$\int_{\Omega_0} \mu_0 = \int_{\Omega_1} \mu_1 = 1$$

so that the same total mass is associated with Ω_0 and Ω_1. We consider diffeomorphisms ϕ from (Ω_0, μ_0) to (Ω_1, μ_1) which map one density to the other in the sense that

$$\mu_0 = |D\phi| \, \mu_1 \circ \phi, \tag{7}$$

which we will call the *mass preservation* (MP) property, and write $\phi \in MP$. Equation (7) is called the *Jacobian equation*. Here $|D\phi|$ denotes the determinant of the Jacobian map $D\phi$. In particular, Equation (7) implies, for example, that if a small region in Ω_0 is mapped to a larger region in Ω_1, then there must be a corresponding decrease in density in order for the mass to be preserved. A mapping ϕ that satisfies this property may thus be thought of as defining a redistribution of a mass of material from one distribution μ_0 to another distribution μ_1.

There may be many such mappings, and we want to pick out an optimal one in some sense. Thus, we define the L^p Kantorovich–Wasserstein metric as follows:

$$d_p(\mu_0, \mu_1)^p := \inf_{\phi \in MP} \int \|\phi(x) - x\|^p \mu_0(x)\,dx. \tag{8}$$

An *optimal MP map*, when it exists, is one which minimizes this integral. This functional is seen to place a penalty on the distance the map ϕ moves each bit of material, weighted by the material's mass.

The case $p = 2$ has been extensively studied. The L^2 Monge–Kantorovich problem has been studied in statistics, functional analysis, and the atmospheric sciences; see [2, 4] and the references therein. A fundamental theoretical result is that *there is a unique optimal $\phi \in MP$ transporting μ_0 to μ_1, and that this ϕ is characterized as the gradient of a convex function w*, i.e., $\phi = \nabla w$. Note that from Equation (7), we have that w satisfies the *Monge–Ampère* equation

$$|Hw|\,\mu_1 \circ (\nabla w) = \mu_0,$$

where $|Hw|$ denotes the determinant of the Hessian Hw of w.

In summary, the Kantorovich–Wasserstein metric defines the distance between two mass densities, by computing the cheapest way to transport the mass from one domain to the other with respect to the functional given in (8), the optimal transport map in the $p = 2$ case being the gradient of a certain function. This will be denoted by ϕ_{MK}. The novelty of this result is that like the Riemann mapping theorem in the plane, the procedure singles out a particular map with preferred geometry.

4. Optimal mass transport and optimal control

The L^2 Monge–Kantorovich problem may also be given an optimal control formulations in the following manner [4]. Consider

$$\inf \int \int_0^1 \mu(t, x)\|\nabla g(t, x)\|^2\,dt\,dx \tag{9}$$

over all time varying densities μ and functions g satisfying

$$\frac{\partial \mu}{\partial t} + \operatorname{div}(\mu \nabla g) = 0, \tag{10}$$

$$\mu(0, \cdot) = \mu_0, \ \mu(1, \cdot) = \mu_1.$$

The functional is the kinetic energy while $u = \nabla g$ represents velocity. One may show that this infimum is attained for some $\mu_{\min}$ and $g_{\min}$. We set $u_{\min} = \nabla g_{\min}$. Further, the flow $X = X(x, t)$ corresponding to the minimizing velocity field $u_{\min}$ defined by

$$X(x, 0) = x, \ X_t = u_{\min}(X(x, t), t) \tag{11}$$

is given simply as

$$X(x, t) = x + t\,(\phi_{MK}(x) - x). \tag{12}$$

Note that when $t = 0$, X is the identity map and when $t = 1$, it is the solution ϕ_{MK} to the Monge–Kantorovich problem. This analysis provides appropriate justification for using (12) to *define* our continuous warping map X between the densities μ_0 and μ_1.

5. Riemannian structure on probability densities

Define

$$\mathcal{D} := \left\{ \mu \geq 0 : \int \mu = 1 \right\},$$

the space of densities. The tangent space at a given point μ may be identified with

$$T_\mu \mathcal{D} \cong \{u : \int u = 0\}.$$

Thus inspired by the Benamou and Brenier [4], given two "points," $\mu_0, \mu_1 \in \mathcal{D}$, the geodesic (Wasserstein) distance is:

$$\inf_{\mu, g} \left\{ \int \int_0^1 \mu(t, x) \|\nabla g(t, x)\|^2 \, dt \, dx \right.$$

$$\text{subject to } \frac{\partial \mu}{\partial t} + \text{div}(\mu \nabla g) = 0,$$

$$\left. \mu(0, \cdot) = \mu_0, \ \mu(1, \cdot) = \mu_1 \right\} \tag{13}$$

In other words, we look at all curves in $\mathcal{D}$ connecting μ_0 and μ_1, and take the shortest one with respect to the Wasserstein metric.

This leads us to give $\mathcal{D}$ a Riemannian structure, which will induce this Wasserstein distance. This idea is due to Jordan *et al.* [12, 17]. Namely, under suitable assumptions on differentiability for $\mu \in \mathcal{D}$, and $u \in T_\mu \mathcal{D}$, one solves the Poisson equation

$$u = -\text{div}(\mu \nabla g). \tag{14}$$

This allows us to identify the tangent space with functions up to additive constant. Thus, for any given u we denote the solution of (14) by g_u. Then given, $u_1, u_2 \in T_\mu \mathcal{D}$, we can define the inner product

$$\langle u_1, u_2 \rangle_\mu := \int \mu \nabla g_{u_1} \cdot \nabla g_{u_2}. \tag{15}$$

An integration by parts argument, shows that this inner product will exactly induce the Wasserstein distance given by Equation (13). Note also that

$$\langle u, u \rangle_\mu = \int \mu \nabla g_u \cdot \nabla g_u$$

$$= -\int g_u \text{div}(\mu \nabla g_u) \text{ (integration by parts)} = \int u g_u. \tag{16}$$

6. Linear heat equation

We now describe a beautiful connection Boltzmann entropy and the linear heat equation [12]. As in Section 2, we define the following:

$$S := - \int \mu \log \mu.$$

Also as above, we evaluate S along a 1-parameter family in $\mathcal{D}$, $\mu(t)$ (we drop the dependence on the spatial variables x in the notation), and take the derivative with respect to t:

$$\frac{dS}{dt} = - \int (\mu_t \log \mu + \mu_t) = - \int (\mu_t \log \mu), \tag{17}$$

since $\int \mu = 1$. Now noting the characterization of the Wasserstein norm from Equation (16), we see that the steepest gradient direction (with respect to the Wasserstein metric) is given by

$$\mu_t = \mathrm{div}(\mu \nabla \log \mu) = \Delta \mu,$$

which is the **linear heat equation**. The diffusivity coefficient here is normalized to 1.

If we substitute $\mu_t = \Delta \mu$ into Equation (17), and integrate by parts, we get

$$\frac{dS}{dt} = \int \frac{\|\nabla \mu\|^2}{\mu}, \tag{18}$$

which is the **Fisher information metric**.

7. Transportation geometry and power spectra

In an analogous manner we explore implications of using the Wasserstein geometry on the space of power spectra of stochastic processes that, for the purposes of this paper, are typically normalized to have integral 1. The principal reason for our use of the Wasserstein geometry is that the corresponding metrics are weak*-continuous and thus capable of quantifying distances between spectral lines as well as absolutely continuous power spectra equally well [8]. In view of this fact, similarities that emerge between this and a metric derived in [7] that is not weak*-continuous, appear quite intriguing.

In this section, μ represents a spectral density of a discrete-time random process, though extension to being a non-absolutely-continuous spectral measure is straightforward. Thus, μ is taken to be non-negative on the unit circle – herein identified with $[0, 1)$. Therefore, $\mu(0) = \mu(1)$. The integral $\int \mu$ which is the variance of the random process is normalized to be 1. In this context, we consider the (differential) *entropy* on $\mathcal{D}$:

$$S_d := - \int \log \mu. \tag{19}$$

As before, we evaluate S_d along a 1-parameter family and take the derivative

$$\frac{dS_d}{dt} = -\int \frac{\mu_t}{\mu}. \tag{20}$$

The steepest gradient direction with respect to the Wasserstein metric is now given by

$$\mu_t = -\text{div}\left(\mu \nabla \frac{1}{\mu}\right) = \text{div}\left(\frac{\nabla \mu}{\mu}\right) = \frac{\Delta \mu}{\mu} - \frac{\|\nabla \mu\|^2}{\mu^2}. \tag{21}$$

This is a **nonlinear heat equation**. We specialize to one dimension and we can write Equation (21) explicitly using partials with respect to this one dimension

$$\mu_t = \frac{\mu \mu_{xx} - (\mu_x)^2}{\mu^2}.$$

Upon substitution into (20), and integration by parts we obtain that

$$\frac{dS_d}{dt} = \int \frac{\|\nabla \mu\|^2}{\mu^3} = \int \frac{(\mu_x)^2}{\mu^3}. \tag{22}$$

Thus, equation (22) defines yet another statistical metric derived from the spectral entropy S_d via Wasserstein.

Turning to time-series and their respective power spectra, we choose to view μ_0, μ_1 as power spectral distributions (hence, possibly unnormalized). Metrics between statistical models are essential in signal processing, system identification, and control. A classical theory for statistical models goes back to the work of C.R. Rao, R.A. Fisher and more recently Amari [1] and is known as "information geometry". In this, a possible starting point for the Fisher metric is the comparison of two probability distributions in the context of source coding where their dissimilarity quantifies the degradation of coding efficiency for a code based on one and then applied to source following the other. In a completely analogous manner [7] we compare power spectra in the context of optimal estimation: if we let $\{\mathbf{u}(k),\ k \in \mathbb{Z}\}$ be a discrete-time, zero-mean, weakly stationary stochastic process taking values in $\mathbb{C}$, the least-variance linear prediction

$$\min\left\{\mathcal{E}\{|\mathbf{p}(0)|^2\} : \mathbf{p}(0) = \mathbf{u}(0) - \sum_{k>0} P_k \mathbf{u}(-k),\ P_k \in \mathbb{C}\right\}, \tag{23}$$

where $\mathcal{E}$ denotes expectation, can be expressed in terms of the power spectrum μ of the stochastic process via the Szegö-Kolmogorov formula [14, p. 369]

$$\exp\left(\int \log \mu\right) = \exp(-S_d), \tag{24}$$

with S_d the differential entropy in (19).

We compare two power spectra by selecting prediction coefficients optimizing (23) based on one of the two spectra, and then evaluating the variance of the prediction error on the other. Considering that the two spectra differ little from one another, e.g., $\mu_1 = \mu_0 + \mu_x \delta x$ the degradation of prediction-error variance is quadratic in the perturbation and gives rise to a Riemannian metric. This metric, about the point $\mu_0 =: \mu$, takes the form

$$\int \frac{\mu_x^2}{\mu^2} - \left(\int \frac{\mu_x}{\mu} \right)^2. \tag{25}$$

Aside from a normalization this can be written as

$$\int \frac{\mu_x^2}{\mu^2}, \tag{26}$$

This exact same expression (26) can also be obtained as a measure of dissimilarity by comparing how close the power spectrum of the innovation process $\mathbf{p}(k)$ $(k \in \mathbb{Z})$ is from being constant across frequencies (see [11]). The analogy between (26), (25), (21) and the Fisher information metric

$$\int \frac{(\mu_x)^2}{\mu}, \tag{27}$$

is rather evident. Yet, their qualitative differences remain to be studied.

8. Porous medium equation and image processing

Following the earlier recipe, one can define alternative "information quantities"

$$S_g = - \int f(\mu),$$

where f is a suitable differentiable increasing function. Then the exact computation given earlier in Sections 5 and 6 shows that the corresponding gradient flow with respect to the Wasserstein metric is

$$\mu_t = \operatorname{div}(\mu \nabla f'(\mu)). \tag{28}$$

Moreover, we get then that

$$\frac{dS_g}{dt} = \int \mu \| f'(\mu) \|^2.$$

As in [17], if we take $f(x) = \frac{1}{n-1} x^n$, $n \geq 0$, then equation (28) becomes

$$\mu_t = \Delta \mu^n. \tag{29}$$

In one spatial variable this reduces to

$$\mu_t = (\mu^n)_{xx}.$$

Now we can turn this into an equation of the form

$$u_t = (u_{xx})^n \tag{30}$$

as follows. Simply apply $-\Delta^{-1}$ to both sides, and set $u = -\Delta\mu$. Interestingly, for $n = 1/3$, we get the one-dimensional **affine invariant heat equation** [3] of great popularity in image processing. While this equation is known not to be derivable via any L^2-based gradient flow, we have just shown that it can be derived as such via the Wasserstein geometric structure.

9. Unbalanced densities

General distributions (histograms, power spectra, spatio-temporal energy densities, images) may not necessarily be normalized to have the same integral. Thus, it is imperative to devise appropriate metrics and theory. The purpose of such a theory is to provide ways for "interpolating" data in the form of distributions, viewed as points in a suitable metric space. The first candidate is the space of L^2-integrable functions. However, as we will note next, geodesics are simply linear intervals and fail to have a number of desirable properties [8, 10]. In particular, the "linear average" of two *unimodal* distributions is typically *bimodal*. Thus, important features are typically "written over". It is instructive for us to consider this first.

One can show that

$$d_{L^2}(\mu_0, \mu_1)^2 = \inf_{\mu, v} \int \int_0^1 |\partial_t \mu(t, x)|^2 \, dt \, dx \qquad (31)$$

over all time varying densities μ and vector fields v satisfying

$$\frac{\partial \mu}{\partial t} + \mathrm{div}(\mu v) = 0, \qquad (32)$$

$$\mu(0, \cdot) = \mu_0, \ \mu(1, \cdot) = \mu_1.$$

The optimality condition for the path is then given by

$$\partial_{tt} \mu(t, x) = 0,$$

which gives as optimal path the "interval" $(t \in [0, 1])$

$$\mu(t, x) = [\mu_1(x) - \mu_0(x)]t + \mu_0(x).$$

Our claim about bi-modality of a mix of two unimodal distributions is evident. On the other hand, OMT geodesics represent nonlinear mixing and have a considerably different character [10].

Yet, the L^2 problem can be used in conjunction with OMT in case of unbalanced mass distributions. Indeed, given the two unbalanced densities μ_0 and μ_1 it is natural to seek a distribution $\tilde{\mu}_1$ the closest density to μ_1 in the L^2 sense, which minimizes the Wasserstein distance $d_{\mathrm{wass}}(\mu_0, \tilde{\mu}_1)^2$. The L^2 perturbation can be interpreted as "noise." One can then show that this problem amounts to minimizing

$$\inf_{\mu, v, \tilde{\mu}_1} \int \int_0^1 \mu(t, x) \|v\|^2 \, dt \, dx + \alpha/2 \int |\mu_1(x) - \tilde{\mu}_1(x)|^2 \, dx \qquad (33)$$

over all time varying densities μ and vector fields v satisfying

$$\frac{\partial \mu}{\partial t} + \mathrm{div}(\mu v) = 0, \qquad (34)$$

$$\mu(0, \cdot) = \mu_0, \ \mu(1, \cdot) = \tilde{\mu}_1.$$

This idea has been taken further in [8, 10] where the two end points μ_0, μ_1 are allowed to be perturbed slightly into $\tilde{\mu}_0, \tilde{\mu}_1$ while the perturbation equalizes their integrals and is accounted for in the metric. This leads to a *modified Monge–Kantorovich problem* which is best expressed in terms of its dual. Recall that the original OMT-problem of transferring (balanced) μ_0 into μ_1 has the following dual (see, e.g., [17])

$$\max_{\phi(x)+\psi(y)\leq\rho(x,y)} \int \phi(x)\mu_0(x)dx + \int \psi(y)\mu_1(y)dy, \tag{35}$$

which for the case $\rho(x,y) = |x-y|$ can be shown to be

$$\max_{\|\phi\|_{\mathrm{Lip}}\leq 1} \int \phi(\mu_0 - \mu_1)$$

with $\|\phi\|_{\mathrm{Lip}} = \sup \frac{|g(x)-g(y)|}{|x-y|}$ the Lipschitz norm. Replacing μ_0, μ_1 by $\tilde{\mu}_0, \tilde{\mu}_1$ in the OMT problem while penalizing the magnitude of the errors $\|\mu_i - \tilde{\mu}_i\|$ leads to the following metric:

$$\inf_{\tilde{\mu}_0(\Omega)=\tilde{\mu}_1(\Omega)} d_1(d\tilde{\mu}_0, d\tilde{\mu}_1) + \kappa \sum_{i=1}^{2} d_{\mathrm{TV}}(d\mu_i, d\tilde{\mu}_i). \tag{36}$$

This metric in fact "interpolates" the total variation (which is not weak* continuous) and the Wasserstein distance. The above expression has the interesting physical interpretation where μ_i's are noisy version of the $\tilde{\mu}_i$'s. It has also a considerable practical significance since it allows comparing distributions of unequal mass in a natural manner. Further, *it is weak* continuous* in contrast to most "distances" used in the signal processing literature (Itakura–Saito, logarithmic spectral deviation, and Kullback–Leibler distance; see [9]). The dual formulation (36) is particularly simple:

$$d_{\mathrm{unbalanced}}(\mu_0, \mu_1) := \max_{\substack{\|\phi\|_{\mathrm{Lip}}\leq 1 \\ \|\phi\|_\infty \leq c}} \int \phi(\mu_0 - \mu_1). \tag{37}$$

The constant c depends on the penalty κ. Interestingly, geodesics of this metric average nicely several of the features of the end points [8, 10]. We also note parallel work on the geometry for unbalanced mass distributions using substantially different tools in [5, 6].

10. Conclusions

The optimal mass transport problem has wide ranging ramifications for statistical estimation, image analysis, information geometry, and control. The span of ideas presented here draws together natural metrics in probability, statistical prediction theory, and physics. Several new metrics may be derived in a similar manner. Indeed, following [4] again, we can devise geodesic distances analogous to *action integrals* (for pressureless fluid flow). Such integrals may be taken in the form:

$$\inf \int \int_0^1 \mu(t,x)h(v(t,x))\, dt\, dx \tag{38}$$

over all time-varying densities μ and vector fields v satisfying

$$\frac{\partial \mu}{\partial t} + \text{div}(\mu v) = 0, \tag{39}$$

$$\mu(0, \cdot) = \mu_0, \ \mu(1, \cdot) = \mu_1.$$

Here h is a strictly convex even function. (In the present paper and in [4], $h(v) = \|v\|^2/2$.) One can show that this leads to the flow

$$\mu_t = \text{div}(\mu \nabla h^* (\nabla f'(\mu))),$$

where h^* is the Legendre transform of h. It will be interesting to further unify the study of the earlier fundamental concepts of information and prediction theory along the above lines.

Acknowledgment

This work was supported in part by grants from AFOSR and ARO. Further, this project was supported by grants from the National Center for Research Resources (P41-RR-013218) and the National Institute of Biomedical Imaging and Bioengineering (P41-EB-015902) of the National Institutes of Health. This work is part of the National Alliance for Medical Image Computing (NAMIC), funded by the National Institutes of Health through the NIH Roadmap for Medical Research, Grant U54 EB005149. Information on the National Centers for Biomedical Computing can be obtained from http://nihroadmap.nih.gov/bioinformatics.

References

[1] Amari, S. and Nagaoka, H., *Methods of Information Geometry, Memoirs of AMS* **191**, 2007.

[2] S. Angenent, S. Haker, and A. Tannenbaum, "Minimizing flows for the Monge–Kantorovich problem," *SIAM J. Math. Analysis*, vol. 35, pp. 61–97, 2003.

[3] S. Angenent, G. Sapiro, and A. Tannenbaum, "On the affine invariant heat equation for nonconvex curves," *Journal of the American Mathematical Society* **11** (1998), pp. 601–634.

[4] J.-D. Benamou and Y. Brenier, "A computational fluid mechanics solution to the Monge–Kantorovich mass transfer problem," *Numerische Mathematik* **84** (2000), pp. 375–393.

[5] J.D. Benamou, "Numerical resolution of an unbalanced mass transport problem," *Mathematical Modelling and Numerical Analysis*, **37(5)**, 851–862, 2003.

[6] A. Figalli,"The optimal partial transport problem," *Archive for rational mechanics and analysis*, **195(2)**: 533–560, 2010.

[7] T. Georgiou, "Distances and Riemannian metrics for spectral density functions," *IEEE Trans. Signal Processing* **55** (2007), pp. 3995–4004.

[8] T.T.Georgiou, J. Karlsson, and S. Takyar,"Metrics for power spectra: an axiomatic approach," *IEEE Trans. on Signal Processing*, **57(3)**: 859–867, March 2009.

[9] R.M. Gray, A. Buzo, A.H. Gray, and Y. Matsuyama, "Distortion measures for speech processing," *IEEE Trans. on Acoustics, Speech, and Signal Proc.*, **28(4)**, pp. 367–376, 1980.

[10] X. Jiang, Z.Q. Luo and T.T. Georgiou, "Geometric Methods for Spectral Analysis," *IEEE Trans. on Signal Processing,* to appear, 2012.

[11] Xianhua Jiang, Lipeng Ning, and Tryphon T. Georgiou, "Distances and Riemannian metrics for multivariate spectral densities," `arXiv:1107.1345v1`.

[12] R. Jordan, D. Kinderlehrer, and F. Otto, "The variational formulation of the Fokker-Planck equation," *SIAM J. Math. Anal.* **29** (1998), pp. 1–17.

[13] L.V. Kantorovich, "On a problem of Monge," *Uspekhi Mat. Nauk.* **3** (1948), pp. 225–226.

[14] P. Masani, "Recent trends in multivariable prediction theory," in Krishnaiah, P.R., Editor, *Multivariate Analysis*, pp. 351–382, Academic Press, 1966.

[15] D. McQuarrie, *Statistical Mechanics* (2nd edition), University Science Books, 2000.

[16] S. Rachev and L. Rüschendorf, *Mass Transportation Problems*, Volumes I and II, Probability and Its Applications, Springer, New York, 1998.

[17] C. Villani, *Topics in Optimal Transportation,* Graduate Studies in Mathematics, vol. 58, AMS, Providence, RI, 2003.

Emmanuel Tannenbaum
Department of Chemistry
Ben-Gurion University of the Negev, Israel
e-mail: `emmanuel.tannenbaum@gmail.com`

Tryphon Georgiou
Department of Electrical and Computer Engineering
University of Minnesota
Minneapolis, MN, USA
e-mail: `tryphon@ece.umn.edu`

Allen Tannenbaum
Departments of Electrical & Computer
and Biomedical Engineering
Boston University, Boston, MA, USA
e-mail: `tannenba@bu.edu`

Operator Theory:
Advances and Applications, Vol. 222, 325–349

LMI Representations of Convex Semialgebraic Sets and Determinantal Representations of Algebraic Hypersurfaces: Past, Present, and Future

Victor Vinnikov

To Bill Helton, on the occasion of his 65th birthday

Abstract. 10 years ago or so Bill Helton introduced me to some mathematical problems arising from semidefinite programming. This paper is a partial account of what was and what is happening with one of these problems, including many open questions and some new results.

Mathematics Subject Classification. Primary: 14M12, 90C22; secondary: 14P10, 52A20.

Keywords. Semidefinite programming (SDP), linear matrix inequality (LMI), convex semialgebraic sets, determinantal representations of polynomials or of hypersurfaces, kernel sheaves, real zero (RZ) polynomials, rigidly convex algebraic interiors, interlacing polynomials, hyperbolic polynomials.

1. Introduction

Semidefinite programming (SDP) is probably the most important new development in optimization in the last two decades. The (primal) semidefinite programme is to minimize an affine linear functional ℓ on $\mathbb{R}^d$ subject to a linear matrix inequality (LMI) constraint

$$A_0 + x_1 A_1 + \cdots + x_d A_d \geq 0;$$

here $A_0, A_1, \ldots, A_d \in \mathbb{SR}^{n \times n}$ (real symmetric $n \times n$ matrices) for some n and $Y \geq 0$ means that $Y \in \mathbb{SR}^{n \times n}$ is positive semidefinite (has nonnegative eigenvalues or equivalently satisfies $y^\top Y y \geq 0$ for all $y \in \mathbb{R}^n$). This can be solved efficiently, both theoretically (finding an approximate solution with a given accuracy ϵ in a time that is polynomial in $\log(1/\epsilon)$ and in the input size of the problem) and in many concrete situations, using interior point methods. Notice that semidefinite programming is a far reaching extension of linear programming (LP) which corresponds to the case when the real symmetric matrices $A_0, A_1, \ldots, A_d$ commute (i.e., are simultaneously diagonalizable). The literature on the subject is quite vast, and we only mention the pioneering book [40], the surveys [52] and [39], and the book [51] for applications to systems and control.

One very basic mathematical question is which convex sets arise as feasibility sets for SDP? In other words, *given a convex set C, do there exist $A_0, A_1, \ldots, A_d \in \mathbb{SR}^{n \times n}$ for some n such that*

$$C = \{x = (x_1, \ldots, x_d) \in \mathbb{R}^d : A_0 + x_1 A_1 + \cdots + x_d A_d \geq 0\}? \tag{1}$$

We refer to (1) as a *LMI representation* of C^1. Sets having a LMI representation are also called *spectrahedra*. This notion was introduced and studied in [49], and the above question – which convex sets admit a LMI representation, i.e., are spectrahedra – was formally posed in [45]. A complete answer for $d = 2$ was obtained in [28], though there are still outstanding computational questions, see [29, 46, 47]; for $d > 2$, no answer is known, though the recent results of [6, 43, 42] shed some additional light on the problem. It is the purpose of this paper to survey some aspects of the current state of the affairs.

Since a real symmetric matrix is positive semidefinite if and only if all of its principal minors are nonnegative, the set on the right-hand side of (1) coincides with the set where all the principal minors of $A_0 + x_1 A_1 + \cdots + x_d A_d$ are nonnegative. Therefore if a convex set C admits a LMI representation then C is a *basic closed semialgebraic set* (i.e., a set defined by finitely many nonstrict polynomial inequalities). However, as shown in [28], C is in fact much more special: it is a *rigidly convex algebraic interior*, i.e., an *algebraic interior* whose *minimal defining polynomial* satisfies the *real zero (RZ)* condition with respect to any point in the interior of C. Furthermore, LMI representations are (essentially) *positive real symmetric determinantal representations* of certain multiples of the minimal defining polynomial of C. This reduces the question of the existence (and a construction) of LMI representations to an old problem of algebraic geometry – we only mention here the classical paper [12] and refer to [5], [13, Chapter 4], and [32] for a detailed bibliography – but with two additional twists: first, we require positivity; second, there is a freedom provided by allowing multiples of the given polynomial.

This paper is organized as follows. In Section 2 we define rigidly convex sets and *RZ* polynomials, and explain why LMI representations are determinantal representations. In Section 3 we discuss some of what is currently known and

---

[1]We can also consider a (complex) self-adjoint LMI representation of C, meaning that $A_0, A_1, \ldots, A_d \in \mathbb{HC}^{n \times n}$ (complex hermitian $n \times n$ matrices) for some n. If $A = B + iC \in \mathbb{HC}^{n \times n}$ with $B, C \in \mathbb{R}^{n \times n}$, and we set $\tilde{A} = \begin{bmatrix} B & -C \\ C & B \end{bmatrix} \in \mathbb{SR}^{2n \times 2n}$, then $A \geq 0$ if and only if $\tilde{A} \geq 0$ and $\det \tilde{A} = (\det A)^2$. So a self-adjoint LMI representation gives a real symmetric LMI representation as defined in the main text with the size of matrices doubled and the determinant of the linear matrix polynomial squared, see [49, Section 1.4] and [43, Lemma 2.14].

unknown about determinantal representations, with a special emphasis on positive real symmetric determinantal representations. In Section 4 we review some of the ways to (re)construct a determinantal representation starting from its kernel sheaf, especially the construction of the adjoint matrix of a determinantal representation that goes back to [12] and was further developed in [53, 3, 32]. In Section 5 we show how this construction yields positive self-adjoint determinantal representations in the case $d = 2$ by using a RZ polynomial that *interlaces* the given RZ polynomial. This provides an alternative proof of the main result of [28] (in a slightly weaker form since we obtain a representation that is self-adjoint rather than real symmetric) which is constructive algebraic in that it avoids the use of theta functions.

We have concentrated in this paper on the non-homogenous setting (convex sets) rather than on the homogeneous setting (convex cones). In the homogeneous setting, RZ polynomials correspond to *hyperbolic polynomials* and rigidly convex algebraic interiors correspond to their *hyperbolicity cones*, see, e.g., [17, 18, 36, 44, 21, 7, 50]. Theorem 3 then provides a solution the Lax conjecture concerning homogeneous hyperbolic polynomials in three variables, see [38], whereas Conjecture 5, which may be called the generalized Lax conjecture, states that *any hyperbolicity cone is a semidefinite slice*, i.e., equals the intersection of the cone of positive semidefinite matrices with a linear subspace.

Finally, the LMI representation problem considered here is but one of the several important problems of this kind arising from SDP. Other major problems have to do with lifted LMI representations (see [35, 25, 26]) and with the free noncommutative setting (see [24, 23]).

2. From LMI representations of convex sets to determinantal representations of polynomials

2.1. A closed set C in $\mathbb{R}^d$ is called an *algebraic interior* [28, Section 2.2] if there is a polynomial $p \in \mathbb{R}[x_1, \ldots, x_d]$ such that C equals the closure of a connected component of

$$\{x \in \mathbb{R}^d \colon p(x) > 0\}.$$

In other words, there is a $p \in \mathbb{R}[x_1, \ldots, x_d]$ which vanishes on the boundary ∂C of C and such that $\{x \in C \colon p(x) > 0\}$ is connected with closure equal to C. (Notice that in general p may vanish also at some points in the interior of C; for example, look at $p(x_1, x_2) = x_2^2 - x_1^2(x_1 - 1)$.) We call p a defining polynomial of C. It is not hard to show that if C is an algebraic interior then a minimal degree defining polynomial p of C is unique (up to a multiplication by a positive constant); we call it a *minimal defining polynomial* of C, and it is simply a reduced (i.e., without multiple irreducible factors) polynomial such that the real affine hypersurface

$$\mathcal{V}_p(\mathbb{R}) = \{x \in \mathbb{R}^d \colon p(x) = 0\} \tag{2}$$

equals the Zariski closure $\overline{\partial C}^{\mathrm{Zar}}$ of the boundary ∂C in $\mathbb{R}^d$ (normalized to be positive at an interior point of C). Any other defining polynomial q of C is given by $q = ph$ where h is an arbitrary polynomial which is strictly positive on a dense connected subset of C. An algebraic interior is a semialgebraic set (i.e., a set defined by a finite boolean combination of polynomial inequalities) since it is the closure of a connected component of a semialgebraic set.

Let now C be a convex set in $\mathbb{R}^d$ that admits a LMI representation (1). We will assume that $\mathrm{Int}\, C \neq \emptyset$; it turns out that by restricting the LMI representation (i.e., the matrices $A_0, A_1, \ldots, A_d$) to a subspace of $\mathbb{R}^n$, one can assume without loss of generality that $A_0 + x_1 A_1 + \cdots + x_d A_d > 0$ for one and then every point of $\mathrm{Int}\, C$ ($Y > 0$ means that $Y \in \mathbb{SR}^{n \times n}$ is positive definite, i.e., Y has strictly positive eigenvalues or equivalently satisfies $y^\top Y y > 0$ for all $y \in \mathbb{R}^n$, $y \neq 0$). It is then easy to see that C is an algebraic interior with defining polynomial $\det(A_0 + x_1 A_1 + \cdots + x_d A_d)$. Conversely, if C is an algebraic interior with defining polynomial $\det(A_0 + x_1 A_1 + \cdots + x_d A_d)$, and $A_0 + x_1 A_1 + \cdots + x_d A_d > 0$ for one point of $\mathrm{Int}\, C$, then it follows easily that (1) is a LMI representation of C. (See [28, Section 2.3] for details.)

Let $q(x) = \det(A_0 + x_1 A_1 + \cdots + x_d A_d)$, let $x^0 = (x_1^0, \ldots, x_d^0) \in \mathrm{Int}\, C$, and let us normalize the LMI representation by $A_0 + x_1^0 A_1 + \cdots + x_d^0 A_d = I$. We restrict the polynomial q to a straight line through x^0, i.e., for any $x \in \mathbb{R}^d$ we consider the univariate polynomial $q_x(t) = q(x^0 + tx)$. Because of our normalization, we can write

$$q_x(t) = \det(I + t(x_1 A_1 + \cdots + x_d A_d)),$$

and since all the eigenvalues of the real symmetric matrix $x_1 A_1 + \cdots + x_d A_d$ are real, we conclude that $q_x \in \mathbb{R}[t]$ has only real zeroes.

A polynomial $p \in \mathbb{R}[x_1, \ldots, x_d]$ is said to satisfy the *real zero (RZ)* condition with respect to $x^0 \in \mathbb{R}^d$, or to be a RZ_{x^0} *polynomial*, if for all $x \in \mathbb{R}^d$ the univariate polynomial $p_x(t) = p(x^0 + tx)$ has only real zeroes. It is clear that a divisor of a RZ_{x^0} polynomial is again a RZ_{x^0} polynomial. We have thus arrived at the following result of [28].

Theorem 1. *If a convex set C with $x^0 \in \mathrm{Int}\, C$ admits a LMI representation, then C is an algebraic interior whose minimal defining polynomial p is a RZ_{x^0} polynomial. (1) is a LMI representation of C (that is positive definite on $\mathrm{Int}\, C$) if and only if $A_0 + x_1^0 A_1 + \cdots + x_d^0 A_d > 0$ and*

$$\det(A_0 + x_1 A_1 + \cdots + x_d A_d) = p(x)h(x),$$

where $h \in \mathbb{R}[x_1, \ldots, x_d]$ satisfies $h > 0$ on $\mathrm{Int}\, C$.

2.2. The definition of a RZ_{x^0} polynomial has a simple geometric meaning ([28, Section 3]). Assume for simplicity that p is reduced (i.e., without multiple irreducible factors) of degree m. Then p is a RZ_{x^0} polynomial if and only if a general straight line through x^0 in $\mathbb{R}^d$ intersects the corresponding real affine hypersurface

$\mathcal{V}_p(\mathbb{R})$ (see (2)) in m distinct points. Alternatively, every straight line through x^0 in the real projective space $\mathbb{P}^d(\mathbb{R})$ intersects the projective closure $\mathcal{V}_P(\mathbb{R})$ of $\mathcal{V}_p(\mathbb{R})$,

$$\mathcal{V}_P(\mathbb{R}) = \{[X] \in \mathbb{R}^d \colon P(X) = 0\}, \tag{3}$$

in exactly m points counting multiplicities. Here we identify as usual the d-dimensional real projective space $\mathbb{P}^d(\mathbb{R})$ with the union of $\mathbb{R}^d$ and of the hyperplane at infinity $X_0 = 0$, so that the affine coordinates $x = (x_1, \ldots, x_d)$ and the projective coordinates $X = (X_0, X_1, \ldots, X_d)$ are related by $x_1 = X_1/X_0, \ldots, x_d = X_d/X_0$; we denote by $[X] \in \mathbb{P}^d(\mathbb{R})$ the point with the projective coordinates X; and we let $P \in \mathbb{R}[X_0, X_1, \ldots, X_d]$ be the homogenization of p,

$$P(X_0, X_1, \ldots, X_d) = X_0^m p(X_1/X_0, \ldots, X_d/X_0). \tag{4}$$

Notice that if $X = (1, x)$ and $X^0 = (1, x^0)$,

$$P(X + sX^0) = (s+1)^m p(x_0 + (s+1)^{-1}(x - x^0)). \tag{5}$$

It turns out that if p is a RZ_{x^0} polynomial with $p(x^0) > 0$, and if x' belongs to the interior of the closure of the connected component of x^0 in $\{x \in \mathbb{R}^d \colon p(x) > 0\}$, then $p(x') > 0$ and p is also a $RZ_{x'}$ polynomial ([28, Section 5.3]). We call an algebraic interior $\mathcal{C}$ whose minimal defining polynomial satisfies the RZ condition with respect to one and then every point of $\operatorname{Int}\mathcal{C}$ a *rigidly convex* algebraic interior.

As simple examples, we see that the circle $\{(x_1, x_2) \colon x_1^2 + x_2^2 \le 1\}$ is a rigidly convex algebraic interior, while the "flat TV screen" $\{(x_1, x_2) \colon x_1^4 + x_2^4 \le 1\}$ is not. *Theorem 1 tells us that a necessary condition for $\mathcal{C}$ to admit a LMI representation is that $\mathcal{C}$ is a rigidly convex algebraic interior, and the size n of the matrices in a LMI representation is greater than or equal to the degree m of a minimal defining polynomial p of $\mathcal{C}$.*

Rigidly convex algebraic interiors are always convex sets ([28, Section 5.3]). They are also basic closed semialgebraic sets, as follows ([41, Remark 2.6] following [50]). Let p be a minimal defining polynomial of a rigidly convex algebraic interior $\mathcal{C}$, of degree m, and let $x^0 \in \operatorname{Int}\mathcal{C}$. We set

$$P_{x^0}^{(k)}(X) = \frac{d^k}{ds^k} P(X + sX^0)\big|_{s=0}, \quad p_{x^0}^{(k)}(x) = P_{x^0}^{(k)}(1, x_1, \ldots, x_d), \tag{6}$$

where P is the homogenization of p (see (4)) and $X^0 = (1, x^0)$; $p_{x^0}^{(k)}$ is called the *kth Renegar derivative* of p with respect to x^0. Then $p_{x^0}^{(k)}$ is a RZ_{x^0} polynomial with $p_{x^0}^{(k)}(x^0) > 0$ for all $k = 1, \ldots, m-1$. The rigidly convex algebraic interiors $\mathcal{C}^{(k)}$ containing x^0 with minimal defining polynomials $p_{x^0}^{(k)}$ (i.e., the closures of the connected components of x^0 in $\{x \in \mathbb{R}^d \colon p_{x^0}^{(k)}(x) > 0\}$) are increasing: $\mathcal{C} = \mathcal{C}^{(0)} \subseteq \mathcal{C}^{(1)} \subseteq \cdots \subseteq \mathcal{C}^{(m-1)}$, and

$$\mathcal{C} = \{x \in \mathbb{R}^d \colon p(x) \ge 0, p_{x^0}^{(1)}(x) \ge 0, \ldots, p_{x^0}^{(m-1)}(x) \ge 0\}. \tag{7}$$

RZ polynomials can be also characterized by a very simple global topology of the corresponding real projective hypersurface $\mathcal{V}_P(\mathbb{R})$ (see (3); readers who prefer can assume that the corresponding real affine hypersurface $\mathcal{V}_p(\mathbb{R})$ is compact in $\mathbb{R}^d$

– this implies that the degree m of p is even – and replace in the following the real projective space $\mathbb{P}^d(\mathbb{R})$ by the affine space $\mathbb{R}^d$). We call $W \subseteq \mathbb{P}^d(\mathbb{R})$ an *ovaloid* if W is isotopic in $\mathbb{P}^d(\mathbb{R})$ to a sphere $S \subset \mathbb{R}^m \subset \mathbb{P}^m(\mathbb{R})$, i.e., there is a homeomorphism F of $\mathbb{P}^d(\mathbb{R})$ with $F(S) = W$, and furthermore F is homotopic to the identity, i.e., there is a homeomorphism H of $[0,1] \times \mathbb{P}^d(\mathbb{R})$ such that $H_t = H|_{\{t\} \times \mathbb{P}^d(\mathbb{R})}$ is a homeomorphism of $\mathbb{P}^d(\mathbb{R})$ for every t, $H_0 = \mathrm{Id}_{\mathbb{P}^d(\mathbb{R})}$, and $H_1 = F$. Notice that $\mathbb{P}^d(\mathbb{R}) \setminus S$ consists of two connected components only one of which is contractible, hence the same is true of $\mathbb{P}^d(\mathbb{R}) \setminus W$; we call the contractible component the *interior* of the ovaloid W, and the non-contractible component the *exterior*. We call $W \subseteq \mathbb{P}^d(\mathbb{R})$ a *pseudo-hyperplane* if W is isotopic in $\mathbb{P}^d(\mathbb{R})$ to a (projective) hyperplane $H \subseteq \mathbb{P}^d(\mathbb{R})$. In the case $d = 2$ we say *oval* and *pseudo-line* instead of ovaloid and pseudo-hyperplane. We then have the following result; we refer to [28, Sections 5 and 7] for proof, discussion, and implications.

Proposition 2. *Let $p \in \mathbb{R}[x_1, \ldots, x_d]$ be reduced of degree m and assume that the corresponding real projective hypersurface $\mathcal{V}_P(\mathbb{R})$ is smooth. Then p satisfies RZ_{x^0} with $p(x^0) \neq 0$ if and only if*

   a. *if $m = 2k$ is even, $\mathcal{V}_P(\mathbb{R})$ is a disjoint union of k ovaloids $W_1, \ldots, W_k$, with W_i contained in the interior of W_{i+1}, $i = 1, \ldots, k-1$, and x^0 lying in the interior of W_1;*

   b. *if $m = 2k + 1$ is odd, $\mathcal{V}_P(\mathbb{R})$ is a disjoint union of k ovaloids $W_1, \ldots, W_k$, with W_i contained in the interior of W_{i+1}, $i = 1, \ldots, k-1$, and x^0 lying in the interior of W_1, and a pseudo-hyperplane W_{k+1} contained in the exterior of W_k.*

Let us denote by $\mathcal{I}$ the interior of W_1, let us normalize p by $p(x^0) > 0$, and let $H_\infty = \{X_0 = 0\}$ be the hyperplane at infinity in $\mathbb{P}^d(\mathbb{R})$. If $\mathcal{I} \cap H_\infty = \emptyset$, then the closure of $\mathcal{I}$ in $\mathbb{R}^d$ is a rigidly convex algebraic interior with a minimal defining polynomial p. If $\mathcal{I} \cap H_\infty \neq \emptyset$, then $\mathcal{I} \setminus \mathcal{I} \cap H_\infty$ consists of two connected components, the closure of each one of them in $\mathbb{R}^d$ being a rigidly convex algebraic interior with a minimal defining polynomial p (if m is even) or p for one component and $-p$ for the other component (if m is odd).

3. Determinantal representations of polynomials: some of the known and of the unknown

3.1. The following is proved in [28, Section 5] (based on the results of [54] and [4], see also [14]).

Theorem 3. *Let $p \in \mathbb{R}[x_1, x_2]$ be a RZ_{x^0} polynomial of degree m with $p(x^0) = 1$. Then there exist $A_0, A_1, A_2 \in \mathbb{S}\mathbb{R}^{m \times m}$ with $A_0 + x_1^0 A_1 + x_2^0 A_2 = I$ such that*

$$\det(A_0 + x_1 A_1 + x_2 A_2) = p(x). \tag{8}$$

We will review the proof of Theorem 3 given in [28] in Section 4 below, and then present in Section 5 an alternate proof for positive self-adjoint (rather than real symmetric) determinantal representations that avoids the transcendental machinery of Jacobian varieties and theta functions (though it still involves, to a certain extent, meromorphic differentials on a compact Riemann surface).

Theorem 3 tells us that a necessary and sufficient condition for $\mathcal{C} \subseteq \mathbb{R}^2$ to admit a LMI representation is that $\mathcal{C}$ is a rigidly convex algebraic interior, and the size of the matrices in a LMI representation can be taken equal to be the degree m of a minimal defining polynomial p of $\mathcal{C}$.

There can be no exact analogue of Theorem 3 for $d > 2$. Indeed, we have

Proposition 4. A general polynomial $p \in \mathbb{C}[x_1, \ldots, x_d]$ of degree m does not admit a determinantal representation

$$\det(A_0 + x_1 A_1 + \cdots + x_d A_d) = p(x), \tag{9}$$

with $A_0, A_1, \ldots, A_d \in \mathbb{C}^{m \times m}$, for $d > 3$ and for $d = 3$, $m \geq 4$.

Since for any fixed $x^0 \in \mathbb{R}^d$ the set of RZ_{x^0} polynomials of degree m with $p(x^0) > 0$ such that the corresponding real projective hypersurface $\mathcal{V}_P(\mathbb{R})$ is smooth is an open subset of the vector space of polynomials over $\mathbb{R}$ of degree m (see [28, Sections 5 and 7] following [44]), it follows that a general RZ_{x^0} polynomial $p \in \mathbb{R}[x_1, \ldots, x_d]$ of degree m with $p(x^0) > 0$ does not admit a determinantal representation (9) with $m \times m$ matrices – even without requiring real symmetry or positivity – for $d > 3$ and for $d = 3$, $m \geq 4$. (For the remaining cases when $d = 3$, the case $m = 2$ is straightforward and the case $m = 3$ is treated in details in [8] when the corresponding complex projective cubic surface $\mathcal{V}_P$ in $\mathbb{P}^3(\mathbb{C})$ is smooth; in both cases there are no positive real symmetric determinantal representations of size m as in Theorem 3, but there are positive self-adjoint determinantal representations of size m, i.e., representations (9) with $m \times m$ self-adjoint matrices such that $A_0 + x_1^0 A_1 + x_2^0 A_2 + x_3^0 A_3 = I$.)

Proposition 4 follows by a simple count of parameters, see [11]. It also follows from Theorem 11 below using the Noether–Lefschetz theory [37, 22, 20], since for a general homogeneous polynomial $P \in \mathbb{C}[X_0, X_1, \ldots, X_d]$ of degree m with $d > 3$ or with $d = 3$, $m \geq 4$, the only line bundles on $\mathcal{V}_P$ are of the form $\mathcal{O}_{\mathcal{V}_P}(j)$ and these obviously fail the conditions of the theorem.

The following is therefore the "best possible" generalization of Theorem 3 to the case $d > 2$.

Conjecture 5. Let $p \in \mathbb{R}[x_1, \ldots, x_d]$ be a RZ_{x^0} polynomial of degree m with $p(x^0) = 1$. Then there exists a RZ_{x^0} polynomial $h \in \mathbb{R}[x_1, \ldots, x_d]$ of degree ℓ with $h(x^0) = 1$ and with the closure of the connected component of x^0 in $\{x \in \mathbb{R}^d : h(x) > 0\}$ containing the closure of the connected component of x^0 in $\{x \in \mathbb{R}^d : p(x) > 0\}$, and $A_0, A_1, \ldots, A_d \in \mathbb{SR}^{n \times n}$, $n \geq m + \ell$, with $A_0 + x_1^0 A_1 + \cdots + x_d^0 A_d = I$, such that

$$\det(A_0 + x_1 A_1 + \cdots + x_d A_d) = p(x) h(x). \tag{10}$$

Notice that is enough to require that h is a polynomial that is strictly positive on the connected component of x^0 in $\{x \in \mathbb{R}^d \colon h(x) > 0\}$, since it then follows from (10) that h is a RZ_{x^0} polynomial with $h(x^0) = 1$ and with the closure of the connected component of x^0 in $\{x \in \mathbb{R}^d \colon h(x) > 0\}$ containing the closure of the connected component of x^0 in $\{x \in \mathbb{R}^d \colon p(x) > 0\}$.

Conjecture 5 tells us that a necessary and sufficient condition for $\mathcal{C} \subseteq \mathbb{R}^d$ to admit a LMI representation is that $\mathcal{C}$ is a rigidly convex algebraic interior.

We can also homogenize (10),

$$\det(X_0 A_0 + X_1 A_1 + \cdots + X_d A_d) = P(X)\tilde{H}(X), \tag{11}$$

where $\tilde{H}(X) = H(X)X_0^{n-m-\ell}$ and P and H are the homogenizations of P and H respectively (see (4)).

3.2. The easiest way to establish Conjecture 5 would be to try taking $h = 1$ in (10) bringing us back to (9); in the homogeneous version, $\tilde{H} = X_0^{n-m}$ in (11). This was the form of the conjecture stated in [28]. It was given further credence by the existence of real symmetric determinantal representations without the requirement of positivity.

Theorem 6. *Let $p \in \mathbb{R}[x_1, \ldots, x_d]$. Then there exist $A_0, A_1, \ldots, A_d \in \mathbb{SR}^{n \times n}$ for some $n \geq m$ such that p admits the determinantal representation (9).*

Theorem 6 was first established in [27] using free noncommutative techniques. More precisely, the method was to take a lifting of p to the free algebra and to apply results of noncommutative realization theory to first produce a determinantal representation with $A_0, A_1, \ldots, A_d \in \mathbb{R}^{n \times n}$ and then to show that it is symmetrizable; see [27, Section 14] for details and references. An alternate proof of Theorem (6) that uses more elementary arguments was given in [48]. As it turns out, determinantal representations also appear naturally in algebraic complexity theory, and a proof of Theorem (6) from this perspective was given in [19].

Unfortunately, the analogue of Theorem 6 for positive real symmetric (or positive self-adjoint) determinantal representations fails. Counterexamples were first established in [6], and subsequently in [43]. Indeed we have

Proposition 7. *A general RZ_{x^0} polynomial $p \in \mathbb{R}[x_1, \ldots, x_d]$ of degree m with $p(x^0) = 1$ does not admit a determinantal representation (9), where A_0, A_1, ..., $A_d \in \mathbb{HC}^{n \times n}$ for some $n \geq m$ with $A_0 + x_1^0 A_1 + \cdots + x_d^0 A_d = I$, for any fixed $m \geq 4$ and d large enough or for any fixed $d \geq 3$ and m large enough.*

Here "large enough" means that $d^3 m^2 < \binom{m+d}{m} - 1$. We refer to [43, Section 3] for details and numerous examples of RZ polynomials that do not admit a positive self-adjoint determinantal representation as in Proposition 7. One simple example is

$$p = (x_1 + 1)^2 - x_2^2 - \cdots - x_d^2 \tag{12}$$

for $d \geq 5$ (for $d = 4$ this polynomial admits a positive self-adjoint determinantal representation but does not admit a positive real symmetric determinantal representation). The proofs are based on the fact that a positive self-adjoint (or real symmetric) determinantal representation of size n always contains, after a unitary (or orthogonal) transformation of the matrices $A_0, A_1, \ldots, A_d$, a direct summand $I_{n-n'} + x_1 0_{n-n'} + \cdots + x_d 0_{n-n'}$ – yielding a determinantal representation of size n' – for relatively small n': one can always take $n' \leq md$ and in many instances one can actually take $n' = m$, see [43, Theorems 2.4 and 2.7]. It would be interesting to compare these results with the various general conditions for decomposability of determinantal representations obtained in [33, 32].

3.3. The next easiest way to establish Conjecture 5 is to try taking h in (10) to be a power of p, $h = p^{r-1}$, so that we are looking for a positive real symmetric determinantal representation of p^r,

$$\det(A_0 + x_1 A_1 + \cdots + x_d A_d) = p(x)^r; \tag{13}$$

in the homogeneous version, $\tilde{H} = P^{r-1} \cdot X_0^{n-mr}$ in (11). If we do not require positivity or real symmetry, then at least for p irreducible, p^r admits a determinantal representation (13) with $A_0, A_1, \ldots, A_d \in \mathbb{C}^{n \times n}$, $n = mr$, for some $r \in \mathbb{N}$; this follows by the theory of matrix factorizations [16], since pI_r can be written as a product of matrices with linear entries, see [2, 30] (and also the references in [43]).

As established in [6], the answer for positive real symmetric determinantal representations is again no. Namely, let p be a polynomial of degree 4 in 8 variables labeled $x_a, x_b, x_c, x_d, x_{a'}, x_{b'}, x_{c'}, x_{d'}$, defined by

$$p = \sum_{S \in \mathcal{B}(V_8)} \prod_{j \in S} (x_j + 1), \tag{14}$$

where $\mathcal{B}(V_8)$ is the set consisting of all 4-element subsets of $\{a, b, c, d, a', b', c', d'\}$ except for

$$\{a, a', b, b'\}, \ \{b, b', c, c'\}, \ \{c, c', d, d'\}, \ \{d, d', a, a'\}, \ \{a, a', c, c'\}.$$

$\mathcal{B}(V_8)$ is the set of bases of a certain matroid V_8 on the set $\{a, b, c, d, a', b', c', d'\}$ called the *Vamos cube*. Then

Theorem 8. *p is RZ with respect to 0, and for all $r \in \mathbb{N}$, the polynomial p^r does not admit a determinantal representation (13) where $A_0, A_1, \ldots, A_d \in \mathbb{SR}^{n \times n}$ for some $n \geq mr$ with $A_0 = I$.*

This follows since on the one hand, V_8 is a half-plane property matroid, and on the other hand, it is not representable over any field, more precisely its rank function does not satisfy Ingleton inequalities. See [6, Section 3] for details. Notice that it turns out that one can take without loss of generality $n = mr$ in Theorem 8, see the paragraph following Proposition 7 above. Notice also that because of the footnote on page 326, it does not matter here whether we are considering real symmetric or self-adjoint determinantal representations.

The polynomial (14) remains so far the only example of a RZ polynomial no power of which admits a positive real symmetric determinantal representation[2]. For instance, we have

Theorem 9. *Let $p \in \mathbb{R}[x_1, \ldots, x_d]$ be a RZ_{x^0} polynomial of degree 2 with $p(x^0) = 1$. Then there exists $r \in \mathbb{N}$ and $A_0, A_1, \ldots, A_d \in \mathbb{SR}^{n \times n}$, $n = mr$, with $A_0 + x_1^0 A_1 + \cdots + x_d^0 A_d = I$, such that p^r admits the determinantal representation (13).*

Theorem 9 has been established in [43] using Clifford algebra techniques. More precisely, one associates to a polynomial $p \in \mathbb{R}[x_1, \ldots, x_d]$ of degree m a unital *-algebra as follows. Let $\mathbb{C}\langle z_1, \ldots, z_d \rangle$ be the free *-algebra on d generators, i.e., $z_1, \ldots, z_d$ are noncommuting self-adjoint indeterminates. For the homogenization P of p, we can write

$$P(-x_1 z_1 - \cdots - x_d z_d, x_1, \ldots, x_d) = \sum_{k \in \mathbb{Z}_+^d, |k|=m} q_k(z) x^k,$$

for some $q_k \in \mathbb{C}\langle z_1, \ldots, z_d \rangle$, where $k = (k_1, \ldots, k_d)$, $|k| = k_1 + \cdots + k_d$, and $x^k = x_1^{k_1} \cdots x_d^{k_d}$. We define the *generalized Clifford algebra associated with* p to be the quotient of $\mathbb{C}\langle z_1, \ldots, z_d \rangle$ by the two-sided ideal generated by $\{q_k\}_{|k|=m}$. It can then be shown that at least if p is irreducible, p^r admits a self-adjoint determinantal representation (13) of size mr with $A_0 = I$ for some $r \in \mathbb{N}$ if and only if the generalized Clifford algebra associated with p admits a finite-dimensional unital *-representation. In case $m = 2$ and p is an irreducible RZ_0 polynomial, the generalized Clifford algebra associated with p turns out to be "almost" the usual Clifford algebra, yielding the proof of Theorem 9. For details and references, see [43, Sections 4 and 5]. It would be interesting to investigate the generalized Clifford algebra associated with the polynomial (14).[3]

A new obstruction to powers of p admitting a positive real symmetric determinantal representation has been recently discovered in [42]. It is closely related to the question of how to test a polynomial for the RZ condition, see [29]. For any monic polynomial $f \in \mathbb{R}[t]$ of degree m with zeroes $\lambda_1, \ldots, \lambda_m$, let us define the matrix $H(f) = [h_{ij}]_{i,j=1,\ldots,m}$ by $h_{ij} = \sum_{k=1}^d \lambda_k^{i+j-2}$; notice that h_{ij} are actually polynomials in the coefficients of f. $H(f)$ is called the *Hermite matrix* of f, and it is positive semidefinite if and only if all the zeroes of f are real. Given $p \in \mathbb{R}[x_1, \ldots, x_d]$ of degree m with $p(x^0) = 1$, we now consider $H(\check{p}_x)$ where $\check{p}_x(t) = t^m p(x^0 + t^{-1}x)$; it is a polynomial matrix that we call the *Hermite matrix* of p with respect to x^0 and denote $H(p; x^0)$. p is a RZ_{x^0} polynomial if and only if $H(p; x^0)(x) \geq 0$

---

[2] *Added in proof.*
Peter Brändén noticed (see http://www-e.uni-magdeburg.de/ragc/talks/branden.pdf) that one can use the symmetry of the polynomial (14) to produce from it a RZ polynomial in 4 variables no power of which admits a positive real symmetric determinantal representation.
[3] *Added in proof.*
Tim Netzer recently reported (see http://www-e.uni-magdeburg.de/ragc/talks/netzer.pdf) that for an irreducible RZ_0 polynomial p with $p(0) = 1$, Conjecture 5 holds (with $x^0 = 0$) if and only if -1 is not a sum of hermitian squares in the generalized Clifford algebra associated with p.

for all $x \in \mathbb{R}^d$. Now, it turns out that if there exists $r \in \mathbb{N}$ such that p^r admits a determinantal representation (13) with $A_0, A_1, \ldots, A_d \in \mathbb{SR}^{n \times n}$, $n = mr$, and $A_0 + x_1^0 A_1 + \cdots + x_d^0 A_d = I$, then $H(p; x^0)$ can be factored: $H(p; x^0) = Q^\top Q$ for some polynomial matrix Q, i.e., $H(p; x^0)$ is a *sum of squares*. We notice that $H(p; x^0)$ can be reduced by homogeneity to a polynomial matrix in $d-1$ variables, implying that the sum of squares decomposition (factorization) is not an obstruction in the case $d = 2$, but it is in the case $d > 2$. In particular, there is numerical evidence that for the polynomial p of (14) the Hermite matrix (with respect to 0) is not a sum of squares. We refer to [42] for details. It would be very interesting to use these ideas in the case $d = 2$ to obtain a new proof of a weakened version of Theorem 3 that gives a positive real symmetric determinantal representation of p^r (of size mr) for some $r \in \mathbb{N}$.

3.4. There have been so far no attempts to pursue Conjecture 5 with other choices of h than 1 or a power of p. Two natural candidates are products of (not necessarily distinct) linear forms (that are nonnegative on the closure of the connected component of x^0 in $\{x \in \mathbb{R}^d \colon p(x) > 0\}$), and products of powers of Renegar derivatives of p with respect to x^0 (see (6)).

Conjecture 5 is a reasonable generalization of Theorem 3 for the purposes of LMI representations of convex sets (provided the solution gives a good hold of the extra factor h and of the size n). It is less satisfactory as a means of describing or generating RZ polynomials. The following alternative conjecture, that was proposed informally by L. Gurvits, might be more useful for that purpose. It is based on the fact that we have two systematic ways of generating RZ polynomials: positive real symmetric (or self-adjoint) determinantal representations and Renegar derivatives.

Conjecture 10. *Let $p \in \mathbb{R}[x_1, \ldots, x_d]$ be a RZ_{x^0} polynomial of degree m with $p(x^0) = 1$. Then there exist $k \in \mathbb{Z}_+$, a RZ_{x^0} polynomial $q \in \mathbb{R}[x_1, \ldots, x_d]$ of degree $m + k$ such that $p = q_{x^0}^{(k)}$, and $A_0, A_1, \ldots, A_d \in \mathbb{SR}^{(m+k) \times (m+k)}$ with $A_0 + x_1^0 A_1 + \cdots + x_d^0 A_d = I$, such that*

$$\det(A_0 + x_1 A_1 + \cdots + x_d A_d) = q(x).$$

4. Determinantal representations of homogeneous polynomials and sheaves on projective hypersurfaces

The kernel of a determinantal representation of a homogeneous polynomial is a sheaf on the corresponding projective hypersurface from which the representation itself can be reconstructed. We consider here the ways to do so that use the duality between the kernel and the left kernel; this gives the only known approaches to the proof of Theorem 3. For a different way using the resolution of the kernel sheaf see [5]; we refer also to the bibliography in [5, 32] and to [13, Chapter 4] and the references therein for more about this old topic in algebraic geometry.

4.1. Let $P \in \mathbb{C}[X_0, X_1, \ldots, X_d]$ $(d > 1)$ be a reduced (i.e., without multiple irreducible factors) homogeneous polynomial of degree m, and let

$$\mathcal{V}_P = \{[X] \in \mathbb{C}^d \colon P(X) = 0\} \tag{15}$$

be the corresponding complex projective hypersurface. Notice that when P is a polynomial over $\mathbb{R}$, $\mathcal{V}_P$ is naturally endowed with an antiholomorphic involution τ (the complex conjugation or the Galois action of $\mathrm{Gal}(\mathbb{C}/\mathbb{R})$) and the set of fixed points of τ is exactly the real projective hypersurface $\mathcal{V}_P(\mathbb{R})$ as in (3). Let

$$\det(X_0 A_0 + X_1 A_1 + \cdots + X_d A_d) = P(X),$$
$$A_\alpha = [A_{\alpha,ij}]_{i,j=1,\ldots,m} \in \mathbb{C}^{m \times m}, \ \alpha = 0, 1, \ldots, d, \tag{16}$$

be a determinantal representation of P, and let

$$U = X_0 A_0 + X_1 A_1 + \cdots + X_d A_d, \quad V = [V_{ij}]_{i,j=1,\ldots,m} = \mathrm{adj}\, U, \tag{17}$$

where $\mathrm{adj}\, Y$ denotes the adjoint matrix of a $m \times m$ matrix Y, i.e., the matrix whose (i,j) entry is $(-1)^{i+j}$ times the determinant of the matrix obtained from Y by removing the jth row and the ith column, so that $Y \cdot \mathrm{adj}\, Y = \det Y \cdot I$. Notice that

$$\det V = P^{m-1}, \tag{18}$$
$$\mathrm{adj}\, V = P^{m-2} \cdot U. \tag{19}$$

Notice also that using the formula for the differentiation of a determinant and row expansion,

$$\frac{\partial P}{\partial X_\alpha} = \sum_{l,k=1}^{n} A_{\alpha,lk} V_{kl}. \tag{20}$$

In particular, $V(X)$ is not zero for a smooth point $[X]$ of the hypersurface $\mathcal{V}_P$, so that $V(X)$ has rank 1 there and $U(X)$ has rank $m - 1$.

4.2. We restrict our attention now to the case $d = 2$, i.e., $\mathcal{V}_P$ is a projective plane curve. Let us assume for a starter that P is irreducible and that $\mathcal{V}_P$ is smooth – we will explain how to relax this assumption in Section 4.7 below. Then we conclude that $\mathcal{L}([X]) = \ker U(X)$ is a one-dimensional subspace of $\mathbb{C}^m$ for all points $[X]$ on $\mathcal{V}_P$, and these subspaces glue together to form a line bundle $\mathcal{L}$ on $\mathcal{V}_P$; more precisely, $\mathcal{L}$ is a subbundle of the trivial rank m vector bundle $\mathcal{V}_P \times \mathbb{C}^m$ whose fiber at the point $[X]$ equals $\mathcal{L}([X])$. It is convenient to twist and define $\mathcal{E} = \mathcal{L}(m-1)$. More algebraically, $\mathcal{E}$ is determined by the exact sequence of sheaves on $\mathcal{V}_P$

$$0 \longrightarrow \mathcal{E} \longrightarrow \mathcal{O}_{\mathcal{V}_P}^{\oplus m}(m-1) \overset{U}{\longrightarrow} \mathcal{O}_{\mathcal{V}_P}^{\oplus m}(m) \longrightarrow \mathrm{coker}(U) \longrightarrow 0, \tag{21}$$

where U denotes the operator of right multiplication by the matrix acting on columns. The following are some of the properties of the kernel line bundle.

1. The determinantal representation is determined up to a natural equivalence (multiplication on the left and on the right by constant invertible matrices) by the isomorphism class of the line bundle $\mathcal{E}$.

2. The columns $F_j = [V_{ij}]_{i=1,\ldots,m}$ of the adjoint matrix V form a basis for the space $H^0(\mathcal{E}, \mathcal{V}_P)$ of global sections of $\mathcal{E}$.
3. $\mathcal{E}$ satisfies $h^0(\mathcal{E}(-1)) = h^1(\mathcal{E}(-1)) = 0$.

See [9, 53, 5] for details. By the Riemann–Roch theorem, $\mathcal{E}(-1)$ is a line bundle of degree $g - 1$ on $\mathcal{V}_P$ (where g denotes the genus), and it is general in that it has no global sections, i.e., it lies on the complement of the theta divisor in the Jacobian of $\mathcal{V}_P$.

There is a similarly defined line bundle $\mathcal{L}_\ell$ on $\mathcal{V}_P$ with fibres $\mathcal{L}_\ell([X]) = \ker_\ell U(X)$, where $\ker_\ell$ denotes the left kernel of a matrix (a subspace of $\mathbb{C}^{1 \times m}$); we set, analogously, $\mathcal{E}_\ell = \mathcal{L}_\ell(m - 1)$. $\mathcal{E}_\ell$ is defined by an exact sequence similar to (21) except that U is now acting as the operator of left multiplication by the matrix on rows. The rows $G_i = [V_{ij}]_{j=1,\ldots,m}$ of the adjoint matrix V form a basis for the space $H^0(\mathcal{E}_\ell, \mathcal{V}_P)$ of global sections of $\mathcal{E}_\ell$. There is furthermore a nondegenerate pairing $\mathcal{E} \times \mathcal{E}_\ell \to \mathcal{K}_{\mathcal{V}_P}(2)$ (here $\mathcal{K}_{\mathcal{V}_P} \cong \mathcal{O}_{\mathcal{V}_P}(m - 3)$ is the canonical line bundle on $\mathcal{V}_P$), i.e., $\mathcal{E}_\ell(-1)$ is isomorphic to the Serre dual $(\mathcal{E}(-1))^* \otimes \mathcal{K}_{\mathcal{V}_P}$ of $\mathcal{E}(-1)$, which is key to the reconstruction of the determinantal representation from the corresponding line bundle.

Notice that if P is a polynomial over $\mathbb{R}$ and the determinantal representation is self-adjoint then $\mathcal{E}_\ell \cong \mathcal{E}^\tau$, whereas if the determinantal representation is real symmetric, then $\mathcal{E}_\ell \cong \mathcal{E}^\tau \cong \mathcal{E}$. (In fact, in the real symmetric case the line bundle $\mathcal{E}$ is defined over $\mathbb{R}$ which is a somewhat stronger condition than $\mathcal{E}^\tau \cong \mathcal{E}$ but the two actually coincide if $\mathcal{V}_P(\mathbb{R}) \neq \emptyset$, see [54] and the references there.)

4.3.   There are two ways to define the pairing $\mathcal{E} \times \mathcal{E}_\ell \to \mathcal{K}_{\mathcal{V}_P}(2)$. One way, originating in multivariable operator theory and multidimensional system theory, simply pairs the right and left kernels of the matrix $U(X)$ against appropriate linear combinations of the coefficient matrices A_0, A_1, A_2; see [3]. This leads to explicit formulae for the coefficient matrices in terms of theta functions, given a line bundle $\mathcal{E}(-1)$ on $\mathcal{V}_P$ with $h^0(\mathcal{E}(-1)) = h^1(\mathcal{E}(-1)) = 0$, see [4, Theorems 4.1 and 5.1]. It is obvious from these formulae that choosing $\mathcal{E}(-1)$ with $(\mathcal{E}(-1))^* \otimes \mathcal{K}_{\mathcal{V}_P} \cong \mathcal{E}(-1)^\tau \cong \mathcal{E}(-1)$ (i.e., $\mathcal{E}(-1)$ is a real theta characteristic on $\mathcal{V}_P$) yields a real symmetric determinantal representation (at least in the case $\mathcal{V}_P(\mathbb{R}) \neq \emptyset$). [28, Section 4] verifies (using the tools developed in [54]) that in case the dehomogenization $p(x_1, \ldots, x_d) = P(1, x_1, \ldots, x_d)$ of the original polynomial P is RZ, appropriate choices of $\mathcal{E}(-1)$ will yield a positive determinantal representation (to be more precise, the positivity is "built in" [28, (4.1)–(4.3)]). "Appropriate choices" means that the line bundle $\mathcal{E}(-1)$ of degree $g - 1$ (more precisely, its image under the Abel–Jacobi map) has to belong to a certain distinguished real g-dimensional torus T_0 in the Jacobian of $\mathcal{V}_P$, see [54, Sections 3 and 4]; accidentally, this already forces $\mathcal{E}(-1)$ to be in the complement of the theta divisor, i.e., the condition $h^0(\mathcal{E}(-1)) = h^1(\mathcal{E}(-1)) = 0$ becomes automatic. It is interesting to notice that recent computational advances in theta functions on Riemann surfaces make this approach possibly suitable for computational purposes, see [47].

4.4. Another way to define the pairing $\mathcal{E} \times \mathcal{E}_\ell \to \mathcal{K}_{\mathcal{V}_P}(2)$ is more algebraic and goes back to the classical paper [12]; it uses the adjoint matrix V of the determinantal representation. This leads to the following construction of the determinantal representation given a line bundle $\mathcal{E}(-1)$ on $\mathcal{V}_P$ with $h^0(\mathcal{E}(-1)) = h^1(\mathcal{E}(-1)) = 0$, see [12, 53, 3]. Take bases $\{F_1, \ldots, F_m\}$ and $\{G_1, \ldots, G_m\}$ for the spaces of global sections of $\mathcal{E}$ and of $\mathcal{E}_\ell$, respectively, where $\mathcal{E}_\ell(-1) := (\mathcal{E}(-1))^* \otimes \mathcal{K}_{\mathcal{V}_P}$ is the Serre dual. Then $V_{ij} := \langle F_j, G_i \rangle$ is a global section of $\mathcal{K}_{\mathcal{V}_P}(2) \cong \mathcal{O}_{\mathcal{V}_P}(m-1)$, hence a homogeneous polynomial in X_0, X_1, X_2 of degree $m - 1$. It can be shown that the matrix $V = [V_{ij}]_{i,j=1,\ldots,m}$ has rank 1 on $\mathcal{V}_P$, implying that (18) holds, up to a constant factor c, and that every entry of adj V is divisible by P^{m-2}. We can now define a matrix U of linear homogeneous forms by (19), and it will be a determinantal representation of P, up to the constant factor c^{m-1}. It remains only to show that the constant factor is not zero, i.e., that $\det V$ is not identically zero. This follows by choosing the bases for the spaces of global sections adapted to a straight line, so that V becomes diagonal along that line, and uses essentially the condition $h^0(\mathcal{E}(-1)) = h^1(\mathcal{E}(-1)) = 0$.

It is quite straightforward that if $\mathcal{E}$ satisfies $(\mathcal{E}(-1))^* \otimes \mathcal{K}_{\mathcal{V}_P} \cong (\mathcal{E}(-1))^\tau \cong \mathcal{E}(-1)$ we obtain a real symmetric determinantal representation (at least in the case $\mathcal{V}_P(\mathbb{R}) \neq \emptyset$, since we really need $\mathcal{E}$ to be defined over $\mathbb{R}$), whereas if $(\mathcal{E}(-1))^* \otimes \mathcal{K}_{\mathcal{V}_P} \cong \mathcal{E}(-1)^\tau$ we obtain a self-adjoint determinantal representation.

4.5. The above procedure can be written down more explicitly in terms of divisors and linear systems. We recall that for a homogeneous polynomial $F \in \mathbb{C}[X_0, X_1, X_2]$, the divisor (F) of F on $\mathcal{V}_P$ is the formal sum of the zeroes of F on $\mathcal{V}_P$ with the orders of the zeroes as coefficients (the order of the zero equals also the intersection multiplicity of the curves $\mathcal{V}_Q$ and $\mathcal{V}_P$ – here Q can have multiple irreducible factors so that the curve $\mathcal{V}_Q$ can have multiple components, i.e., it may be a non-reduced subscheme of $\mathbb{P}^2$ over $\mathbb{C}$).

Let $Q \in \mathbb{C}[X_0, X_1, X_2]$ be an auxiliary homogeneous polynomial of degree $m-1$, together with a decomposition $(Q) = D + D_\ell$, $\deg D = \deg D_\ell = m(m-1)/2$. We assume that D and D_ℓ satisfy the condition that $D - (L)$ or equivalently $D_\ell - (L)$ is not linearly equivalent to an effective divisor on $\mathcal{V}_P$, where L is a linear form.

Take a basis $\{V_{11}, \ldots, V_{m1}\}$ of the vector space of homogeneous polynomials of degree $m - 1$ that vanish on D, with $V_{11} = Q$, and a basis $\{V_{11}, \ldots, V_{1m}\}$ of the vector space of homogeneous polynomials of degree $m - 1$ that vanish on D_ℓ. Write $(V_{i1}) = D + D_{\ell,i}$ and $(V_{1j}) = D_j + D_\ell$, where $D_1 = D$ and $D_{\ell,1} = D_\ell$. Define homogeneous polynomials V_{ij} of degree $m - 1$ for $i > 1$ and $j > 1$ by $(V_{ij}) = D_j + D_{\ell,i}$. We then set $V = [V_{ij}]_{i,j=1,\ldots,m}$, and obtain a determinantal representation U of P by (19).

To be able to obtain a real symmetric determinantal representation of a polynomial P over $\mathbb{R}$, we need $\mathcal{V}_Q$ to be a real contact curve of $\mathcal{V}_P$, i.e., to be defined by a polynomial Q over $\mathbb{R}$ and to have even intersection multiplicity at all points of intersection (in this case $D = D_\ell$ is uniquely determined). To be able to obtain a self-adjoint determinantal representation we need $\mathcal{V}_Q$ to be a real curve that is contact to $\mathcal{V}_P$ at all real points of intersection (in this case the real points of D and of $D_\ell = D^\tau$ are uniquely determined whereas the non-real points can be shuffled between the two).

4.6. Unlike the approach of Section 4.3, the approach of Sections 4.4–4.5 does not produce directly the coefficient matrices of the determinantal representation, so it is not clear a priori how to obtain a real symmetric or self-adjoint representation that is positive. A delicate calculation with differentials carried out in [54, Sections 4–6] shows that this will happen exactly in case the original polynomial p is RZ and $\mathcal{E}(-1)$ (more precisely, its image under the Abel–Jacobi map) belongs to the distinguished real g-dimensional torus T_0 in the Jacobian of $\mathcal{V}_P$. We will obtain a corresponding result in terms of the auxiliary curve $\mathcal{V}_Q$ in Section 5 below by elementary methods.

4.7. We consider now how to relax the assumption that $\mathcal{V}_P$ is irreducible and smooth. A full analysis of determinantal representations for a general reduced polynomial P involves torsion free sheaves of rank 1 on a possibly reducible and singular curve; see [32] and the references therein. However one can get far enough to obtain a full proof of Theorem 3 by considering a restricted class of determinantal representations.

Let $\nu\colon \tilde{\mathcal{V}}_P \to \mathcal{V}_P$ be the normalization or equivalently the desingularization. $\tilde{\mathcal{V}}_P$ is a disjoint union of smooth complex projective curves (or compact Riemann surfaces) corresponding to the irreducible factors of P (the irreducible components of $\mathcal{V}_P$) and

$$\nu\big|_{\tilde{\mathcal{V}}_P \setminus \nu^{-1}((\mathcal{V}_P)_{\mathrm{sing}})} : \tilde{\mathcal{V}}_P \setminus \nu^{-1}((\mathcal{V}_P)_{\mathrm{sing}}) \to \mathcal{V}_P \setminus (\mathcal{V}_P)_{\mathrm{sing}}$$

is a (biregular or complex analytic) isomorphism, where $(\mathcal{V}_P)_{\mathrm{sing}}$ denotes the set of singular points of $\mathcal{V}_P$. Let $\lambda \in (\mathcal{V}_P)_{\mathrm{sing}}$; we assume that λ lies in the affine plane $\mathbb{C}^2 \subseteq \mathbb{P}^2(\mathbb{C})$ (otherwise we just choose different affine coordinates near λ). For every $\mu \in \nu^{-1}(\lambda)$ (i.e., for every branch of $\mathcal{V}_P$ at λ), the differential

$$\nu^*\left(\frac{dx_1}{\partial p/\partial x_2}\right) = -\nu^*\left(\frac{dx_2}{\partial p/\partial x_1}\right)$$

on $\tilde{\mathcal{V}}_P$ has a pole at μ; we denote the order of the pole by m_μ. We define

$$\Delta_\lambda = \sum_{\mu \in \nu^{-1}(\lambda)} m_\mu \mu$$

(the adjoint divisor of λ), and

$$\Delta = \sum_{\lambda \in (\mathcal{V}_P)_{\mathrm{sing}}} \Delta_\lambda \tag{22}$$

(the adjoint divisor, or the divisor of singularities, of $\mathcal{V}_P$); see, e.g., [1, Appendix A2].

A determinantal representation U of P is called *fully saturated* (or $\widetilde{\mathcal{V}}_P/\mathcal{V}_P$ saturated) if all the entries of the adjoint matrix V vanish on the adjoint divisor: $(\nu^* V_{ij}) \geq \Delta$ for all $i, j = 1, \ldots, m$. This is a somewhat stronger condition than being a *maximal* (or *maximally generated*) determinantal representation, which means that for every $\lambda \in (\mathcal{V}_P)_{\text{sing}}$, $\dim \ker U(\lambda)$ has the maximal possible dimension equal to the multiplicity of λ on $\mathcal{V}_P$. We refer to [32, 33] for details. If P is reducible than a fully saturated determinantal representation always decomposes, up to equivalence, as a direct sum of determinantal representations of the irreducible factors of P; hence we can assume that P is irreducible.

For a fully saturated determinantal representation U of P, we can define a line bundle $\widetilde{\mathcal{L}}$ on $\widetilde{\mathcal{V}}_P \setminus \nu^{-1}((\mathcal{V}_P)_{\text{sing}})$ with fibres $\widetilde{\mathcal{L}}([X]) = \ker U(X)$ and then extend it uniquely to all of $\widetilde{\mathcal{V}}_P$; we then define $\widetilde{\mathcal{E}} = \widetilde{\mathcal{L}}(m-1)(-\Delta)$, see [3] – here $\widetilde{\mathcal{L}}(m-1) = \widetilde{\mathcal{L}} \otimes \nu^* \mathcal{O}_{\mathcal{V}_P}(m-1)$. Alternatively, we can define $\widetilde{\mathcal{E}} = \nu^* \mathcal{E}$, where the sheaf $\mathcal{E}$ on $\mathcal{V}_P$ is still defined by (21), see [32]. We introduce similarly the left kernel line bundle $\widetilde{\mathcal{E}}_\ell$. Most of Sections 4.2–4.4 and 4.6 now carry over for a fully saturated determinantal representation U of P and line bundles $\widetilde{\mathcal{E}}$ and $\widetilde{\mathcal{E}}_\ell$ on $\widetilde{\mathcal{V}}_P$; notice that the canonical line bundle on $\widetilde{\mathcal{V}}_P$ is given by $\mathcal{K}_{\widetilde{\mathcal{V}}_P} \cong \nu^* \mathcal{O}_{\mathcal{V}_P}(m-3)(-\Delta)$.

In Section 4.5, we have to take the auxiliary polynomial Q to vanish on the adjoint divisor: $(\nu^* Q) \geq \Delta$, with a decomposition $(\nu^* Q) = D + D_\ell + \Delta$. We then take a basis $\{V_{11}, \ldots, V_{m1}\}$ of the vector space of homogeneous polynomials of degree $m-1$ that vanish on D and on the adjoint divisor, with $V_{11} = Q$, and a basis $\{V_{11}, \ldots, V_{1m}\}$ of the vector space of homogeneous polynomials of degree $m-1$ that vanish on D_ℓ and on the adjoint divisor; we write $(V_{i1}) = D + D_{\ell,i} + \Delta$ and $(V_{1j}) = D_j + D_\ell + \Delta$, where $D_1 = D$ and $D_{\ell,1} = D_\ell$; and we define homogeneous polynomials V_{ij} of degree $m-1$ for $i > 1$ and $j > 1$ by $(V_{ij}) = D_j + D_{\ell,i} + \Delta$.

4.8. The recent work [32] extends the construction of the adjoint matrix of a determinantal representation outlined in Section 4.4 to the most general higher-dimensional situation. Let $P = P_1^{r_1} \cdots P_k^{r_k} \in \mathbb{C}[X_0, X_1, \ldots, X_d]$, where $P_1, \ldots, P_k$ are (distinct) irreducible polynomials, and let

$$\mathcal{V}_P = \operatorname{Proj} \mathbb{C}[X_0, X_1, \ldots, X_d]/\langle P \rangle \tag{23}$$

be the corresponding closed subscheme of $\mathbb{P}^n$ over $\mathbb{C}$; of course $\mathcal{V}_P$ is in general highly non-reduced. Let U be a determinantal representation of P as in (16)–(17); we define the kernel sheaf $\mathcal{E}$ on $\mathcal{V}_P$ by the exact sequence (21), as before. $\mathcal{E}$ is a torsion-free sheaf on $\mathcal{V}_P$ of multirank $(r_1, \ldots, r_k)$ (these notions have to be somewhat carefully defined), and we have

$$h^0(\mathcal{E}(-1)) = h^{d-1}(\mathcal{E}(1-d)) = 0, \quad h^i(\mathcal{E}(j)) = 0, \, i = 1, \ldots, d-2, \, j \in \mathbb{Z}. \tag{24}$$

Conversely,

Theorem 11. *Let $\mathcal{E}$ be a torsion-free sheaf on $\mathcal{V}_P$ of multirank $(r_1, \ldots, r_k)$ satisfying the vanishing conditions (24); then $\mathcal{E}$ is the kernel sheaf of a determinantal representation of P.*

As in Section 4.4, Theorem 11 is proved by taking bases of $H^0(\mathcal{E}, \mathcal{V}_P)$ and of $H^0(\mathcal{E}_\ell, \mathcal{V}_P)$, $\mathcal{E}_\ell = \mathcal{E}^* \otimes \omega_{\mathcal{V}_P}(d)$ (here $\omega_{\mathcal{V}_P} = \mathcal{O}_{\mathcal{V}_P}(m - d - 1)$ is the dualizing sheaf), pairing these bases to construct a matrix V of homogeneous polynomials of degree $m - 1$, and then defining the determinantal representation U by (19); there are quite a few technicalities, especially because the scheme is non-reduced. For P a polynomial over $\mathbb{R}$, the determinantal representation can be taken to be self-adjoint if (and only if) $\mathcal{E}^\tau \cong \mathcal{E}^* \otimes \omega_{\mathcal{V}_P}(d)$ where τ is again the complex conjugation. It should be also possible to characterize real symmetric determinantal representations. (Complex symmetric determinantal representations correspond to $\mathcal{E} \cong \mathcal{E}^* \otimes \omega_{\mathcal{V}_P}(d)$.)

Theorem 11 provides a new venue for pursuing Conjecture 5. To make it effective requires progress in two directions:

1. Given a reduced homogeneous polynomial P, characterize large classes of homogeneous polynomials $\tilde{H}$ such that the scheme $\mathcal{V}_{P\tilde{H}}$ admits torsion free sheaves of correct multirank satisfying the vanishing conditions (24).

2. If p is RZ, characterize positive real symmetric or self-adjoint determinantal representations of P in terms of the kernel sheaf $\mathcal{E}$. This is interesting not only for the general conjecture but also for special cases, compare the recent paper [10] dealing with singular nodal quartic surfaces in $\mathbb{P}^3$. It could be that the results of Section 5 below admit some kind of a generalization.

5. Interlacing RZ polynomials and positive self-adjoint determinantal representations

5.1. Let $p \in \mathbb{R}[x_1, \dots, x_d]$ be a reduced (i.e., without multiple factors) RZ_{x^0} polynomial of degree m with $p(x^0) \neq 0$, and let P be the homogenization of p (see (4)). Let $Q \in \mathbb{R}[X_0, X_1, \dots, X_d]$ be a homogeneous polynomial of degree $m - 1$ that is relatively prime with P. We say that Q *interlaces* P if for a general $X \in \mathbb{R}^{d+1}$, there is a zero of the univariate polynomial $Q(X + sX^0)$ in the open interval between any two zeroes of the univariate polynomial $P(X + sX^0)$, where $X^0 = (1, x^0)$. Alternatively, for any $X \in \mathbb{R}^{d+1}$,

$$s_1 \leq s_1' \leq s_2 \leq \cdots \leq s_{m-1} \leq s_{m-1}' \leq s_m, \tag{25}$$

where $s_1, \dots, s_m$ are the zeroes of $P(X + sX^0)$ and $s_1', \dots, s_{m-1}'$ are zeroes of $Q(X + sX^0)$, counting multiplicities. Notice (see (5)) that we can consider instead the zeroes of the univariate polynomials $\check{p}_x(t) = t^m p(x_0 + t^{-1}x)$ and $\check{q}_x(t) = t^{m-1}\tilde{q}(x_0 + t^{-1}x)$ for a general or for any $x \in \mathbb{R}^d$, where $\tilde{q}(x) = Q(1, x_1, \dots, x_d)$. It follows that $\tilde{q}$ is a RZ_{x^0} polynomial with $\tilde{q}(x^0) \neq 0$, and (upon normalizing $p(x^0) > 0$, $\tilde{q}(x^0) > 0$) the closure of the connected component of x^0 in $\{x \in \mathbb{R}^d : \tilde{q}(x) > 0\}$ contains the closure of the connected component of x^0 in $\{x \in \mathbb{R}^d : p(x) > 0\}$. The degree of $\tilde{q}$ is either $m - 1$ (in which case Q is the homogenization of $\tilde{q}$) or $m - 2$ (in which case Q is the homogenization of $\tilde{q}$ times X_0).

Geometrically, let $\mathcal{L}$ be a general straight line through $[X^0]$ in $\mathbb{P}^d(\mathbb{R})$. Then Q interlaces P if and only if any there is an intersection of $\mathcal{L}$ with the real projective hypersurface $\mathcal{V}_Q(\mathbb{R})$ in any open interval on $\mathcal{L} \setminus [X^0]$ between two intersections of $\mathcal{L}$ with the real projective hypersurface $\mathcal{V}_P(\mathbb{R})$. If Q does not contain X_0 as a factor, we can consider instead of $\mathcal{L} \setminus [X^0]$ the two open rays $\mathcal{L}_\pm$ starting at x^0 of a general straight line through x^0 in $\mathbb{R}^d$ and their intersections with the real affine hypersurfaces $\mathcal{V}_{\check{q}}(\mathbb{R})$ and $\mathcal{V}_p(\mathbb{R})$.

An example of a polynomial Q interlacing P is the first directional derivative $P_{x^0}^{(1)}$, see (6) (in this case $\tilde{q} = p_{x^0}^{(1)}$ is the first Renegar derivative).

It is not hard to see that (upon normalizing $p(x^0) > 0$) the definition of interlacing is independent of the choice of a point x^0 in a rigidly convex algebraic interior with a minimal defining polynomial p. In case the real projective hypersurfaces $\mathcal{V}_P(\mathbb{R})$ and $\mathcal{V}_Q(\mathbb{R})$ are both smooth, the interlacing of polynomials simply means the interlacing of ovaloids, see Proposition 2. More precisely, in this case Q interlaces P if and only if

a. If $m = 2k$ is even and $\mathcal{V}_P(\mathbb{R}) = W_1 \coprod \cdots \coprod W_k$ and $\mathcal{V}_Q(\mathbb{R}) = W_1' \coprod \cdots \coprod W_k'$ are the decompositions into connected components, then the ovaloid W_i' is contained in the "shell" obtained by removing the interior of the ovaloid W_i from the closure of the interior of the ovaloid W_{i+1}, $i = 1, \ldots, k-1$, and the pseudo-hyperplane W_k' is contained in the closure of the exterior of the ovaloid W_k;

b. If $m = 2k + 1$ is odd and $\mathcal{V}_P(\mathbb{R}) = W_1 \coprod \cdots \coprod W_k \coprod W_{k+1}$ and $\mathcal{V}_Q(\mathbb{R}) = W_1' \coprod \cdots \coprod W_k'$ are the decompositions into connected components, then the ovaloid W_i' is contained in the "shell" obtained by removing the interior of the ovaloid W_i from the closure of the interior of the ovaloid W_{i+1}, $i = 1, \ldots, k-1$, and the ovaloid W_k' is contained in the closure of the exterior of the ovaloid W_k and the pseudo-hyperplane W_{k+1} is contained in the closure of the exterior of W_k'.

Interlacing can be tested via the Bezoutiant, similarly to testing the RZ condition via the Hermite matrix. For polynomials $f, g \in \mathbb{R}[t]$ with f of degree m and g of degree at most m, we define the Bezoutiant of f and g, $B(f,g) = [b_{ij}]_{i,j=1,\ldots,m}$, by the identity

$$\frac{f(t)g(s) - f(s)g(t)}{t - s} = \sum_{i,j=0}^{m-1} b_{ij} t^i s^j;$$

notice that the entries of $B(f,g)$ are polynomials in the coefficients of f and of g. The nullity of $B(f,g)$ equals the number of common zeroes of f and of g (counting multiplicities), and (assuming that the degree of g is at most $m-1$), $B(f,g) > 0$ if and only if f has only real and distinct zeroes and there is a zero of g in the open interval between any two zeroes of f; see, e.g., [34]. Given $p \in \mathbb{R}[x_1, \ldots, x_d]$ a reduced polynomial of degree m with $p(x^0) \neq 0$, with homogenization P, and $Q \in \mathbb{R}[X_0, X_1, \ldots, X_d]$ a homogeneous polynomial of degree $m-1$ that is relatively prime with P, we now consider $B(\check{p}_x, \check{\tilde{q}}_x)$, where $\check{p}_x, \check{\tilde{q}}_x$ are as before; it is a polynomial matrix that we call the Bezoutiant of P and Q with respect to x^0 and denote $B(P, Q; x^0)$. We see that p is a RZ_{x^0} polynomial and Q interlaces P if and only if $B(P, Q; x^0)(x) \geq 0$ for all $x \in \mathbb{R}^d$.

5.2. Before stating and proving the main result of this section, we make some preliminary observations.

Let $P \in \mathbb{C}[X_0, X_1, \ldots, X_d]$ be a reduced homogeneous polynomial of degree m with the corresponding complex projective hypersurface $\mathcal{V}_P$ (see (15), and let U be a determinantal representation of P with the adjoint matrix V as in (17). Since $\dim \ker U(X) = 1$ for a general point $[X]$ of any irreducible component of $\mathcal{V}_P$, the rows of V are proportional along $\mathcal{V}_P$ and so are the columns. An immediate consequence is that no element of V can vanish along $\mathcal{V}_P$: otherwise, because of the proportionality of the rows, a whole row or a whole column of V would vanish along $\mathcal{V}_P$, hence be divisible by P, hence be identically 0 (since all the elements have degree $m - 1$ which is less than the degree of P), implying that $\det V$ is identically 0, a contradiction. Another consequence is that every minor of order 2 in V, $V_{ij} V_{kl} - V_{kj} V_{il}$, vanishes along $\mathcal{V}_P$.

Lemma 12. *Let* $F_j = [V_{ij}]_{i=1,\ldots,m}$, $j = 1, \ldots, m$, *and* $G_i = [V_{ij}]_{j=1,\ldots,m}$, $i = 1, \ldots, m$, *be the columns and the rows of the adjoint matrix* V, *respectively, let* $X^0 = (X_0^0, X_1^0, \ldots, X_d^0) \in \mathbb{C}^{d+1} \setminus \{0\}$, *and let*

$$P'_{X^0}(X) = \frac{d}{ds} P(X + sX^0)\big|_{s=0} = \sum_{\alpha=0}^{d} X_\alpha^0 \frac{\partial P}{\partial X_\alpha}(X) \tag{26}$$

be the directional derivative. Then

$$G_i \, U(X^0) \, F_j = V_{ij} \, P'_{X^0} \tag{27}$$

along $\mathcal{V}_P$.

The result follows immediately by substituting (20) into (26) to calculate the directional derivative in terms of the entries of the adjoint matrix and of the coefficient matrices of the determinantal representation, and using the vanishing of the minors of order 2 in V along $\mathcal{V}_P$. A version of (27) was established in [54, Corollary 5.8] in case $d = 2$ and $\mathcal{V}_P$ is smooth (the proof given there works verbatim for fully saturated determinantal representations, see Section 4.7, when $\mathcal{V}_P$ is possibly singular and/or reducible) using essentially the pairing between the kernel and the left kernel alluded to in Section 4.3.

Assume now that the dehomogenization $p(x_1, \ldots, x_d) = P(1, x_1, \ldots, x_d)$ is a RZ_{x^0} polynomial with $p(x^0) \neq 0$, let $X^0 = (1, x^0)$, and let U be a self-adjoint determinantal representation. Let $\mathcal{L}$ be a straight line through $[X^0]$ in $\mathbb{P}^d(\mathbb{R})$ intersecting $\mathcal{V}_P(\mathbb{R})$ in m distinct points $[X^1], \ldots, [X^m]$. Then we have

Lemma 13. $U(X^0) > 0$ *if and only if the compression of* $U(X^0)$ *to* $\ker U(X^i)$ *is positive definite for* $i = 1, \ldots, m$.

This is just a special case of [54, Proposition 5.5]: the statement there is for $d = 2$ but the proof for general d is exactly the same (it amounts to restricting the determinantal representation U to the straight line $\mathcal{L}$, and looking at the canonical form of the resulting hermitian matrix pencil). We give a direct argument in our situation.

Proof of Lemma 13. Choose $X \in \mathbb{R}^{d+1}$ so that $\mathcal{L} \setminus [X^0] = \{X - sX^0\}_{s \in \mathbb{R}}$. Then $X^i = X - s_i X^0$, where s_i, $i = 1, \ldots, m$, are the zeroes of the univariate polynomial $P(X - sX^0)$, i.e., the eigenvalues of the generalized eigenvalue problem

$$\left(U(X) - sU(X^0) \right) v = 0.$$

The corresponding eigenspaces are precisely $\ker U(X^i)$; since there are m distinct eigenvalues, these eigenspaces span all of $\mathbb{C}^m$,

$$\mathbb{C}^m = \ker U(X^1) \dotplus \cdots \dotplus \ker U(X^m).$$

The lemma now follows since the different eigenspaces are orthogonal with respect to $U(X^0)$: if $v_i \in \ker U(X^i)$, $v_j \in \ker U(X^j)$, $i \neq j$, then

$$s_i v_j^* U(X^0) v_i = v_j^* U(X) v_i = s_j v_j^* U(X^0) v_i$$

(since $s_j \in \mathbb{R}$), implying that $v_j^* U(X^0) v_i = 0$ (since $s_i \neq s_j$). □

We notice that Lemma 13 remains true for non-reduced polynomials P provided the determinantal representation U is *generically maximal* (or *generically maximally generated*) [32]: if $P = P_1^{r_1} \cdots P_k^{r_k}$, where $P_1, \ldots, P_k$ are distinct irreducible polynomials, this means that $\dim \ker U(X) = r_i$ at a general point $[X]$ of $\mathcal{V}_{P_i}$, $i = 1, \ldots, k$. Since positive self-adjoint determinantal representations are always generically maximal, this may open the possibility of generalizing Theorem 14 below to the non-reduced setting.

Theorem 14. *Let $p \in \mathbb{R}[x_1, \ldots, x_d]$ be an irreducible RZ_{x^0} polynomial of degree m with $p(x^0) \neq 0$, let P be the homogenization of p, and let $X^0 = (1, x^0)$. Let U be a self-adjoint determinantal representation of P with adjoint matrix V, as in (17). Then $U(X^0)$ is either positive or negative definite if and only if the polynomial V_{jj} interlaces P; here j is any integer between 1 and m.*

Proof. The fact that $U(X^0) > 0$ implies the interlacing follows immediately from Cauchy's interlace theorem for eigenvalues of Hermitian matrices, see, e.g., [31]. We provide a unified proof for both directions.

Let $\mathcal{L}$ be a straight line through $[X^0]$ in $\mathbb{P}^d(\mathbb{R})$ intersecting $\mathcal{V}_P(\mathbb{R})$ in m distinct points $X^1, \ldots, X^m$ none of which is a zero of V_{jj}. Lemma 12 implies that for any $[X] \in \mathcal{V}_P(\mathbb{R})$,

$$F_j(X)^* \, U(X^0) \, F_j(X) = P'_{X^0}(X) \, V_{jj}(X).$$

Lemma 13 then shows that $U(X^0)$ is positive or negative definite if and only if $P'_{X^0} V_{jj}$ has the same sign (positive or negative, respectively) at X^i for $i = 1, \ldots, m$.

Similarly to the proof of Lemma 13, let us choose $X \in \mathbb{R}^{d+1}$ so that $\mathcal{L} \setminus [X^0] = \{X + sX^0\}_{s \in \mathbb{R}}$, so that $X^i = X + s_i X^0$, where $s_1 < \cdots < s_m$ are the zeroes of the univariate polynomial $P(X + sX^0)$. It follows from Rolle's Theorem that $\dfrac{d}{ds}P(X + sX^0) = P'_{X^0}(X + sX^0)$ has exactly one zero in each open interval (s_i, s_{i+1}), $i = 1, \ldots, m - 1$, hence has opposite signs at s_i and at s_{i+1}. Therefore $U(X^0)$ is positive or negative definite if and only if $V_{jj}(X + sX^0)$ has opposite signs at s_i and at s_{i+1}, i.e., if and only if V_{jj} interlaces P. $\qquad\square$

It would be interesting to find an analogue of Theorem 14 for other signatures of a self-adjoint determinantal representation, similarly to [54, Section 5].

Combining Theorem 14 with the construction of determinantal representations that was sketched in Section 4.5 (see also Section 4.7 for the extension of the construction to the singular case) then yields the following result.

Theorem 15. *Let $p \in \mathbb{R}[x_1, x_2]$ be an irreducible RZ_{x^0} polynomial of degree m with $p(x^0) = 1$, let P be the homogenization of p, let $\nu\colon \widetilde{\mathcal{V}}_P \to \mathcal{V}_P$ be the desingularization of the corresponding complex projective curve, and let Δ be the adjoint divisor on $\widetilde{\mathcal{V}}_P$. Let $Q \in \mathbb{R}[X_0, X_1, X_2]$ be a homogeneous polynomial of degree $m - 1$ that interlaces P and that vanishes on the adjoint divisor: $(\nu^*Q) \geq \Delta$. Then there exist $A_0, A_1, A_2 \in \mathbb{HC}^{m \times m}$ with $A_0 + x_1^0 A_1 + x_2^0 A_2 = I$ such that $\det(A_0 + x_1 A_1 + x_2 A_2) = p(x)$ and such that the first principal minor of $A_0 + x_1 A_1 + x_2 A_2$ equals $Q(1, x_1, x_2)$.*

We emphasize that the determinantal representation $A_0 + x_1 A_1 + x_2 A_2$ is given by an explicit algebraic construction starting with P and Q. Theorem 15 implies a version of Theorem 3 for positive self-adjoint determinantal representations since there certainly exist interlacing polynomials vanishing on the adjoint divisor: we can take the directional derivative $Q = P_{x'}^{(1)}$ for any interior point x' of the rigidly convex algebraic interior containing x^0 with a minimal defining polynomial p. The two basic open questions here are:

1. "How many" positive self-adjoint determinantal representations does one obtain starting with directional derivatives as above?
2. What other methods are there to produce interlacing polynomials (vanishing on the adjoint divisor)?

Proof of Theorem 15. It is not hard to see that Q interlacing P implies that $\mathcal{V}_Q$ is contact to $\mathcal{V}_P$ at real points of intersection, and that we can write $(\nu^*Q) = D + D^\tau + \Delta$. It only remains to show that $D - (L)$ is not linearly equivalent to an effective divisor, where L is a linear form.

Notice that τ lifts to an antiholomorphic involution on the desingularization (this was already implied when we wrote, e.g., D^τ). Furthermore, the fact that p is a RZ polynomial, implies that $\widetilde{\mathcal{V}}_P$ is a compact real Riemann surface of dividing type, i.e., $\widetilde{\mathcal{V}}_P \setminus \widetilde{\mathcal{V}}_P(\mathbb{R})$ consists of two connected components interchanged by τ, where $\widetilde{\mathcal{V}}_P(\mathbb{R})$ is the fixed point set of τ, see [54] and the references therein and [28]. We orient $\widetilde{\mathcal{V}}_P(\mathbb{R})$ as the boundary of one of these two connected components.

It is now convenient to change projective coordinates so that $[X^0] = [1, x^0]$ becomes $[0, 0, 1]$. It is not hard to see that in the new coordinates, both the meromorphic differential $\nu^* dx_1$ and the function $\nu^* \dfrac{Q(1, x_1, x_2)}{\partial p / \partial x_2}$ have constant sign (are either everywhere nonnegative or everywhere nonpositive) on $\widetilde{\mathcal{V}}_P(\mathbb{R})$. It follows that so is the meromorphic differential $\omega = \nu^* \dfrac{Q(1, x_1, x_2) dx_1}{\partial p / \partial x_2}$. We have (see, e.g., [1, Appendix A2]) $(\omega) = (Q) - \Delta - 2(X_0) = D + D^\tau - 2(X_0)$. If there existed a rational function f and an effective divisor E on $\widetilde{\mathcal{V}}_P$ so that $(f) + D - (X_0) = E$, we would have obtained that $(f \omega f^\tau) = E + E^\tau$, i.e., $f \omega f^\tau$ is a nonzero holomorphic differential that is everywhere nonnegative or everywhere nonpositive on $\widetilde{\mathcal{V}}_P(\mathbb{R})$, a contradiction since its integral over $\widetilde{\mathcal{V}}_P(\mathbb{R})$ has to vanish by Cauchy's Theorem. $\qquad\square$

Notice that this proof is essentially an adaptation of [54, Proposition 4.2] which is itself an adaptation of [15]; it would be interesting to find a more elementary argument.

Acknowledgment

Apart from my joint work with Bill Helton, a lot of what is described here is based on earlier joint work with Joe Ball, as well as on more recent collaboration with Dmitry Kerner. It is a pleasure to thank Didier Henrion, Tim Netzer, Daniel Plaumann, and Markus Schweighofer for many useful discussions.

References

[1] E. Arbarello, M. Cornalba, P.A. Griffiths, and J. Harris. *Geometry of Algebraic Curves: Volume I.* Springer, New York, 1985.

[2] J. Backelin, J. Herzog, and H. Sanders. Matrix factorizations of homogeneous polynomials. *Algebra – some current trends* (Varna, 1986), pp. 1–33. Lecture Notes in Math., 1352, Springer, Berlin, 1988.

[3] J.A. Ball and V. Vinnikov. Zero-pole interpolation for matrix meromorphic functions on an algebraic curve and transfer functions of 2D systems. *Acta Appl. Math.* 45:239–316, 1996.

[4] J.A. Ball and V. Vinnikov. Zero-pole interpolation for meromorphic matrix functions on a compact Riemann surface and a matrix Fay trisecant identity. *Amer. J. Math.* 121:841–888, 1999.

[5] A. Beauville. Determinantal hypersurfaces. *Mich. Math. J.* 48:39–64, 2000.

[6] P. Brändén. Obstructions to determinantal representability. *Advances in Math.*, to appear (arXiv:1004.1382).

[7] H.H. Bauschke, O. Güler, A.S. Lewis, and H.S. Sendov. Hyperbolic polynomials and convex analysis. *Canad. J. Math.* 53:470–488, 2001.

[8] A. Buckley and T. Košir. Determinantal representations of smooth cubic surfaces. *Geom. Dedicata* 125:115–140, 2007.

[9] R.J. Cook and A.D. Thomas. Line bundles and homogeneous matrices. *Quart. J. Math. Oxford Ser.* (2) 30:423–429, 1979.

[10] A. Degtyarev and I. Itenberg. On real determinantal quartics. Preprint (arXiv:1007.3028).

[11] L.E. Dickson. Determination of all general homogeneous polynomials expressible as determinants with linear elements. *Trans. Amer. Math. Soc.* 22:167–179, 1921.

[12] A. Dixon. Note on the reduction of a ternary quartic to a symmetrical determinant. *Proc. Cambridge Phil. Soc.* 11:350–351, 1900–1902
(available at http://www.math.bgu.ac.il/k̃ernerdm/).

[13] I. Dolgachev. *Classical Algebraic Geometry: A Modern View.* Cambridge Univ. Press, to appear
(available at http://www.math.lsa.umich.edu/ idolga/topics.pdf).

[14] B.A. Dubrovin. Matrix finite zone operators. *Contemporary Problems of Mathematics* (Itogi Nauki i Techniki) 23, pp. 33–78 (1983) (Russian).

[15] J.D. Fay. *Theta Functions On Riemann Surfaces.* Lecture Notes in Math. 352, Springer, Berlin, 1973.

[16] D. Eisenbud. Homological algebra on a complete intersection, with an application to group representations. *Trans. Amer. Math. Soc.* 260:35–64, 1980.

[17] L. Gårding. Linear hyperbolic partial differential equations with constant coefficients. *Acta Math.* 85:2–62, 1951.

[18] L. Gårding. An inequality for hyperbolic polynomials. *J. Math. Mech.* 8:957–965, 1959.

[19] B. Grenet, E. Kaltofen, P. Koiran, and N. Portier. Symmetric Determinantal Representation of Formulas and Weakly Skew Circuits. *Randomization, Relaxation, and Complexity in Polynomial Equation Solving* (edited by L. Gurvits, P. Pébay, and J.M. Rojas), pp. 61–96. Contemporary Mathematics 556, Amer. Math. Soc., 2011 (arXiv:1007.3804).

[20] P. Griffiths and J. Harris. On the Noether–Lefschetz theorem and some remarks on codimension two cycles. *Math. Ann.* 271:31–51, 1985.

[21] O. Güler. Hyperbolic polynomials and interior point methods for convex programming. *Mathematics of Operations Research* 22:350–377, 1997.

[22] R. Hartshorne. *Ample subvarieties of algebraic varieties.* Lecture Notes in Math., 156, Springer-Verlag, Heidelberg, 1970.

[23] J.W. Helton and S.A. McCullough. Every convex free basic semi-algebraic set has an LMI representation. Preprint [http://arxiv.org/abs/0908.4352]

[24] J.W. Helton, S. McCullough, M. Putinar, and V. Vinnikov. Convex matrix inequalities versus linear matrix inequalities. *IEEE Trans. Automat. Control* 54:952–964, 2009.

[25] J.W. Helton and J. Nie. Sufficient and necessary conditions for semidefinite representability of convex hulls and sets. *SIAM J. Optim.* 20:759–791, 2009.

[26] J.W. Helton and J. Nie. Semidefinite representation of convex sets. *Math. Program.* 122:21–64, 2010.

[27] J.W. Helton, S.A. McCullough, and V. Vinnikov. Noncommutative convexity arises from linear matrix inequalities. *J. Funct. Anal.* 240:105–191, 2006.

[28] J.W. Helton and V. Vinnikov. Linear Matrix Inequality Representation of Sets. *Comm. Pure Appl. Math.* 60:654–674, 2007.

[29] D. Henrion. Detecting rigid convexity of bivariate polynomials. *Linear Algebra Appl.* 432:1218–1233, 2010.

[30] J. Herzog, B. Ulrich, and J. Backelin. Linear Cohen–Macaulay modules over strict complete intersections. *J. Pure Appl. Algebra* 71:187–202, 1991.

[31] S.-G. Hwang. Cauchy's interlace theorem for eigenvalues of Hermitian matrices. *Amer. Math. Monthly* 111:157–159, 2004.

[32] D. Kerner and V. Vinnikov. On the determinantal representations of singular hypersurfaces in $\mathbb{P}^n$. *Advances in Math.*, to appear (arXiv:0906.3012).

[33] D. Kerner and V. Vinnikov. On decomposability of local determinantal representations of hypersurfaces. Preprint (arXiv:1009.2517).

[34] M.G. Krein and M.A. Naimark. *The method of symmetric and Hermitian forms in the theory of the separation of the roots of algebraic equations.* Kharkov, 1936. English translation (by O. Boshko and J.L. Howland): *Lin. Mult. Alg.* 10:265–308, 1981.

[35] J.B. Lasserre. Convex sets with semidefinite representation. *Math. Program.* 120:457–477, 2009.

[36] P. Lax. Differential equations, difference equations and matrix theory. *Comm. Pure Appl. Math.* 11:175–194, 1958.

[37] S. Lefschetz. *L'analysis situs et la géometrie algébrique.* Gauthier-Villars, Paris, 1924.

[38] A.S. Lewis, P.A. Parrilo, and M.V. Ramana. The Lax conjecture is true. *Proc. Amer. Math. Soc.* 133:2495–2499, 2005.

[39] A. Nemirovskii. Advances in convex optimization: conic programming. *Proceedings of the International Congress of Mathematicians (ICM)* (Madrid, 2006), Vol. I (Plenary Lectures), pp. 413–444. Eur. Math. Soc., Zürich, 2007 (available at http://www.icm2006.org/proceedings/).

[40] Y. Nesterov and A. Nemirovskii. *Interior-point polynomial algorithms in convex programming.* SIAM Studies in Applied Mathematics, 13, Society for Industrial and Applied Mathematics (SIAM), Philadelphia, PA, 1994.

[41] T. Netzer, D. Plaumann, and M. Schweighofer. Exposed faces of semidefinitely representable sets. *SIAM J. Optim.* 20:1944–1955, 2010.

[42] T. Netzer, D. Plaumann, and A. Thom. Determinantal representations and the Hermite matrix. Preprint (arXiv:1108.4380).

[43] T. Netzer, A. Thom: Polynomials with and without determinantal representations. Preprint (arXiv:1008.1931).

[44] W. Nuij. A note on hyperbolic polynomials. *Math. Scand.* 23:69–72, 1968.

[45] P. Parrilo and B. Sturmfels. Minimizing polynomial functions. *Algorithmic and quantitative real algebraic geometry* (Piscataway, NJ, 2001), pp. 83–99. DIMACS Ser. Discrete Math. Theoret. Comput. Sci., 60, Amer. Math. Soc., Providence, RI, 2003.

[46] D. Plaumann, B. Sturmfels, and C. Vinzant. Quartic Curves and Their Bitangents. Preprint (arXiv:1008.4104).

[47] D. Plaumann, B. Sturmfels, and C. Vinzant. Computing Linear Matrix Representations of Helton-Vinnikov Curves. This volume (arXiv:1011.6057).

[48] R. Quarez. Symmetric determinantal representations of polynomials. Preprint (http://hal.archives-ouvertes.fr/hal-00275615_v1/).

[49] M. Ramana and A.J. Goldman. Some geometric results in semidefinite programming. *J. Global Optim.* 7:33–50, 1995.

[50] J. Renegar. Hyperbolic programs, and their derivative relaxations. *Found. Comput. Math.* 6:59–79, 2006.

[51] R.E. Skelton, T. Iwasaki, and K.M. Grigoriadis. *A Unified Algebraic Approach to Linear Control Design.* Taylor & Francis, 1997.

[52] L. Vandenberghe and S. Boyd. Semidefinite programming. *SIAM Rev.* 38:49–95, 1996.

[53] V. Vinnikov. Complete description of determinantal representations of smooth irreducible curves. *Lin. Alg. Appl.* 125:103–140, 1989.

[54] V. Vinnikov. Self-adjoint determinantal representions of real plane curves. *Math. Ann.* 296:453–479, 1993.

Victor Vinnikov
Department of Mathematics
Ben-Gurion University of the Negev
Beer-Sheva, Israel, 84105
e-mail: `vinnikov@math.bgu.ac.il`

Operator Theory:
Advances and Applications, Vol. 222, 351–368

Some Analysable Instances of μ-synthesis

N.J. Young

To Bill Helton, inspiring mathematician and friend

Abstract. I describe a verifiable criterion for the solvability of the 2×2 spectral Nevanlinna-Pick problem with two interpolation points, and likewise for three other special cases of the μ-synthesis problem. The problem is to construct an analytic 2×2 matrix function F on the unit disc subject to a finite number of interpolation constraints and a bound on the cost function $\sup_{\lambda \in \mathbb{D}} \mu(F(\lambda))$, where μ is an instance of the structured singular value.

Mathematics Subject Classification. Primary: 93D21, 93B36;
secondary: 32F45, 30E05, 93B50, 47A57.

Keywords. Robust control, stabilization, analytic interpolation, symmetrized bidisc, tetrablock, Carathéodory distance, Lempert function.

1. Introduction

It is a pleasure to be able to speak at a meeting in San Diego in honour of Bill Helton, through whose early papers, especially [31], I first became interested in applications of operator theory to engineering. I shall discuss a problem of Heltonian character: a hard problem in pure analysis, with immediate applications in control engineering, which can be addressed by operator-theoretic methods. Furthermore, the main advances I shall describe are based on some highly original ideas of Jim Agler, so that San Diego is the ideal place for my talk.

The μ-synthesis problem is an interpolation problem for analytic matrix functions, a generalization of the classical problems of Nevanlinna-Pick, Carathéodory-Fejér and Nehari. The symbol μ denotes a type of cost function that generalizes the operator and H^∞ norms, and the μ-synthesis problem is to construct an analytic matrix function F on the unit disc satisfying a finite number of interpolation conditions and such that $\mu(F(\lambda)) \leq 1$ for $|\lambda| < 1$. The precise definition of μ is in Section 4 below, but for most of the paper we need only a familiar special case of μ – the spectral radius of a square matrix A, which we denote by $r(A)$.

The purpose of this lecture is to present some cases of the μ-synthesis problem that are amenable to analysis. I shall summarize some results that are scattered through a number of papers, mainly by Jim Agler and me but also several others of my collaborators, without attempting to survey all the literature on the topic. I shall also say a little about recent results of some specialists in several complex variables which bear on the matter and may lead to progress on other instances of μ-synthesis.

Although the cases to be described here are too special to have significant practical applications, they do throw some light on the μ-synthesis problem. More concretely, the results below could be used to provide test data for existing numerical methods and to illuminate the phenomenon (known to engineers) of the numerical instability of some μ-synthesis problems.

We are interested in citeria for μ-synthesis problems to be solvable. Here is an example. We denote by $\mathbb{D}$ and $\mathbb{T}$ the open unit disc and the unit circle respectively in the complex plane $\mathbb{C}$.

Theorem 1. *Let $\lambda_1, \lambda_2 \in \mathbb{D}$ be distinct points, let W_1, W_2 be nonscalar 2×2 matrices of spectral radius less than 1 and let $s_j = \operatorname{tr} W_j$, $p_j = \det W_j$ for $j = 1, 2$. The following three statements are equivalent:*

(1) *there exists an analytic function $F : \mathbb{D} \to \mathbb{C}^{2\times 2}$ such that*

$$F(\lambda_1) = W_1, \quad F(\lambda_2) = W_2 \quad \text{and} \quad r(F(\lambda)) \leq 1 \quad \text{for all } \lambda \in \mathbb{D};$$

(2)
$$\max_{\omega \in \mathbb{T}} \left| \frac{(s_2 p_1 - s_1 p_2)\omega^2 + 2(p_2 - p_1)\omega + s_1 - s_2}{(s_1 - \bar{s}_2 p_1)\omega^2 - 2(1 - p_1 \bar{p}_2)\omega + \bar{s}_2 - s_1 \bar{p}_2} \right| \leq \left| \frac{\lambda_1 - \lambda_2}{1 - \bar{\lambda}_2 \lambda_1} \right|;$$

(3)
$$\left[\frac{\overline{(2 - \omega s_i)}(2 - \omega s_j) - \overline{(2\omega p_i - s_i)}(2\omega p_j - s_j)}{1 - \bar{\lambda}_i \lambda_j} \right]_{i,j=1}^{2} \geq 0 \text{ for all } \omega \in \mathbb{T}.$$

The paper is organised as follows. Section 2 contains the definition of the spectral Nevanlinna-Pick problem, sketches the ideas that led to Theorem 1 – reduction to the complex geommetry of the symmetrized bidisc $\mathbb{G}$, the associated "magic functions" Φ_ω and the calculation of the Carathéodory distance on $\mathbb{G}$ – and fills in the final details of the proof of Theorem 1 using the results of [11]. It also discusses ill-conditioning and the possibility of generalization of Theorem 1. In Section 3 there is an analogous solvability criterion for a variant of the spectral Nevanlinna-Pick problem in which the two interpolation points coalesce (Theorem 10). In Section 4, besides the definition of μ and μ-synthesis, there is some motivation and history. Important work by H. Bercovici, C. Foiaş and A. Tannenbaum is briefly described, as is Bill Helton's alternative approach to robust stabilization problems. In Section 5 we consider an instance of μ-synthesis other than the spectral radius. Here we can only obtain a solvability criterion in two very special circumstances (Theorems 11 and 12). The paper concludes with some speculations in Section 6.

We shall denote the closed unit disc in the complex plane by Δ.

.

2. The spectral Nevanlinna-Pick problem

A particularly appealing special case of the μ-synthesis problem is the *spectral Nevanlinna-Pick problem*:

Problem SNP *Given distinct points $\lambda_1, \ldots, \lambda_n \in \mathbb{D}$ and $k \times k$ matrices $W_1, \ldots, W_n$, construct an analytic $k \times k$ matrix function F on $\mathbb{D}$ such that*

$$F(\lambda_j) = W_j \quad \text{for } j = 1, \ldots, n \tag{1}$$

and

$$r(F(\lambda)) \leq 1 \quad \text{for all } \lambda \in \mathbb{D}. \tag{2}$$

When $k = 1$ this is just the classical Nevanlinna-Pick problem, and it is well known that a suitable F exists if and only if a certain $n \times n$ matrix formed from the λ_j and W_j is positive (this is *Pick's Theorem*). We should very much like to have a similarly elegant solvability criterion for the case that $k > 1$, but strenuous efforts by numerous mathematicians over three decades have failed to find one.

About 15 years ago Jim Agler and I devised a new approach to the problem in the case $k = 2$ based on operator theory and a dash of several complex variables ([5] to [13]). Since interpolation of the eigenvalues fails, how about interpolation of the coefficients of the characteristic polynomials of the W_j, or in other words of the elementary symmetric functions of the eigenvalues? This thought brought us to the study of the complex geometry of a certain set $\Gamma \subset \mathbb{C}^2$, defined below. By this route we were able to analyse quite fully the simplest then-unsolved case of the spectral Nevanlinna-Pick problem: the case $n = k = 2$. For the purpose of engineering application this is a modest achievement, but it nevertheless constituted progress. It had the merit of revealing some unsuspected intricacies of the problem, and may yet lead to further discoveries.

2.1. The symmetrized bidisc Γ

We introduce the notation

$$\Gamma = \{(z + w, zw) : z, w \in \Delta\}, \tag{3}$$
$$\mathbb{G} = \{(z + w, zw) : z, w \in \mathbb{D}\}.$$

Γ and $\mathbb{G}$ are called the *closed* and *open symmetrized bidiscs* respectively. Their importance lies in their relation to the sets

$$\Sigma \stackrel{\text{def}}{=} \{A \in \mathbb{C}^{2 \times 2} : r(A) \leq 1\},$$
$$\Sigma^o \stackrel{\text{def}}{=} \{A \in \mathbb{C}^{2 \times 2} : r(A) < 1\}.$$

Σ and its interior Σ^o are sometimes called "spectral unit balls", though the terminology is misleading since they are not remotely ball-like, being unbounded and non-convex. Observe that, for a 2×2 matrix A,

$$A \in \Sigma \Leftrightarrow \text{ the zeros of the polynomial } \lambda^2 - \text{tr } A\lambda + \det A \text{ lie in } \Delta$$
$$\Leftrightarrow \text{tr } A = z + w, \det A = zw \text{ for some } z, w \in \Delta.$$

We thus have the following simple assertion.

Proposition 2. *For any* $A \in \mathbb{C}^{2 \times 2}$

$$A \in \Sigma \text{ if and only if } (\operatorname{tr} A, \det A) \in \Gamma,$$
$$A \in \Sigma^o \text{ if and only if } (\operatorname{tr} A, \det A) \in G.$$

Consequently, if $F : \mathbb{D} \to \Sigma$ is analytic and satisfies the equations (1) above, where $k = 2$, then $h \overset{\text{def}}{=} (\operatorname{tr} F, \det F)$ is an analytic map from $\mathbb{D}$ to Γ satisfying the interpolation conditions

$$h(\lambda_j) = (\operatorname{tr} W_j, \det W_j) \text{ for } j = 1, \dots, n. \tag{4}$$

Let us assume that none of the target matrices W_j is a scalar multiple of the identity. On this hypothesis it is simple to show the converse [16] by similarity transformation of the W_j to companion form.

Proposition 3. *Let* $\lambda_1, \dots, \lambda_n$ *be distinct points in* $\mathbb{D}$ *and let* $W_1, \dots, W_n$ *be non-scalar* 2×2 *matrices. There exists an analytic map* $F : \mathbb{D} \to \mathbb{C}^{2 \times 2}$ *such that equations* (1) *and* (2) *hold if and only if there exists an analytic map* $h : \mathbb{D} \to \Gamma$ *that satisfies the conditions* (4).

We have therefore (in the case $k = 2$) reduced the given analytic interpolation problem for Σ-valued functions to one for Γ-valued functions (the assumption on the W_j is harmless, since any constraint for which W_j is scalar may be removed by the standard process of Schur reduction).

Why is it an advance to replace Σ by Γ? For one thing, of the two sets, the geometry of Γ is considerably the less rebarbative. Σ is an unbounded, non-smooth 4-complex-dimensional set with spikes shooting off to infinity in many directions. Γ is somewhat better: it is compact and only 2-complex-dimensional, though Γ too is non-convex and not smoothly bounded. But the true reason that Γ is amenable to analysis is that there is a 1-parameter family of linear fractional functions, analytic on G, that has special properties *vis-à-vis* Γ. For ω in the unit circle $\mathbb{T}$ we define

$$\Phi_\omega(s, p) = \frac{2\omega p - s}{2 - \omega s}. \tag{5}$$

We use the variables s and p to suggest "sum" and "product". The Φ_ω determine G in the following sense.

Proposition 4. *For every* $\omega \in \mathbb{T}$, Φ_ω *maps* G *analytically into* $\mathbb{D}$. *Conversely, if* $(s, p) \in \mathbb{C}^2$ *is such that* $|\Phi_\omega(s, p)| < 1$ *for all* $\omega \in \mathbb{T}$, *then* $(s, p) \in G$.

Both statements can be derived from the identity

$$|2 - z - w|^2 - |2zw - z - w|^2 = 2(1 - |z|^2)|1 - w|^2 + 2(1 - |w|^2)|1 - z|^2.$$

See [11, Theorem 2.1] for details.

There is an analogous statement for Γ, but there are some subtleties. For one thing Φ_ω is undefined at $(2\bar{\omega}, \bar{\omega}^2) \in \Gamma$ when $\omega \in \mathbb{T}$.

Proposition 5. *For every $\omega \in \mathbb{T}$, Φ_ω maps $\Gamma \setminus \{(2\bar{\omega}, \bar{\omega}^2)\}$ analytically into Δ. Conversely, if $(s, p) \in \mathbb{C}^2$ is such that $|\Phi_\omega(rs, r^2 p)| < 1$ for all $\omega \in \mathbb{T}$ and $0 < r < 1$ then $(s, p) \in \Gamma$.*

In the second statement of the proposition the parameter r is needed: it does not suffice that $|\Phi_\omega(s, p)| \leq 1$ for all $\omega \in \mathbb{T}$ (in the case that $p = 1$ the last statement is true if and only if $s \in \mathbb{R}$, whereas for $(s, p) \in \Gamma$, of course $|s| \leq 2$).

We found the functions Φ_ω by applying Agler's theory of families of operator tuples [5, 6]. We studied the family $\mathcal{F}$ of commuting pairs of operators for which Γ is a spectral set, and its dual cone $\mathcal{F}^\perp$ (that is, the collection of hereditary polynomials that are positive on $\mathcal{F}$). Agler had previously done the analogous analysis for the bidisc, and shown that the dual cone was generated by just two hereditary polynomials; this led to his celebrated realization theorem for bounded analytic functions on the bidisc. On incorporating symmetry into the analysis we found that the cone $\mathcal{F}^\perp$ had the 1-parameter family of generators $1 - \Phi_\omega^\vee \Phi_\omega$, $\omega \in \mathbb{T}$. From this fact many conclusions follow: see [13] for more on these ideas.

Operator theory played an essential role in our discovery of the functions Φ_ω. Once they are known, however, the geometry of $\mathbb{G}$ and Γ can be developed without the use of operator theory.

2.2. A necessary condition

Suppose that F is a solution of the spectral Nevanlinna-Pick problem (1), (2) with $k = 2$. Let us write $s_j = \operatorname{tr} W_j$, $p_j = \det W_j$ for $j = 1, \ldots, n$. For any $\omega \in \mathbb{T}$ and $0 < t < 1$ the composition

$$\mathbb{D} \xrightarrow{tF} \Sigma^o \xrightarrow{(\operatorname{tr}, \det)} \mathbb{G} \xrightarrow{\Phi_\omega} \mathbb{D}$$

is an analytic self-map of $\mathbb{D}$ under which

$$\lambda_j \mapsto \Phi_\omega(ts_j, t^2 p_j) = \frac{2\omega t^2 p_j - ts_j}{2 - \omega ts_j} \quad \text{for } j = 1, \ldots, n.$$

Thus, by Pick's Theorem,

$$\left[\frac{1 - \overline{\Phi_\omega(ts_i, t^2 p_i)} \Phi_\omega(ts_j, t^2 p_j)}{1 - \bar{\lambda}_i \lambda_j} \right]_{i,j=1}^n \geq 0. \tag{6}$$

On conjugating this matrix inequality by $\operatorname{diag}\{2 - \omega ts_j\}$ and letting $\alpha = t\omega$ we obtain the following necessary condition for the solvability of a 2×2 spectral Nevanlinna-Pick condition [5, Theorem 5.2].

Theorem 6. *If there exists an analytic map $F : \mathbb{D} \to \Sigma$ satisfying the equations*

$$F(\lambda_j) = W_j \quad \text{for } j = 1, \ldots, n$$

and

$$r(F(\lambda)) \leq 1 \quad \text{for all } \lambda \in \mathbb{D}$$

then, for every α such that $|\alpha| \leq 1$,

$$\left[\frac{\overline{(2-\alpha s_i)}(2-\alpha s_j) - |\alpha|^2 \overline{(2\alpha p_i - s_i)}(2\alpha p_j - s_j)}{1 - \overline{\lambda_i}\lambda_j}\right]_{i,j=1}^{n} \geq 0 \qquad (7)$$

where

$$s_j = \operatorname{tr} W_j, \qquad p_j = \det W_j \quad for \ j = 1, \dots, n.$$

In the case that the W_j all have spectral radius strictly less than one, the condition (7) holds for all $\alpha \in \Delta$ if and only if it holds for all $\alpha \in \mathbb{T}$, and hence the condition only needs to be checked for a one-parameter pencil of matrices. It is of course less simple than the classical Pick condition in that it comprises an infinite collection of algebraic inequalities, but it is nevertheless checkable in practice with the aid of standard numerical packages. Its major drawback is that it is *not* sufficient for solvability of the 2×2 spectral Nevanlinna-Pick problem.

Example 7. Let $0 < r < 1$ and let

$$h(\lambda) = \left(2(1-r)\frac{\lambda^2}{1+r\lambda^3}, \frac{\lambda(\lambda^3 + r)}{1+r\lambda^3}\right).$$

Let $\lambda_1, \lambda_2, \lambda_3$ be any three distinct points in $\mathbb{D}$ and let $h(\lambda_j) = (s_j, p_j)$ for $j = 1, 2, 3$. We can prove [3] that, in any neighbourhood of (s_1, s_2, s_3) in $(2\mathbb{D})^3$, there exists a point (s_1', s_2', s_3') such that $(s_j', p_j) \in \mathbb{G}$, the Nevanlinna-Pick data

$$\lambda_j \mapsto \Phi_\omega(s_j', p_j), \quad j = 1, 2, 3,$$

are solvable for all $\omega \in \mathbb{T}$, but the Nevanlinna-Pick data

$$\lambda_j \mapsto \Phi_{m(\lambda_j)}(s_j', p_j), \quad j = 1, 2, 3,$$

are unsolvable for some Blaschke factor m. It follows that the interpolation data

$$\lambda_j \mapsto (s_j', p_j), \quad j = 1, 2, 3,$$

satisfy the necessary condition of Theorem 6 for solvability, and yet there is no analytic function $h : \mathbb{D} \to \Gamma$ such that $h(\lambda_j) = (s_j', p_j)$ for $j = 1, 2, 3$.

Hence, if we choose nonscalar 2×2 matrices W_1, W_2, W_3 such that

$$(\operatorname{tr} W_j, \det W_j) = (s_j, p_j),$$

then the spectral Nevanlinna-Pick problem with data $\lambda_j \mapsto W_j$ satisfies the necessary condition of Theorem 6 and yet has no solution.

See also [22] for another example.

2.3. Two points and two-by-two matrices

When $n = k = 2$ the condition in Theorem 6 *is* sufficient for the solvability of the spectral Nevanlinna-Pick problem.

We shall now prove the main theorem from Section 1. Recall the statement:

Theorem 1.1. *Let* $\lambda_1, \lambda_2 \in \mathbb{D}$ *be distinct points, let* W_1, W_2 *be nonscalar* 2×2 *matrices of spectral radius less than 1 and let* $s_j = \operatorname{tr} W_j$, $p_j = \det W_j$ *for* $j = 1, 2$. *The following three statements are equivalent:*

(1) *there exists an analytic function* $F : \mathbb{D} \to \mathbb{C}^{2 \times 2}$ *such that*

$$F(\lambda_1) = W_1, \qquad F(\lambda_2) = W_2$$

and

$$r(F(\lambda)) \le 1 \quad \text{for all } \lambda \in \mathbb{D};$$

(2)

$$\max_{\omega \in \mathbb{T}} \left| \frac{(s_2 p_1 - s_1 p_2)\omega^2 + 2(p_2 - p_1)\omega + s_1 - s_2}{(s_1 - \bar{s}_2 p_1)\omega^2 - 2(1 - p_1 \bar{p}_2)\omega + \bar{s}_2 - s_1 \bar{p}_2} \right| \le \left| \frac{\lambda_1 - \lambda_2}{1 - \bar{\lambda}_2 \lambda_1} \right|; \tag{8}$$

(3)

$$\left[\frac{\overline{(2 - \omega s_i)}(2 - \omega s_j) - \overline{(2\omega p_i - s_i)}(2\omega p_j - s_j)}{1 - \bar{\lambda}_i \lambda_j} \right]_{i,j=1}^{2} \ge 0 \tag{9}$$

for all $\omega \in \mathbb{T}$.

The proof depends on some elementary notions from the theory of invariant distances. A good source for the general theory is [35], but here we only need the following rudiments.

We denote by d the pseudohyperbolic distance on the unit disc $\mathbb{D}$:

$$d(\lambda_1, \lambda_2) = \left| \frac{\lambda_1 - \lambda_2}{1 - \bar{\lambda}_2 \lambda_1} \right| \quad \text{for } \lambda_1, \lambda_2 \in \mathbb{D}.$$

For any domain $\Omega \in \mathbb{C}^n$ we define the *Lempert function* $\delta_\Omega : \Omega \times \Omega \to \mathbb{R}^+$ by

$$\delta_\Omega(z_1, z_2) = \inf d(\lambda_1, \lambda_2) \tag{10}$$

over all $\lambda_1, \lambda_2 \in \mathbb{D}$ such that there exists an analytic map $h : \mathbb{D} \to \Omega$ such that $h(\lambda_1) = z_1$ and $h(\lambda_2) = z_2$. We define[1] the *Carathéodory distance* $C_\Omega : \Omega \times \Omega \to \mathbb{R}^+$ by

$$C_\Omega(z_1, z_2) = \sup d(f(z_1), f(z_2)) \tag{11}$$

over all analytic maps $f : \Omega \to \mathbb{D}$. If Ω is bounded then C_Ω is a metric on Ω.

It is not hard to see (by the Schwarz-Pick Lemma) that $C_\Omega \le \delta_\Omega$ for any domain Ω. The two quantities C_Ω, δ_Ω are not always equal – the punctured disc provides an example of inequality. The question of determining the domains Ω for which $C_\Omega = \delta_\Omega$ is one of the concerns of invariant distance theory.

---

[1]Conventionally the definition of the Carathéodory distance contains a $\tanh^{-1}$ on the right-hand side of (11). For present purposes it is convenient to omit the $\tanh^{-1}$.

Proof. Let $z_j = (s_j, p_j) \in \mathbb{G}$.

(1)$\Leftrightarrow$(2) In view of Proposition 3 we must show that the inequality (8) is equivalent to the existence of an analytic $h : \mathbb{D} \to \Gamma$ such that $h(\lambda_j) = z_j$ for $j = 1, 2$. By definition of the Lempert function $\delta_\mathbb{G}$, such an h exists if and only if

$$\delta_\mathbb{G}(z_1, z_2) \le d(z_1, z_2).$$

By [11, Corollary 5.7] we have $\delta_\mathbb{G} = C_\mathbb{G}$, and by [11, Theorem 1.1 and Corollary 3.4],

$$C_\mathbb{G}(z_1, z_2) = \max_{\omega \in \mathbb{T}} d(\Phi_\omega(z_1), \Phi_\omega(z_2)) \tag{12}$$

$$= \max_{\omega \in \mathbb{T}} \left| \frac{(s_2 p_1 - s_1 p_2)\omega^2 + 2(p_2 - p_1)\omega + s_1 - s_2}{(s_1 - \bar{s}_2 p_1)\omega^2 - 2(1 - p_1 \bar{p}_2)\omega + \bar{s}_2 - s_1 \bar{p}_2} \right|.$$

Thus the desired function h exists if and only if the inequality (8) holds.

(2)$\Leftrightarrow$(3) By equation (12), the inequality (8) is equivalent to

$$d(\Phi_\omega(z_1), \Phi_\omega(z_2)) \le d(\lambda_1, \lambda_2) \quad \text{for all } \omega \in \mathbb{T}.$$

By the Schwarz-Pick Lemma, this inequality holds if and only if, for all $\omega \in \mathbb{T}$, there exists a function f_ω in the Schur class such that $f_\omega(\lambda_j) = \Phi_\omega(z_j)$ for $j = 1, 2$. By Pick's Theorem this in turn is equivalent to the relation

$$\left[\frac{1 - \bar{\Phi}_\omega(z_i)\Phi_\omega(z_j)}{1 - \bar{\lambda}_i \lambda_j} \right]_{i,j=1}^2 \ge 0.$$

Conjugate by $\operatorname{diag}\{2 - \omega s_1, 2 - \omega s_2\}$ to obtain (2)$\Leftrightarrow$(3). □

Remark 8. If one removes the hypothesis that W_1, W_2 be nonscalar from Theorem 1 one can still give a solvability criterion. If both of the W_j are scalar matrices then the problem reduces to a scalar Nevanlinna-Pick problem. If $W_1 = cI$ and W_2 is nonscalar then the corresponding spectral Nevanlinna-Pick problem is solvable if and only if

$$r((W_2 - cI)(I - \bar{c}W_2)^{-1}) \le d(\lambda_1, \lambda_2)$$

(see [7, Theorem 2.4]). This inequality can also be expressed as a somewhat cumbersome algebraic inequality in c, s_2, p_2 and $d(\lambda_1, \lambda_2)$ [7, Theorem 2.5(2)].

2.4. Ill-conditioned problems

The results of the preceding subsection suggest that solvability of spectral Nevanlinna-Pick problems depends on the derogatory structure of the target matrices – that is, in the case of 2×2 matrices, on whether or not they are scalar matrices. It is indeed so, and in consequence problems in which a target matrix is close to scalar can be very ill-conditioned.

Example 9. [7, Example 2.3] Let $\beta \in \mathbb{D} \setminus \{0\}$ and, for $\alpha \in \mathbb{C}$ let

$$W_1(\alpha) = \begin{bmatrix} 0 & \alpha \\ 0 & 0 \end{bmatrix}, \quad W_2 = \begin{bmatrix} 0 & \beta \\ 0 & \frac{2\beta}{1+\beta} \end{bmatrix}.$$

Consider the spectral Nevanlinna-Pick problem with data $0 \mapsto W_1(\alpha)$, $\beta \mapsto W_2$. If $\alpha = 0$ then the problem is not solvable. If $\alpha \neq 0$, however, by Proposition 3 the problem is solvable if and only if there exists an analytic function $f : \mathbb{D} \to \Gamma$ such that

$$f(0) = (0,0) \quad \text{and} \quad f(\beta) = \frac{2\beta}{1+\beta}.$$

It may be checked [8] that

$$f(\lambda) = \left(\frac{2(1-\beta)\lambda}{1-\beta\lambda}, \frac{\lambda(\lambda-\beta)}{1-\beta\lambda} \right)$$

is such a function. Thus the problem has a solution F_α for any $\alpha \neq 0$. Consider a sequence (α_n) of nonzero complex numbers tending to zero: the functions F_{α_n} cannot be locally bounded, else they would have a cluster point, which would solve the problem for $\alpha = 0$. If α is, say, 10^{-100} then *any* numerical method for the spectral Nevanlinna-Pick problem is liable to run into difficulty in this example.

2.5. Uniqueness and the construction of interpolating functions

Problem SNP *never* has a unique solution. If F is a solution of Problem SNP then so is $P^{-1}FP$ for any analytic function $P : \mathbb{D} \to \mathbb{C}^{k \times k}$ such that $P(\lambda)$ is nonsingular for every $\lambda \in \mathbb{D}$ and $P(\lambda_j)$ is a scalar matrix for each interpolation point λ_j. There are always many such P that do not commute with F, save in the trivial case that F is scalar. Nevertheless, the solution of the corresponding interpolation problem for Γ *can* be unique. Consider again the case $n = k = 2$ with W_1, W_2 nonscalar. By Theorem 1, the problem is solvable if and only if inequality (8) holds. In fact it is solvable *uniquely* if and only if inequality (8) holds with equality. This amounts to saying that each pair of distinct points of $\mathbb{G}$ lies on a unique complex geodesic of $\mathbb{G}$, which is true by [12, Theorem 0.3]. (An analytic function $h : \mathbb{D} \to \mathbb{G}$ is a *complex geodesic* of $\mathbb{G}$ if h has an analytic left-inverse). Moreover, in this case the unique analytic function $h : \mathbb{D} \to \mathbb{G}$ such that $h(\lambda_j) = (s_j, p_j)$ for $j = 1, 2$ can be calculated explicitly as follows [11, Theorem 5.6].

Choose an $\omega_0 \in \mathbb{T}$ such that the maximum on the left-hand side of (8) is attained at ω_0. Since equality holds in (8), we have

$$d(\Phi_{\omega_0}(z_1), \Phi_{\omega_0}(z_2)) = d(\lambda_1, \lambda_2),$$

where $z_j = (s_j, p_j)$. Thus Φ_{ω_0} is a Carathéodory extremal function for the pair of points z_1, z_2 in $\mathbb{G}$. It is easy (for example, by Schur reduction) to find the unique Blaschke product p of degree at most 2 such that

$$p(\lambda_1) = p_1, \quad p(\lambda_2) = p_2 \quad \text{and} \quad p(\bar{\omega}_0) = \bar{\omega}_0^2).$$

Define s by

$$s(\lambda) = 2\frac{\omega_0 p(\lambda) - \lambda}{1 - \omega_0 \lambda} \quad \text{for } \lambda \in \mathbb{D}.$$

Then $h \stackrel{\text{def}}{=} (s, p)$ is the required complex geodesic.

Note that h is a rational function of degree at most 2. It can also be expressed in the form of a realization: $h(\lambda) = (\operatorname{tr} H(\lambda), \det H(\lambda))$ where H is a 2×2 function in the Schur class given by

$$H(\lambda) = D + C\lambda(1 - A\lambda)^{-1}B$$

for a suitable unitary 3×3 or 4×4 matrix $\begin{bmatrix} A & B \\ C & D \end{bmatrix}$ given by explicit formulae (see [4], [12, Theorem 1.7]).

2.6. More points and bigger matrices

Our hope in addressing the case $n = k = 2$ of the spectral Nevanlinna-Pick problem was of course that we could progress to the general case. Alas, we have not so far managed to do so. We have some hope of giving a good solvability criterion for the case $k = 2$, $n = 3$, but even the case $n = 4$ appears to be too complicated for our present methods.

The case of two points and $k \times k$ matrices, for any k, looks at first sight more promising. There is an obvious way to generalize the symmetrized bidisc: we define the *open symmetrized polydisc* $\mathbb{G}_k$ to be the domain

$$\mathbb{G}_k = \{(\sigma_1(z), \ldots, \sigma_k(z)) : z \in \mathbb{D}^k\} \subset \mathbb{C}^k$$

where σ_m denotes the elementary symmetric polynomial in $z = (z^1, \ldots, z^k)$ for $1 \leq m \leq k$. Similarly one defines the closed symmetrized polydisc Γ_k. As in the case $k = 2$, one can reduce Problem SNP to an interpolation problem for functions from $\mathbb{D}$ to Γ_k under mild hypotheses on the target matrices W_j (specifically, that they be nonderogatory). However, the connection between Problem SNP and the corresponding interpolation problems for Γ_k are more complicated for $k > 2$, because there are more possibilities for the rational canonical forms of the target matrices [37]. The analogues for Γ_k of the Φ_ω were described by D.J. Ogle [39] and subsequently other authors, e.g., [23, 29]. Ogle generalized to higher dimensions the operator-theoretic method of [6] and thereby obtained a necessary condition for solvability analogous to Theorem 6.

The solvability of Problem SNP when $n = 2$ is generically equivalent to the inequality

$$\delta_{\mathbb{G}_k}(z_1, z_2) \leq d(\lambda_1, \lambda_2)$$

where z_j is the k-tuple of coefficients in the characteristic polynomial of W_j. All we need is an effective formula for $\delta_{\mathbb{G}_k}$. It turns out that this is a much harder problem for $k > 2$. In particular, it is *false* that $\delta_{\mathbb{G}_k} = C_{\mathbb{G}_k}$ when $k > 2$. This discovery [38] was disappointing, but not altogether surprising.

There is another type of solvability criterion for the 2×2 spectral Nevanlinna-Pick problem with general n [10, 14], but it involves a search over a nonconvex set, and so does not count for the purpose of this paper as an analytic solution of the problem. Another paper on the topic is [24].

It is heartening that the study of the complex geometry and analysis of the symmetrized polydisc has been taken up by a number of specialists in several complex variables, including G. Bharali, C. Costara, A. Edigarian, M. Jarnicki, L. Kosinski, N. Nikolov, P. Pflug, P. Thomas and W. Zwonek. Between them they have made many interesting discoveries about these and related domains. There is every hope that some of their results will throw further light on the spectral Nevanlinna-Pick problem.

3. The spectral Carathéodory-Fejér problem

This is the problem that arises from the spectral Nevanlinna-Pick problem when the interpolation points coalesce at 0.

Problem SCF. *Given $k \times k$ matrices $V_0, V_1, \ldots, V_n$, find an analytic function $F :$ $\mathbb{D} \to \mathbb{C}^{k \times k}$ such that*

$$F^{(j)}(0) = V_j \quad \text{for } j = 0, \ldots, n \tag{13}$$

and

$$r(F(\lambda)) \leq 1 \quad \text{for all } \lambda \in \mathbb{D}. \tag{14}$$

This problem also can be converted to an interpolation problem for analytic functions from $\mathbb{D}$ into Γ_k [34, Theorem 2.1], [37]. However, the resulting problem is again hard when $k \geq 2$, and the only truly explicit solution we have is in the case $k = 2, n = 1$ [34, Theorem 1.1].

Theorem 10. *Let*

$$V_m = [v_{ij}^m]_{i,j=1}^2 \quad \text{for } m = 0,1$$

and suppose that V_0 is nonscalar. There exists an analytic function $F : \mathbb{D} \to \mathbb{C}^{2 \times 2}$ such that

$$F(0) = V_0, \quad F'(0) = V_1 \quad \text{and} \quad r(F(\lambda)) < 1 \text{ for all } \lambda \in \mathbb{D} \tag{15}$$

if and only if

$$\max_{|\omega|=1} \left| \frac{(s_1 p_0 - s_0 p_1)\omega^2 + 2\omega p_1 - s_1}{\omega^2 (s_0 - \bar{s}_0 p_0) - 2\omega(1 - |p_0|^2) + \bar{s}_0 - s_0 \bar{p}_0} \right| \leq 1, \tag{16}$$

where

$$s_0 = \operatorname{tr} V_0, \quad p_0 = \det V_0,$$

$$s_1 = \operatorname{tr} V_1, \quad p_1 = \begin{vmatrix} v_{11}^0 & v_{12}^1 \\ v_{21}^0 & v_{22}^1 \end{vmatrix} + \begin{vmatrix} v_{11}^1 & v_{12}^0 \\ v_{21}^1 & v_{22}^0 \end{vmatrix}.$$

The proof of this theorem in [34] again depends on the calculation in [11] of the Carathéodory metric on $\mathbb{G}$, but this time on the infinitesimal version $c_{\mathbb{G}}$ of the metric: the left-hand side of inequality (16) is the value of $c_{\mathbb{G}}$ at (s_0, p_0) in the direction (s_1, p_1). This fact is [11, Corollary 4.4], but unfortunately there is an ω missing in the statement of Corollary 4.4. The proof shows that the correct formula is as in (16). An important step is the proof that the infinitesimal Carathéodory and Kobayashi metrics on $\mathbb{G}$ coincide.

The ideas behind Theorem 10 can be used to find solutions of Problem SCF: see [34, Section 6]. The ideas can also be used to derive a necessary condition for the spectral Carathéodory-Fejér problem (13), (14) in the case that $n = 1$ and $k > 2$ [34, Theorem 4.1], but there is no reason to expect this condition to be sufficient.

4. The structured singular value

The *structured singular value* of a matrix relative to a space of matrices was introduced by J.C. Doyle and G. Stein in the early 1980s [25, 26] and was denoted by μ. It is a refinement of the usual operator norm of a matrix and is motivated by the problem of the robust stabilization of a plant that is subject to structured uncertainty. Initially, in the H^∞ approach to robustness, the uncertainty of a plant was modelled by a meromorphic matrix function (on a disc or half-plane) that is subject to an L^∞ bound but is otherwise completely unknown. The problem of the simultaneous stabilization of the resulting collection of plant models could then be reduced to some classical analysis and operator theory, notably to the far-reaching results of Adamyan, Arov and Krein from the 1970s [30].

In practice one may have some structural information about the uncertainty in a plant – for example, that certain entries are zero. By incorporating such structural information one should be able to achieve a less conservative stabilizing controller. The structured singular value was devised for this purpose. A good account of these notions is in [27, Chapter 8]. Unfortunately, the behaviour of μ differs radically from that of the operator norm – for one thing, μ is not in general a norm at all, and none of the relevant classical theorems (such as Pick's theorem) or methods appear to extend to the corresponding questions for μ. This provides a challenge for mathematicians: we should help out our colleagues in engineering by creating an AAK-type theory for μ.

For any $A \in \mathbb{C}^{k \times \ell}$ and any subspace E of $\mathbb{C}^{\ell \times k}$ we define the structured singular value $\mu_E(A)$ by

$$\frac{1}{\mu_E(A)} = \inf\{\|X\| : X \in E,\, 1 - AX \text{ is singular}\} \tag{17}$$

with the understanding that $\mu_E(A) = 0$ if $1 - AX$ is always nonsingular.

Two instances of the structured singular value are the operator norm $\|.\|$ (relative to the Euclidean norms on $\mathbb{C}^k$ and $\mathbb{C}^\ell$) and the spectral radius r. If we take $E = \mathbb{C}^{\ell \times k}$ then we find that $\mu_E(A) = \|A\|$. On the other hand, if $k = \ell$ and we choose E to be the space of scalar multiples of the identity matrix, then $\mu_E(A) = r(A)$. These two special μs are in a sense extremal: it is always the case, for any E, that $\mu_E(A) \leq \|A\|$. If $k = \ell$ and E contains the identity matrix, then $\mu_E(A) \geq r(A)$. A comprehensive discussion of the properties of μ can be found in [40].

Here is a formulation of the *μ-synthesis problem* [26, 27].

Given positive integers k, ℓ, a subspace E of $\mathbb{C}^{\ell \times k}$ and analytic functions A, B, C on $\mathbb{D}$ of types $k \times \ell, k \times k$ and $\ell \times \ell$ respectively, construct an analytic function $F : \mathbb{D} \to \mathbb{C}^{k \times \ell}$ of the form

$$F = A + BQC \quad \text{for some analytic} \quad Q : \mathbb{D} \to \mathbb{C}^{k \times \ell} \tag{18}$$

such that

$$\mu_E(F(\lambda)) \leq 1 \quad \text{for all} \quad \lambda \in \mathbb{D}. \tag{19}$$

The condition (18), that F be expressible in the form $A + BQC$ for some analytic Q, can be regarded as an interpolation condition on F. In the event that $k = \ell$, B is the scalar polynomial

$$B(\lambda) = (\lambda - \lambda_1) \ldots (\lambda - \lambda_n) I$$

with distinct zeros $\lambda_j \in \mathbb{D}$ and C is constant and equal to the identity, then F is expressible in the form $A + BQC$ if and only if

$$F(\lambda_1) = A(\lambda_1), \ldots, F(\lambda_n) = A(\lambda_n).$$

With this choice of B and C, if we take E to be the space of scalar matrices, we obtain precisely the spectral Nevanlinna-Pick problem. If we now replace B by the polynomial λ^n, we get the spectral Carathéodory-Fejér problem.

In engineering applications μ-synthesis problems arise after some analysis is carried out on the plant model to produce the A, B and C in condition (18), and the resulting B and C will not usually be scalar functions. Nevertheless, explicit pointwise interpolation conditions provide a class of easily-formulated test cases, and it is arguable that such problems are the *hardest* cases of μ-synthesis.

Conditions of the form (18) are said to be of *model matching type* [30].

The most sustained attempt to develop an AAK-type theory for the structured singular value in full generality is due to H. Bercovici, C. Foiaş and A. Tannenbaum ([15] to[21]). They have a far-reaching theory: *inter alia* they have constructed many illuminating examples, found properties of extremal solutions and obtained a type of solvability criterion for μ-synthesis problems. The criterion results from a combination of the Commutant Lifting Theorem with the application of similarity transformations. To apply the criterion to a concrete spectral Nevanlinna-Pick problem one must solve an optimization problem over a high-dimensional unbounded and non-convex set. We can certainly hope that this is not the last word on the subject of solvability. Despite the achievements of Bercovici, Foiaş and Tannenbaum, there is still plenty of room for further study of μ-synthesis.

One of their examples [18, Section 7, Example 5] exhibits an important fact about the spectral Nevanlinna-Pick problem: *diagonalization does not work*. It shows that diagonalization of the target matrices W_j in Problem SNP by similarity transformations, even when possible, does not help solve the problem. One could hope that if the W_j were diagonal one might be able to decouple the problem into a series of scalar interpolation problems, but they show that such a hope is vain.

Bill Helton himself, along with collaborators, has developed an alternative approach to the refinement of H^∞ control; his viewpoint is set out in [32]. His part in the introduction of the results of Adamyan, Arov, Krein and other operator-theorists into robust control theory in the early 1980s is well known. He subsequently worked extensively (with Orlando Merino, Trent Walker and others) during the 1990s on the more delicate optimization problems that arise from refinements of the basic H^∞ picture of modelling uncertainty. As in the μ approach, the aim is to incorporate more subtle specifications and robustness conditions into methods for controller design. He developed a very flexible formulation of such problems as optimization problems over spaces of vector-valued analytic functions on the disc, and devised an algorithm for their numerical solution – see [33] and several other papers. The authors proved convergence results and described numerical trials. However, the spectral Nevanlinna-Pick problem cannot be satisfactorily treated by the Helton scheme. Although it can be cast in the basic problem formulation [32, Chapter 2], solution algorithms require smoothness properties (of the function "Γ") which the spectral radius does not possess.

5. The next case of μ

After the two extremes $\mu = \|.\|_{H^\infty}$ and $\mu = r$ the next natural case to consider is the one in which, in (17), $k = \ell$ and E is the space $\mathrm{Diag}(k)$ of diagonal matrices. For the rest of this section μ will denote $\mu_{\mathrm{Diag}(2)}$ and we shall study the following problem:

Given distinct points $\lambda_1,\ldots,\lambda_n \in \mathbb{D}$ and 2×2 matrices $W_1,\ldots,W_n$, construct an analytic function $F : \mathbb{D} \to \mathbb{C}^{2\times2}$ such that

$$F(\lambda_j) = W_j \quad for\ j = 1,\ldots,n \tag{20}$$

and

$$\mu(F(\lambda)) \le 1 \quad for\ all\ \lambda \in \mathbb{D}. \tag{21}$$

For the 2×2 spectral Nevanlinna-Pick problem we had some modest success through reduction to an interpolation problem for Γ-valued functions. In the present case we tried an analogous approach, with still more modest success [1, 2, 41]. The following result is [2, Theorem 9.4 and Remark 9.5(iii)].

Theorem 11. *Let $\lambda_0 \in \mathbb{D}$, $\lambda \ne 0$, let $\zeta \in \mathbb{C}$ and let*

$$W_1 = \begin{bmatrix} 0 & \zeta \\ 0 & 0 \end{bmatrix}, \qquad W_2 = \begin{bmatrix} a & * \\ * & b \end{bmatrix}. \tag{22}$$

Suppose that $|b| \le |a|$ and let $p = \det W_2$. There exists an analytic function $F : \mathbb{D} \to \mathbb{C}^{2\times2}$ such that

$$F(\lambda_1) = W_1, \quad F(\lambda_2) = W_2 \quad and \quad \mu(F(\lambda)) \le 1 \quad for\ all\ \lambda \in \mathbb{D} \tag{23}$$

if and only if $|p| < 1$ *and*

$$\begin{cases} \dfrac{|a - \bar{b}p| + |ab - p|}{1 - |p|^2} \leq |\lambda_0| & \text{if } \zeta \neq 0 \\ |\lambda_0|^4 - (|a|^2 + |b|^2 + 2|ab - p|)|\lambda_0|^2 + |p|^2 \geq 0 & \text{if } \zeta = 0. \end{cases}$$

The stars in the formula for W_2 in (22) denote arbitrary complex numbers.

What is the analog of Γ for this case of μ? To determine whether a 2×2 matrix $A = [a_{ij}]$ satisfies $r(A) \leq 1$ one needs to know only the two numbers $\operatorname{tr} A$ and $\det A$; this fact means that the spectral Nevanlinna-Pick problem can generically be reduced to an interpolation problem for Γ. To determine whether $\mu(A) \leq 1$ one needs to know the three numbers $a_{11}, a_{22}, \det A$. This led us to introduce a domain $\mathbb{E}$ which we call the *tetrablock*:

$$\mathbb{E} = \{x \in \mathbb{C}^3 : 1 - x^1 z - x^2 w + x^3 zw \neq 0 \text{ whenever } |z| \leq 1, |w| \leq 1\}. \quad (24)$$

Its closure is denoted by $\bar{\mathbb{E}}$. The name reflects the fact that the intersection of $\mathbb{E}$ with $\mathbb{R}^3$ is a regular tetrahedron. The domain $\mathbb{E}$ is relevant because $\mu(A) < 1$ if and only if $(a_{11}, a_{22}, \det A) \in \mathbb{E}$. There exists a solution of the 2-point μ-synthesis problem (23) if and only if the corresponding interpolation problem for analytic functions from $\mathbb{D}$ to $\mathbb{E}$ is solvable [2, Theorem 9.2], and accordingly the solvability problem for this μ-synthesis problem is equivalent to the calculation of the Lempert function $\delta_{\mathbb{E}}$. As far as I know no one has yet computed $\delta_{\mathbb{E}}$ for a general pair of points of $\mathbb{E}$, but we did calculate it in the case that one of the points is the origin in $\mathbb{C}^3$, that is, we proved a Schwarz lemma for $\mathbb{E}$. The result is Theorem 11.

Observe that ill-conditioning appears in this instance of μ-synthesis too [2, Remark 9.5(iv)]. If, in Theorem 11, $a = b = p = \frac{1}{2}$ then there exists a solution F_ζ of the problem if and only if

$$|\lambda_0| \geq \begin{cases} \frac{2}{3} & \text{if } \zeta \neq 0 \\ \frac{1}{\sqrt{2}} & \text{if } \zeta = 0 \end{cases}$$

Thus if $\frac{2}{3} < |\lambda_0| < \frac{1}{\sqrt{2}}$, the F_ζ are not locally bounded as $\zeta \to 0$, and so are sensitive to small changes in ζ near 0.

The complex geometry of $\mathbb{E}$ has also proved to be of interest to researchers in several complex variables. To my surprise, it was recently shown [28] that the Lempert function and the Carathéodory distance on $\mathbb{E}$ coincide. This might be a step on the way to the derivation of a formula for $\delta_{\mathbb{E}}$. It would suffice to compute $\delta_{\mathbb{E}}$ in the case that one of the two points is of the form $(0, 0, \lambda)$ for some $\lambda \in [0, 1)$, since every point of $\mathbb{E}$ is the image of such a point under an automorphism of $\mathbb{E}$ [41, Theorem 5.2].

The fourth and final special case of μ-synthesis in this paper is the μ-analog of the 2×2 Carathéodory-Fejér problem:

Given 2×2 matrices $V_0, \ldots, V_n$, construct an analytic function $F : \mathbb{D} \to \mathbb{C}^{2 \times 2}$ such that

$$F^{(j)}(0) = V_j \text{ for } j = 0, \ldots, n \quad \text{and} \quad \mu(F(\lambda)) \leq 1 \quad \text{for all } \lambda \in \mathbb{D}.$$

Again the problem can be reduced to an interpolation problem for $\mathbb{E}$, but the resulting problem has only been solved in an exceedingly special case.

Theorem 12. *Let V_0, V_1 be 2×2 matrices such that*

$$V_0 = \begin{bmatrix} 0 & \zeta \\ 0 & 0 \end{bmatrix}$$

for some $\zeta \in \mathbb{C}$ and $V_1 = [v_{ij}]$ is nondiagonal. There exists an analytic function $F : \mathbb{D} \to \mathbb{C}^{2 \times 2}$ such that

$$F(0) = V_0, \quad F'(0) = V_1 \quad \text{and} \quad \mu(F(\lambda)) \leq 1 \text{ for all } \lambda \in \mathbb{D}$$

if and only if

$$\max\{|v_{11}|, |v_{22}|\} + |\zeta v_{21}| \leq 1.$$

This result follows from [41, Theorem 2.1].

6. Conclusion

Although μ-analysis remains a useful tool, it is fair to say that μ-synthesis, as a major technique for robust control system design, has been something of a disappointment up to now. The trouble is that the μ-synthesis problem is difficult. It is a highly non-convex problem. There do exist heuristic numerical methods for addressing particular μ-synthesis problems, notably a Matlab toolbox [36] based on the "DK algorithm" [27, Section 9.3], but there is no practical solvability criterion, no fast algorithm nor any convergence theorem for any known algorithm. For these reasons engineers have largely turned to other approaches to robust stabilization over the past 20 years. If, however, a satisfactory analytic theory of the problem is developed, engineers' attention may well return to μ-synthesis as a promising design tool. We are still far from having such a theory, but perhaps these special cases and the interest of the several complex variables community may yet lead to one.

References

[1] A.A. Abouhajar, *Function theory related to H^∞ control,* Ph.D. thesis, Newcastle University, 2007.

[2] A.A. Abouhajar, M.C. White and N.J. Young, A Schwarz lemma for a domain related to mu-synthesis, *J. Geometric Analysis* **17** (2007) 717–750.

[3] J. Agler, Z.A. Lykova and N.J. Young, Extremal holomorphic maps and the symmetrised bidisc, submitted.

[4] J. Agler, F.B. Yeh and N.J. Young, Realization of functions into the symmetrized bidisc, in *Reproducing Kernel Spaces and Applications,* ed. D. Alpay, *Operator Theory: Advances and Applications* **143**, Birkhäuser Verlag (2003), 1–37.

[5] J. Agler and N.J. Young, A commutant lifting theorem for a domain in $\mathbb{C}^2$ and spectral interpolation, *J. Functional Analysis* **161** (1999) 452–477.

[6] J. Agler and N.J. Young, Operators having the symmetrized bidisc as a spectral set, *Proc. Edin. Math. Soc.* **43** (2000) 195–210.

[7] J. Agler and N.J. Young, The two-point spectral Nevanlinna-Pick problem, *Integral Equations Operator Theory* **37** (2000) 375–385.

[8] J. Agler and N.J. Young, A Schwarz lemma for the symmetrised bidisc, *Bull. London Math. Soc.* **33** (2001) 175–186.

[9] J. Agler and N.J. Young, A model theory for Γ-contractions, *J. Operator Theory* **49** (2003) 45–60.

[10] J. Agler and N.J. Young, The two-by-two spectral Nevanlinna-Pick problem, *Trans. Amer. Math. Soc.* **356** (2004) 573–585.

[11] J. Agler and N.J. Young, The hyperbolic geometry of the symmetrized bidisc, *J. Geom. Anal.* **14** (2004) 375–403.

[12] J. Agler and N.J. Young, The complex geodesics of the symmetrized bidisc, *International J. Math.* **17** (2006) 375–391.

[13] J. Agler and N.J. Young, The magic functions and automorphisms of a domain,*Complex Analysis and Operator Theory* **2** (2008) 383–404.

[14] H. Bercovici, Spectral versus classical Nevanlinna-Pick interpolation in dimension two, *Electronic Journal of Linear Algebra* **10** (2003), 60–64.

[15] H. Bercovici, C. Foiaş, P.P. Khargonekar and A. Tannenbaum, On a lifting theorem for the structured singular value, *J. Math. Analysis Appl.* **187** (1994) 617–627.

[16] H. Bercovici, C. Foiaş, and A. Tannenbaum, Spectral variants of the Nevanlinna-Pick interpolation problem, commutant lifting theorem, *Signal processing, scattering and operator theory, and numerical methods*, Progr. Systems Control Theory, Vol. 5, Birkhäuser, Boston, 1990, pp. 23–45.

[17] H. Bercovici, C. Foiaş and A. Tannenbaum, On the optimal solutions in spectral commutant lifting theory, *J. Functional Analysis* **101** (1991) 38–49.

[18] H. Bercovici, C. Foiaş and A. Tannenbaum, A spectral commutant lifting theorem, *Trans. Amer. Math. Soc.* **325** (1991) 741–763.

[19] H. Bercovici, C. Foiaş and A. Tannenbaum, On spectral tangential Nevanlinna-Pick interpolation, *J. Math. Analysis Appl.* **155** (1991) 156–176.

[20] H. Bercovici, C. Foiaş and A. Tannenbaum, Structured interpolation theory, *Operator Theory: Advances and Applications* **47** (1992) 195–220.

[21] H. Bercovici, C. Foiaş and A. Tannenbaum, The structured singular value for linear input-output operators, *SIAM J. Control Optimization* **34** (1996).

[22] G. Bharali, Some new observations on interpolation in the spectral unit ball, *Integral Eqns. Operator Theory* **59** (2007) 329–343.

[23] C. Costara, *Le problème de Nevanlinna-Pick spectrale*, Ph.D. thesis, Université Laval, Quebec City, Canada, 2004.

[24] C. Costara, The 2×2 spectral Nevanlinna–Pick problem, *J. London Math. Soc.* **71** (2005) 684–702.

[25] J.C. Doyle, Analysis of feedback systems with structured uncertainty, *IEE Proceedings* **129** (1982) 242–250.

[26] J.C. Doyle and G. Stein, Multivariable feedback design: concepts for a classical/modern synthesis, *IEEE Transactions on Automatic Control*, **26** (1981) 4–16.

[27] G. Dullerud and F. Paganini, *A course in robust control theory: a convex approach*, Texts in Applied Mathematics **36**, Springer (2000).

[28] A. Edigarian, L. Kosinski and W. Zwonek, The Lempert theorem and the tetrablock, arXiv:1006.4883.

[29] A. Edigarian and W. Zwonek, Geometry of the symmetrised polydisc, *Archiv Math.*, **84** (2005) 364–374.

[30] B.A. Francis, *A Course in H_∞ Control Theory*, Lecture Notes in Control and Information Sciences No. 88, Springer Verlag, Heidelberg, 1987.

[31] J.W. Helton, Orbit structure of the Möbius transformation semigroup action on H^∞ (broadband matching), *Adv. in Math. Suppl. Stud.* **3**, Academic Press, New York (1978), 129–197.

[32] J.W. Helton, *Operator theory, analytic functions, matrices, and electrical engineering* CBMS Regional Conference Series in Mathematics No. 68, AMS, Providence RI, 1987.

[33] J.W. Helton, O. Merino and T. Walker, Algorithms for optimizing over analytic functions, *Indiana Univ. Math. J.* **42** (1993) 839–874.

[34] H-N. Huang, S. Marcantognini and N.J. Young, The spectral Carathéodory-Fejér problem, *Integral Equations and Operator Theory* **56** (2006) 229–256.

[35] M. Jarnicki and P. Pflug, Invariant distances and metrics in complex analysis revisited, *Dissertationes Math. (Rozprawy Mat.)* **430** (2005) 1–192.

[36] *Matlab μ-Analysis and Synthesis Toolbox*, The Math Works Inc., Natick, Massachusetts, http://www.mathworks.com/products/muanalysis/~.

[37] N. Nikolov, P. Pflug and P.J. Thomas, Spectral Nevanlinna-Pick and Carathéodory-Fejér problems, to appear in *Indiana Univ. Math. J.*, arXiv:1002.1706.

[38] N. Nikolov, P. Pflug and W. Zwonek, The Lempert function of the symmetrized polydisc in higher dimensions is not a distance, *Proc. Amer. Math. Soc.* **135** (2007) 2921–2928.

[39] D. Ogle, Operator and Function Theory of the Symmetrized Polydisc, Ph.D. thesis, Newcastle University (1999), http://www.maths.leeds.ac.uk/ nicholas/abstracts/ogle.html.

[40] A. Packard and J.C. Doyle, The complex structured singular value, *Automatica* **29** (1993) 71–109.

[41] N.J. Young, The automorphism group of the tetrablock, *J. London Math. Soc.* **77** (2008) 757–770.

N.J. Young
Department of Pure Mathematics
Leeds University
Leeds LS2 9JT, England

and

School of Mathematics and Statistics
Newcastle University
Newcastle upon Tyne NE1 7RU, England
e-mail: N.J.Young@leeds.ac.uk

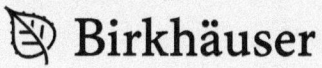 **Birkhäuser** | **www.birkhauser-science.com**

Operator Theory: Advances and Applications (OT)

This series is devoted to the publication of current research in operator theory, with particular emphasis on applications to classical analysis and the theory of integral equations, as well as to numerical analysis, mathematical physics and mathematical methods in electrical engineering.

Edited by
Joseph A. Ball (Blacksburg, VA, USA), Harry Dym (Rehovot, Israel),
Marinus A. Kaashoek (Amsterdam, The Netherlands), Heinz Langer (Vienna, Austria),
Christiane Tretter (Bern, Switzerland)